Practical Numerical Analysis

Practical Numerical Analysis

Gwynne A. Evans
De Montfort University, Leicester, UK

JOHN WILEY & SONS
Chichester • New York • Brisbane • Toronto • Singapore

Baffins Lane, Chichester,
West Sussex PO19 IUD, England

National 01243 779777
International (+44) 1243 779777

Other Wiley Editorial Offices

John Wiley & Sons, Inc., 605 Third Avenue,
New York, NY 10158-0012, USA

Jacaranda Wiley Ltd, 33 Park Road, Milton,
Queensland 4064, Australia

John Wiley & Sons (Canada) Ltd, 22 Worcester Road,
Rexdale, Ontario M9W 1L1, Canada

John Wiley & Sons (SEA) Pte Ltd, 37 Jalan Pemimpin #05-04,
Block B, Union Industrial Building, Singapore 2057

Library of Congress Cataloging-in-Publication Data

Evans, Gwynne.
Practical numerical analysis / Gwynne A. Evans.
p. cm.
Includes bibliographical references and indexes.
ISBN 0 471 95535 3 (alk. paper)
1. Numerical analysis. I. Title.
QA297.E93 1995
519.4 — dc20

British Library Cataloguing in Publication Data

A catalogue record for this book is available from the British Library

ISBN 0 471 95535 3

Typeset in 10/12pt Times by Laser Words, Madras
Printed and bound in Great Britain by Biddles Ltd, Guildford and King's Lynn

This book is printed on acid-free paper responsibly manufactured from sustainable forestation, for which at least two trees are planted for each one used for paper production.

To Margaret, our children,
our animals and impending grandchild

Contents

Preface

Over a period of twenty-five years, I have had the pleasure to teach Numerical Analysis to all three undergraduate years at Loughborough University, and to the M.Sc. groups. These lectures constitute the basis of this book. Two courses which were separate from the conventional numerical teaching are also covered here. These are variational methods and optimization techniques. There is an inevitable bias in the choice of material due to my own conception of the subject, but this has been heavily diluted by the combined influence of colleagues and students whose criticisms have helped to improve the quality of the subject matter, and bring in new ideas.

The ordering of the material presents some difficulties in many books on Numerical Analysis, in that pre-requisites are necessary for some topics. In particular, it is convenient to choose a topic in Chapter 1 which acts as an introduction to the philosophy of the subject which underlies the whole work. Authors have commonly used the solution of non-linear equations as such a starting point, and indeed lecturers also find this topic amenable for the first lectures. I have also employed this approach. The problem then arises as to where to cover sets of non-linear equations. Some authors place this topic in a later chapter after sets of linear equations on which there is some dependence. In this book, I have kept sets of non-linear equations to the first chapter with the proviso that these sections can be omitted at the first reading. Topics are not then split across the book.

I have also taken advantage of this first chapter to include some techniques which have broad application across the subject. These include some analytic and linear algebra preliminaries, and methods used to accelerate sequence convergence.

One advantage of having taught a subject over a long period is the wealth of exercises generated as hand-outs and examination papers. Some of these appear in the exercises at the ends of each chapter. Selected exercises have answers and comments appearing in Appendix A. Readers will find a range of problems in these exercises ranging from problems soluble by hand or by using a simple calculator, to more extensive computer-orientated examples which would form mini projects in their known right. It is easily forgotten that when I first taught this subject, electronic calculators had not been constructed, and all computation was carried out either on main-frame machines with considerable restriction of use, or by electro-mechanical machines which could take several seconds to complete a division, and had no built-in sets of standard functions. Hence the exercises reflect the steady introduction of computing devices and their ease of availability and use. In all exercises, the object is to provide examples which illustrate the methods and associated theory rather than to provide large-scale number crunching exercises. Clearly much of the practical application of Numerical Analysis leads to large-scale computing, but this in general requires no further understanding than is given by the simpler illustrative examples provided. Where special difficulties do arise, as in very large sets of linear algebraic equations, then coverage of the area is provided, and pointers to further reading are made. Indeed in most chapters it is impossible to cover the wealth of material which is sufficient for a complete book in its own right. The further reading will lead the reader to some of these specialist books and extended references to specific papers can be found in these secondary sources.

The chapters are ordered differently from the courses given, in the obvious respect that in each topic, all three years of material falls successively into the relevant chapter. On the other hand, it seems educationally sensible to cover each of the major topics in each year of a course. Many of the Loughborough students took their third year in industry, and a broad coverage in their first two years gave them a good basis for this training before they returned for the final year.

Hence Chapter 1 is of an introductory nature but uses non-linear equations as a starting topic to help to set the scene for later chapters. The analytic and linear algebra preliminaries are without proof but the further reading list contains references for pursuing these important background topics in more detail. Chapter 2 also has an element of introduction, being devoted to the archaic topic of finite differences and the risky process of numerical differentiation. Emphasis is made on deriving results which are of later use.

Chapters 3 and 4 contain the work on numerical linear algebra, split into the solution of sets of linear equations in Chapter 3 and finding eigenvalues and eigenvectors in Chapter 4. In the original course the contents of Chapter 3 were taught in years one and two, and those in Chapter 4 in years two and three. This is a particular area in which reference is made to further reading, and inevitably some choice of topics has been made.

Chapter 5 and 6 link together, and cover approximation theory and quadrature respectively. Space has restricted Padé Theory to limited coverage only, and the aim has been to motivate the quadrature work via approximations. The quadrature chapter is a condensed version of my previous book 'Practical Numerical Integration', with many of the specialist topics omitted.

The door is then open to cover the huge topic of ordinary differential equations in Chapter 7. This is one of the longest chapters, and attempts to survey the current state of the subject. Methods for solving stiff systems and the relation between graph theory and Runge–Kutta methods are pursued.

A related topic follows in Chapter 8 where integral equations are tackled, including some recent work on non-linear problems. Variational methods are introduced here to deal with eigenvalue problems defined in the integral equation context. Following this theme, Chapter 9 is devoted to partial differential equations, with both finite difference and finite element methods being discussed. Again space restricts more than an introductory coverage of the finite element method.

Finally in Chapter 10 a survey of the methods of optimization is made, where again it is recognized that complete books have been written on this topic alone. The chapter is split into two parts, the first part being devoted to unconstrained optimization and the second to constrained problems. The latter introduces linear programming and Kuhn–Tucker theory.

Answers to selected exercises include just straight numerical results to act as a guide to the reader, as well as more detailed comments on a proof or derivation. Some questions include the answer as part of the question and are omitted, and others follow closely the work in the text and again draw no comment at this stage. Numerical results usually indicate the method used and the level of effort rather than giving numerically correct values to the problem. Hence a solution may give the results of applying Euler's method with a particular step size rather than the numerically correct values for the original differential equation. Some emphasis is placed on students validating their own calculations as they would with a real problem, rather than relying on provided answers at every turn.

Throughout the text, reference is made to the original work for most of the methods. Selectivity is always a problem here, and in some cases a secondary source is quoted which will lead the reader to a vast specialised bibliography for that topic. Because authors' names play a vital role in naming methods, a separate author index is provided. Where reference to a topic or author takes on many pages, the index refers to the first page of the relevant set. The topics for further reading are also included in the author index.

The author remains very grateful to various colleagues for support in the preparation of this book, not least to those at John Wiley who have been helpful and approachable at all stages. The start of the preparation occurred at Loughborough University on an Apple Mac-based TEX system. Algorithms, answers and the preparation of diagrams were completed on the University main-frame HP computers. UNIRAS was used specifically for diagrams. Part way through the preparation, with the basis of all ten chapters written, I was delighted to be promoted to a Chair of Mathematics at De Montfort University. It was in this new and developing environment that the final stages were completed. Again thanks must go to a second set of colleagues who prepared the technical equipment for the work to continue. After some qualms, the text was transfered to the HP system at De Montfort, and the diagrams were processed through the network on the original Loughborough system. Particular names include Helen Ramsey and David Ireland at John Wiley's; Professor C. Storey at Loughborough University's Mathematics Department; David Hogg and Graham Gerrard at the Computer Centre, Loughborough University; and in the closing stages, Paul Luker, Ian Vlaeminke and David Chiddick at De Montfort University. Throughout the long hours of proof-reading I was nearly always accompanied by my canine friend Ben who must know more about numerical analysis than any other dog.

1

Introduction to Numerical Processes and Non-linear Equations

1.1 History and introduction

Numerical processes have their roots in the early development of numbers themselves. Some of the earliest computations were employed to find values for π, and to compute areas and volumes. The Babylonians appear to have used the area of a trapezium to calculate irregularly shaped areas, and Bryson, one of Pythagoras' pupils, used the method of exhaustion in around 450 BC which effectively exhausts the area between a curve and an inscribed polygon by successively increasing the number of sides of the polygon. In both cases, the idea of splitting an area into small regions of known area as an approximation to the full area is a typical ploy of numerical analysis. It was Archimedes in 225 BC who was able to place π in the range $\frac{22}{7} > \pi > \frac{223}{71}$ by the method of exhaustion. In a similar vein, Kepler (1615) discretized the volume of a wine barrel into a series of shallow cylinders as a means of settling an argument with a wine merchant in Linz.

The introduction of the calculus brought in the concept of limit and derivative, with immediate discrete analogues which developed into numerical techniques. Because of the lack of any computing aids, many of the developments were very much side-lines when they were first found, and their significance has only been apparent recently. For example, Newton in the Principia (1686) investigates the solution of a cubic, but it is a more general application of this rule which is used for the solution of non-linear equations. Names associated with methods seem to have little historical relevance. For example, Lagrange's interpolation formula appears first in 'Meditationes Algebricae' by E. Waring, the Lucasian Professor of mathematics at Cambridge in 1770. The significance of many results remained hidden for decades. For example Chebyshev polynomials were known in the 1870's, but their relevance to approximation and the resulting algorithms in quadrature occurred in the 1960's.

The missing link was the ability to compute. The Edinburgh mathematician Napier in 'Mirifici logarithmorum canonis descripto' in 1614 opened the possibility of hand calculation, and the London mathematics teacher Richard Delamain designed a circular slide rule in 1630. The clergyman and teacher of mathematics Oughtred also produced 'Circles of Proportion' in 1632, the first accepted practical slide rule. Without computational aids, numerical analysis could not develop. Foundations were laid as part of other mathematical developments, 'Methodus nova integralum valores per approximationem inveniendi' by Gauss (1886) gave us Gaussian quadrature: Weierstrass (1885) established that any continuous function could be approximated to arbitrary accuracy by a polynomial of sufficiently high degree. This is a fundamental concept of much numerical work.

It was the force of war which pushed forward the manufacture of an electronic programmable computing machine such as the University of Pennsylvania's ENIAC in 1946. Now it became possible to test out methods on practical problems. The concept of rounding errors and resulting stability problems arose, and the modern subject of numerical analysis was born.

At first computing machines were physically large with a single process being executed. Hence 'batch running' with programmes submitted in succession using a physical medium such as punched cards was common. Machines which could cope with multiple users each sharing the main processor and storage became commonplace in the later 1970's and early 1980's as computer power increased. Both computing speed and storage have become relatively cheap and available through networking and hardware advances, so that now enormous volumes of arithmetic processes can be carried out on vast stores of data using machines which fit under a desk.

Progress is now towards machines in which many processors work simultaneously on the same problem. This is the parallel environment, and allows gains in real time to be made. Extensions to computer languages are required to allow the benefits of parallel working to be made. There is always some overhead in synchronizing simultaneous computations, and accessing common or independent storage. Machines which simply handle vector processes in a parallel manner can give large gains in computational time for numerical problems in linear algebra in particular. This whole area continues to be a subject for current research.

There are many libraries of numerical software and packages which are considered as aids to mathematical thinking. The NAG library is a traditional collection of programmes to solve almost any problem in numerical analysis. Each area of the subject has a chapter, and flow diagrams are used to direct the user to the most suitable method. Even with a tool of this sort, a knowledge of how and why a particular technique works makes the choice of rule less a matter of trust in the library writer's recommendation. When things go wrong it is easier to see what has happened when not operating in a theoretical blackness. The desk-top packages which now carry out algebraic manipulations include Mathematica, DERIVE, MAPLE and Matlab. These packages and some of their capabilities will be mentioned throughout this book, though the primary object is to provide the background behind these methods.

Any student of mathematics soon encounters the need to practically generate numerical values as a solution to a given problem. Almost every branch of mathematics has a numerical counterpart in which numerical values are required. At school level, students meet simultaneous linear algebraic equations which they solve by elimination methods. If the number of unknowns rises beyond say five, it becomes increasingly tedious to generate algebraic solutions and at this stage a numerical process takes over. In this case a computer programme is written to simulate closely the elimination process. When equations are non-linear, algebraic means are not generally available as in the simple example

$$x = e^{-x}. \tag{1.1.1}$$

In such a case a numerical technique is required which progresses in stages towards the required result according to some rule or algorithm. Such a process is an iteration.

Just as it is very easy to produce a non-linear equation whose roots are not easily written explicitly, it is also easy to find examples of integrals whose values cannot be expressed in closed form. Indeed, transcendental functions are often defined in just this way, leaving the numerical problem of evaluation to be solved separately. Hence the integral

$$\int_0^{\pi/2} \sin^{1/2} x \, dx \tag{1.1.2}$$

is such an example, the integral in fact being an elliptic function. Equally, differential equations soon become intractable analytically with

$$\frac{d^2 y}{dx^2} = xy \tag{1.1.3}$$

being a particularly simple example with no solution in elementary functions. The solutions are in this case the Airy functions $\mathrm{Ai}(x)$ and $\mathrm{Bi}(x)$. Once more a numerical process is required to compute values of y for various x given suitable boundary conditions, and tables of these functions are obtained by numerical processes on various analytic representations including differential equations and integral forms.

It is the purpose of this book to address some of these problems, and cover the associated mathematics. There seem to be three main themes which need attention in tackling this subject. The first and obvious requirement is to invent a method to apply to the problem. The inventive phase is achieved sometimes by mathematical argument, sometimes using experience or intuition or using some form of lateral thinking. It is only after a method is found that the mathematical part really starts to establish the validity of the process. The practical implementation of this method will be an algorithm or rule of computation which will manifest itself as a programme on a computing machine. The object here is not to use a specific commercial language but to give algorithms in a well-defined form which will allow easy conversion to a specific language. There will be a few exceptions in this book to this rule when the precise form of an algorithm is required, most likely where the originality lies primarily in the algorithm itself. Fortran and Pascal have proved suitable languages for this work.

The story does not end with the production of a method, as in most cases the method will be approximate in some sense. Hence the second main theme is to analyse the method with a view to establishing just what the error in the process is. This may involve finding rates of convergence of iterative methods, finding conditions under which methods converge at all, or establishing the size of a truncation error after chopping short an infinite series. In effect this aspect is the study of the ideal performance of the method on a perfectly accurate computer.

One of the big stumbling blocks in numerical analysis is the further errors which arise because of the use of real computing machines in which all floating point numbers have a finite representation and are inevitably only represented to a specific number of figures. Hence all computed values carry with them an underlying rounding error, and if that error grows in the execution of the algorithm it is possible that the final error could ultimately completely dominate the true result and no correct figures be generated. On an exact machine, the algorithm would be subject only to its truncation error, and would work satisfactorily, but on a real machine the rounding error is magnified and so gives wrong results. This is known as ill-conditioning and seems to arise all too often in numerical processes.

Hence the subject has a third theme, the analysis of the stability of the process. The question as to whether small rounding errors introduced in the input data, and similar errors introduced during the computation, result in large errors in the final values, is answered. It is clear that the invention of a method is only a small part of the process of solving a given problem. It is the analysis of truncation and rounding errors which give most work. A good way to illustrate these effects is to consider the simple approximation

$$\frac{f(x+h)-f(x)}{h} \simeq f'(x) \tag{1.1.4}$$

for varying h. This formula is a simple numerical differentiator based directly on the analytic definition. By using a Taylor expansion it is easy to see that the truncation error is of order $hf''(x)/2$ and so as $h \to 0$ the approximation will become closer and closer to $f'(x)$ if the computations are carried out exactly. On a real machine the situation is different because as h gets smaller the finite length arithmetic cannot resolve $x+h$ from x to full accuracy. Hence $f(x+h)-f(x)$ cannot be computed accurately. In other words because $f(x+h)$ and $f(x)$ lie close together, subtractive cancellation occurs. Hence if as an example $f(x)$ is taken to be e^x and if x is taken to be unity, then the results in Table 1.1 are obtained for $f'(x)$ as h gets small.

For moderate h, the error falls off with decreasing h as the dominant truncation error behaves as $hf''(1)/2$, the true value being 2.718 28 to six figures. Around $h = 10^{-4}$ the error begins to rise, as these elementary calculations were carried out to a seven-digit precision. Hence at 10^{-4} the values of $f(x+h)$ and $f(x)$ are 2.718 28 and 2.718 55 respectively, causing a loss of four digits on subtraction. Therefore the rounding error is now dominating the situation, and progressive loss of accuracy occurs as h becomes ever smaller until h is near to machine precision when all of the significant digits are lost.

Table 1.1

h	f'
10^{-2}	2.731 94
10^{-3}	2.719 88
10^{-4}	2.720 36
10^{-5}	2.717 97
10^{-6}	2.622 60

Another example which demonstrates rounding error instability is the use of a truncated Taylor series for the evaluation of $\sin x$.

Suppose the truncated series:

$$\sin x = x - \frac{x^3}{3!} + \frac{x^5}{5!} - \cdots + (-1)^{m+1}\frac{x^{2m-1}}{(2m-1)!} \tag{1.1.5}$$

is used to evaluate $\sin x$. The series alternates and the possibility of subtractive cancellation is present. A range of x values and the corresponding sum is illustrated in Table 1.2.

Summation continues until the truncation error is below the machine accuracy of 10^{-7}. The errors are then entirely due to rounding and can be seen to increase rapidly for the larger x values. It is easy to analyse the cause. The general term $T(m)$ is of the form $x^m/m!$ or $x^m/\Gamma(m+1)$. Hence this term has a turning point when its derivative with respect to m has a zero. But this derivative is

$$T'(m) = (x^m \ln(x) - x^m \psi(m+1))/\Gamma(m+1) \tag{1.1.6}$$

where $\psi(x)$ is the Psi or digamma function. Hence the turning point occurs when

$$\psi(m+1) = \ln(x) \tag{1.1.7}$$

and asymptotically $\psi(x) \sim \ln x + 1/2x + \cdots$, giving the maximum term at the value $x \sim m+1$ for sufficiently large m. Both the number of terms (m) used, and the size of the maximum term are given in Table 1.2. Clearly the sum itself has modulus less than unity and yet the maximum term grows to order $(m+1)^m/m!$ which at $x = 15$ is as high as -3.3×10^5. These extra digits must cancel to give the final value and the number of digits in the maximum term is a measure of the loss of accuracy in

Table 1.2

x	Sum	True value	m	Max term
1.0	8.414 71(1)	8.414 71(−1)	6	1.0
5.0	−9.589 24(−1)	−9.589 24(−1)	13	2.604 17(+1)
10.0	−5.440 61(−1)	−5.440 21(−1)	20	2.755 73(+3)
15.0	6.435 53(−1)	6.502 88(−1)	27	−3.348 65(+5)

the final sum. This is well illustrated in the table with, for example, the expectation of a loss of three digits, and indeed only four being obtained when $x = 10$.

These examples show the need for extreme care in the selection of a method. It is not just the truncation error in a process which is important, but also the possibly well-hidden effect of rounding errors. Throughout the chapters which follow these problems will be pursued in a range of topics from linear algebra to ordinary differential equations.

1.2 Mathematical preliminaries

A number of results and theorems from analysis, linear algebra and applied mathematics are used extensively in numerical analysis, and these are outlined here for completeness. Firstly from analysis, a number of theorems which are used throughout the subject are stated below. The proofs of these results can be found in almost any standard work on analysis, for example Apostol (1974).

(1) Rolle's theorem states that if f is a continuous function on the interval $a \le x \le b$ and if $f'(x)$ is itself continuous on $a < x < b$, then if $f(a) = 0$ and $f(b) = 0$, then $f'(\xi) = 0$ for at least one point ξ such that $a < \xi < b$.

(2) The differential mean value theorem states that if $f(x)$ is continuous for $a \le x \le b$ and if $f'(x)$ is continuous for $a < x < b$, then $f(b) - f(a) = (b - a)f'(\xi)$ for at least one ξ in the interval $a < \xi < b$.

(3) The first integral mean value theorem states that if $f(x)$ is continuous for $a \le x \le b$ then

$$\int_a^b f(x)\,\mathrm{d}x = (b - a)f(\xi) \tag{1.2.1}$$

for at least one point ξ in the interval $a < \xi < b$.

(4) The second integral mean value theorem states that if $f(x)$ is continuous on the interval $a \le x \le b$ and if the function $g(x)$ is of constant sign in the interval (a, b), then

$$\int_a^b f(x)g(x)\mathrm{d}x = f(\xi)\int_a^b g(x)\mathrm{d}x \tag{1.2.2}$$

for at least one ξ such that $a < \xi < b$.

(5) An important result is Taylor's theorem. Let $f(x)$ be such that its $(n + 1)$th derivative exists and is continuous on the interval (a, b), then for any two points x and y belonging to (a, b)

$$\begin{aligned} f(x) = f(y) + (x - y)f'(y) + \frac{(x - y)^2}{2!}f''(y) + \cdots \\ + \frac{(x - y)^n f^{(n)}(y)}{n!} + R_{n+1}(x, y) \end{aligned} \tag{1.2.3}$$

where $R_{n+1}(x, y)$ is the remainder term. This remainder term can be expressed in a number of ways. In differential form it has the value

$$R_{n+1}(x, y) = \frac{(x-y)^{n+1} f^{(n+1)}(\xi)}{(n+1)!} \tag{1.2.4}$$

where ξ lies in the interval (a, b), and in integral form the remainder can be expressed as

$$R_{n+1}(x, y) = \frac{1}{n!} \int_y^x (x-s)^n f^{(n+1)}(s)\, ds \tag{1.2.5}$$

Moreover, equation (1.2.4) can be obtained from (1.2.5) by a simple application of the second integral mean value theorem as $(x-s)^n$ does not change sign in the interval (y, x).

(6) The concept of a supremum or least upper bound will be employed from time to time, and needs careful definition. Let $\mathcal{A}$ be the set of real numbers. Then if there exists a real number y such that $y \in \mathcal{A}$ implies that $a \leq y$ for any $a \in \mathcal{A}$ then y is an upper bound for $\mathcal{A}$. Such a set is bounded above.

Let $\mathcal{A}$ be a set bounded above. Suppose that for any $y \in \mathcal{A}$

(i) y is an upper bound for $\mathcal{A}$.
(ii) If z is an upper bound for $\mathcal{A}$ then $y \leq z$.

Then y is a least upper bound or supremum of $\mathcal{A}$.

(7) A set of standard functions with wide-ranging application throughout numerical analysis is the set of Chebyshev polynomials. It is convenient to consider their main properties at this stage rather than have a piecemeal account spread in many chapters. Chebyshev polynomials are defined by

$$T_r(x) = \cos(r \arccos x) = \cos r\theta \tag{1.2.6}$$

where $x = \cos\theta$. As such they are easy to write down using the multi-angle formulae for $\cos r\theta$. Hence

$$\begin{aligned}
T_0 &= 1\\
T_1 &= x\\
T_2 &= 2x^2 - 1\\
T_3 &= 4x^3 - 3x\\
T_4 &= 8x^4 - 8x^2 + 1\\
T_5 &= 16x^5 - 20x^3 + 5x\\
T_6 &= 32x^6 - 48x^4 + 18x^2 - 1\\
T_7 &= 64x^7 - 112x^5 + 56x^3 - 7x\\
T_8 &= 128x^8 - 256x^6 + 160x^4 - 32x^2 + 1
\end{aligned}$$

$$T_9 = 256x^9 - 576x^7 + 432x^5 - 120x^3 + 9x$$
$$T_{10} = 512x^{10} - 1280x^8 + 1120x^6 - 400x^4 + 50x^2 - 1 \qquad (1.2.7)$$

are the polynomials to order x^{10}. Further values are given in Appendix D. These polynomial coefficients are easily computed because the trigonometric identity

$$\cos(r+1)\theta = 2\cos\theta\cos r\theta - \cos(r-1)\theta \qquad (1.2.8)$$

can be immediately rewritten in terms of Chebyshev polynomials as

$$T_{r+1}(x) = 2xT_r(x) - T_{r-1}(x) \qquad (1.2.9)$$

yielding a three-term recurrence relation. The set of Chebyshev polynomials is one of the classical sets of polynomials which are orthogonal over a given interval with respect to a given weight function. In this case the orthogonality condition is

$$\int_{-1}^{1} \frac{T_r(x)T_s(x)}{\sqrt{1-x^2}} = \begin{cases} 0 & r \neq s \\ \frac{\pi}{2} & r = s \neq 0 \\ \pi & r = s = 0 \end{cases} \qquad (1.2.10)$$

making the set of Chebyshev polynomials orthogonal over the interval $[-1, 1]$ with the weight function $1/\sqrt{1-x^2}$. The orthogonality condition is easy to establish using the substitution $x = \cos\theta$ to reduce the integral to a simple trigonometric form. How orthogonal polynomials are generated in general, given a weight function and an interval, will be dealt with in the approximation theory work of Chapter 5. Like other orthogonal polynomials, the Chebyshev polynomials not only satisfy a three-term recurrence relation but also satisfy a second-order differential equation. Hence $y = T_r(x)$ is a solution of

$$(1-x^2)y'' - xy' + r^2y = 0 \qquad (1.2.11)$$

with the other solution being a Chebyshev function of the second kind which is denoted by $y = U_r(x)$.

One of the main uses of Chebyshev polynomials is in the formation of approximations of other functions. It will be shown in Chapter 5 that such approximations are particularly accurate, and so the functions find application in approximation theory: Chapter 5; in differentiation: Chapter 2, and in numerical integration in Chapter 6. One way of expressing a function $f(x)$ as a series of Chebyshev polynomials in the form

$$f(x) = {\sum_{r=0}^{N}}'' a_r T_r(x) \qquad (1.2.12)$$

where the notation $''$ indicates that the first and last terms of the series are halved, is to use a discrete orthogonalization result such as

$$ {\sum_{s=0}^{N}}'' \cos\frac{\pi i s}{N} \cos\frac{\pi j s}{N} = \begin{cases} 0 & i \neq j \\ N & i = j = 0 \text{ or } N \\ \frac{N}{2} & i = j \neq 0 \text{ or } N \end{cases} \qquad (1.2.13)$$

This result can be proved by writing

$$\cos\frac{\pi is}{N}\cos\frac{\pi js}{N}=\left(\cos\frac{\pi s(i-j)}{N}+\cos\frac{\pi s(i+j)}{N}\right)\Big/2 \tag{1.2.14}$$

and applying the formula

$$\sum_{k=0}^{N}\cos kx=\cos\frac{nx}{2}\sin\frac{(n+1)x}{2}\operatorname{cosec}\frac{x}{2} \tag{1.2.15}$$

which itself follows easily if the expansion

$$\cos kx\sin\tfrac{1}{2}x=\left(\sin x\left(k+\tfrac{1}{2}\right)-\sin x\left(k-\tfrac{1}{2}\right)\right)/2 \tag{1.2.16}$$

is summed over the range $k=0,N$, cancellation occurring except for the first and last terms which yield the required result. There are other discrete orthogonalization results of this type involving products of sin's and other arguments. Some of these will be used and discussed in Chapter 6 on numerical integration.

Hence multiplying both sides of (1.2.12) by $\cos\pi s\hat{r}/N$ and summing from $s=0$ to N gives

$$\sum_{s=0}^{N}{}'' f(x_s)\cos\frac{\pi s\hat{r}}{N}=\sum_{s=0}^{N}{}''\sum_{r=0}^{N}{}'' a_rT_r(x_s)\cos\frac{\pi s\hat{r}}{N} \tag{1.2.17}$$

Making the specific choice $x_s=\cos\pi s/N$ gives $T_r(x_s)=\cos\pi rs/N$ and the right-hand side of (1.2.17) is precisely in the form of equation (1.2.13) and only the term which multiplies $a_{\hat{r}}$ is non-zero, having the coefficient $N/2$. The use of the double prime notation allows this coefficient to be written in one form, whether or not $\hat{r}$ is 0 or N or some intermediate value. Hence the useful expansion in (1.2.12) has coefficients a_r given by

$$a_r=\frac{2}{N}\sum_{s=0}^{N}{}'' f\left(\cos\frac{\pi s}{N}\right)\cos\frac{\pi rs}{N} \tag{1.2.18}$$

(8) A number of results from linear algebra make up the second group of preliminaries and will be required in the work to follow. Again, further details can be found in any standard work such as Finkbeiner (1960). Firstly a set of linear algebraic equations of the form

$$\mathbf{Ax}=\lambda\mathbf{x} \tag{1.2.19}$$

will possess only the trivial solution $\mathbf{x}=\mathbf{0}$ unless the condition

$$\det(\mathbf{A}-\lambda\mathbf{x})=0 \tag{1.2.20}$$

holds. This condition yields a polynomial in λ of degree n, the order of the matrix, and is the characteristic equation for the eigenvalues λ. Finding such eigenvalues constitutes a major problem in numerical analysis and is covered in Chapter 4.

(9) A matrix $\mathbf{A}$ is said to be positive definite if its eigenvalues are positive real numbers. Equivalent definitions include all the principal minors of $\mathbf{A} > 0$, and $\mathbf{x}^{\mathbf{T}}\mathbf{A}\mathbf{x} > 0$ for all vectors $\mathbf{x} \neq 0$.

(10) The transpose of a matrix $\mathbf{A}$ with elements a_{ij} is written as $\mathbf{A}^T$ and has the elements a_{ji}, that is rows and columns have been interchanged. If $\mathbf{A}^T = \mathbf{A}$ then the matrix is symmetric. Symmetric matrices have real eigenvalues which makes the numerical problem in this case much easier. The inverse matrix $\mathbf{A}^{-1}$ satisfies

$$\mathbf{A}\mathbf{A}^{-1} = \mathbf{I} \tag{1.2.21}$$

If

$$\mathbf{A}\mathbf{A}^T = \mathbf{A}^T\mathbf{A} = \mathbf{I} \tag{1.2.22}$$

then $\mathbf{A}$ is orthogonal.

If the matrix $\mathbf{A}$ has complex elements the equivalent of transpose is the Hermitian transpose $\mathbf{A}^H$ in which the new elements are $\bar{a}_{ji}$, that is the complex conjugates of the original elements transposed. In the real case the definition reduces to that in (10) for transpose. If further

$$\mathbf{A}\mathbf{A}^H = \mathbf{A}^H\mathbf{A} = \mathbf{I} \tag{1.2.23}$$

the matrix is said to be unitary, which is the equivalent of orthogonal in the real case.

(11) A further useful concept is the Gram–Schmidt Process. This states that in any finite-dimensional inner product space V, $\exists$ a basis of V consisting of mutually orthogonal vectors. These vectors can be constructed directly and the process has applications in optimization in Chapter 10.

Define the inner product of two vectors by the notation $(\mathbf{x}, \mathbf{y})$ which might be exemplified by the scalar product $x_1y_1 + \cdots + x_ny_n$. The construction is made by supposing the result is true for dimension k. Let V have dimension $k+1$ and let S be any subspace of dimension k. Then $\exists$ a basis $\mathbf{u_1}, \ldots, \mathbf{u_k}$ for S which is orthogonal that is $(\mathbf{u_i}, \mathbf{u_j}) = 0$, $i \neq j$. Let $\mathbf{v} \notin S$ then let

$$\mathbf{a} = \sum_{i=1}^{k} \frac{(\mathbf{v}, \mathbf{u}_i)}{(\mathbf{u}_i, \mathbf{u}_i)}\mathbf{u}_i = \sum_{i=1}^{k} c_i\mathbf{u}_i \tag{1.2.24}$$

Then $\mathbf{b} = \mathbf{v} - \mathbf{a}$ is a vector orthogonal to each of the $\mathbf{u}_i$'s as

$$\begin{aligned}
(\mathbf{b}, \mathbf{u}_i) &= (\mathbf{v} - \mathbf{a}, \mathbf{u}_i) \\
&= (\mathbf{v}, \mathbf{u}_i) - \sum_{j=1}^{k} c_j(\mathbf{u}_j, \mathbf{u}_i) \\
&= (\mathbf{v}, \mathbf{u}_i) - c_i(\mathbf{u}_i, \mathbf{u}_i) \\
&= (\mathbf{v}, \mathbf{u}_i) - \frac{(\mathbf{v}, \mathbf{u}_i)}{(\mathbf{u}_i, \mathbf{u}_i)}(\mathbf{u}_i, \mathbf{u}_i) = 0
\end{aligned} \tag{1.2.25}$$

as required. In effect this process takes the vector **v** and splits it into components or projections along each of the orthogonal directions, leaving just the extra new orthogonal vector **b**.

The third group of preliminaries comes from applied mathematics and concerns the representation of combinations of partial derivatives as vector operators. Such operators are used freely to represent the equations of mathematical physics such as those of fluid mechanics and those of electromagnetic theory (see for example Acheson (1990) and Atkin (1962)).

(12) The vector operators div, grad and curl are defined in Cartesian co-ordinates (x, y, z) by

$$\text{(i)} \quad \operatorname{grad}\phi = \left\{\frac{\partial\phi}{\partial x}, \frac{\partial\phi}{\partial y}, \frac{\partial\phi}{\partial z}\right\} = \nabla\phi \tag{1.2.26}$$

and operate on a scalar to yield a vector.

$$\text{(ii)} \quad \operatorname{div}\mathbf{a} = \frac{\partial a_x}{\partial x} + \frac{\partial a_y}{\partial y} + \frac{\partial a_z}{\partial z} = \nabla.\mathbf{a} \tag{1.2.27}$$

which operates on a vector to yield a scalar.

$$\text{(iii)} \quad \operatorname{curl}\mathbf{a} = \begin{vmatrix} \mathbf{i} & \mathbf{j} & \mathbf{k} \\ \frac{\partial}{\partial x} & \frac{\partial}{\partial y} & \frac{\partial}{\partial z} \\ a_x & a_y & a_z \end{vmatrix} = \nabla \times \mathbf{a} \tag{1.2.28}$$

which operates on a vector to yield a vector, **i**, **j** and **k** being unit vectors in the directions of x, y and z respectively.

It is common to require these operators in different co-ordinate systems. If the metric of a general co-ordinate system is

$$\mathrm{d}s^2 = h_1^2\mathrm{d}u^2 + h_2^2\mathrm{d}v^2 + h_3^2\mathrm{d}w^2 \tag{1.2.29}$$

where ds is the small increment of length generated as the co-ordinates u, v and w vary by small increments du, dv and dw. Then in cylindrical polar co-ordinates, $u = r$, $v = \theta$ and $w = z$ with $h_1 = 1$, $h_2 = r$ and $h_3 = 1$, and in spherical polars $u = r$, $v = \theta$ and $w = \phi$ with $h_1 = 1$, $h_2 = r$ and $h_3 = r\sin\theta$. The operators in a general system of co-ordinates are now

$$\operatorname{grad}\phi = \left\{\frac{1}{h_1}\frac{\partial\phi}{\partial u_1}, \frac{1}{h_2}\frac{\partial\phi}{\partial u_2}, \frac{1}{h_3}\frac{\partial\phi}{\partial u_3}\right\} \tag{1.2.30}$$

$$\operatorname{div}\mathbf{a} = \frac{1}{h_1h_2h_3}\left\{\frac{\partial}{\partial u_1}(h_2h_3a_1) + \frac{\partial}{\partial u_2}(h_1h_3a_2) + \frac{\partial}{\partial u_3}(h_1h_2a_3)\right\} \tag{1.2.31}$$

$$\operatorname{curl}\mathbf{a} = \frac{1}{h_1h_2h_3}\begin{vmatrix} h_1\mathbf{i} & h_2\mathbf{j} & h_3\mathbf{k} \\ \frac{\partial}{\partial u_1} & \frac{\partial}{\partial u_2} & \frac{\partial}{\partial u_3} \\ h_1a_1 & h_2a_2 & h_3a_3 \end{vmatrix} \tag{1.2.32}$$

(13) An important result which relates surface and volume integrals is the divergence theorem:

If a closed region V is bounded by a simple closed surface S then

$$\int_S \mathbf{F}.\mathrm{d}\mathbf{S} = \int_V \operatorname{div} \mathbf{F}\, \mathrm{d}V \tag{1.2.33}$$

which is used freely in the derivations of many of the partial differential equations considered in Chapter 9.

1.3 Non-linear equations

This section treats one of the fundamental problems of numerical analysis, namely the numerical solution of non-linear equations. The section has been placed in the first chapter as the solution of non-linear equations employs many of the concepts which will recur throughout the work. On the other hand the ideas involved are relatively simple and so the first-time reader can obtain a feel for the topic. On a first reading, it may be advisable to omit Section 1.4 on Bairstow's method and Section 1.6 on sets of non-linear equations. These have been included at this stage to avoid fragmentation and spreading of topics throughout other chapters.

The problem here can be simply written down as

$$f(x) = 0 \tag{1.3.1}$$

where values of x which satisfy equation (1.3.1) are required. There may well be many such solutions, and part of the problem can be to isolate a particular solution, or find all the solutions in a given interval. In this introductory chapter attention will be confined to obtaining an accurate root having been given the approximate region in which it lies. Two simple examples of such equations are

$$e^x = \frac{1}{x} \tag{1.3.2}$$

and

$$x^3 - 9x^2 + 4 = 0 \tag{1.3.3}$$

and a range of methods will be studied, each refining the numerical process used. As each method is developed, the three parts of numerical work mentioned in the first paragraph will become very evident. There will be some intuitive meanderings to derive a method. Once derived, strict analysis will determine such factors as convergence rates, and finally shortcomings due to rounding errors and stability will be covered. There is always an assessment of a method covering practical issues of reliability and convenience, and ranging through to speed of convergence and stability.

As a start, consider a simple approach to the solution of equation (1.3.1). If the equation is re-arranged in the form

$$x = \phi(x) \tag{1.3.4}$$

then it is possible to consider the associated iteration

$$x_{n+1} = \phi(x_n). \tag{1.3.5}$$

Hence this process generates a sequence of values x_n and it is true that if this sequence converges then

$$x_{n+1} \to x_n \to x \tag{1.3.6}$$

where x then satisfies equation (1.3.4). Hence with example (1.3.2), by setting,

$$x_{n+1} = e^{-x_n} \tag{1.3.7}$$

and using $x_0 = 1$ as a starting point, the sequence

$$\{1.0, 0.37, 0.69, 0.50, 0.61, 0.54, 0.58, 0.56, 0.57, 0.565, \ldots\}$$

is obtained and slow convergence is observed. A similar slow convergence is obtained with the rearrangement

$$x_{n+1} = \frac{x_n^3 + 4}{9x_n} \tag{1.3.8}$$

for example (1.3.3) to yield

$$\{1.0, 0.55, 0.83, 0.61, 0.77, 0.64, \ldots\}$$

and indeed not all rearrangements will yield convergent processes. An analysis of the iteration is easily made by considering the error

$$\varepsilon_n = x_n - x_{n-1} \tag{1.3.9}$$

and hence

$$\begin{aligned} \varepsilon_{n+1} &= x_{n+1} - x_n = \phi(x_{n+1}) - \phi(x_n) \\ &= \phi(x_n) + (x_{n+1} - x_n)\phi'(\xi) - \phi(x_n) \end{aligned} \tag{1.3.10}$$

where $\xi \in [x_n, x_{n+1}]$ using the differential mean value theorem from Section 1.2 (2). Immediately (1.3.10) yields:

$$\varepsilon_{n+1} = \varepsilon_n \phi'(\xi) \tag{1.3.11}$$

and if $|\phi'(\xi)| < 1$ then convergence will result as the errors will reduce at every step. In the first example $\phi'(x) = -e^{-x}$ which has a numerical modulus of approximately 0.6 around the point of convergence. The power n for which $(0.6)^n = 0.1$ will give the number of iterations required to further the iteration one more decimal place. In this case $n = 4.5$, yielding around five iterations per decimal, and this is borne out in the example.

An assessment of this method would be dominated by the two obvious shortcomings. The method is somewhat ad hoc in that it requires a rearrangement of the equation in such a way that the convergence criterion is satisfied, and even when such a rearrangement is complete, the convergence rate can be very slow. There is a clear need for a method with much more guarantee of success.

The first method to be considered introduces a very basic idea in numerical work and yields a highly robust method. The method is called the bisection method and is very simple indeed. Suppose it is known that $f(a) < 0$ and $f(b) > 0$ or vice versa. Then evaluating the function $f(x)$ at the midpoint $c = (a+b)/2$ and considering the sign of the result will determine which side of c the required zero lies. Hence if the sign of $f(c)$ is the same as the sign of $f(a)$ then the zero lies between c and b. Alternatively if the sign of $f(c)$ is the same as that of $f(b)$ then the zero lies between a and c. The process is then repeated with the new limits. After M steps the interval length is $(b-a)/2^M$ and hence if n digits are required the number of iterations must satisfy

$$10^{-n} = (b-a)/2^M \tag{1.3.12}$$

to give the number of iterations as

$$M = \frac{\ln(b-a) + n \ln 10}{\ln 2} \tag{1.3.13}$$

or approximately $\ln 10/\ln 2$ iterations per figure, which is roughly three per figure.

As a simple example of this process consider the solution of the non-linear equation

$$f(x) = e^{-x} - \sin(x) \tag{1.3.14}$$

with initial interval from 0.0 to 1.5 or approximately $\pi/2$. There will be other intersections for larger x values. The resulting intervals are shown in Table 1.3, as the bisection process develops.

Table 1.3

a	b
0.0000	1.5000
0.0000	0.7500
0.3750	0.7500
0.5625	0.7500
0.5625	0.6563
0.5625	0.6094
0.5859	0.6094
0.5859	0.5977
0.5859	0.5918
0.5859	0.5889
0.5874	0.5889
0.5881	0.5889
0.5885	0.5889
0.5885	0.5886
0.5885	0.5885

It can be seen that the rough rule of three iterations per figure holds good with 14 iterations for four figures, and the accepted value is taken when a and b agree to the required accuracy. As long as the initial interval contains the required root, this method will be very rugged. But like the direct iterative method, convergence is fairly slow, though unlike direct iteration the rate of convergence is not subject to the vagaries of the function $f(x)$ itself. More rapid convergence with no corresponding loss of reliability is required.

Two such related methods rely on simple geometrical ideas. These are usually known as the secant method and the method of false position. The geometry is illustrated in Figure 1.1.

Suppose the points x_0 and x_1 straddle the root at x. These points can be determined by a crude initial tabulation. In the method of false position, a line is drawn from the point $(x_0, f(x_0))$ to the point $(x_1, f(x_1))$, cutting the x-axis at the new point x_2. This point is intuitively nearer to x than x_1. The process is repeated to generate an iterative sequence. The line joining the general point x_n to the original point x_0 has the equation

$$y - f(x_0) = \frac{f(x_n) - f(x_0)}{x_n - x_0}(x - x_0) \tag{1.3.15}$$

and the point of intersection with the x-axis gives $y = 0$ and $x = x_{n+1}$ and the update formula

$$x_{n+1} = \frac{x_0 f(x_n) - x_n f(x_0)}{f(x_n) - f(x_0)} \tag{1.3.16}$$

which is the method of false position or regula falsi.

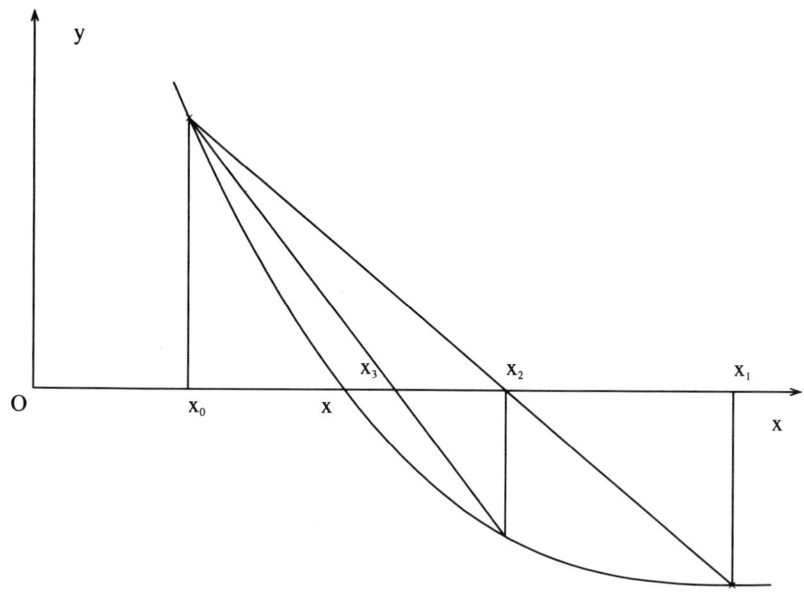

Figure 1.1

An obvious modification of the method is to update both x_n and x_0 so that the most recent values are used. This results in the secant method with the update formula

$$x_{n+1} = \frac{x_{n-1}f(x_n) - x_n f(x_{n-1})}{f(x_n) - f(x_{n-1})} \tag{1.3.17}$$

Now for example applying formula (1.3.16) to example (1.3.3) gives the iteration

$$x_{n+1} = \frac{x_n^3 - 9x_n^2 + 4 - (-4)x_n}{x_n^3 - 9x_n^2 + 8} \tag{1.3.18}$$

where x_0 has been set to unity. The iterates are now

$$\{0.0, 0.5, 0.659, 0.688, 0.693, \ldots\}$$

with more rapid convergence than direct iteration, and no hit and miss rearrangement required.

To investigate the convergence of the method of false position, it is convenient to rearrange the equation as

$$x_{n+1} = x_n + \frac{f(x_n)(x_0 - x_n)}{f(x_n) - f(x_0)} \tag{1.3.19}$$

and again using equation (1.3.9) for ε gives

$$\begin{aligned}\varepsilon_{n+1} &= \varepsilon_n + \frac{f(x_n + \varepsilon_n)(x_0 - x - \varepsilon_n)}{f(x + \varepsilon_n) - f(x_0)} \\ &= \varepsilon_n + \frac{\varepsilon_n f'(x)(x_0 - x)}{-f(x_0)} + O(\varepsilon^2) \\ &= \varepsilon_n \left(1 - \frac{f'(x)(x_0 - x)}{f(x_0)}\right) + O(\varepsilon^2)\end{aligned} \tag{1.3.20}$$

by expanding by Taylor's series to first order. Noting now that the ratio

$$\frac{(x_0 - x)}{f(x_0)} \tag{1.3.21}$$

is approximately $1/f'(x)$ as $f(x) = 0$ it is clear that

$$\left|1 - \frac{f'(x)(x_0 - x)}{f(x_0)}\right| < 1 \tag{1.3.22}$$

and so convergence is established. Convergence of this form is said to be linear or of order 1.

For the secant method, a similar approach is used to yield:

$$\begin{aligned}\varepsilon_{n+1} &= \varepsilon_{n-1}\varepsilon_n \frac{f''(x)}{f'(x)} + \text{higher terms} \\ &= A\varepsilon_{n-1}\varepsilon_n + \text{higher terms}\end{aligned} \tag{1.3.23}$$

To discover what order of convergence this relation represents, a value of m is required such that

$$\varepsilon_{n+1} = K\varepsilon_n^m \tag{1.3.24}$$

Then

$$\varepsilon_{n-1} = K^{-1/m}\varepsilon_n^{1/m} \tag{1.3.25}$$

to leave

$$\varepsilon_{n+1} = A\varepsilon_n K^{-1/m}\varepsilon_n^{1/m} \tag{1.3.26}$$

and hence

$$1 + \frac{1}{m} = m \tag{1.3.27}$$

resulting in $m = (1 + \sqrt{5})/2 = 1.618$. This method has an order of convergence somewhat better than linear and yet less good than quadratic. This is sometimes termed super-linear.

There is also a possibility of rounding error problems with this method as both the denominator and the numerator of equation (1.3.17) will involve the difference of two very similar numbers as convergence is approached. An implementation of the method would at first result in quite rapid convergence, but as the point x is reached the rounding errors will limit the level of accuracy which can be ultimately obtained. This effect is magnified if the difference in the neighbouring f values is second order near to x, as will be the case if $f'(x)$ is small. Observation of the convergence formula (1.3.23) shows that under these circumstances the rate of convergence will also be impaired.

The end product of the secant and false position methods is a common dichotomy in numerical work. The method of false position is stable but slowly convergent, whereas the price paid for improved convergence is a fall in stability in the secant method.

The very commonly used method due to Newton answers some of the above criticisms. The geometry of the method is shown in Figure 1.2.

Instead of a secant being drawn as in the previous methods, a tangent to the curve $y = f(x)$ at the current point is used. Where this tangent cuts the real axis is taken to be the next point in the iteration. Hence the tangent at $x = x_0$ has the equation

$$y - f(x_0) = f'(x_0)(x - x_0) \tag{1.3.28}$$

and at the point of intersection with the real axis $y = 0$ and

$$x_1 = x_0 - \frac{f(x_0)}{f'(x_0)} \tag{1.3.29}$$

or in general

$$x_{n+1} = x_n - \frac{f(x_n)}{f'(x_n)} \tag{1.3.30}$$

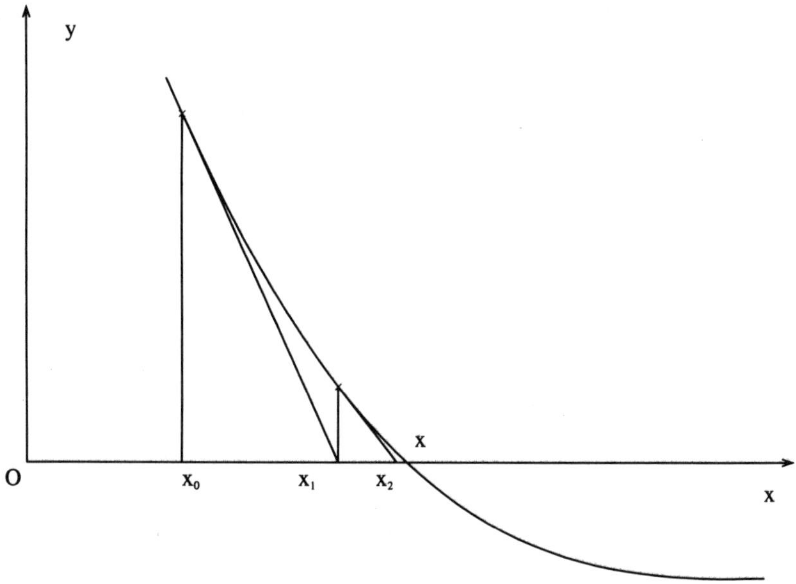

Figure 1.2

This is Newton's method and requires a knowledge of $f'(x)$. The method appears to converge quickly as is illustrated in the example $x^3 - 9x^2 + 4 = 0$ for which the iteration is

$$x_{n+1} = x_n - \frac{x_n^3 - 9x_n^2 + 4}{3x_n^2 - 18x_n} \tag{1.3.31}$$

to yield the iterates

$$\{1.0,\ 0.733\,33,\ 0.694\,872\,79,\ 0.693\,958\,11, \ldots\}$$

The rate of convergence can be easily investigated as before using equation (1.3.9) to give

$$\begin{aligned} \varepsilon_{n+1} &= \varepsilon_n - \frac{f(x+\varepsilon_n)}{f'(x+\varepsilon_n)} \\ &= \varepsilon_n - \frac{f(x) + \varepsilon_n f'(x) + \frac{\varepsilon_n^2}{2} f''(x) + \cdots}{f'(x) + \varepsilon_n f''(x) + \cdots} \\ &= \frac{\varepsilon_n f'(x) + \varepsilon_n^2 f''(x) - \varepsilon_n f'(x) - \frac{\varepsilon_n^2}{2} f''(x) + \cdots}{f'(x) + \varepsilon_n f''(x) + \cdots} \\ &= \varepsilon_n^2 \frac{f''(x)}{2f'(x)} + O(\varepsilon_n^3) \end{aligned} \tag{1.3.32}$$

by using Taylor's expansion and remembering that $f(x) = 0$. This convergence is quadratic and the number of correct figures will double at each iteration. This effect can be observed in the example.

Should it be impossible to find $f'(x)$ in an analytic form, approximations to $f'(x)$ may be used based on the formula

$$f'(x) = \frac{f(x+h) - f(x)}{h} \tag{1.3.33}$$

with h not taken to be either too large and giving a large truncation error or too small when the rounding errors will become significant as seen earlier in equation (1.1.4).

There is an alternative approach to deriving Newton's method which does not rely on geometric ideas, but rather uses analytic methods. It is this approach which allows an easy generalization to many variables. Firstly in one variable consider an iterate x_n being found and the next iterate x_{n+1} being required. Then Taylor's series yields the formula

$$f(x_{n+1}) = f(x_n) + (x_{n+1} - x_n)f'(x_n) + O((x_{n+1} - x_n)^2) \tag{1.3.34}$$

and forcing $f(x_{n+1}) = 0$ leaves an infinite degree polynomial for the new point x_n. Clearly a good approximation to the solution of this polynomial near to the point of convergence will be that obtained from just the linear terms in (1.3.34). Immediately Newton's method is obtained in the form of equation (1.3.30).

The error analysis of equation (1.3.32) suggests that if $f'(x) = 0$ then quadratic convergence is lost. Indeed if $f'(x_n)$ becomes zero during the iterations, then the tangent is 'horizontal', and there is no finite intersection with the real axis. This situation can arise accidently during an iteration, and the usual practical approach is to programme an 'escape clause', that if $f'(x_n)$ becomes less than some threshold, then a fixed step is taken in the direction of convergence as seen from the previous iterates. The process can then be restarted. If the fixed step yields another small $f'(x_n)$, then halve the step and try again. If some agreed number of such steps does not resolve the problem, then maybe the $f(x)$ is itself close to a straight line parallel to the x-axis, and the problem being solved has no sensible solution.

Another way that $f'(x) = 0$ becomes a problem is if there are multiple roots such as

$$f(x) = (x-1)^2 h(x) = 0 \tag{1.3.35}$$

for some function $h(x)$ with simple roots not at $+1$. The standard form of Newton's method will not now yield quadratic convergence. If Newton's method is thought of as $x_{n+1} = g(x_n)$ with $g(x) = x - f(x)/f'(x)$, then the linear convergence condition (1.3.11) gives

$$g'(x) = f(x)f''(x)/f'(x)^2 \tag{1.3.36}$$

which yields zero and indicates at least quadratic convergence. However for a double root, say, $f(x) = f'(x) = 0$, an application of l'Hôpital's rule gives $g'(x) =$

$\frac{1}{2}$. Hence convergence still occurs but now it is linear convergence. It is then clear that if $g(x)$ is adjusted to $g(x) = x - 2f(x)/f'(x)$ then $g'(x) = 0$ once again and at least quadratic convergence is re-established. In general with a root of multiplicity p the iteration

$$x_{n+1} = x_n - pf(x_n)/f'(x_n) \tag{1.3.37}$$

will yield quadratic convergence. This formula is, however, not very useful in practice. Given some non-linear equation $f(x) = 0$, there will be no a priori knowledge as to the multiplicity of the roots. Hence there is no guide as to which p to use. Obviously if quadratic convergence is seen not to be occurring, then alarm bells may ring, but by then the root is known by the resulting linear convergence. Even then what p to use is not clear. Care has to be taken in inventing algorithms like this with little more than theoretical interest.

There are some interesting further methods which link together a number of ideas. Given a sequence of iterates, say $x_n,\ x_{n+1}, \ldots$, it is tempting to fit these values to a simple curve and extrapolate from this curve to the next member of the sequence. Assumptions as to the behaviour of the sequence need to be made to obtain mathematical formulae for the update rule. One very simple assumption is to suppose that the iteration in question is converging linearly.

Hence if $e_n = x - x_n$ is the error, x being the converged value, then linear convergence gives $e_{n+1} = \lambda e_n$. Hence if three successive iterates are known then

$$\begin{aligned} x - x_{n+1} &= \lambda(x - x_n) \\ x - x_{n+2} &= \lambda(x - x_{n+1}) \end{aligned} \tag{1.3.38}$$

and λ can be eliminated to yield an estimate of x. The resulting formula is

$$\begin{aligned} x &= \frac{x_{n+2}x_n - x_{n+1}^2}{x_{n+2} - 2x_{n+1} + x_n} \\ &= x_n - \frac{(x_{n+1} - x_n)^2}{(x_{n+2} - 2x_{n+1} + x_n)} \end{aligned} \tag{1.3.39}$$

and is called Aitken's extrapolation formula.

The power of this rule can be illustrated by considering a somewhat hybrid process for solving $x = f(x)$ based on the direct iteration method of equations (1.3.4) and following. In that discussion, emphasis was placed on condition (1.3.11) for convergence. Used with the Aitken method the convergence condition can be ignored using the following device. Use direct iteration to compute x_{n+1} and x_{n+2} from the current x_n, irrespective of whether the process converges or not! Then apply Aitken's formula to get x. Use this x as the new x_n and repeat. Hence the rule is

$$\begin{aligned} x_{n+1} &= f(x_n) \\ x_{n+2} &= f(x_{n+1}) \end{aligned}$$

$$x = \frac{f(x_{n+1})x_n - f^2(x_n)}{f(x_{n+1}) - 2f(x_n) + x_n} \tag{1.3.40}$$

If x_{n+1} and x_{n+2} are substituted out, and if x is then renamed x_{n+1} the update rule becomes:

$$x_{n+1} = \frac{x_n f(f(x_n)) - f^2(x_n)}{f(f(x_n)) - 2f(x_n) + x_n} \tag{1.3.41}$$

This rule is called Steffenson's method, but the form in equation (1.3.41) is that for the solution of $x = f(x)$. It is more conventional in the literature to generate the form for the solution of $F(x) = 0$ which is easily derived from (1.3.41) by substituting $f(x) = x + F(x)$ which gives after a few lines of manipulation the more standard form of Steffenson's method:

$$x_{n+1} = x_n - \frac{F(x_n)}{F(x_n + F(x_n)) - F(x_n)} F(x_n) \tag{1.3.42}$$

This formula can be derived from (1.3.30) by using a estimate for the derivative $f'(x)$ in the form

$$x_{n+1} = x_n - \left[\frac{F(x_n + h) - F(x_n)}{h}\right]^{-1} F(x_n) \tag{1.3.43}$$

and substituting $h = f(x_n)$. Further an error analysis along the lines of (1.3.32) establishes that for simple roots the formula converges quadratically, but does not involve any evaluation of derivatives. This can be illustrated by considering the example

$$f(x) = e^{-x} - \sin x \tag{1.3.44}$$

An equivalent direct iteration is

$$x_{n+1} = -\ln(\sin(x_n)) \tag{1.3.45}$$

which is easily shown to be divergent. Nevertheless, computing two iterates and using Aitken yields the set of results in Table 1.4. Rapid convergence is seen despite the underlying sequence being divergent.

Table 1.4

k	x_1	x_2	x_3
0	0.500 000 0000	0.735 166 6863	0.399 417 2343
x	0.596 867 7651		
1	0.596 867 7651	0.576 156 3470	0.607 327 6688
x	0.588 599 8355		
2	0.588 599 8355	0.588 432 2295	0.588 683 3591
x	0.588 532 7484		
3	0.588 532 7484	0.588 532 7374	0.588 532 7538
x	0.588 532 74398		

This particular topic illustrates well the three areas of study which constitute numerical analysis. In the first, various approaches are used to invent methods for the problem in hand. There are no set rules here. Sometimes geometric ideas are fruitful (the method of false position), sometimes analytic (Newton derived from a Taylor expansion). Steffenson uses a link between two different ideas. In each case the second topic is then pursued rigorously. The theoretical convergence rate of the process can be investigated, the underlying assumption being that all arithmetic is carried out exactly. These investigations yield shortcomings in methods such as Newton no longer being a quadratically convergent method for multiple roots. The third area is stability when rounding errors occur. This is less apparent as a problem in non-linear equations, but the secant method results in subtractive cancellation near to the point of convergence with an associated loss of precision.

It is convenient to complete the topic of non-linear equations in this chapter rather than to return to the topic later. There are two remaining topics which merit consideration. The first is the particular problem of finding the roots of a real polynomial, in which the problem of finding the complex pairs of roots arises. A method to tackle this problem is Bairstow's method which will be covered next. The second topic is that of sets of non-linear equations which will also be treated here.

Finally a further general purpose topic will be covered. This is the subject of sequence accelerators of which Aitken's extrapolation formula is a simple example. Accelerators are used in quite a number of different applications later in the book, and it is convenient to consider the topic amongst the 'preliminaries'. The first-time reader may well be advised to skip these sections and return to them as necessary at a later time.

1.4 Bairstow's method

A very common problem in non-linear equations is to find the roots of a polynomial. Clearly any of the above methods could be employed, but these all involve only real arithmetic. Newton's method, for example, can be shown to converge in the complex field, but there is always the practical difficulty in an algorithm, that if any real value arises in the iteration then the following iterates must all be real. The real axis acts as a 'black hole' and locks the iterations into themselves. Algorithmic trickery can be used in such cases to restart the method off the axis and re-introduce complex numbers, but this can be very tedious if the required root is near to the real axis. A modification of Newton's method which easily yields complex roots for a polynomial is the method of Bairstow (1914). This prejudges the next paragraph on sets of equations, as finding roots in a complex plane is in effect solving two real problems. However, the concern here is for real polynomials with possibly complex pairs of roots.

The principle involved is again simple and effective, though some algebraic manipulation is required. Suppose $f(z)$ is written as

$$f(z) = (z^2 - pz - l)q(z) + rz + s \tag{1.4.1}$$

then the problem of finding the roots of $f(z)$ can be re-phrased as the problem of finding p and l such that r and s are made zero. Obviously r and s are functions of p and l, and so the solution is well-posed in that there are two unknowns, namely p and l and two equations $r(p,l) = 0$ and $s(p,l) = 0$. Hence the process involves iterating for the coefficients of the quadratic factor, rather than for the roots themselves. These coefficients will be real, and the corresponding roots then follow as simple roots of the resulting quadratic. The algebraic aspect of the derivation is how $r(p,l)$ and $s(p,l)$ can be evaluated in an efficient manner, and indeed how can their derivatives be found in order to complete a Newton-like rule. This is done using the remainder theorem. Suppose $q(z)$ and $f(z)$ have the forms

$$f(z) = a_n z^n + \cdots + a_0 \tag{1.4.2}$$

$$q(z) = q_n z^{n-2} + \cdots + q_2 \tag{1.4.3}$$

then equating coefficients of z^i in (1.4.1) gives the recurrence relations

$$\begin{aligned} q_n &= a_n \\ q_{n-1} &= a_{n-1} + pq_n \\ q_s &= a_s + pq_{s+1} + lq_{s+2} \qquad s = n-2, \ldots, 2 \\ r &= a_1 + pq_2 + lq_3 \end{aligned} \tag{1.4.4}$$

Formal continuation of these relations gives in addition

$$\begin{aligned} q_1 &= a_1 + pq_2 + lq_3 \\ q_0 &= a_0 + pq_1 + lq_2 \end{aligned} \tag{1.4.5}$$

which gives $r = q_1$ and $s = q_0 - pq_1 = q_0'$, say.

Hence the evaluation of r and s is related to being able to evaluate $f(\alpha + \mathrm{i}\beta)$ where the roots of $z^2 + pz - l$ are $\alpha + \mathrm{i}\beta$ and $\alpha - \mathrm{i}\beta$, as

$$f(\alpha + \mathrm{i}\beta) = (r\alpha + s) + \mathrm{i}r\beta \tag{1.4.6}$$

In a similar manner $f'(\alpha + \mathrm{i}\beta)$ can be found so paving the way forward to applying Newton's method. Let

$$q(z) = (z^2 - pz - l)T(z) + T_1 z + T_0' \tag{1.4.7}$$

with

$$t(z) = T_{n-2} z^{n-4} + T_{n-3} z^{n-5} + \cdots + T_2 \tag{1.4.8}$$

then again equating coefficients gives the relations

$$\begin{aligned} T_{n-2} &= q_n \\ T_{n-3} &= q_{n-1} + pT_{n-2} \end{aligned}$$

$$T_s = q_{s+2} + pT_{s+1} + lT_{s+2} \qquad s = n-4, \ldots, 0$$
$$T'_0 = T_0 - pT_1 \tag{1.4.9}$$

Differentiating (1.4.1) gives

$$f'(z) = (z^2 - pz - l)q'(z) + (2z - p)q(z) = q_1 \tag{1.4.10}$$

and hence

$$f'(\alpha + \mathrm{i}\beta) = (2\alpha + 2\mathrm{i}\beta - p)q(\alpha + \mathrm{i}\beta) + q_1 \tag{1.4.11}$$

but equation (1.4.7) gives

$$q(\alpha + \mathrm{i}\beta) = T_1\alpha + T'_0 + \mathrm{i}T_1\beta \tag{1.4.12}$$

and $2\alpha = p$ to give finally

$$f'(\alpha + \mathrm{i}\beta) = 2\mathrm{i}\beta(T_1\alpha + T'_0 + \mathrm{i}T_1\beta) + q_1 \tag{1.4.13}$$

Newton's method can now be invoked. Using the Taylor series approach, the object is to find p and l which force $r(p, l)$ and $s(p, l)$ to zero to first order, that is ignoring second derivative terms. Hence in terms of the q_i's

$$(q_1z + q'_0) + \frac{\partial}{\partial p}(q_1z + q'_0)\delta p + \frac{\partial}{\partial l}(q_1z + q'_0)\delta l = 0 \tag{1.4.14}$$

is the first order expansion as a zero of $q_1z + q'_0$ is required. δp and δl are the increments which are needed to update p and l at each iteration. From equation (1.4.1)

$$\frac{\partial f}{\partial l} = -q(z) + (z^2 - pz - l)\frac{\partial q}{\partial l} + \frac{\partial q_1}{\partial l}z + \frac{\partial q'_0}{\partial l} \tag{1.4.15}$$
$$\frac{\partial f}{\partial p} = -zq(z) + (z^2 - pz - l)\frac{\partial q}{\partial p} + \frac{\partial q_1}{\partial p}z + \frac{\partial q'_0}{\partial p} \tag{1.4.16}$$

and only arithmetic modulo $(z^2 - pz - l)$ need be considered, hence

$$\frac{\partial q_1}{\partial l}z + \frac{\partial q'_0}{\partial l} = q(z) = T_1z + T'_0 \tag{1.4.17}$$
$$\frac{\partial q_1}{\partial p}z + \frac{\partial q'_0}{\partial p} = zq(z) = z(pT_1 + T'_0) + lT_1 \tag{1.4.18}$$

ignoring the factor $z^2 - pz - l$. Hence the equations for the increments δp and δl are

$$q_1 + (pT_1 + T'_0)\delta p + T_1\delta l = 0 \tag{1.4.19}$$
$$q'_0 + lT_1\delta p + T'_0\delta l = 0 \tag{1.4.20}$$

which give the update formulae

$$\delta p = (T_1 q_0 - T_0 q_1)/D \tag{1.4.21}$$

$$\delta l = (M q_1 - T_0 q_0)/D \tag{1.4.22}$$

where $M = lT_1 + pT_0$ and $D = T_0^2 - MT_1$. Initial values for p and l are used to start the process, say $p^{(0)}$ and $l^{(0)}$, and each step uses the update

$$p^{(n+1)} = p^{(n)} + \delta p \tag{1.4.23}$$

$$l^{(n+1)} = l^{(n)} + \delta l \tag{1.4.24}$$

with δp and δl being given by the recurrence relations (1.4.4) and (1.4.9) and the formulae (1.4.21) and (1.4.22). Once a factor has been found it can be divided out and the new polynomial of degree two less can be processed in the same way, and so on until either a quadratic or linear factor remains.

The process is illustrated by the example

$$z^4 - 8z^3 + 24z^2 + 336z + 120 = 0 \tag{1.4.25}$$

using the initial values $p = -4.5$ and $l = -1.5$. The first step gives

$$\begin{array}{lll} q_4 = 1 & q_3 = -25/2 & q_2 = 315/4 \\ q_1 = 3/8 & q_0 = 3/16 & q_0' = 15/8 \end{array} \tag{1.4.26}$$

and

$$\begin{array}{lll} T_2 = 1 & T_1 = -17 & T_0 = 615/4 \\ M = -5331/8 & D = 559\,479/16 & \end{array} \tag{1.4.27}$$

to give $\delta p = -0.003\,48$ and $\delta l = -0.007\,97$ and the iterated factor $z^2 + 4.503z + 1.508$. Repeated steps gives convergence to the factor $z^2 + 4.5049z + 1.5226$.

Once factors are found, either by Bairstow's process for a quadratic factor or by Newton's method for a real linear factor, these known factors are divided out to yield a lower degree polynomial with all the original roots except the known ones. This process is generally known as deflation and will recur in Chapter 4 on eigenvalue problems. For simple polynomials the deflation process is no more than division and can be implemented quite succinctly. Firstly for a linear factor, let

$$p(x) = a_n x^n + \cdots + a_0 \tag{1.4.28}$$

then $q(x)$ is required where

$$p(x) = (x - x_0) q(x) \tag{1.4.29}$$

and

$$q(x) = b_n x^{n-1} + \cdots + b_1 \tag{1.4.30}$$

Comparing coefficients then yields the recurrence relation

$$b_k = a_k + b_{k+1}x_0 \tag{1.4.31}$$

for the new coefficients. This process is Horner's method of synthetic division.

In the quadratic case

$$p(x) = (x^2 - px - l)q(x) \tag{1.4.32}$$

with

$$q(x) = b_n x^{n-2} + \cdots + b_2 \tag{1.4.33}$$

to give the recurrence relation

$$b_k = a_k + pb_{k+1} + lb_{k+2} \tag{1.4.34}$$

A little care needs to be exercised in the use of these deflations in a repeated manner. The first division will be by a factor which is itself approximate, having been found by iteration only to some determined accuracy. Hence the deflated polynomial will have coefficients with this error built in. The roots of a polynomial can be very sensitive to its coefficients, and hence small errors now introduced into the deflated coefficients can magnify into the roots obtained at this stage. Clearly successive dividing out can exasperate the problem until after a few steps there is little relation between the deflated roots and those of the original polynomial. One safeguard is to accept the deflated roots on a temporary basis and then return to the original and use the deflated roots as starting values to confirm that no problems have arisen. Any small loss of accuracy will then be recovered, and if severe ill-conditioning has occurred then further investigation can be made.

1.5 QD algorithm

The QD algorithm is basically a technique to find the poles of an analytic function given its Taylor expansion. It is a simple modification to obtain a method for the extraction of zeros from a polynomial or indeed to find the eigenvalues of a matrix. The derivation is based on the Hankel determinants $H_k^{(n)}$ for the function $f(z) = \sum a_n z^n$ defined by $H_0^{(n)} = 1$ and

$$H_k^{(n)} = \begin{vmatrix} a_n & a_{n+1} & \cdots & a_{n+k-1} \\ a_{n+1} & a_{n+2} & \cdots & a_{n+k} \\ \vdots & \vdots & & \vdots \\ a_{n+k-1} & a_{n+k} & \cdots & a_{n+2k-2} \end{vmatrix} \tag{1.5.1}$$

with $n = 0, 1, 2, \ldots$ and $k = 1, 2, \ldots$. A number of properties of these determinants will be quoted here, the proofs being given in Henrici (1958). The first result is the recurrence relation linking Hankel determinants which has the form

$$(H_k^{(n)})^2 - H_k^{(n-1)}H_k^{(n+1)} + H_{k+1}^{(n-1)}H_{k-1}^{(n+1)} = 0 \tag{1.5.2}$$

The second result concerns the poles of $f(z)$. Suppose the poles of $f(z)$ are at $z_i = 1/\lambda_i$ with respective order m_j, then as $n \to \infty$

$$H_m^{(n+1)}/H_m^{(n)} \to \lambda_1^{m_1}\lambda_2^{m_2}\cdots\lambda_k^{m_k} \tag{1.5.3}$$

and

$$m = \sum_{j=1}^{k} m_j \tag{1.5.4}$$

Hence these results constitute a device to compute the poles of a function, except that the computation of the Hankel determinants explicitly is tedious. The QR algorithm avoids this tedium.

Define the 'quotients', $q_k^{(n)}$ as

$$q_k^{(n)} = \frac{H_k^{(n+1)} H_{k-1}^{(n)}}{H_k^{(n)} H_{k-1}^{(n+1)}} \tag{1.5.5}$$

then (1.5.3) implies that $q_k^{(n)} \to \lambda_k$ as $n \to \infty$. In order to avoid the explicit computation of the Hankel determinants, the differences, $e_k^{(n)}$, are defined by

$$e_k^{(n)} = \frac{H_{k+1}^{(n)} H_{k-1}^{(n+1)}}{H_k^{(n)} H_k^{(n+1)}} \tag{1.5.6}$$

with $e_0^{(n)} = 0$ and $n = 0, 1, 2, \ldots$ and $k = 1, 2, \ldots$. The definitions of q in (1.5.5), e in (1.5.6) and the relation between the Hankel determinants in (1.5.2) gives the simple relations

$$q_{k+1}^{(n)} e_k^{(n)} = q_k^{(n+1)} e_k^{(n+1)} \tag{1.5.7}$$

$$q_k^{(n)} + e_k^{(n)} = q_k^{(n+1)} + e_{k-1}^{(n+1)} \tag{1.5.8}$$

which define the QD scheme. The scheme is usually laid out in the following manner:

$$\begin{array}{ccccc}
q_1^{(0)} & & & & \\
 & e_1^{(0)} & & & \\
q_1^{(1)} & & q_2^{(0)} & & \\
 & e_1^{(1)} & & e_2^{(0)} & \\
q_1^{(2)} & & q_2^{(1)} & & \ddots \\
 & e_1^{(2)} & & e_2^{(1)} & \\
q_1^{(3)} & & q_2^{(2)} & & \\
 & e_1^{(3)} & & e_2^{(2)} & \\
q_1^{(4)} & & q_2^{(3)} & & \\
\vdots & \vdots & \vdots & \vdots &
\end{array} \tag{1.5.9}$$

The scheme defined in equations (1.5.7) and (1.5.8) can be used to complete the array in (1.5.9) either by being given the first column and then proceeding from left to right, or by being given the top diagonal and then working downwards. The first strategy proves unstable and so methods are based on the second downward progression. Hence the next stage is to determine these initial elements.

To find the zeros of a polynomial

$$P(z) = c_0 z^N + c_1 z^{N-1} + \cdots + c_{N-1} z + c_N, \qquad c_i \neq 0; \quad i = 0, 1, \ldots, N \tag{1.5.10}$$

the equivalent problem of finding the poles of the function $f(z)$ where

$$f(z) = \frac{1}{P(z)} = \sum a_n z^n \tag{1.5.11}$$

is considered. To obtain a link with the top diagonal of the QD scheme, Wronski's identity

$$\frac{1}{f(z)} = \sum_{k=0}^{\infty} \frac{(-1)^{(1/2)k(k+1)}}{(a_0)^{k+1}} H_k^{(-k+2)} z^k \tag{1.5.12}$$

is used with

$$\frac{1}{f(z)} = \sum_{k=0}^{N} c_{N-k} z^k \tag{1.5.13}$$

where $H_k^{(n)}$ is defined for negative n by considering $a_n = 0$ for $n < 0$. Comparing coefficients gives

$$H_k^{(-k+2)} = \begin{cases} (-1)^{(1/2)k(k+1)} (a_0)^{k+1} c_k, & k = 0, 1, \ldots, N \\ 0, & k > N \end{cases} \tag{1.5.14}$$

But $a_0 = c_0^{-1}$ and

$$H_k^{(-k+1)} = (-1)^{(1/2)k(k+1)} (a_0)^k \tag{1.5.15}$$

$$H_k^{(-k)} = 0 \tag{1.5.16}$$

to give the formulae

$$q_k^{(-k+1)} = \begin{cases} \frac{-c_1}{c_0}, & k = 1 \\ 0, & k > 1 \end{cases} \tag{1.5.17}$$

$$e_k^{(-k+1)} = \begin{cases} \frac{c_{k+1}}{c_k}, & k = 1, 2, \ldots, N-1 \\ 0, & k \geq N \end{cases} \tag{1.5.18}$$

for the computation of the leading diagonal. The condition that no c_i should be zero in (1.5.10) which would cause overflow in (1.5.17) and (1.5.18) if violated, presents no problem, as a linear transformation of the form $\hat{z} = z + A$ for a suitable A can be used which will remove the zero elements. After the process is complete the resulting roots can be corrected by the shift A to recover the roots of the original equation.

To see this process working in practice, consider finding the roots of the polynomial

$$(x-3)(x+4)(x^2+x+2) = x^4+2x^3-9x^2-10x-24 \tag{1.5.19}$$

Then the QD scheme for this problem is shown in Table 1.5. The leading calculations are given in detail and the computations were continued to the 30th iteration which appears at the end of the table.

It can be seen that the first two columns of q's converge gently to the correct roots at $x = 3$ and $x = -4$. Corresponding to the converging q's are the intermediate columns of errors or e's which are moving towards zero. On the other hand, column $e^{(3)}$ is not tending to zero and suggests a complex pair of roots. The quadratic factor, $x^2 - rx + s$, representing these roots is found from the relations:

$$r = q_m^{(3)} + q_m^{(4)} \tag{1.5.20}$$

$$s = q_{m-1}^{(3)} q_m^{(4)} \tag{1.5.21}$$

to give

$$r = -0.293\,89 - 0.706\,11 = -1.000\,00$$

$$s = (-2.832\,427) \times (-0.706\,108) = 1.999\,99 \tag{1.5.22}$$

which are the correct coefficients for the quadratic factor. Clearly formulae (1.5.20) and (1.5.21) can be used anywhere that an e column is not converging.

To place this algorithm in context, it can be seen that convergence is slow compared with quadratically converging algorithms such as Newton. The strength of the process is that it isolates the roots and hence allows more rapidly convergent methods to be applied from the starting points provided from QD. This process can

Table 1.5

$e^{(0)}$	$q^{(1)}$	$e^{(1)}$	$q^{(2)}$	$e^{(2)}$	$q^{(3)}$	$e^{(3)}$	$q^{(4)}$	$e^{(4)}$
	2.000		0.0		0.0		0.0	
0		−4.500		1.111		2.400		0
	−6.5		5.611		1.289		−2.400	
0		3.885		2.552		−4.469		0
	−2.615		1.982		−3.435		2.069	
0		−2.943		−4.424		2.692		0
	−5.559		4.4827		−3.014		−0.6225	
⋮				⋮				⋮
⋮				⋮				⋮
	−3.999		2.999		−2.8324		1.8324	
0		−1.64(−3)		−1.11(−9)		2.53854		0
	−4.0007		3.0007		−0.29389		−0.70611	

be extended to finding eigenvalues of matrices directly from the matrix elements, (Henrici (1958, 1963)).

1.6 Sets of non-linear equations

In this paragraph, natural extensions of the one-dimensional case are made to give effective multi-dimensional methods for non-linear equations. Now the problem is to solve the vector equations

$$\begin{aligned} f_1(x_1, \ldots, x_n) &= 0 \\ f_2(x_1, \ldots, x_n) &= 0 \\ \vdots \quad & \quad \vdots \\ \vdots \quad & \quad \vdots \\ f_n(x_1, \ldots, x_n) &= 0 \end{aligned} \tag{1.6.1}$$

or

$$\mathbf{f}(\mathbf{x}) = 0 \tag{1.6.2}$$

or in rearranged form as

$$\mathbf{x} = \mathbf{G}(\mathbf{x}) \tag{1.6.3}$$

It is now obvious that a direct iteration analogous to equation (1.3.5) can be applied to give

$$\mathbf{x}^{(k+1)} = \mathbf{G}(\mathbf{x}^{(k)}) \tag{1.6.4}$$

which will suffer from the same problems as in one dimension. The iteration requires an ad hoc setting up manipulation, the iteration may not converge and if it does the convergence is only linear. Nevertheless effective iterations can be obtained such as in the following example.

$$\begin{aligned} 3x_1 - \cos(x_2 x_3) - \tfrac{1}{2} &= 0 \\ x_1^2 - 81(x_2 + 0.1)^2 + \sin x_3 + 1.06 &= 0 \\ e^{-x_1 x_2} + 20x_3 + \tfrac{1}{3}(10\pi - 3) &= 0 \end{aligned} \tag{1.6.5}$$

with $-1 \leq x_1 \leq 1$. A manipulation which gives a satisfactory direct iteration formula is

$$\begin{aligned} x_1 &= \tfrac{1}{3} \cos x_2 x_3 + \tfrac{1}{6} \\ x_2 &= \tfrac{1}{9}\sqrt{x_1^2 + \sin x_3 + 1.06} - 0.1 \\ x_3 &= -\tfrac{1}{20} e^{-x_1 x_2} - \tfrac{1}{60}(10\pi - 3) \end{aligned} \tag{1.6.6}$$

which yields the iteration in Table 1.6.

Table 1.6

k	x_1	x_2	x_3
0	0.1	0.1	0.1
1	0.4999	0.0094	−0.5231
2	0.49999	0.000 025	−0.523 36
3	0.50000	0.000 0123	−0.523 598

The true solution is $\{0.5,\ 0.0,\ -\pi/6\}$ which confirms the fairly rapid convergence in this carefully chosen example.

To overcome the criticisms already levelled at this approach, by analogy with the one-dimensional work, a Taylor expansion can be used to generate the multi-dimensional equivalent of Newton's method. Hence the expansion

$$\mathbf{f}(\mathbf{x}^{(k)}) = \mathbf{f}(\mathbf{x}^{(k-1)}) + \frac{\partial \mathbf{f}}{\partial x_1}(x_1^{(k)} - x_1^{(k-1)}) + \cdots + \frac{\partial \mathbf{f}}{\partial x_n}(x_n^{(k)} - x_n^{(k-1)}) + \cdots O(x_i^{(k)} - x_i^{(k-1)})^2 \tag{1.6.7}$$

is considered with $\mathbf{f}(\mathbf{x}^{(k)}) = 0$ and the higher terms ignored to give

$$0 = \mathbf{f}(\mathbf{x}^{(k-1)}) + \mathbf{J}(\mathbf{x}^{(k-1)})\mathbf{x}^{(k)} - \mathbf{J}(\mathbf{x}^{(k-1)})\mathbf{x}^{(k-1)} \tag{1.6.8}$$

where $\mathbf{J}$ is the Jacobian matrix

$$\mathbf{J} = \begin{pmatrix} \frac{\partial \mathbf{f_1}}{\partial x_1} & \cdots & \cdots & \frac{\partial \mathbf{f_1}}{\partial x_n} \\ \vdots & & & \vdots \\ \vdots & & & \vdots \\ \frac{\partial \mathbf{f_n}}{\partial x_1} & \cdots & \cdots & \frac{\partial \mathbf{f_n}}{\partial x_n} \end{pmatrix} \tag{1.6.9}$$

Hence the update formula involves solving a set of linear algebraic equations for the new values $\mathbf{x}^{(k)}$ according to the equations

$$\mathbf{J}(\mathbf{x}^{(k-1)})\mathbf{x}^{(k)} = \mathbf{f}(\mathbf{x}^{(k-1)}) - \mathbf{J}(\mathbf{x}^{(k-1)})\mathbf{x}^{(k-1)} \tag{1.6.10}$$

and these are solved by one of the methods of Chapter 3 such as Gaussian elimination in Section 3.2.

In the example of equation (1.6.5) the Jacobian matrix has the specific elements:

$$\mathbf{J} = \begin{pmatrix} 3 & x_3 \sin x_2x_3 & x_2 \sin x_2x_3 \\ 2x_1 & -162(x_2 + 0.1) & \cos x_3 \\ -x_2e^{-x_1x_2} & -x_1e^{-x_1x_2} & 20 \end{pmatrix} \tag{1.6.11}$$

and the corresponding iterates are given in Table 1.7.

This example, like its one-dimensional counterpart demonstrates quadratic convergence with a doubling of the number of correct digits at each iteration.

Again the obvious problem will be the difficulty of obtaining analytic derivatives to form $\mathbf{J}$ in some cases, especially if the defining non-linear equations are not

Table 1.7

k	x_1	x_2	x_3
0	0.1	0.1	0.1
1	0.5000	0.0194	−0.5215
2	0.500 00	0.001 589	−0.523 557
3	0.500 000	0.000 012 44	−0.523 598 45
4	0.500 0000	0.000 000	−0.523 598 77

analytic in form but are derived from say another numerical process. Estimates of the partial derivatives using

$$\frac{\partial f_j}{\partial x_i} \sim \frac{f_j(x_i + h) - f_j(x_i)}{h} \tag{1.6.12}$$

can be effective remembering not to allow h to be either too small or too large. A complete way around this difficulty is to use what is known as a quasi-Newton method in which $\mathbf{J}$ is replaced with an approximation $\mathbf{A}_1$ which satisfies

$$\mathbf{A}_1(\mathbf{x}^{(1)} - \mathbf{x}^{(0)}) = \mathbf{f}(\mathbf{x}^{(1)} - \mathbf{x}^{(0)}) \tag{1.6.13}$$

Equation (1.6.13) does not define $\mathbf{A}_1$ uniquely as there is no restriction on the variation of $\mathbf{A}_1$ along any vector orthogonal to $(\mathbf{x}^{(1)} - \mathbf{x}^{(0)})$. Hence the condition

$$\mathbf{A}_1\mathbf{z} = \mathbf{J}(\mathbf{x}^{(0)})\mathbf{z} \tag{1.6.14}$$

can be satisfied for any $\mathbf{z}$ for which $(\mathbf{x}^{(1)} - \mathbf{x}^{(0)})^T\mathbf{z} = 0$. Now $\mathbf{A}_1$ is determined according to

$$\mathbf{A}_1 = \mathbf{J}(\mathbf{x}^{(0)}) + \frac{[\mathbf{f}(\mathbf{x}^{(1)}) - \mathbf{f}(\mathbf{x}^{(0)}) - \mathbf{J}(\mathbf{x}^{(0)})(\mathbf{x}^{(1)} - \mathbf{x}^{(0)})](\mathbf{x}^{(1)} - \mathbf{x}^{(0)})^T}{||\mathbf{x}^{(1)} - \mathbf{x}^{(0)}||_2^2} \tag{1.6.15}$$

to yield a complete method defined by the formulae

$$\mathbf{A}_i = \mathbf{A}_{i-1} + \frac{(\mathbf{y}_i - \mathbf{A}_{i-1}\mathbf{s}_i)\mathbf{s}_i^T}{||\mathbf{s}_i||_2^2} \tag{1.6.16}$$

with

$$\mathbf{x}^{(i+1)} = \mathbf{x}^{(i)} - \mathbf{A}_i^{-1}\mathbf{f}(\mathbf{x}^{(i)}) \tag{1.6.17}$$

$$\mathbf{y}_i = \mathbf{f}(\mathbf{x}^{(i)}) - \mathbf{f}(\mathbf{x}^{(i-1)}) \tag{1.6.18}$$

and

$$\mathbf{s}_i = \mathbf{x}^{(i)} - \mathbf{x}^{(i-1)} \tag{1.6.19}$$

This method is due to Broyden (1965) and will be referred to again in Chapter 10 where optimization is considered. The application of this quasi-Newton method to example (1.6.5) is illustrated in Table 1.8.

Table 1.8

k	x_1	x_2	x_3
0	0.1	0.1	0.1
1	0.499 86	0.0194	−0.521 52
2	0.499 9863	0.008 73	−0.523 1746
3	0.500 0066	0.000 00867	−0.523 69818

Good convergence is observed, but unlike Newton's method the matrix **J** is not directly computed. The price paid is a more involved algorithm and a little less rapid convergence.

The general area of non-linear equation solving is well represented in packages and libraries. For example sets of equations can be handled in MAPLE with simple linguistic features such as

$$fsolve(\{x^2 - 2y = 1; x + y^2 = 2\}, \{x, y\})$$

which allows the two quoted non-linear equations to be solved for x and y.

1.7 Sequence accelerators

This chapter is concluded with a study of methods which take a given sequence, and generate a second sequence which has accelerated convergence. The extrapolation formula of Aitken used in Steffenson's method for non-linear equations is a simple example. These processes can be used with any numerical technique in which a sequence of converging, or even diverging values, has resulted.

The first of the accelerator methods to be considered will be that of Shanks (1955) who considered the sequence denoted by

$$A = \{A_r\} \qquad r = 0, 1, 2, \ldots \tag{1.7.1}$$

The root of the method is to try to fit the elements of the sequence to a pre-defined form which has free constants. These constants are then computed, and the limit of the pre-defined form gives an estimate of the required limit. This process is applied in a local manner, and the sequence of estimated limits itself forms a sequence, and it is this sequence which should converge more rapidly than the original sequence, but to the same limit. The pre-defined form used by Shanks is

$$A_r = B_{kn} + \sum_{i=1}^{k} \alpha_{in} q_{in}^r \qquad n - k \le r \le n + k; \quad n \ge k \tag{1.7.2}$$

and $q_{in} \neq 1$ or 0. The term B_{kn} is called the local base and for $|q_{in}| \le 1$ it is clear that B_{kn} will be the required limit as $r \to \infty$. Hence there are $2k + 1$ non-linear equations for the free constants B_{kn}, α_{in} and q_{in}, and the requirement is to compute B_{kn}. The special form of these non-linear equations allows them to be solved using the following result. If the quantities p_s are defined as

$$p_s = \sum_{i=1}^{k} a_i q_i^s \qquad s = 0, 1, \ldots, 2k-1 \tag{1.7.3}$$

then the q_i's are roots of

$$\begin{vmatrix} 1 & q & \cdots & q^{k-1} & q^k \\ p_0 & p_1 & \cdots & p_{k-1} & p_k \\ p_1 & p_2 & \cdots & p_k & p_{k+1} \\ \vdots & \vdots & & & \vdots \\ p_{k-1} & \cdots & \cdots & \cdots & p_{2k-1} \end{vmatrix} = 0 \tag{1.7.4}$$

If now the difference of successive equations in (1.7.2) is taken and if p_s is substituted for ΔA_{n-k+s}, and a_i for $a_{in}[q_{in}^{n-k+1} - q_{in}^{n-k}]$ then equation (1.7.4) yields the result that the q_{in}'s are the roots of

$$\begin{vmatrix} 1 & q & \cdots & q^k \\ \Delta A_{n-k} & & \cdots & \Delta A_n \\ \vdots & & & \vdots \\ \Delta A_{n-1} & & \cdots & \Delta A_{n+k-1} \end{vmatrix} = 0 \tag{1.7.5}$$

or

$$c_n q^k + c_{n-1} q^{k-1} + \cdots + c_{n-k} = 0. \tag{1.7.6}$$

where the c_j are the minors obtained by expanding (1.7.5) by its first row. If the expression in equation (1.7.2) is substituted into the expression

$$\frac{c_{n-k}A_{n-k} + \cdots + c_{n-1}A_{n-1} + c_n A_n}{c_{n-k} + \cdots + c_{n-1} + c_n} \tag{1.7.7}$$

then using equation (1.7.6) yields

$$\frac{B_{kn}(c_{n-k} + \cdots + c_n) + 0 + \cdots + 0}{c_{n-k} + \cdots + c_n} = B_{kn}. \tag{1.7.8}$$

But expression (1.7.7) is the expansion by the first rows of the ratio of the determinants

$$B_{kn} = \frac{\begin{vmatrix} A_{n-k} & \cdots & A_{n-1} & A_n \\ \Delta A_{n-k} & \cdots & \Delta A_{n-1} & \Delta A_n \\ \Delta A_{n-k+1} & \cdots & \Delta A_n & \Delta A_{n+1} \\ \vdots & & & \vdots \\ \Delta A_{n-1} & \cdots & \Delta A_{n+k-2} & \Delta A_{n+k-1} \end{vmatrix}}{\begin{vmatrix} 1 & \cdots & 1 & 1 \\ \Delta A_{n-k} & \cdots & \Delta A_{n-1} & \Delta A_n \\ \Delta A_{n-k+1} & \cdots & \Delta A_n & \Delta A_{n+1} \\ \vdots & & & \vdots \\ \Delta A_{n-1} & \cdots & \Delta A_{n+k-2} & \Delta A_{n+k-1} \end{vmatrix}} \tag{1.7.9}$$

so yielding a form for B_{kn} as required. The simplest case arises with determinants of order two and yields the transformation

$$B_{1n} = \frac{A_{n-1}A_{n+1} - A_n^2}{A_{n+1} - 2A_n + A_{n-1}} \tag{1.7.10}$$

which is the familiar Aitken extrapolation formula. This formula can be applied to successive sequences of three terms to generate the new sequence $\{B_{1n}, B_{2n}, \ldots\}$ which is usually denoted by the operator e_1 where

$$e_1(A) = \{B_{1n}\} \qquad n \geq 1 \tag{1.7.11}$$

and more generally the operator e_k is defined by

$$e_k(A) = \{B_{kn}\} \qquad n \geq k \tag{1.7.12}$$

A further operator called the diagonal operator e_d is defined by

$$e_d(A) = \{B_{nn}\}. \tag{1.7.13}$$

These operators can be applied to produce a table of new sequences, the first column being the original sequence, the second being e_1, the third e_2, and so on. The sequence from the operator e_d then appears along the top of the display. This table is sometimes called a Shanks table and can be extended downwards until convergence is obtained. An alternative strategy is the repeated use of the simple transformation by applying e_1 to the second column to get the third, giving columns with the transformations e_1, e_1^2, and so on. The effectiveness of these methods is well illustrated by the example

$$A_n = -\sum_{m=1}^{n} \frac{(-2)^m}{m} \tag{1.7.14}$$

which is the Taylor expansion for $\ln(1 + x)$ which diverges for $x = 2$. Yet application of the above transformations gives the impressive results of Table 1.9, where rapid convergence is noted to the true value of ln(3) which is 1.098 612 2886.

The implementation of the Shanks accelerators can be made very succinct using the ε-algorithm, the link between this algorithm and the determinantal form of the

Table 1.9

k	A_k	e_d
1	2.0	
2	0.0	1.142 857
3	2.666 67	1.114 492
4	−1.333 33	1.098 8056
5	5.066 6667	1.098 6256
6	−5.6000	1.098 613 23

Shanks' sequences being made by Wynn (1956). The algorithm is very simple and uses the recurrence relations

$$\varepsilon_0^{(-1)} = 0 \tag{1.7.15}$$

$$\varepsilon_n^{(0)} = A_n \tag{1.7.16}$$

$$\varepsilon_n^{(p)} = \varepsilon_{n+1}^{(p-2)} + [\varepsilon_{n+1}^{(p-1)} - \varepsilon_n^{(p-1)}]^{-1} \tag{1.7.17}$$

and in terms of these ε's the accelerated sequences are given by

$$B_{kn} = \varepsilon_{n-k}^{2k} \tag{1.7.18}$$

to give a typical display of the form

$$\begin{array}{cccccc}
\varepsilon_0^{(-1)} & & & & & \\
\varepsilon_1^{(-1)} & \varepsilon_0^{(0)} & & & & \\
\varepsilon_2^{(-1)} & \varepsilon_1^{(0)} & \varepsilon_0^{(1)} & & & \\
\varepsilon_3^{(-1)} & \varepsilon_2^{(0)} & \varepsilon_1^{(1)} & \varepsilon_0^{(2)} & & \\
\varepsilon_4^{(-1)} & \varepsilon_3^{(0)} & \varepsilon_2^{(1)} & \varepsilon_1^{(2)} & \varepsilon_0^{(3)} & \\
\varepsilon_5^{(-1)} & \varepsilon_4^{(0)} & \varepsilon_3^{(1)} & \varepsilon_2^{(2)} & \varepsilon_1^{(3)} & \varepsilon_0^{(4)} \\
 & & & & & \\
 & A_n & & B_{1n} & & B_{2n}
\end{array} \tag{1.7.19}$$

where the columns representing the sequences $\{A_n\}$, $\{B_{1n}\}$ and $\{B_{2n}\}$ are marked below the relevant column. This display can be built up row by row as the convergence rate dictates. A start is made with three A values and $\varepsilon_0^{(2)}$ is found. Two more A values are added and the table extended through to $\varepsilon_0^{(4)}$. As soon as suitable agreement occurs in one of the new sequences the process can be terminated.

As an example consider the series

$$\arctan x = x - \frac{x^3}{3} + \frac{x^5}{5} - \frac{x^7}{7} + \cdots \tag{1.7.20}$$

for $x = 1$ with a view to computing $\pi/4$. The ε-table then has the form:

$$\begin{array}{cccccc}
0 & & & & & \\
0 & 1 & & & & \\
0 & \frac{2}{3} & -3.0 & & & \\
0 & 0.866\,67 & 5.0 & 0.791\,666 & & \\
0 & 0.723\,809 & -7.0 & 0.783\,333 & -114.9999 & \\
0 & 0.834\,921 & 9.0 & 0.786\,309 & 329.055\,38 & 0.785\,585
\end{array} \tag{1.7.21}$$

giving good converence to $\pi/4 = 0.785\,398\,16$ in only a very few steps.

It is not difficult to generate a sequence which will defeat the Shanks transformations. For example, $A_r = 1/r^2$, will not have the form which was predetermined in equation (1.7.2). To improve the effectiveness of the transformations, Levin (1973) fitted the sequence in a different way. The assumption here is that

$$A_r = T_{kn} + R_k(r) \qquad n \le r \le n+k \tag{1.7.22}$$

with

$$R_k(r) = R_1(r) \sum_{i=0}^{k-1} \gamma_{in} f_i(r) \tag{1.7.23}$$

where the constants γ_{in} are free to force the fit to the given sequence. The functions $R_1(r)$ and $f_i(r)$ are chosen to give a range of different methods, and to simplify the final form. Hence the equations

$$A_r = T_{kn} + R_r \sum_{i=0}^{k-1} \gamma_{in} f_i(r) \qquad n \le r \le n+k \tag{1.7.24}$$

form a set of linear algebraic equations to determine the unknown constants, where R_r has been used for $R_1(r)$. In particular Cramer's rule can be applied to give the following determinantal form for T_{kn}, namely

$$T_{kn} = \frac{\begin{vmatrix} A_n & A_{n+1} & \cdots & A_{n+k} \\ R_n f_0(n) & R_{n+1} f_0(n+1) & \cdots & R_{n+k} f_0(n+k) \\ R_n f_1(n) & R_{n+1} f_1(n+1) & \cdots & R_{n+k} f_1(n+k) \\ \vdots & \vdots & & \vdots \\ R_n f_{k-1}(n) & R_{n+1} f_{k-1}(n+1) & \cdots & R_{n+k} f_{k-1}(n+k) \end{vmatrix}}{\begin{vmatrix} 1 & 1 & \cdots & 1 \\ R_n f_0(n) & R_{n+1} f_0(n+1) & \cdots & R_{n+k} f_0(n+k) \\ R_n f_1(n) & R_{n+1} f_1(n+1) & \cdots & R_{n+k} f_1(n+k) \\ \vdots & \vdots & & \vdots \\ R_n f_{k-1}(n) & R_{n+1} f_{k-1}(n+1) & \cdots & R_{n+k} f_{k-1}(n+k) \end{vmatrix}} \tag{1.7.25}$$

By now taking $f_i(r) = r^{-i}$ and by dividing successive columns by $R_n, R_{n+1}, \ldots, R_{n+k}$, an expansion by the first row gives

$$T_{kn} = \frac{\sum_{j=0}^{k} (-1)^j \binom{k}{j} \left(\frac{n+j}{n+k}\right)^{k-1} \frac{A_{n+j}}{R_{n+j}}}{\sum_{j=0}^{k} (-1)^j \binom{k}{j} \left(\frac{n+j}{n+k}\right)^{k-1} \frac{1}{R_{n+j}}} \tag{1.7.26}$$

The so-called t-transformation is then obtained by taking $R_r = a_r = \Delta A_{r-1}$, and the operator t_k analogous to e_k is defined as

$$t_k(A) = \{T_{kn}\} \tag{1.7.27}$$

with the corresponding diagonal form

$$t_d(A) = \{T_{n1}\} \tag{1.7.28}$$

Two modifications of the t-transformation are derived by differing choices of R_r. Hence if $R_r = ra_r$ the u-transformation is produced with the formula

$$U_{kn} = \frac{\sum_{j=0}^{k} (-1)^j \binom{k}{j} \left(\frac{n+j}{n+k}\right)^{k-2} \frac{A_{n+j}}{a_{n+j}}}{\sum_{j=0}^{k} (-1)^j \binom{k}{j} \left(\frac{n+j}{n+k}\right)^{k-2} \frac{1}{a_{n+j}}} \tag{1.7.29}$$

with the corresponding operators defined as

$$u_k(A) = \{U_{kn}\} \tag{1.7.30}$$

and

$$u_d(A) = \{U_{n1}\}. \tag{1.7.31}$$

It is now possible to demonstrate these transformations on the very example which defeats Shanks, namely

$$A_n = \sum_{m=1}^{n} m^{-2} \tag{1.7.32}$$

which has the well-known sum $\pi^2/6$. Applications of t_d and u_d are shown in Table 1.10.

It is clear that rapid convergence is obtained despite the very slow convergence of the original partial sums.

The second modification builds in a simple form of the Shanks transformation to the t-transform. The simple Aitken extrapolation rule in equation (1.7.10) can be written as

$$\begin{aligned} e_1(A)_n &= \frac{A_{n-1}a_{n+1} - A_n a_n}{a_{n+1} - a_n} \\ &= A_n + \frac{a_n a_{n+1}}{a_n - a_{n+1}} \end{aligned} \tag{1.7.33}$$

Table 1.10

k	A_k	$t_d(A)_{2k+1}$	$u_d(A)_{2k+1}$
1	1.0	1.544 9074	1.643 518 51
2	1.25	1.597 3147	1.644 951 38
3	1.361 111	1.617 1562	1.644 933 94
4	1.423 611	1.626 7522	1.644 934 666

and hence the ν-transformation is derived using

$$R_r = \frac{a_r a_{r+1}}{a_{r+1} - a_r} \tag{1.7.34}$$

which gives the definition

$$V_{kn} = \frac{\sum_{j=0}^{k} (-1)^j \binom{k}{j} \left(\frac{n+j}{n+k}\right)^{k-1} \frac{a_{n+j+1} - a_{n+j}}{a_{n+j} a_{n+j+1}} A_{n+j}}{\sum_{j=0}^{k} (-1)^j \binom{k}{j} \left(\frac{n+j}{n+k}\right)^{k-1} \frac{a_{n+j+1} - a_{n+j}}{a_{n+j} a_{n+j+1}}} \tag{1.7.35}$$

and again the related operators are defined by

$$\nu_k(A) = \{V_{kn}\} \tag{1.7.36}$$

and

$$\nu_d(A) = \{V_{n1}\}. \tag{1.7.37}$$

These transformations prove to be most powerful tools in quadrature, especially in that the forms of the sequences to be accelerated can be, to some degree, manufactured to ensure good acceleration when one or other of the above methods is employed. Such applications will be pursued in detail in Chapter 6.

Further reading

Analysis
Apostol T. M. (1974) *Mathematical Analysis*, Addison-Wesley, Reading, Massachusetts.
Linear algebra
Finkbeiner D. T. (1960) *Introduction to Matrices and Linear Transformations*, W. H. Freeman, San Francisco.
Non-linear equations
Ortega J. and Rheinboldt W. (1970) *Iterative Solution of Non-linear Equations in Several Variables*, Academic Press, New York.
Householder A. S. (1970) *The Numerical Treatment of a Single Non-linear Equation*, McGraw-Hill, New York.
Applied mathematics
Atkin R. H. (1962) *Theoretical Electromagnetism*, Heinemann, London.
Acheson D. J. (1990) *Elementary Fluid Mechanics*, Clarendon Press, Oxford.
Bourne D. E. and Kendall P. C. (1977) *Vector Analysis and Cartesian Tensors*, Nelson, Sunbury-on-Thames.

Exercises

(1.1) Prove that

$$\int_{-1}^{1} \frac{T_r(x)T_s(x)}{\sqrt{1-x^2}} = \begin{cases} 0 & r \neq s \\ \frac{\pi}{2} & r = s \neq 0 \\ \pi & r = s = 0 \end{cases}$$

(1.2) Show that $y = T_r(x)$ satisfies

$$(1 - x^2)y'' - xy' + r^2 y = 0$$

(1.3) Prove that

$$\sum_{s=0}^{N}{}'' \sin\frac{\pi i s}{N}\sin\frac{\pi j s}{N} = \begin{cases} 0 & i \neq j \\ N & i = j = 0 \text{ or } N \\ \frac{N}{2} & i = j \neq 0 \text{ or } N \end{cases}$$

(1.4) If the iteration

$$x_{n+1} = \frac{3x_n^4 + 10}{4x_n^3 - 1}$$

converges to x, find the non-linear equation whose root is x. Find two roots of this equation by experimenting with different starting values.

(1.5) Define the terms rounding error and truncation error. Give an illustrative example of each. The formula

$$y = [120(\sin x - x) + 20x^3]/x^5$$

is used to compute y from x. An error δx is made in x. Find the resulting error δy in y from first principles ignoring any terms of order $(\delta x)^2$ or smaller. Show that for small x

$$\delta y = 10\,\delta x/x$$

and hence demonstrate that ill-conditioning occurs for small x.

(1.6) Apply the bisection process to the non-linear equation $x^3 - 2 = 0$ to find $\sqrt[3]{2}$ to one decimal place. Use the initial interval $(1, 2)$. By considering the rate of convergence of the method, calculate the number of steps required for 16 digit accuracy.

(1.7) Determine which of the iterations

$$\text{(a)}\quad y_{n+1} = \tfrac{1}{10}\tan\left[\tfrac{1}{2}\ln 100 + \tfrac{1}{2}\pi - \tfrac{1}{2}\ln(1 + 100y_n^2)\right]$$

$$\text{(b)}\quad y_{n+1} = \tfrac{1}{10}\sqrt{\exp[\ln 100 + \pi - 2\arctan 10y_n] - 1}$$

converges to the root of

$$\tfrac{1}{2}\ln\left[\tfrac{1}{100} + y^2\right] + \arctan 10y = \tfrac{1}{2}\pi$$

near to $x = 1$.

(1.8) Show that if the iteration $x_{n+1} = \sin 2x_n$ converges, then it converges to the value of x at which the curve $y = \tan x/x^2$ has a minimum. Given that this minimum lies between $x = 0.8$ and $x = 0.9$ show that the iteration does converge and complete two steps of the iteration using $x_0 = 0.9599$. Repeat the solution with a quadratically convergent method.

(1.9) Apply the secant method, the method of false position and Newton's method to the non-linear equations:

(i) $x^4 - x = 10$
(ii) $e^{-x} = \sin x$
(iii) $x = \frac{1}{2} + \sin x$.

Also apply the methods to find $\sqrt{2}$ and $\sqrt[3]{2}$ and to find two roots near to $x = 2$ of the equation

$$x^4 - 5x^3 - 12x^2 + 76x - 79 = 0$$

(1.10) Find the smallest positive root of

$$1 - x + \frac{x^2}{(2!)^2} - \frac{x^3}{(3!)^2} + \frac{x^4}{(4!)^2} - \cdots = 0$$

(1.11) A goat herder in the Middle East has the misfortune to own a circular field. He wishes to graze a goat on one half of the area of the field only, by tethering the goat to a point on the circumference of the field. How long must the tether be in terms of the radius of the field?

(1.12) Apply Steffenson's method to find the roots of the non-linear equations

$$\text{(a)} \quad x = \cos x$$

$$\text{(b)} \quad e^{-x} = \sin x$$

(1.13) Find all the roots of the quartic

$$2x^4 - 3x^3 + x^2 - 7x + 8 = 0$$

(1.14) Show that the iteration

$$x_{n+1} = \frac{1}{4}\left[3x_n + \frac{3a}{2x_n} - \frac{x_n^3}{2a}\right]$$

is a third-order process for finding $\sqrt{a}$. Use the iteration to find $\sqrt{10}$ starting at $x_0 = 3$. Compare the process with the usual quadratically convergent Newton iteration

$$x_{n+1} = (x_n + a/x_n)/2$$

(1.15) Use Newton's method to solve the pair of simultaneous non-linear equations

$$f_1(x, y) = \sin(4x/y) - 2y$$

$$f_2(x, y) = e^{-x} - \sqrt{y}$$

(1.16) Use Bairstow's method to find the roots of

$$z^4 - 8z^3 + 24z^2 + 336z + 120 = 0$$

with the trial factor $(z^2 + \frac{9}{2}z + \frac{3}{2}) = 0$ in the first instance.

(1.17) Solve the problem in exercise (1.16) using the QD algorithm. It becomes apparent that the QD algorithm would make a good starting place to find close initial estimates of roots and quadratic factors.

(1.18) Use the QD algorithm after a suitable linear change of variable to find the roots of the polynomial

$$x^6 - 4x^4 + 3x^3 + 4x - 1 = 0$$

(1.19) Use a combination of QD, Newton and Bairstow to generate an effective method for finding the roots of the polynomial

$$x^6 + 2x^5 - 4x^4 + 3x^3 - x^2 + 4x - 2 = 0$$

(1.20) Use Shanks' accelerator method through the ε algorithm to evaluate

$$\frac{\sin \pi x}{\pi x} = \prod_{n=1}^{\infty} \left(1 - \frac{x^2}{n^2}\right)$$

with $x = \frac{1}{4}$ taking the process to B_{22}.

(1.21) Use Shanks' accelerator to evaluate the infinite range oscillatory integral

$$\int_0^{\infty} e^{-x} \sin 100x \, dx$$

by taking the sequence of integrals

$$\int_0^{\pi/100} \quad \int_{\pi/100}^{\pi/50} \quad \int_{\pi/50}^{3\pi/100}$$

and so on. Hence each cycle of the infinite range integral forms a term of a sequence. Evaluate the integrals over each cycle analytically and use sufficient terms to get convergence.

(1.22) Use Shanks' process to evaluate the singular integral

$$\int_0^1 \frac{e^{-x^{1/2}}}{x^{1/2}} \, dx$$

by forming the sequence

$$A_i = \int_{1/(2^{i+1})}^{1} \frac{e^{-x^{1/2}}}{x^{1/2}} \, dx$$

Compare the outcome of using $i = 0, 1, 2, 3, 4$ with the true value. Evaluate the sequence integrals analytically.

2

Differences, Interpolation and Differentiation

2.1 Finite differences

A fundamental requirement in numerical work is the ability to obtain intermediate values from a tabulated definition of a function. This ability was particularly important in the pre-computer era, but still plays a major role, as many numerical methods generate values at finite steps which is in reality just a table. Moreover, an interpolation rule may itself be differentiated or integrated to give differentiation and integration rules. This chapter is split into two major sections, the first dealing with the more archaic topic of finite differences, and the second moving on to more general interpolation methods. The importance of the finite difference section is the use of the notation in other areas, and it seems reasonable to cover the topic at this point.

This chapter also includes various forms of curve fitting based on least squares, where the data is not fitted exactly as in the interpolation methods, but statistical or experimental error is expected in the data and the fitting curve is meant to smoothen out these errors in some sense. In general there will be many more data points than free coefficients in the fitting curve so that the problem is over-specified. With interpolation, on the other hand, a quadratic with three free coefficients will be fitted to three points, for example, with no over specification. The discussion begins with finite differences and their associated operators.

A notation which once played a major role in numerical analysis before computing machinery was available is that of finite differences. In the days of hand calculations, much numerical work was performed on the basis of tabular values. A table would be set up of say $f(x)$ against x with the x's equally spaced so that

$$x_{i+1} = x_i + h \tag{2.1.1}$$

where h is the step size. The following operators can then be defined. The forward difference operator, Δ, is defined as

$$\Delta f(x) = f(x+h) - f(x) \tag{2.1.2}$$

whereas the backward difference operator, ∇, is defined as

$$\nabla f(x) = f(x) = f(x-h) \tag{2.1.3}$$

and finally the central difference operator, δ, is defined as

$$\delta f(x) = f(x + h/2) - f(x - h/2) \tag{2.1.4}$$

In addition to these difference operators, the following operators can be defined which will allow useful formulae to be derived which link the various operators together. The first of these extra operators is the averaging operator, μ, defined by

$$\mu f(x) = \tfrac{1}{2}[f(x + h/2) + f(x - h/2)]. \tag{2.1.5}$$

The main use of this operator is with central differences. Clearly only even order central differences are tabulated, the odd order differences involving values midway between the tabulated entries. By combining odd order central differences with the operator μ, tabular values are returned. For use in generating interpolation formulae the shifting operator E is defined as

$$Ef(x) = f(x + h), \tag{2.1.6}$$

and similarly for use in differentiation and integration formula respectively the operators D and J are defined by

$$Df(x) = f'(x) \tag{2.1.7}$$

$$Jf(x) = \int_x^{x+h} f(t)\,\mathrm{d}t \tag{2.1.8}$$

The difference operators operating on a table of values of $f(x)$ can be easily enumerated using a difference table, and with these values interpolation, differentiation and integration formulae can be found. A typical difference table is shown in Table 2.1.

In terms of these tabulated values, the definitions of the difference operators yields for example that $\Delta f_2 = 0.021\,89 = \nabla f_3$ and $\Delta^2 f_1 = -0.009\,63 = \nabla^2 f_2$,

Table 2.1

x		f		Δ	Δ^2	Δ^3	Δ^4
x_1	1.2	f_1	0.932 04				
				0.03152			
x_2	1.3	f_2	0.963 56		−0.009 63		
				0.02189		−0.000 22	
x_3	1.4	f_3	0.985 45		−0.009 85		0.000 11
				0.01204		−0.000 11	
x_4	1.5	f_4	0.997 49		−0.009 96		0.000 08
				0.00208		−0.000 03	
x_5	1.6	f_5	0.999 57		−0.009 99		0.000 12
				−0.00791		0.000 09	
x_6	1.7	f_6	0.991 66		−0.009 90		
				−0.01781			
x_7	1.8	f_7	0.973 85				

and so any formulae with tabulated differences can be easily computed. Hence in terms of forward differences the equivalent table is shown in Table 2.2.

The equivalent table for backward differences is shown in Table 2.3.

The central differences are a little more difficult in that odd order differences involve values such as $f(x+h/2)$ and $f(x-h/2)$ which do not appear in the table. However the even order differences such as

$$\delta^2 f(x) = f(x+h) - 2f(x) + f(x-h) \tag{2.1.9}$$

are tabulated as are the half differences such as

$$\delta f(x+h/2) = \delta f_{1/2} = f(x+h) - f(x) \tag{2.1.10}$$

The result is a central difference table equivalent to the previous forward and backward differences illustrated in Table 2.4.

Many simple relations hold between the various operators defined above. It is immediate from the definitions that

$$\Delta = E - 1 \tag{2.1.11}$$

$$\nabla = 1 - E^{-1} \tag{2.1.12}$$

$$\delta = E^{1/2} - E^{-1/2} \tag{2.1.13}$$

$$\mu = \tfrac{1}{2}(E^{1/2} + E^{-1/2}) \tag{2.1.14}$$

Table 2.2

x	f	Δ	Δ^2	Δ^3
x_0	f_0			
		Δf_0		
x_1	f_1		$\Delta^2 f_0$	
		Δf_1		$\Delta^3 f_0$
x_2	f_2		$\Delta^2 f_1$	
		Δf_2		
x_3	f_3			

Table 2.3

x	f	∇	∇^2	∇^3
x_{n-3}	f_{n-3}			
		∇f_{n-2}		
x_{n-2}	f_{n-2}		$\nabla^2 f_{n-1}$	
		∇f_{n-1}		$\nabla^3 f_n$
x_{n-1}	f_{n-1}		$\nabla^2 f_n$	
		∇f_n		
x_n	f_n			

Table 2.4

x	f	δ	δ^2	δ^3	δ^4
x_{-2}	f_{-2}				
		$\delta f_{-3/2}$			
x_{-1}	f_{-1}		$\delta^2 f_{-1}$		
		$\delta f_{-1/2}$		$\delta^3 f_{-1/2}$	
x_0	f_0		$\delta^2 f_0$		$\delta^4 f_0$
		$\delta f_{1/2}$		$\delta^3 f_{1/2}$	
x_1	f_1		$\delta^2 f_1$		
		$\delta f_{3/2}$			
x_2	f_2				

where the inverse operator E^{-1} is defined as

$$E^{-1} f(x) = f(x-h) \tag{2.1.15}$$

which then yields correctly that $EE^{-1} = 1$, where $1f(x) = f(x)$ defines a unit operator.

The differentiation and integration operators are related by the algebra

$$\begin{aligned} DJf(x) &= \frac{\mathrm{d}}{\mathrm{d}x}\int_x^{x+h} f(t)\,\mathrm{d}t \\ &= f(x+h) - f(x) = \int_x^{x+h} \frac{\mathrm{d}f(t)}{\mathrm{d}t}\,\mathrm{d}t = JDf(x) \end{aligned}$$

to give the result

$$DJ = JD = \Delta \tag{2.1.16}$$

Hence an interpolation formula can be found from

$$E^s = (1+\Delta)^s = (1-\nabla)^{-s}$$

by a formal expansion to yield Newton's forward and backward interpolation formulae namely:

$$E^s f(x) = f(x+sh)$$

$$= \left(1 + s\Delta + \frac{s(s-1)}{1.2}\Delta^2 + \cdots + {}^sC_k\Delta^k + \cdots\right) f(x) \tag{2.1.17}$$

$$= \left(1 + s\nabla + \frac{s(s+1)}{1.2}\nabla^2 + \cdots + \frac{s(s+1)\cdots(s+k-1)}{k!}\nabla^k + \cdots\right) f(x). \tag{2.1.18}$$

By using equation (2.1.13), formulae for interpolation using central differences are derivable, now based on the expansion of

$$E^s = (1 + E^{1/2}\delta)^s \tag{2.1.19}$$

which gives

$$E^s = 1 + sE^{1/2}\delta + \frac{s(s-1)}{1.2}E\delta^2 + \cdots + \frac{s(s-1)\cdots(s-2n+1)}{(2n)!}E^n\delta^{2n} + + \frac{s(s-1)\cdots(s-2n)}{(2n+1)!}E^{(2n+1)/2}\delta^{2n+1} + \cdots \tag{2.1.20}$$

However $E^n\delta^{2n}f(x) = \delta^{2n}f(x_0+2nh)$, and so a move of origin to $\hat{x}_0 = x_0 + (n-1)h$, which is equivalent to putting $s = t + n - 1$ gives

$$\frac{s(s-1)\cdots(s-2n+1)}{(2n)!}\delta^{2n}f(x_0+2nh) = \frac{(t+n-1)(t+n-2)\cdots t\cdots(t-(n-1))(t-n)}{(2n)!}\delta^{2n}f(\hat{x}_0) = \frac{t(t^2-1^2)(t^2-2^2)\cdots(t^2-(n-1)^2)(t-n)}{(2n)!}\delta^{2n}f(\hat{x}_0). \tag{2.1.21}$$

Similarly using $s = t + n$ for the odd terms gives

$$\frac{t(t^2-1^2)(t^2-2^2)\cdots(t^2-n^2)}{(2n+1)!}\delta^{2n+1}f\left(\hat{x}_0 + \frac{h}{2}\right) \tag{2.1.22}$$

Using these moves of origin, and renaming $\hat{x}_0$ as x_0 then yields the Gauss forward central difference formula:

$$f(x_0+th) = f(x_0) + t\delta f(x_0+h/2) + \frac{t(t-1)}{2!}\delta^2 f(x_0) + \frac{t(t^2-1^2)}{3!}\delta^3 f(x_0+h/2) + \cdots + \frac{t(t^2-1^2)\cdots(t^2-(n-1)^2)(t-n)}{(2n)!}\delta^{2n}f(x_0) + \frac{t(t^2-1^2)\cdots(t^2-n^2)}{(2n+1)!}\delta^{2n+1}f(x_0+h/2) + \cdots \tag{2.1.23}$$

where it is noted that all the differences are tabulated, as the odd central differences are required at $f(x_0+h/2)$ rather than $f(x_0)$. A similar process based on the expansion of

$$E^s = (1 - E^{1/2}\delta)^{-s} \tag{2.1.24}$$

gives the Gauss backward central difference formula:

$$f(x_0+th) = f(x_0) + t\delta f(x_0-h/2) + \frac{t(t+1)}{2!}\delta^2 f(x_0) + \frac{t(t^2-1^2)}{3!}\delta^3 f(x_0-h/2) + \cdots$$

$$+\frac{t(t^2-1^2)\cdots(t^2-(n-1)^2)(t+n)}{(2n)!}\delta^{2n}f(x_0)$$

$$+\frac{t(t^2-1^2)\cdots(t^2-n^2)}{(2n+1)!}\delta^{2n+1}f(x_0-h/2)+\cdots \quad (2.1.25)$$

A third formula in this series is obtained by taking the mean of the two Gauss formulae (2.1.23) and (2.1.25). The general odd and even terms then become

$$\frac{s(s^2-1^2)\cdots(s^2-(m-1)^2)[(s-m)+(s+m)]}{2.(2m)!}\delta^{2m}f_0 \quad (2.1.26)$$

and

$$\frac{s(s^2-1^2)\cdots(s^2-m^2)}{2(2m+1)!}(\delta^{2m+1}f_{1/2}+\delta^{2m+1}f_{-1/2}) \quad (2.1.27)$$

and using

$$\mu\delta^{2n+1}f_0=\tfrac{1}{2}(\delta^{2n+1}f_{1/2}+\delta^{2n+1}f_{-1/2}) \quad (2.1.28)$$

yields Stirling's formula:

$$f_s=f(x_0+sh)=f_0+s\mu\delta f_0+\frac{s^2}{2!}\delta^2 f_0+\frac{s(s^2-1^2)}{3!}\mu\delta^3 f_0+\cdots$$

$$+\frac{s^2(s^2-1^2)\cdots(s^2-(m-1)^2)}{(2m)!}\delta^{2m}f_0$$

$$+\frac{s^2(s^2-1^2)\cdots(s^2-m^2)}{(2m+1)!}\mu\delta^{2m+1}f_0+\cdots \quad (2.1.29)$$

Again this formula has a form which uses tabulated difference operators, now achieved using the averaging operator μ with any odd order central differences.

To obtain formulae involving the operators D and J, a formal form of Taylor's series gives:

$$p(x+h)=p(x)=hp'(x)+\frac{h^2}{2!}p''(x)+\cdots+\frac{h^n}{n!}p^{(n)}(x)+\cdots \quad (2.1.30)$$

which can be expressed as

$$E=1+hD+\frac{h^2}{2!}D^2+\cdots \quad (2.1.31)$$

to yield the formal expression

$$E=e^{hD} \quad (2.1.32)$$

which rearranges to give

$$hD=\ln(E)-\ln(1+\Delta)=-\ln(1-\nabla). \quad (2.1.33)$$

Hence the differentiation formulae

$$f'(x) = \frac{1}{h}\left(\Delta - \frac{1}{2}\Delta^2 + \frac{1}{3}\Delta^3 - \cdots\right) f(x) \tag{2.1.34}$$

and

$$f'(x) = \frac{1}{h}\left(\nabla + \frac{1}{2}\nabla^2 + \frac{1}{3}\nabla^3 + \cdots\right) f(x) \tag{2.1.35}$$

As an example of the use of these formulae consider a table of $y = \tan x$ with known derivative $y' = \sec^2 x$. The resulting difference table is shown in Table 2.5.

Now the forward and backward differentiation formulae can be applied to compute the value of y' at $x = \pi/4$ or 45° where the true value is $y' = 2$. The forward rule gives

$$\begin{aligned} y' &= \frac{180}{\pi}\left(0.0355 - \frac{1}{2}(0.0014) - \frac{1}{3}(0.0001)\right) \\ &= \frac{180}{\pi}(0.0348) = 1.994 \end{aligned} \tag{2.1.36}$$

and the corresponding backward formula yields

$$\begin{aligned} y' &= \frac{180}{\pi}\left(0.034\,3000 + \frac{1}{2}(0.001\,10)\right) \\ &= \frac{180}{\pi}(0.0348) = 1.994 \end{aligned} \tag{2.1.37}$$

where in both cases the differences are picked from the relevant part of the table. It is clear again that forward differences are more applicable at the start of a table and backward differences at the end.

Table 2.5

$x°$	f	Δ	Δ^2	Δ^3	Δ^4
42	0.9004				
		0.0321			
43	0.9325		0.0011		
		0.0322		0.0000	
44	0.9657		0.0011		0.0001
		0.0343		0.00011	
45	1.0000		0.0012		0.0002
		0.0355		0.00003	
46	1.0355		0.0014		−0.0001
		0.0369		−0.00009	
47	1.0724		0.0013		
		0.0382			
48	1.1106				

From equation (2.1.13) a corresponding central difference formula can be derived. Treating (2.1.13) as a quadratic in $E^{1/2}$ yields the solution for δ in the form

$$E^{1/2} = \tfrac{1}{2}\delta + \left(1 + \tfrac{1}{4}\delta^2\right)^{1/2} \tag{2.1.38}$$

and then the formula analogous to equation (2.1.33) is

$$hD = 2\ln\left(\left(1 + \tfrac{1}{4}\delta^2\right)^{1/2} + \tfrac{1}{2}\delta\right) = 2\,\text{arcsinh}\,\tfrac{1}{2}\delta \tag{2.1.39}$$

Now using

$$\mu^2 = 1 + \delta^2/4 \tag{2.1.40}$$

gives the expansion

$$\begin{aligned} f'(x) &= \frac{2}{h}\,\text{arcsinh}\,\frac{\delta}{2} f(x) \\ &= \frac{2\mu}{h}\,\frac{\text{arcsinh}\,\frac{\delta}{2}}{\sqrt{1+\delta^2/4}} f(x) \\ &= \frac{\mu}{h}\left(\delta - \frac{1^2}{3!}\delta^3 + \frac{1^2.2^2}{5!}\delta^5 - \cdots\right) f(x) \end{aligned} \tag{2.1.41}$$

where (2.1.40) has been used to engineer an expansion which involves the tabulated difference operators of the form $\mu\delta^{2n+1}$. From Table 2.5 the equivalent result to (2.1.36) and (2.1.37) is

$$\begin{aligned} y' &= \frac{180}{\pi}\left(\left(\frac{0.0343 + 0.0355}{2}\right) - \frac{1}{3!}\left(\frac{0.0001 + 0.0002}{2}\right) + \cdots\right) \\ &= \frac{180}{\pi}(0.0348) = 1.994 \end{aligned} \tag{2.1.42}$$

again agreeing with the previous values.

Formulae for integration can equally be derived from the relationships involving E. Hence

$$\begin{aligned} \int_{x_0}^{x_0+rh} f(x)\,\mathrm{d}x &= (1 + E + E^2 + \cdots + E^{r-1}) J f(x_0) \\ &= \frac{E^r - 1}{E - 1} J f_0 \end{aligned} \tag{2.1.43}$$

Then using equations (2.1.11, 2.1.16) and (2.1.33) gives the formal form:

$$\int_{x_0}^{x_0+rh} f(x)\,\mathrm{d}x = h\left[\frac{(1+\Delta)^r - 1}{\Delta}\right]\left[\frac{\Delta}{\ln(1+\Delta)}\right] f_0 \tag{2.1.44}$$

The expansion of the right-hand side in terms of differences is somewhat irregular but the first few terms can be found using

$$\frac{\Delta}{\ln(1+\Delta)} = \sum_{j=0}^{\infty} c_j \Delta^j \tag{2.1.45}$$

where $c_0 = 1$, $c_1 = \frac{1}{2}$, $c_2 = -\frac{1}{12}$, $c_3 = \frac{1}{24}$ and $c_4 = -\frac{19}{720}$. Hence for different values of r, expansions can be written down for integration with the most commonly used results being for $r = 1$,

$$\int_{x_0}^{x_0+h} f(x)\,\mathrm{d}x = h\left(1 + \frac{\Delta}{2} - \frac{1}{12}\Delta^2 + \frac{1}{24}\Delta^3 - \frac{19}{720}\Delta^4 + \frac{3}{160}\Delta^5 + \cdots\right) f_0 \tag{2.1.46}$$

and for $r = 2$,

$$\int_{x_0}^{x_0+2h} f(x)\,\mathrm{d}x = 2h\left(1 + \Delta + \frac{1}{6}\Delta^2 + 0\Delta^3 - \frac{1}{180}\Delta^4 + \frac{1}{180}\Delta^5 + \cdots\right) f_0 \tag{2.1.47}$$

Taking the first two terms of the first rule gives the trapezoidal rule, whereas the first three terms of the second rule give Simpson's rule. The extra terms give corrections to these well-known integration rules which will be considered in more detail in Chapter 6. The corresponding backward formulae can also be used to give the rules: for $r = 1$,

$$\int_{x_0-h}^{x_0} f(x)\,\mathrm{d}x = h\left(1 - \frac{\nabla}{2} - \frac{1}{12}\nabla^2 - \frac{1}{24}\nabla^3 - \frac{19}{720}\nabla^4 - \frac{3}{160}\nabla^5 + \cdots\right) f_n \tag{2.1.48}$$

and for $r = -1$,

$$\int_{x_0}^{x_0+h} f(x)\,\mathrm{d}x = h\left(1 + \frac{1}{2}\nabla + \frac{5}{12}\nabla^2 + \frac{3}{8}\nabla^3 + \frac{251}{720}\nabla^4 + \frac{95}{288}\nabla^5 + \cdots\right) f_n \tag{2.1.49}$$

Once again there is an equivalent set of central difference formulae based on the equations:

$$\begin{aligned}\int_{x_0-rh}^{x_0+rh} f(x)\,\mathrm{d}x &= (E^{-r} + \cdots + E^{r-1})Jf_0 \\ &= \frac{E^r + E^{-r}}{E-1}Jf_0 = \frac{e^{rhD} - e^{-rhD}}{\Delta}\Delta D^{-1} f_0 \\ &= 2\frac{\sinh rhD}{D} f_0\end{aligned} \tag{2.1.50}$$

and (2.1.50) expands to give:

$$\int_{x_0-rh}^{x_0+rh} f(x)\,\mathrm{d}x = 2rh\left[1 + \frac{r^2(hD)^2}{6} + \frac{r^4(hD)^4}{120} + \cdots\right] \tag{2.1.51}$$

But there is an expansion for hD using:

$$hD = 2\,\text{arcsinh}\,\frac{\delta}{2}$$
$$= \frac{1}{h}\left(\delta - \frac{1^2}{2^2.3!}\delta^3 + \frac{1^2.3^2}{2^4.5!}\delta^5 - \frac{1^2.3^2.5^2}{2^6.7!}\delta^7 + \cdots\right) \tag{2.1.52}$$

which combines with the expansion in equation (2.1.51) to yield the particular formulae, firstly with $r = 1$,

$$\int_{x_0-h}^{x_0+h} f(x)\,\mathrm{d}x = 2h\left(1 + \frac{1}{6}\delta^2 - \frac{1}{180}\delta^4 + \frac{1}{1512}\delta^6 - \cdots\right) f_0 \tag{2.1.53}$$

and with $r = 2$,

$$\int_{x_0-2h}^{x_0+2h} f(x)\,\mathrm{d}x = 4h\left(1 + \frac{2}{3}\delta^2 + \frac{7}{90}\delta^4 - \frac{2}{945}\delta^6 - \cdots\right) f_0 \tag{2.1.54}$$

Equation (2.1.53) can be written in terms of ordinates to give

$$\int_{x_0-h}^{x_0+h} f(x)\,\mathrm{d}x = \frac{h}{3}(f_{-1} + 4f_0 + f_1) - \frac{h}{90}\left(\delta^4 - \tfrac{5}{42}\delta^6 + \cdots\right) f_0 \tag{2.1.55}$$

which is again the well-known Simpson's rule with a corrective term.

These same difference operators can be used to derive the Euler–Maclaurin summation formula. Returning to equation (2.1.16), and using the results in equation (2.1.11) and (2.1.32) gives

$$hp(x) = \left(\frac{hD}{e^{hD} - 1}\right) J\,p(x). \tag{2.1.56}$$

But the term in the brackets can be expanded as the series

$$\frac{hD}{e^{hD} - 1} = \sum_{i=0}^{\infty} \frac{B_i}{i!}(hD)^i \tag{2.1.57}$$

where B_i are the Bernoulli numbers. These are defined and tabulated in Abramowitch and Stegun (1965) in terms of generating functions. For reference purposes the first few are $B_3 = B_5 = \cdots = 0$ and $B_0 = 1$, $B_1 = -\frac{1}{2}$, $B_2 = \frac{1}{6}$, $B_4 = -\frac{1}{30}$, ... Hence

$$hp(x) = \sum_{i=0}^{\infty} \frac{B_i}{i!} h^i D^i J\,p(x)$$
$$= \int_x^{x+h} p(t)\mathrm{d}t + \sum_{i=1}^{\infty} \frac{B_i}{i!} h^i D^i J\,p(x) \tag{2.1.58}$$

and now using (2.1.16) to replace $D^i J p(x)$ by $D^{i-1}[p(x+h) - p(x)]$ gives

$$p(x) = \frac{1}{h}\int_x^{x+h} p(t)\mathrm{d}t + \sum_{i=1}^{\infty} \frac{B_i}{i!} h^{i-1}[p^{(i-1)}(x+h) - p^{(i-1)}(x)] \tag{2.1.59}$$

By putting $x = x_0, x_1, \ldots, x_{r-1}$ where $x_r = x_0 + (r-1)h$ and then summing with the notation $p_k = p(x_k)$ yields

$$\sum_{k=0}^{r-1} p_k = \frac{1}{h}\int_{x_0}^{x_r} p(x)\mathrm{d}x + \sum_{i=1}^{\infty} \frac{B_i}{i!} h^{i-1}[p_r^{(i-1)} - p_0^{(i-1)}] \tag{2.1.60}$$

When $p(x)$ is not a polynomial, there is an additional error term and the series terminates at $i = m$, say. In this form, the formula is

$$\sum_{k=0}^{r} f_k = \frac{1}{h}\int_{x_0}^{x_r} f(x)\mathrm{d}x + \frac{1}{2}(f_0 + f_r) + \sum_{i=1}^{m} \frac{B_{2i}}{(2i)!} h^{2i-1}[f_r^{(2i-1)} - f_0^{(2i-1)}] + E_m \tag{2.1.61}$$

where

$$E_m = r\frac{B_{2m+2}h^{2m+2}}{(2m+2)!} f^{(2m+2)}(\xi) \tag{2.1.62}$$

with $x_0 < \xi < x_r$. The importance of this formula is often not appreciated, and yet some authors have put effort into generating new quadrature rules which rely entirely on the properties of this series. The details for individual methods will appear in Chapter 6, and only the philosophy will be outlined here. The Euler–Maclaurin formula is basically a trapezoidal rule quadrature with an error term represented as an asymptotic series. This series has coefficients which are the difference between the higher derivatives evaluated at the end-points of the interval of integration. By considering transformations which force these end-point derivatives to be zero, considerable enhancement of the normal accuracy of the trapezoidal rule occurs, and some high order quadrature rules ensue. There is no question of an infinite order formula even if all the higher end-point derivatives are zero because the error series is asymptotic and has to be considered for only a finite number of terms, the last term being a special remainder term. Nevertheless some important methods can be derived, in particular in the area of singular quadratures.

Many other formulae using finite differences were in common use before the general use of computing machines. It is more normal at present to make free use of the notation, but little use of the specific formulae of the type generated above. Indeed equivalent formulae will be offered for all the above processes in the relevant chapters to follow. Here attention is confined to the specific area of interpolation.

2.2 Lagrange interpolation

Consider the following simple minded approach to interpolation based on fitting polynomials to the tabulated data. Table 2.6 gives three values for the function $f(x)$.

A simple quadratic fit can be made to the three tabulated points by setting

$$f(x) = A + Bx + Cx^2 \tag{2.2.1}$$

which gives three linear equations for the constants A, B and C. These equations are:

$$\begin{aligned} 0.0873 &= A + 5B + 25C \\ 0.1746 &= A + 10B + 100C \\ 0.2621 &= A + 15B + 225C \end{aligned} \tag{2.2.2}$$

which solve to yield the quadratic

$$f(x) = 0.0873 + 0.0872x + 0.0001x^2 \tag{2.2.3}$$

If this fitting process is extended to higher degree fits, the set of linear equations becomes less well conditioned (see Chapter 3) with a resulting loss of accuracy in the computed constants. The question then arises as to whether the direct process can be circumvented so that the solution of a set of linear algebraic equations is avoided. This alternative is called Lagrange interpolation and depends on the properties of the function $\Pi(x)$ which is defined by

$$\Pi(x) = (x - x_0)(x - x_1)\cdots(x - x_{n-1})(x - x_n) \tag{2.2.4}$$

This function is clearly zero for $x = x_i$, $i = 0, n$ and with this fact in mind, suppose a formula of the type

$$y(x) = \sum_{k=0}^{n} l_k(x) f(x_k) + E(x) \tag{2.2.5}$$

is being sought with the interpolation property that $y(x_i) = f(x_i)$ for $i = 0, n$. Hence the function $l_k(x)$ needs to be a polynomial of degree n, and whenever $x = x_i$, the function needs to be zero except when $i = k$ when it needs to be unity. Now $\Pi(x)$ is zero for all the relevant x_i's, and has degree $n + 1$. By omitting the

Table 2.6

x	$f(x)$
5	0.0873
10	0.1746
15	0.2621

term $(x - x_k)$, the zero requirements are met, and scaling will resolve the problem of forcing a unit value at $x = x_k$. Hence, what are termed Lagrange coefficients are defined by

$$l_k(x) = \frac{(x - x_0)\cdots(x - x_{k-1})(x - x_{k+1})\cdots(x - x_n)}{(x_k - x_0)\cdots(x_k - x_{k-1})(x_k - x_{k+1})\cdots(x_k - x_n)}$$

$$= \frac{\Pi(x)}{(x - x_k)\Pi'(x_k)} \tag{2.2.6}$$

and then the formula (2.2.5) yields Lagrange's interpolation rule with a polynomial being generated which passes through the required points, but does not require the solution of linear equations. Computation of interpolated values can be made directly from (2.2.5) without knowledge of the specific coefficients of the fitting polynomial which are now less easy to obtain.

This approach also yields an error bound on the interpolation formula. This derivation seems to cause some concern to readers new to the subject. One point has to be realized. No error formula will be completely specified. If it were, the error would be known exactly and a perfect numerical method would result by combining the original method with the completely specified error. Hence in error analysis there will always be some form of indeterminancy. Usually this allows bounds to be specified for the error, even though the precise error is unknown. It is useful to know that a result is correct to m figures even if no more precision in the error can be achieved. Hence the following analysis aims to get the form of the error together with a suitable indeterminancy. The argument revolves round the function $F(x)$ defined by

$$F(x) = f(x) - y(x) - K\Pi(x) \tag{2.2.7}$$

Both $f(x) - y(x)$ and $\Pi(x)$ vanish at the points $x = x_0,\ x_1, \ldots, x_n$ and so this is also true of $F(x)$ for any K. Moreover $\Pi(x)$ only vanishes at these points and therefore K can be chosen to force $F(x)$ to vanish at one further point, say $\bar{x}$. Let the closed interval I be limited by the smallest and largest of the values $x_0, \ldots, x_n, \bar{x}$, then $F(x)$ vanishes at least $n + 2$ times in I. By Rolle's Theorem, $F'(x)$ vanishes at least $n + 1$ times, and continuing to differentiate, $F^{(n+1)}(x)$ vanishes at least once in I say at the point ξ. This value ξ gives the required indeterminacy as its exact value is not known. However the relationship

$$0 = f^{(n+1)}(\xi) - y^{(n+1)}(\xi) - K\Pi^{(n+1)}(\xi) \tag{2.2.8}$$

holds to yield a value for K in terms of ξ. Note that $y^{(n+1)}(x) = 0$ as $y(x)$ is a polynomial of degree n so that its $(n + 1)$th derivative vanishes, and $\Pi^{(n+1)}(x)$ is easy to write down as the highest power in the expansion of Π is $x^{(n+1)}$ which differentiates $n + 1$ times to give $(n + 1)!$, all the lower terms giving zero. Hence the error in Lagrange's interpolation formula is

$$E(x) = \frac{f^{(n+1)}(\xi)\Pi(x)}{(n+1)!} \tag{2.2.9}$$

It is worthwhile stopping to consider the practical value of this formula. The presence of ξ is no great difficultly as ξ lies in I and so an upper bound on the error can be set by using an upper bound on $|f^{(n+1)}(x)|$ for $x \in I$. The real problem is in determining the $(n+1)$th derivative of f. Even for quite simple functions the high derivatives can become increasingly tedious to obtain. Hence error estimates of the type in equation (2.2.9) give a theoretical grounding to the method and allow university examiners to produce awkward questions in specific cases, but have limited practical application.

As an example consider Table 2.7 of $f(x) = \log_{10} x$ with the problem of computing $f(2.4)$ by Lagrange interpolation.

Then to investigate the error, the derivative:

$$f^{(m)}(x) = \frac{1}{\ln 10}\frac{(m-1)!}{x^m} \tag{2.2.10}$$

is required and Lagrange's interpolation from equation (2.2.5) gives

$$\begin{aligned} y(2.4) &= \frac{(0.6)(1.6)}{(1)(2)}0.3010 + \frac{(0.4)(-1.6)}{(1)(-1)}0.4771 + \frac{(0.4)(-0.6)}{(2)(1)}0.6021 \\ &= 0.48 \times 0.3010 + 0.64 \times 0.4771 - 0.12 \times 0.6021 \\ &= 0.3776 \end{aligned} \tag{2.2.11}$$

whereas the true value is 0.3802 with an actual error of 0.0026. On the other hand using (2.2.10) gives

$$|f^{(m)}(x)| < \frac{2}{2^3 \ln 10} = \frac{1}{4 \ln 10} \tag{2.2.12}$$

and the error formula (2.2.9) gives

$$|E(x)| \le \frac{1}{4 \ln 10}\frac{(0.4)(-0.6)(-1.6)}{6} = 0.0070 \tag{2.2.13}$$

which is a valid bound.

Table 2.7

x	$f(x)$
2	0.3010
3	0.4771
4	0.6021

2.3 Hermite interpolation

Lagrange's interpolation generates an interpolatory polynomial which agrees with $f(x)$ at a given set of interpolatory x values. It would appear that a more accurate interpolator could be found if agreement at the nodes x_i were forced to occur for both $f(x)$ and $f'(x)$. Such a formula is Hermite's interpolation formula. By analogy with equation (2.2.5) for Lagrange, a formula of the form

$$y(x) = \sum_{k=1}^{m} h_k(x) f(x_k) + \sum_{k=1}^{m} \bar{h}_k(x) f'(x_k) + E(x) \tag{2.3.1}$$

is now sought, with $h_k(x)$ and $\bar{h}_k(x)$ being polynomials of degree $2m - 1$ to yield sufficient degrees of freedom to satisfy the fitting of f and f'. The four conditions needed to be satisfied by h_k and $\bar{h}_k$ are

$$y(x_j) = f(x_j)$$

which leads to

$$h_i(x_j) = \delta i j \quad \text{and} \quad \bar{h}_i(x_j) = 0 \tag{2.3.2}$$

and

$$y'(x_j) = f'(x_j)$$

to give

$$h_i'(x_j) = 0 \quad \text{and} \quad \bar{h}_i'(x_j) = \delta_{ij} \tag{2.3.3}$$

where δ_{ij} has the usual value of 0 if $i \neq j$ and the value of 1 if $i = j$.

It is clear that $[l_i(x)]^2$ is a polynomial of degree $2m - 2$ which comes close to satisfying some of the conditions in (2.3.2) and (2.3.3). By trying

$$h_i(x) = r_i(x)[l_i(x)]^2 \tag{2.3.4}$$

and

$$\bar{h}_i(x) = s_i(x)[l_i(x)]^2 \tag{2.3.5}$$

with r_i and s_i being linear functions of x, the required degree of polynomial is attained and four degrees of freedom remain to complete the satisfaction of (2.3.2) and (2.3.3). Indeed these conditions result in the relations

$$r_i(x_i) = 1 \qquad r_i'(x_i) + 2l_i'(x_i) = 0 \tag{2.3.6}$$

$$s_i(x_i) = 0 \qquad s_i'(x_i) = 1 \tag{2.3.7}$$

The linear forms which satisfy these conditions are

$$r_i(x) = 1 - 2l_i'(x_i)(x - x_i) \tag{2.3.8}$$

$$s_i(x) = x - x_i \tag{2.3.9}$$

which gives Hermite's interpolation formula as

$$y(x) = \sum_{k=1}^{m} h_k(x) f(x_k) + \sum_{k=1}^{m} \bar{h}_k(x) f'(x_k) + E(x) \tag{2.3.10}$$

with

$$h_i(x) = [1 - 2l_i'(x_i)(x - x_i)][l_i(x)]^2 \tag{2.3.11}$$

and

$$\bar{h}_i(x) = (x - x_i)[l_i(x)]^2 \tag{2.3.12}$$

and as in the formula of Lagrange the problem of determining $E(x)$ remains. A similar approach to the Lagrange case yields a suitable error form. Now it is recognized that both $E(x)$ and $\Pi^2(x)$ vanish together with their first derivatives at each of the m points $x_1, x_2, \ldots, x_m$. Hence consider the function

$$F(x) = f(x) - y(x) - K[\Pi(x)]^2 \tag{2.3.13}$$

Then $F(x)$ vanishes with its first derivatives at the same m points. But $\Pi^2(x)$ only vanishes at these m points, which implies that K can be chosen to force $F(x)$ to vanish at a further point, say $\bar{x}$. If I is the smallest closed interval containing all the points $x_1, \ldots, x_m, \bar{x}$ then $F(x)$ vanishes at $m+1$ points in I. Likewise $F'(x)$ vanishes at m different points in I, each intermediate with the original $m+1$ points. But $F'(x)$ also vanishes at the m points $x_1, \ldots, x_m$ and hence in total $F'(x)$ vanishes $2m$ times in I. By the Rolle theorem, argument $F''(x)$ vanishes $2m-1$ times in I, and continuing, $F^{(2m)}(x)$ vanishes once in I, say at $x = \xi$. Hence there is now sufficient information to fix K in terms of the indeterminate value ξ to give

$$0 = F^{(2m)}(\xi) = f^{(2m)}(\xi) - K(2m)! \tag{2.3.14}$$

remembering that $y(x)$ is a polynomial of degree $2m-1$ which differentiates $2m$ times to give zero. Hence the Hermite interpolation formula error is

$$E(x) = \frac{f^{(2m)}(\xi)}{(2m)!}[\Pi(x)]^2 \tag{2.3.15}$$

The formula and its associated accuracy can be illustrated by extending the example in Table 2.7 to include derivatives as shown in Table 2.8.

Table 2.8

x	$f(x)$	$f'(x)$
2	0.301 030	0.217 147
3	0.477 121	0.144 765
4	0.602 060	0.108 574

The Lagrange coefficients in (2.2.6) give

$$\begin{aligned} l_1'(x_1) &= -\tfrac{3}{2} \\ l_2'(x_2) &= 0 \\ l_3'(x_3) &= \tfrac{3}{2} \end{aligned} \tag{2.3.16}$$

and then using formulae (2.3.11) and (2.3.12) gives

$$\begin{aligned} h_1(x) &= (3x-5)l_1^2(x) \\ h_2(x) &= l_2^2(x) \\ h_3(x) &= (13-3x)l_3^2(x) \\ \bar{h}_1(x) &= (x-2)l_1^2(x) \\ \bar{h}_2(x) &= (x-3)l_2^2(x) \\ \bar{h}_3(x) &= (x-4)l_3^2(x) \end{aligned} \tag{2.3.17}$$

Hence the interpolated value at $x = 2.4$ is given by

$$\begin{aligned} y(2.4) &= 0.506\,88 \times 0.301\,030 + 0.4096 \times 0.477\,121 + 0.083\,52 \times 0.602\,060 \\ &\quad + 0.092\,16 \times 0.217\,147 - 0.245\,76 \times 0.144\,765 - 0.023\,04 \times 0.108\,574 \\ &= 0.380\,232 \end{aligned} \tag{2.3.18}$$

and the associated actual error is 0.000 0209. On the other hand the error bound (2.3.15) with the derivative (2.2.10) gives

$$\begin{aligned} E(x) &= (x-x_1)^2(x-x_2)^2(x-x_3)^2\frac{f^{(6)}(\xi)}{6!} \\ &= (0.4)^2(0.6)^2(1.6)^2\frac{0.814}{6!} \\ &= 0.000\,1667 \end{aligned} \tag{2.3.19}$$

which yields a sensible upper bound.

The importance of Hermite interpolation lies much more with other branches of numerical analysis than in its own right. As with finite differences, any interpolation formula effectively generates a polynomial approximation to a given function $f(x)$. Polynomials can be integrated and differentiated analytically, and so the interpolation formulae lead naturally to differentiation and, more importantly, integration rules. In particular Hermite's interpolation rule is one basis for the generation of the important Gauss quadrature rules of Chapter 6.

2.4 Interpolation by splines

In a sense, the use of splines is connected to the previous topic of Hermite interpolation. With Hermite interpolation, increased accuracy was achieved by the introduction of gradients. With spline fitting, the main interval is sub-divided, and in the simplest case, a cubic is fitted in each of the sub-intervals. Continuity is imposed both on the fitting function itself and on its first derivative at the joins of the sub-intervals. These conditions yield sufficient equations to fix the coefficients of the various cubics, with two spare degrees of freedom which are used to vary the end-point conditions.

More specifically, the interval of interest $[a, b]$ is broken up into n sub-intervals by a set of abscissae $a = x_0 < x_1 < \cdots < x_n = b$ which are called knots or nodes. A spline $S(x)$ of degree m is a piecewise polynomial of degree m on each of the subintervals and is arranged to have continuous derivatives on the interval $[a, b]$ to order $m - 1$, that is $S(x) \in \mathcal{C}^{m-1}$. The spline is forced to interpolate the data points and so satisfy

$$S(x_i) = y_i \tag{2.4.1}$$

where the set $\{y_i\}$ are the experimental function values at the sampling points x_i. The extra degrees of freedom are taken up in satisfying the continuity conditions of the higher derivatives to yield a set of banded linear equations for the spline coefficients. The most commonly used spline is the cubic spline with $m \leq 3$, and from an approximation point of view this spline minimizes the oscillations in the fit between the interpolating points (Holladay 1957). This is unlike a polynomial fit which is well-known to introduce spurious loops when fitting irregular data, a feature which has proved irritating to software writers who are manufacturing graphical packages for drawing smooth curves.

By defining

$$x_i - x_{i-1} = h_i \qquad i = 1, 2, \ldots, n \tag{2.4.2}$$

and setting

$$S''(x) = m_{i-1}(x_i - x)/h_i + m_i(x - x_{i-1})/h_i \tag{2.4.3}$$

on the ith interval, forces $S''(x)$ to be linear and equal to m_i at x_i. Hence $S(x)$ itself is a cubic of the form

$$S(x) = \frac{m_{i-1}(x_i - x)^3}{6h_i} + \frac{m_i(x - x_{i-1})^3}{6h_i} + a_i(x_i - x) + b_i(x - x_{i-1}) \tag{2.4.4}$$

on the ith interval. The free constants are now determined using the end point conditions $S(x_{i-1}) = y_{i-1}$ and $S(x_i) = y_i$, which fix a_i and b_i, and then imposing the continuity conditions on $S'(x)$ at each point $x = x_i$. The first two conditions yield

$$a_i = \left(y_{i-1} - \frac{m_{i-1}h_i^2}{6} \right) \Big/ h_i \tag{2.4.5}$$

and

$$b_i = \left(y_i - \frac{m_i h_i^2}{6} \right) \Big/ h_i \tag{2.4.6}$$

and hence the derivative of $S(x)$ becomes

$$S'(x) = \frac{-m_{i-1}(x_i - x)^2 + m_i(x - x_{i-1})^2}{2h_i} + \frac{y_i - y_{i-1}}{h_i} - (m_i - m_{i-1})h_i/6 \tag{2.4.7}$$

on the ith interval. It is now straightforward to impose the continuity condition by equating the two values for $S'(x_i)$ from the neighbouring intervals to yield

$$h_i m_{i-1}/6 + (h_i + h_{i+1})m_i/3 + h_{i+1}m_{i+1}/6 = (y_{i+1} - y_i)/h_{i+1} - (y_i - y_{i-1})/h_i$$
$$\text{for} \quad i = 1, 2, \ldots, n-1 \tag{2.4.8}$$

This tridiagonal set of linear equations is then written succinctly as

$$\alpha_i m_{i-1} + 2m_i + \beta_i m_{i+1} = d_i \qquad i = 1, 2, \ldots, n-1 \tag{2.4.9}$$

where

$$\beta_i = \frac{h_{i+1}}{h_i + h_{i+1}} \tag{2.4.10}$$

and

$$\alpha_i = 1 - \beta_i \tag{2.4.11}$$

The right-hand sides d_i are defined by

$$d_i = \frac{6[(y_{i+1} - y_i)/h_{i+1} - (y_i - y_{i-1})/h_i]}{h_i + h_{i+1}} \tag{2.4.12}$$

To complete the set of equations two additional conditions are required and these can be written in the same form as equation (2.4.9) to give

$$2m_0 + \beta_0 m_1 = d_0$$
$$\alpha_n m_{n-1} + 2m_n = d_n \tag{2.4.13}$$

and it is a matter of choice how these extra constants are fixed. The natural spline is obtained by setting $\beta_0 = d_0 = \alpha_n = d_n = 0$, but it is also feasible to fix the first derivative of $S(x)$ at the end points a and b, or indeed use the interior points as nodes and then specify $S(x)$ itself at the end points. The resulting tridiagonal equation is efficiently solved by the simple device of eliminating m_0 from the second equation, m_1 from the third and so on to m_{n-1} from the last. This will leave two equations at the end in the two unknowns m_{n-1} and m_n which can be solved for m_n, leaving a simple back substitution to yield a recurrence relation for the complete set of m_i's.

This process can be illustrated by considering once again the example of Table 2.7. Setting up the natural spline with two sub-intervals (2, 3) and (3, 4) giving $n = 2$ and $h_i = 1$ for $i = 1, 2$, gives the single equation

$$m_0/6 + 2m_1/3 + m_2/6 = \log 4 - 2\log 3 + \log 2 \tag{2.4.14}$$

by using (2.4.8). The choice of the natural spline gives $m_0 = m_2 = 0$ to leave $m_1 = -0.076\,7288$. Now (2.4.4) with (2.4.5) and (2.4.6) give

$$S_1 = (x-2)^3 \tfrac{1}{6} m_1 + (3-x)y_0 + \left(y_1 - \tfrac{1}{6}m_1\right)(x-2) \tag{2.4.15}$$

$$S_2 = (4-x)^3 \tfrac{1}{6} m_1 + (4-x)\left(y_1 - \tfrac{1}{6}m_1\right) + y_2(x-3) \tag{2.4.16}$$

from which it is easy to confirm that $S_1(2) = y_0$, $S_1(3) = S_2(3) = y_1$ and $S_2(4) = y_2$ correctly. In addition the fit has a continuous first derivative at $x = 3$. Hence

$$\begin{aligned} S_1(2.4) &= 0.4^3 \tfrac{1}{6} m_1 + 0.6 \log 2 + 0.4\left(\log 3 - \tfrac{1}{6}m_1\right) \\ &= -0.000\,818\,440 + 0.180\,618 + 0.195\,964 \\ &= 0.375\,763 \end{aligned} \tag{2.4.17}$$

Higher accuracy could be obtained by using the two spare degrees of freedom to fit the first derivative of $f(x)$ at the end points rather than use the natural spline.

2.5 Least squares curve fitting

Interpolation is the study of fitting a polynomial or other function to a set of error-free data. The resulting curve passes through the collocating points, and the number of free constants in the fitting function exactly matches the number of data points to give a well-posed problem.

With curve fitting, experimental or other errors can be expected in the data to be fitted, and now the object is to find a fitting curve which approximates the data in some sense. The classic problem arises in elementary physics when a series of readings is taken from an experiment and plotted on a graph. There is a scatter of points which roughly lie on a straight line, for example. It may be that the gradient of this line is the required physical constant which is the purpose of the experiment. One approach is to gauge the 'best fit' using a ruler and a good eye. The process can be improved using some simple mathematics.

Somehow a curve is required which is a 'minimum' distance away from the given data. Consider first linear curve fitting where the outcome is a straight line fit to the data. Let the data be the set of points $\{x_i, Y_i\}$, with $i = 1, \ldots, n$ for some n. Suppose the x value is free of error and the measured value is y. Then one measure of the 'distance' of a point from the line would be the sums of squares of the differences between the measured Y_i and the values predicted from the fitted straight line. Let this line have the equation $y = mx + c$, then at the point x_i the predicted y is y_i. The error is $y_i - Y_i$, and a suitable measure of this error(or norm) is to take

$$S = \sum_{i=1}^{n} (y_i - Y_i)^2 \tag{2.5.1}$$

By squaring the actual error, any sign differences are absorbed, and if S is minimized over the sum of all the trial points, equations will be obtained for the two free constants m and c which specify the required curve. Hence

$$S = \sum_{i=1}^{n} (mx_i + c - Y_i)^2 \tag{2.5.2}$$

with turning points occurring when

$$\frac{\partial S}{\partial m} = 0 = \sum_{i=1}^{n} 2(mx_i + c - Y_i)x_i \tag{2.5.3}$$

and

$$\frac{\partial S}{\partial c} = 0 = \sum_{i=1}^{n} 2(mx_i + c - Y_i) \tag{2.5.4}$$

which yield two simultaneous equations for m and c

$$m\sum_{i=1}^{n} x_i^2 + c\sum_{i=1}^{n} x_i = \sum_{i=1}^{n} x_i Y_i \tag{2.5.5}$$

$$m\sum_{i=1}^{n} x_i + cn = \sum_{i=1}^{n} Y_i \tag{2.5.6}$$

As a simple example consider the data in Table 2.9.

Then $\sum x_i = 273.1$, $\sum Y_i = 4438$, $\sum x_i^2 = 186\,07.27$ and $\sum x_i Y_i = 254\,932.5$ to give the pair of algebraic equations

$$\begin{aligned} 186\,07.27m + 273.1c &= 254\,932.5 \\ 273.1m + 5c &= 4438 \end{aligned} \tag{2.5.7}$$

with the solution $m = 3.395$ and $c = 702.2$.

Table 2.9

x	Y
20.5	765
32.7	826
51.0	873
73.2	942
95.7	1032

These linear equations are called normal equations. In statistical terms, if the errors in Y have a normal distribution and constant standard deviation, then the line obtained by the above process is the one with the maximum likelihood of occurrence.

There is no reason why this process cannot be extended to more complicated fitting curves with many free parameters. There may well be physical evidence for choosing the general form of the fitting curve. The same process of equating the partial derivatives of S to zero will yield a set of equations for the free parameters, now of course not necessarily a set of linear equations. Recourse to using numerical solutions of sets of equations from Chapter 1 may be needed in this case. There are however many cases when linear equations either still occur or can be contrived to arise. Two examples will illustrate such cases.

Suppose it is required to fit $y = ax^b$, then a contrivance to give linear equations is to take

$$\log y = \log a + b \log x \tag{2.5.8}$$

to give the values in Table 2.10.

Taking $\zeta = \log y$, $c = \log a$ and $\eta = \log x$ gives the linear problem $\zeta = b\eta + c$ with the equations

$$b\sum \eta^2 + c\sum \eta = \sum \eta\zeta$$

$$b\sum \eta + nc = \sum \zeta$$

or

$$0.6807b + 1.3802c = 1.3614$$

$$1.3802b + 4c = 2.7604$$

with the solution $c = 0$ and $b = 2$, and hence $a = 1$ to give the expected fit $y = x^2$. The use of logarithms in this way is very common.

Linear equations are also obtained if the fitting curve is a polynomial of degree m say. As an example of this, consider the fitting of a quadratic to a data set which is shown in Table 2.11.

The normal equations are now given by

$$S = \sum_{i=1}^{n} (ax_i^2 + bx_i + c - y_i)^2 \tag{2.5.9}$$

Table 2.10

y	x	$\log y$	$\log x$	$\log x \log y$	$\log^2 x$
1	1	0.0000	0.0000	0.0000	0.0000
4	2	0.6021	0.3010	0.1812	0.0906
9	3	0.9542	0.4771	0.4552	0.2276
16	4	1.2041	0.6021	0.7250	0.3625
$\sum$		2.7604	1.3802	1.3614	0.6907

Table 2.11

x_i	y_i	x^2	x^3	x^4	xy	x^2y
0	1.1	0	0	0	0	0
1	2.9	1	1	1	2.9	2.9
2	7.1	4	8	16	14.2	28.4
3	12.9	9	27	81	38.7	116.8
4	21.1	16	64	256	84.4	337.6
5	30.8	25	125	625	154.0	770.0
$\sum$ 15	75.9	55	225	979	294.2	1255.7

with derivatives

$$\frac{\partial S}{\partial a} = 0 = 2\sum_{i=1}^{n}(ax_i^2 + bx_i + c - y_i)x_i^2 \tag{2.5.10}$$

$$\frac{\partial S}{\partial b} = 0 = 2\sum_{i=1}^{n}(ax_i^2 + bx_i + c - y_i)x_i \tag{2.5.11}$$

$$\frac{\partial S}{\partial c} = 0 = 2\sum_{i=1}^{n}(ax_i^2 + bx_i + c - y_i) \tag{2.5.12}$$

to give the equations

$$a\sum_{i=1}^{n} x_i^4 + b\sum_{i=1}^{n} x_i^3 + c\sum_{i=1}^{n} x_i^2 = \sum_{i=1}^{n} x_i^2 y_i \tag{2.5.13}$$

$$a\sum_{i=1}^{n} x_i^3 + b\sum_{i=1}^{n} x_i^2 + c\sum_{i=1}^{n} x_i = \sum_{i=1}^{n} x_i y_i \tag{2.5.14}$$

$$a\sum_{i=1}^{n} x_i^2 + b\sum_{i=1}^{n} x_i + cn = \sum_{i=1}^{n} y_i \tag{2.5.15}$$

The specific equations taking the relevant sums from Table 2.11 are

$$979a + 225b + 55c = 1255.7$$

$$225a + 55b + 15c = 294.2$$

$$55a + 15b + 6c = 75.9$$

with solution $c = 1.1803$, $b = 0.8053$ and $a = 1.0312$ to give the fit $y = 1.0312x^2 + 0.8053x + 1.1803$.

For higher degree polynomials, the set of linear equations which results would need to be solved by a numerical method such as Gaussian elimination which is covered in Chapter 3. If the fitting parameters appear linearly in the proposed fit then linear equations will result. Hence the trial fits

$$y = a + b/x + cx \tag{2.5.16}$$

$$y = a \sin x + be^x \tag{2.5.17}$$

would yield linear equations.

Mathematica is an extensive computer package designed to aid mathematical work. Relevant features for interpolation include various forms of polynomial fitting, least squares packages, splines approximations and associated plotting facilities.

2.6 Differentiation

It has already been demonstrated in Chapter 1 that even a simple finite difference approximation to a derivative, leads rapidly to numerical cancellation and instability. Because of this, it is usual to avoid having to use numerical differentiation as much as possible. Fortunately the analytic process of differentiation is widely applicable, and where the use of numerical differentiation might be a temptation, there are often alternative avoidance methods which should be employed. Such an area is the computation of maxima and minima, more usually called optimization in a numerical context. An obvious algorithm is to combine a numerical differentiator with a zero finding routine from Chapter 1. However, more direct approaches avoiding the differentiator are used and will be detailed in Chapter 10. Sometimes it does become necessary to use a differentiator, and under such circumstances care is needed. One such case would be in electromagnetic problems in which solutions for the potential ϕ are found from Laplace's equation (see Chapter 9) and the practical problem requires the physical field which is given by

$$\mathbf{E} = \operatorname{grad} \phi \tag{2.6.1}$$

which therefore requires the derivatives of ϕ.

Another case arises in the numerical solution of integro-differential equations such as the Hartree–Fock equations of quantum mechanics. The basic problem is to solve an integral equation (see Chapter 8) but some of the terms in the integral equation involve derivatives which need to be evaluated numerically from tabulated iterates for the solution. In such a case, it is useful again to have a numerical differentiator available. The finite difference section gives a number of such formulae such as (2.1.34), (2.1.35) and (2.1.41), and indeed it would be equally feasible to use the derivative of a Lagrange interpolation approximation. However, the proposal here is to employ the special properties of the Chebyshev polynomials quoted in Chapter 1. This follows the work of Evans (1979) in which the solution of the Hartree–Fock equations was the motivating aim.

The approximation (1.2.12) to $f(x)$ with corresponding coefficients a_r given by (1.2.18) are used, and the interval is reduced from the general range $[a, b]$ to

$[-1, 1]$ by

$$F(t) = f\left(\frac{b-a}{2}t + \frac{b+a}{2}\right) \tag{2.6.2}$$

Then $F(t)$ is approximated by

$$F(t) = \sum_{r=0}^{M}{}'' a_r T_r(t) + E(t) \tag{2.6.3}$$

with

$$a_r = \frac{2}{M}\sum_{s=0}^{M}{}'' F\left(\cos\frac{\pi s}{M}\right)\cos\frac{\pi rs}{M} \tag{2.6.4}$$

Hence the derivative can be written as:

$$\frac{\mathrm{d}f}{\mathrm{d}x} = \frac{2}{(b-a)}\frac{\mathrm{d}F}{\mathrm{d}t} \sim \frac{2}{(b-a)}\sum_{r=1}^{M}{}' a_r T'_r(t)$$

$$= \frac{2}{(b-a)}\sum_{r=1}^{m}{}' r a_r \sin r\theta / \sin\theta \tag{2.6.5}$$

with

$$t = \cos\theta \text{ and } x = \tfrac{1}{2}(b+a) + \tfrac{1}{2}(b-a)$$

The attractive feature of this approach is that a computable error estimate can be derived, and it is illustrated in the examples which follow that this estimate does indeed agree with the actual errors. The estimate is derived by considering the parallel expansion to (2.6.3)

$$F(t) = \sum_{r=0}^{\infty}{}' A_r T_r(t) \tag{2.6.6}$$

which is a generalized Fourier expansion of $F(t)$. The Fourier coefficients A_r are related to the expansion coefficients a_r by

$$a_r = \frac{2}{M}\sum_{s=0}^{M}{}'' \cos\frac{\pi rs}{M}\sum_{i=0}^{\infty}{}' A_i \cos\frac{\pi is}{M}$$

$$= A_r + A_{2m-r} + A_{2m+r} + A_{4M-r} + A_{4M+r} + \cdots \tag{2.6.7}$$

obtained by simply substituting (2.6.6) into (2.6.4). The error $E'(t)$ in the derivative $\mathrm{d}F/\mathrm{d}t$ is then given by

$$E'(t) = -\sum_{i=1}^{M}\left[\sum_{j=1}^{\infty}(A_{2j,M-i} + A_{2j,M+i})\right]T'_i(t) + \sum_{i=M+1}^{\infty} A_i T'_i(t) \tag{2.6.8}$$

The next stage is to estimate the size of the Fourier coefficients A_r which are given by the conventional integral

$$A_r = \frac{\displaystyle\int_{-1}^{1} \frac{F(t)T_r(t)}{\sqrt{1-t^2}}\,\mathrm{d}t}{\displaystyle\int_{-1}^{1} \frac{T_r^2(t)}{\sqrt{1-x^2}}\,\mathrm{d}t} \tag{2.6.9}$$

and the substitution $x = \cos\theta$ reduces this to

$$A_r = \frac{2}{\pi}\int_0^{\pi} F(\cos\theta)\cos r\theta\,\mathrm{d}\theta \tag{2.6.10}$$

If $F(t)$ were discontinuous by say ΔF at $t = a$ then integration by parts gives:

$$\begin{aligned} A_r &= \frac{2}{\pi}\int_0^{\pi} F(\cos\theta)\mathrm{d}\left(\frac{\sin r\theta}{r}\right) \\ &= \frac{2}{\pi r}\Delta F\sin ra + \frac{2}{\pi r}\int_0^{\pi} F'(\cos\theta)\sin r\theta\sin\theta\,\mathrm{d}\theta \end{aligned} \tag{2.6.11}$$

and so A_r behaves like K/r. If it were $F'(t)$ in which the discontinuity first appeared then a further integration by parts would yield behaviour like K/r^2. Presumably the worst that can happen in the current situation is for the second derivative to be discontinuous, in which case the fall-off in A_r is like K/r^3. Hence regrouping the error formula $E'(t)$ in descending magnitude of the coefficients A_r gives:

$$\begin{aligned} E'(t) = &-A_M T'_M(t) + A_{M+1}[T'_{M+1}(t) - T'_{M-1}(t)] + \cdots \\ &+ A_{2M-1}[T'_{2M-1}(t) - T'_1(t)] + A_{2M}T'_{2M}(t) \\ &+ A_{2M+1}[T'_{2M+1}(t) - T'_1(t)] + \cdots + A_{3M}[T'_{3M}(t) - 2T'_M(t)] \\ &+ A_{3M+1}[T'_{3M+1} - T'(t)_{M-1}(t)] + \cdots + A_{4M-1}[T'_{4M-1}(t) - T'_1(t)] \\ &+ A_{4M}T'_{4M}(t) + A_{4M+1}[T'_{4M+1}(t) - T'_1(t)] + \cdots \\ &+ A_{5M}[T'_{5M}(t) - 2T'_M(t)] + \cdots \end{aligned} \tag{2.6.12}$$

Taking leading terms, and using the inequalities

$$T'_j(t) = \frac{j\sin j\theta}{\sin\theta} \le j^2 \tag{2.6.13}$$

with $t = \cos\theta$ and

$$|T'_{M+i} - T'_{M-i}| = \left|\frac{2i\cos i\theta\sin M\theta + 2M\cos M\theta\sin i\theta}{\sin\theta}\right| \le 4Mi \tag{2.6.14}$$

leads to the simplified estimate

$$|E'(t)| < M^2|A_M| + 4M|A_{M+1}| + 8M|A_{M+2}| + \cdots \tag{2.6.15}$$

with the leading term

$$|E'(t)| \sim A_M M^2 = a_M M^2/2 \tag{2.6.16}$$

using (2.6.7).

The use of this differentiation formula together with the associated error estimate can be well illustrated by considering two test cases which can be made progressively more difficult from a differentiation point of view by the adjustment of either a parameter or the range of interest. The first example is the function

$$f(x) = cxe^{-cx} \tag{2.6.17}$$

on the interval [0, 1] with the range of c given by

$$c = (0.5, 1.0, 2.0, 4.0, 8.0)$$

Clearly, for increasing c, the steeper the curve becomes. The differentiation rule was applied for $M = 5$, 10, 15, and 20 and in Table 2.12 a comparison is made between the actual errors in both the Chebyshev and the Lagrangian approaches, and the estimated error from the above formula. The derivative was computed at $x = 0$, $x = \frac{1}{2}$ and $x = 1$ for the purposes of the comparison.

It can be seen that the low-order Lagrange results are generally better than the Chebyshev values but as the order increases the Chebyshev rule overhauls the Lagrange rule in accuracy. Moreover the Chebyshev rule also has an error estimate which can be seen to be quite reasonable. Obviously it is the nature of differentiation formulae that the effect of rounding will ultimately be the dominant

Table 2.12

c	M	Cheb error at 0	Lag error at 0	Cheb error at 1	Lag error at 1	Cheb error at $\frac{1}{2}$	Lag error at $\frac{1}{2}$	$\lvert E'(t)\rvert$
0.5	5	9.2(−5)	3.9(−6)	9.6(−5)	3.6(−6)	1.7(−5)	2.3(−7)	9.4(−5)
	10	8.7(−11)	7.3(−10)	2.6(−9)	4.4(−9)	4.7(−11)	1.5(−10)	1.1(−10)
	15	1.0(−9)	3.5(−8)	5.4(−9)	4.5(−8)	1.1(−10)	3.2(−10)	1.7(−10)
1.0	5	2.2(−3)	2.0(−4)	2.3(−3)	1.6(−4)	4.2(−4)	1.1(−5)	2.2(−3)
	10	4.9(−10)	2.5(−9)	3.0(−9)	4.9(−9)	1.2(−10)	1.1(−10)	4.5(−10)
	15	6.4(−10)	6.8(−8)	1.1(−8)	1.0(−8)	1.5(−10)	4.3(−10)	4.7(−10)
2.0	5	3.7(−2)	7.8(−3)	4.4(−2)	5.5(−3)	8.1(−3)	4.1(−4)	4.1(−2)
	10	3.6(−7)	7.5(−8)	3.7(−7)	5.6(−8)	3.9(−9)	4.2(−10)	3.7(−7)
	15	2.5(−9)	7.6(−8)	1.0(−8)	5.9(−8)	1.3(−10)	6.3(−10)	1.0(−10)
4.0	5	3.4(−1)	2.0(−1)	5.0(−1)	1.0(−1)	9.7(−2)	8.7(−3)	4.4(−1)
	10	1.3(−4)	6.1(−5)	2.7(−6)	2.0(−7)	1.4(−4)	4.2(−5)	1.3(−4)
	15	3.4(−9)	4.0(−8)	7.4(−9)	1.7(−8)	8.6(−10)	1.9(−10)	7.1(−10)
8.0	5	5.8(−1)	2.6(0)	2.4(0)	6.1(−1)	5.4(−1)	6.8(−2)	2.1(0)
	10	1.7(−2)	2.1(−2)	2.1(−2)	9.7(−3)	7.7(−4)	5.3(−5)	2.0(−2)
	15	5.8(−6)	1.4(−5)	6.3(−6)	8.4(−6)	1.3(−7)	1.2(−9)	6.1(−6)
	20	7.2(−7)	2.0(−7)	1.0(−8)	2.5(−7)	3.1(−10)	2.7(−10)	1.7(−9)

Table 2.13 Errors over $[a, b]$ with $c = (a + b)/2$

a	b	M	Cheb error at a	Lag error at a	Cheb error at c	Lag error at c	$\|E'(t)\|$
1.0	2.0	5	1.2(−2)	6.8(−4)	2.4(−3)	1.1(−4)	1.3(−2)
		10	2.7(−5)	4.2(−5)	2.3(−6)	1.3(−7)	3.2(−5)
		15	3.0(−8)	1.1(−7)	4.2(−9)	6.3(−10)	3.5(−8)
		20	2.5(−8)	1.8(−6)	1.8(−9)	3.9(−10)	1.2(−9)
0.5	1.0	5	2.2(−1)	2.0(−1)	4.4(−2)	7.2(−3)	3.3(−1)
		10	3.4(−4)	1.2(−4)	2.9(−5)	7.6(−7)	3.5(−4)
		15	4.8(−7)	3.6(−6)	6.3(−9)	7.0(−10)	5.3(−7)
		20	1.5(−7)	3.8(−5)	3.2(−9)	7.0(−9)	1.2(−9)
0.1	0.5	5	1.2(1)	8.1(2)	9.9(1)	1.0(1)	3.1(2)
		10	1.6(2)	7.8(2)	1.2(1)	2.8(−1)	7.7(1)
		15	3.3(1)	2.3(2)	1.2(0)	1.1(−2)	2.8(1)
		20	1.3(0)	6.1(1)	6.5(−2)	9.1(−6)	2.5(0)

cause of error, and it can be seen that around nine figure accuracy is the best that can be achieved using 11 digit precision.

A rather different test can be made with the function

$$f(x) = \frac{1}{x} \sin \frac{1}{x} \tag{2.6.18}$$

in which the interval under consideration is extended towards the difficult region round the origin in order to increase the severity of the test. The test intervals, (a, b), used are $(1.0, 2.0)$, $(0.5, 1.0)$ and $(0.1, 0.5)$ and the results are tabulated in Table 2.13.

It can be seen that the severe errors in the final range are adequately estimated, and it must be noted that the errors quoted in this range are not as bad as they first appear, as the derivatives themselves are large. For example at $x = 0.1$ the true derivative is 893.5 and the Chebyshev rule with $m = 20$ gives 894.8 against the Lagrange value of 832.7.

In addition to its interpolatory packages, Mathematica can generate standard derivatives in analytic form using a construct of the form $D[f(x), x]$ which will yield the analytic derivative of a given function $f(x)$ with respect to x. This capability is clearly limited to standard functions, but can be combined with evaluating routines to yield the possibility of semi-analytic approaches to problems.

Further reading

Differences

Hildebrand F. (1958) *Introduction to Numerical Analysis*, Mcgraw-Hill, New York.
Milne-Thomson L. M. (1951) *The Calculus of Finite Differences*, Macmillan, London.

Splines
Ahlberg J., Nilson E. and Walsh J. (1967) *The Theory of Splines and their Application*, Academic Press, New York.
Greville T. N. E. (Ed) (1969) *Theory and Application of Spline Functions*, Academic Press, New York.

Exercises

(2.1). Use Newton's forward difference formula to interpolate for $f(1.05)$ in the table:

x	$f(x)$
1.0	0.8415
1.1	0.8668
1.2	0.8891
1.3	0.9087
1.4	0.9258

(2.2). Establish the operator results

$$DJ = JD = \Delta$$

and hence use Taylor series to establish the result

$$E = e^{hD}$$

where the usual definitions apply. Hence derive the differentiation formula:

$$f'_0 = \frac{1}{h}\left(\Delta - \frac{1}{2}\Delta^2 + \frac{1}{3}\Delta^3 - \cdots\right) f_0$$

Given the table:

x	$f(x)$
0.0	1.0000
0.1	1.1052
0.2	1.2214
0.3	1.3499

use the above formula to find $f'(0.3)$. Derive the corresponding backward formula and again find $f'(0.3)$.

(2.3). A table is being set up for the function e^x on the interval [0, 1]. Let h be the difference between adjacent tabular values. Assuming that the individual tabular values are more accurate than the required error, what is the value of h so that

(i) linear interpolation
(ii) quadratic interpolation

gives interpolated errors of less than 10^{-6}, using Lagrange's formula?

(2.4). Take a set of tables for the function $\sin(x)$, or use a calculator, for the values $x = 1.1, 1.2, 1.3, 1.4$ and 1.5. Apply Lagrange interpolation of varying orders to compute $\sin(1.24)$, and compare your errors with the theoretical bounds.

(2.5). The following is an interpolatory method for finding square roots. Tabulate $f(x) = \sqrt{x}$ at trivially known points such as

x	$f(x)$
1	1
4	2
9	3
16	4

Up to cubic interpolation can now be used to get an intermediate value such as $\sqrt{7}$. Can you develop this idea into an algorithm which yields arbitrary accuracy?

(2.6) Prove that

$$|\Pi(x)| = |(x - x_0) \cdots (x - x_n)|$$

is minimized if $\Pi(x) = 1/2^n T_{n+1}(x)$ on the interval $[-1, 1]$.

Apply an equally spaced Lagrange interpolation to e^{2x} on the interval $[-1, 1]$ collocating at the points $\{-1,\ -\frac{1}{3},\ \frac{1}{3},\ 1\}$. Hence compute e^0 from the Lagrange fit. Repeat the procedure with the four cosine weighted points

$$x_i = \cos\frac{(2i+1)\pi}{2(n+1)}$$

with $n = 3$ and $i = 0,\ 1,\ 2$ and 3. Compare the results to confirm the proof.

(2.7) Tabulate $f(x) = \ln(x)$ for x ranging from 3.0 to 3.5 in steps of 0.1, and hence generate a difference table to include Δ^3 differences. Use Newton's formula to find $\ln(3.15)$, and estimate the error. Repeat with Lagrange's formula with three collocation points, and use the error formula to obtain an upper bound for the error.

(2.8) In the special case of $n = 2$ show that the Newton forward difference interpolation formula and the Lagrange interpolation formula are identical when the collating points are equally spaced.

(2.9) Find the equation of the quadratic polynomial which collocates with the function

$$f(x) = \frac{1}{(x+2)}$$

at the points $x = 0$, 1 and 2. Use both Lagrange interpolation and Newton's difference formula.

(2.10) Use Newton's difference formula and Lagrange's interpolation formula to find $f(0.25)$ given that the f values corresponding to $x = 0(0.1)0.6$ are 1.0, 0.9950, 0.9801, 0.9553, 0.9211, 0.8776 and 0.8253. Is quadratic interpolation adequate to the tabulated accuracy?

(2.11) Take the set of values in exercise (2.4) and complete the interpolation using Hermite's formula. You will need to compute a set of derivatives of $\sin x$ at the collocation points.

(2.12) Take the process in exercise (2.5) and apply Hermite interpolation. Now you will need $f'(x)$, but these values are also trivial for the values of x used.

(2.13) The quantities given in the following table are related by a rule of the form

$$y = ax + be^{-x/2}$$

Obtain a least squares fit, finding a and b explicitly.

x	y
0	0.60
1	1.66
2	2.82
3	4.03
4	5.28

(2.14) Show that the method of least squares applied to the curve

$$y = a\cos x + b\sin x$$

gives rise to the normal equations

$$a\sum_i \cos^2 x_i + b\sum_i \sin x_i \cos x_i - \sum_i y_i \cos x_i = 0$$

$$a\sum_i \sin x_i \cos x_i + b\sum_i \sin^2 x_i \cos x_i - \sum_i y_i \sin x_i = 0$$

for the values a and b.

The following table gives specific values of x and y.

x	y
0.0	4.0
$\pi/6$	5.1
$\pi/4$	5.1
$\pi/3$	4.8
$\pi/2$	3.2

Show that the maximum value of y occurs at $y = (a^2 + b^2)^{1/2}$. Find this maximum for the given data.

(2.15) Apply the numerical differentiator of equation (2.6.5) to find $f'(x)$ at $x = 0.5$ for

$$f(x) = \tan(x)$$

and compare the results for formulae with orders $M = 5, 10, 15$ with the exact error. Try using the finite difference approach to obtain a further comparison, for a suitable range of tabulated values, for example at $x = 0.0, 0.2, 0.4, 0.6, 0.8$.

3

Solution of Algebraic Equations

3.1 Mathematical introduction

The problem being addressed here is the very basic one of solving a set of linear algebraic equations. Such equations are usually represented by the matrix equation

$$\mathbf{Ax} = \mathbf{b} \tag{3.1.1}$$

where the matrix $\mathbf{A}$ is a real $n \times n$ matrix with elements $\{a_{ij}\}$, $\mathbf{b}$ is a real vector, and the vector $\mathbf{x}$ is required. In order to measure the sizes of errors in the processes used to solve this system, the concept of norm is employed. In pure mathematics terms, the norm $||\mathbf{x}||$ of a vector $\mathbf{x}$ must satisfy the following axioms.

(i) $||\mathbf{x}|| > 0$ unless $\mathbf{x} = \mathbf{0}$ when $||\mathbf{x}|| = 0$.
(ii) $||k\mathbf{x}|| = |k|\ ||\mathbf{x}||$ for any complex scalar k.
(iii) $||\mathbf{x} + \mathbf{y}|| \leq ||\mathbf{x}|| + ||\mathbf{y}||$.

These axioms are commonly used in theoretical linear algebra to prove theorems concerning norms and little attention is paid to specific norms except possibly in exercises. However, in numerical analysis, the norms used to measure errors need to be of a practical nature and allow relatively easy computation. One such family of vector norms is defined by

$$||\mathbf{x}||_p = (|x_1|^p + |x_2|^p + \cdots + |x_n|^p)^{1/p} \tag{3.1.2}$$

for values of $p = 1, 2, \ldots$. Indeed as $p \to \infty$ the norm $||\mathbf{x}||_\infty$ becomes

$$\max_i |x_i|$$

which yields a norm satisfying the above axioms. Quite clearly $||\mathbf{x}||_2$ yields the usual Euclidean length.

A number of useful properties of these norms will be used later and these will be derived in this introduction. If the vector $\mathbf{x}$ is defined as

$$\mathbf{x} = \begin{pmatrix} \lambda_1 + i\mu_1 \\ \lambda_2 + i\mu_2 \\ \vdots \\ \lambda_n + i\mu_n \end{pmatrix} \tag{3.1.3}$$

then the Hermitian of $\mathbf{x}$ is defined as

$$\mathbf{x}^H = (\lambda_1 - i\mu_1, \lambda_2 - i\mu_2, \ldots, \lambda_n - i\mu_n) \tag{3.1.4}$$

and clearly for real $\mathbf{x}$, the Hermitian of $\mathbf{x}$ reduces to the transpose of $\mathbf{x}$ which is denoted by $\mathbf{x}^T$. Immediately, from the two definitions it follows that

$$\mathbf{x}^H.\mathbf{x} = ||\mathbf{x}||_2^2 \tag{3.1.5}$$

and

$$|\mathbf{x}^H\mathbf{y}| \leq ||\mathbf{x}||_2||\mathbf{y}||_2. \tag{3.1.6}$$

A similar set of norms and axioms can be laid down for a matrix $\mathbf{A}$ with the norm denoted by $||\mathbf{A}||$ and now satisfying:

(i) $||\mathbf{A}|| > 0$ unless $\mathbf{A} = \mathbf{0}$ when $||\mathbf{A}|| = 0$.
(ii) $||k\mathbf{A}|| = |k|\,||\mathbf{A}||$ for any complex scalar k.
(iii) $||\mathbf{A} + \mathbf{B}|| \leq ||\mathbf{A}|| + ||\mathbf{B}||$ and
(iv) $||\mathbf{AB}|| \leq ||\mathbf{A}||\,||\mathbf{B}||$.

A whole family of practical norms can be derived from the vector norms above. Such norms are called subordinate matrix norms and are defined in terms of the corresponding vector norm by the relation:

$$||\mathbf{A}|| = \sup_{\mathbf{x}\neq\mathbf{0}} \frac{||\mathbf{Ax}||}{||\mathbf{x}||} \tag{3.1.7}$$

which by the second matrix norm axiom can be reduced to

$$||\mathbf{A}|| = \sup_{||\mathbf{x}||=1} ||\mathbf{Ax}|| \tag{3.1.8}$$

This definition does not give a practical evaluation of a matrix norm, and to realize such a form, some specific subordinate norms need to be re-formulated. In fact there are three norms which can be easily computed which are based on the vector norms, namely $||\ ||_1$, $||\ ||_2$ and $||\ ||_\infty$, and these will be derived next.

Consider first $||\ ||_1$, then from the definition of a subordinate matrix norm the following inequalities arise:

$$\begin{aligned}\frac{||\mathbf{Ax}||_1}{||\mathbf{x}||_1} &= \frac{\left|\left|\sum_{j=1}^{n} a_{ij}x_j\right|\right|_1}{||\mathbf{x}||_1}\\ &= \frac{\sum_{j=1}^{n} |x_j| \left(\sum_{i=1}^{n} |a_{ij}|\right)}{\sum_{j=1}^{n} |x_j|} \qquad \text{by the definition of } ||\ ||_1\end{aligned}$$

$$\leq \frac{\sum_{j=1}^{n} |x_j| \left[\max_j \sum_{i=1}^{n} |a_{ij}|\right]}{\sum_{j=1}^{n} |x_j|}$$

$$= \left[\max_j \sum_{i=1}^{n} |a_{ij}|\right]$$

Hence the final expression is an upper bound for $||\mathbf{A}||_1$. Suppose the maximum is attained for $j = K$, then this bound is attained for the vector $\mathbf{x} = \mathbf{e}_K$, since $||\mathbf{e}_K|| = 1$ and

$$||\mathbf{A}\mathbf{e}_K|| = \sum_{i=1}^{n} \left| \sum_{k=1}^{n} a_{ik}\delta_{kK} \right|$$

$$= \sum_{j=1}^{n} |a_{iK}| \tag{3.1.9}$$

Hence the relation

$$||\mathbf{A}||_1 = \max_j \sum_{i=1}^{n} |a_{ij}| \tag{3.1.10}$$

has been established, and this form of the norm is trivially computable.

In a similar vein the $|| \; ||_\infty$ result can be derived by consideration of the ratio:

$$\frac{||\mathbf{A}\mathbf{x}||_\infty}{||\mathbf{x}||_\infty} = \frac{\max_i \left[\sum_{j=1}^{n} |a_{ij}x_j|\right]}{\max_i [|x_i|]} \quad \text{by the definition of } || \; ||_\infty$$

Now let x_k be $\max_i [|x_i|]$ to give:

$$\frac{||\mathbf{A}\mathbf{x}||_\infty}{||\mathbf{x}||_\infty} = \max_i \left[\sum_{j=1}^{n} |a_{ij}| \left|\frac{x_j}{x_k}\right|\right]$$

$$\leq \max_i \left[\sum_{j=1}^{n} |a_{ij}|\right]$$

and now as for $|| \; ||_1$, an upper bound has been obtained. Suppose the maximum occurs for $i = K$, then for $\mathbf{x}$ with components

$$x_k = \begin{cases} \frac{a_{Kk}}{|a_{Kk}|}, & a_{Kk} \neq 0 \\ 0, & a_{Kk} = 0 \quad k = 1, 2, \ldots, n \end{cases} \tag{3.1.11}$$

then $||\mathbf{x}||_\infty = 1$ and for non-trivial $\mathbf{A}$

$$||\mathbf{A}\mathbf{x}||_\infty = \sum_{k=1}^{n} |a_{kK}| \leq ||\mathbf{A}||_\infty \tag{3.1.12}$$

Hence

$$||\mathbf{A}||_\infty = \max_i \sum_{j=1}^{n} |a_{ij}|. \tag{3.1.13}$$

The proof for the third subordinate norm follows by first defining the singular values σ_i of a matrix $\mathbf{A}$. The matrix $\mathbf{A}^H\mathbf{A}$ is Hermitian and

$$\mathbf{x}^H\mathbf{A}^H\mathbf{A}\mathbf{x} = (\mathbf{A}\mathbf{x})^H(\mathbf{A}\mathbf{x}) \geq \mathbf{0} \tag{3.1.14}$$

so that $\mathbf{A}^H\mathbf{A}$ is also positive definite. Hence the eigenvalues of $\mathbf{A}^H\mathbf{A}$ are ≥ 0 and can be denoted by σ_i^2 where these values are ordered according to

$$\sigma_1^2 \geq \sigma_2^2 \geq \cdots \sigma_n^2 \geq 0 \tag{3.1.15}$$

Hence in terms of $\mathbf{A}^H\mathbf{A}$ and using the continuity argument again

$$\begin{aligned} ||\mathbf{A}||_2^2 &= \max_{\mathbf{x}\neq 0} \frac{||\mathbf{A}\mathbf{x}||_2^2}{||\mathbf{x}||_2^2} \\ &= \max_{\mathbf{x}\neq 0} \frac{\mathbf{x}^H\mathbf{A}^H\mathbf{A}\mathbf{x}}{\mathbf{x}^H\mathbf{x}}. \end{aligned} \tag{3.1.16}$$

If now the eigenvectors of $\mathbf{A}^H\mathbf{A}$ are denoted by $\mathbf{u}_1, \ldots \mathbf{u}_n$ then these eigenvectors form a basis of the underlying vector space and hence any vector in the space can be expressed as a linear combination of these vectors to give

$$\mathbf{x} = \sum_{i=1}^{n} \alpha_i \mathbf{u}_i. \tag{3.1.17}$$

Moreover the eigenvectors $\mathbf{u}_i$ form an orthogonal set and satisfy

$$\mathbf{A}^H\mathbf{A}\mathbf{u} = \sigma^2\mathbf{u} \tag{3.1.18}$$

to give

$$\frac{\mathbf{x}^H\mathbf{A}^H\mathbf{A}\mathbf{x}}{\mathbf{x}^H\mathbf{x}} = \frac{\sum_{i=1}^{n} |\alpha_i|^2\sigma_i^2}{\sum_{i=1}^{n} |\alpha_i|^2} \leq \sigma_1^2 \tag{3.1.19}$$

so establishing the third subordinate matrix norm

$$||\mathbf{A}||_2^2 = (\text{maximum eigenvalue of } \mathbf{A}^H\mathbf{A}) \tag{3.1.20}$$

There are many useful properties which link these norms and throw some light onto which norms are most useful. For example $||\mathbf{I}|| = 1$ for any subordinate norm, although there are non-subordinate norms for which this result is false, namely the Euclidean norm

$$||\mathbf{A}||_E = \left(\sum_{i,j} (a_{ij})^2\right)^{1/2} \tag{3.1.21}$$

for which

$$||\mathbf{I}||_E = n^{1/2} \tag{3.1.22}$$

A relation between $|| \; ||_1$ and $|| \; ||_\infty$ follows immediately from their definitions to give

$$||\mathbf{A}||_1 = ||\mathbf{A}^H||_\infty \tag{3.1.23}$$

and three more results follow by the simple application of the definitions.

(i) Since

$$||\mathbf{A}||_E^2 = \text{trace } \mathbf{A}^H\mathbf{A} = \sum_{i=1}^{n} \sigma_i^2 \tag{3.1.24}$$

then the inequalities

$$\sigma_1^2 \leq \sum_{i=1}^{n} \sigma_i^2 \leq n\sigma_1^2$$

result in the equivalent form

$$||\mathbf{A}||_2 \leq ||\mathbf{A}||_E \leq n^{1/2}||\mathbf{A}||_2 \tag{3.1.25}$$

(ii) Starting with $\mathbf{Ax} = \lambda\mathbf{x}$ gives the inequalities

$$|\lambda|||\mathbf{x}|| = ||\lambda\mathbf{x}|| = ||\mathbf{Ax}|| \leq ||\mathbf{A}|| \; ||\mathbf{x}||$$

which yields the very useful relation

$$|\lambda| \leq ||\mathbf{A}|| \tag{3.1.26}$$

(iii) Using result (ii) in

$$||\mathbf{A}||_2^2 = \text{maximum eigenvalue of } \mathbf{A}^H\mathbf{A}$$

gives the inequalities

$$||\mathbf{A}||_2^2 \leq ||\mathbf{A}^H\mathbf{A}||_1 \leq ||\mathbf{A}^H||_1||\mathbf{A}||_1 = ||\mathbf{A}||_\infty||\mathbf{A}||_1 \tag{3.1.27}$$

It appears from these results that the second subordinate norm gives a tighter value than the others, and so it would seem sensible to use this norm in the measure of errors, so reducing the crudeness of the error estimates. With this in mind the spectral radius ρ of a matrix is defined as

$$\rho(\mathbf{A}) = \max_i |\lambda_i| \tag{3.1.28}$$

As often occurs in numerical analysis, the norms which are easiest to compute such as $|| \; ||_1$ and $|| \; ||_\infty$ give less tight bounds than the more difficult to compute norm, namely $|| \; ||_2$. This second norm requires the computation of an eigenvalue, which is the topic of Chapter 4. Obviously high accuracy is not required, and the requirement for a maximum eigenvalue, helps to make this computation possible by quite simple methods.

3.2 Gaussian elimination

Having set the mathematical background which will be used to analyse the rounding error growth in the solution of sets of linear equations, it is convenient to introduce some methods for the solution of the problem. Gaussian elimination is simply an organized form of school-type elimination in which elements are eliminated until only one remains. A process of back-substitution then yields the other values. In order that the process should be programmable, the steps are formalized, and therefore do not include 'short-cuts' which depend on particular 'nice factors' appearing in the arithmetic. The main practical issue is to be certain that any divisions in the process should be by non-zero values so avoiding accidental failure. This end is achieved by a process called partial pivoting.

The elimination process can be considered in two equivalent ways. Either it can be looked on as the application of a set of elementary row operations, or as the pre-multiplication of the matrix $\mathbf{A}$ by a sequence of so-called elementary matrices. The elementary row operations preserve the equality of the linear equations and hence consist of:

(i) interchanging any pair of rows,
(ii) taking a multiple of one row from another.

The multiple in (ii) is so arranged to force a required element to become zero. The elements are made zero in turn, working down the columns from the first, and treating elements below the diagonal. The result is then an upper triangular matrix from which the unknown $\mathbf{x}$ values follow by a simple process of substitution. The multiple in (ii) will be the value of the element being made zero divided by the diagonal element in the same column. Hence to avoid this diagonal element being zero, a process called partial pivoting finds the largest element in modulus beneath the diagonal in the column in which zeros are being placed, and interchanges the row containing that element with the row containing the corresponding diagonal element. The new diagonal element is called the pivotal element, and if the rank of $\mathbf{A}$ is n then this pivot cannot be zero. Indeed if the rank is not n, then the number of steps completed before the first zero pivot arises yields the rank of the matrix. This will be expanded further when the wider uses of Gaussian elimination are discussed.

It is easier to see this process in action than to describe it in words, and hence an example worked by hand is illustrative. Consider the solution of $\mathbf{Ax} = \mathbf{b}$ with

$$\mathbf{A} = \begin{pmatrix} 2 & 3 & -1 \\ 4 & 4 & -3 \\ 2 & -3 & 1 \end{pmatrix}$$

and $\mathbf{b} = (5, 3, -1)^T$. This set of equations can be laid out as the display:

$$\begin{array}{rrrr} 2 & 3 & -1 & 5 \\ 4 & 4 & -3 & 3 \\ 2 & -3 & 1 & -1 \end{array}$$

The first step is to carry out partial pivoting and select the element of largest modulus in the current leading column, namely column one. This is the element of value four and hence rows one and two are interchanged to put the pivotal element in the current leading row:

$$\begin{array}{rrrrr} & \underline{4} & 4 & -3 & 3 \\ -\frac{1}{2} & 2 & 3 & -1 & 5 \\ -\frac{1}{2} & 2 & -3 & 1 & -1 \end{array}$$

The pivotal element is underlined and the current leading row will now be used to eliminate the leading elements below the diagonal by adding multiples of row one to rows two and three. These multiples are shown next to the corresponding row and for row i are simply $-a_{ir}/a_{rr}$ at the rth step with $r = 1$ at this stage. It is the avoidance of an accidental division by zero at this stage, and successive similar stages, which is the purpose of the partial pivoting. The elimination then yields

$$\begin{array}{rrrr} \underline{4} & 4 & -3 & 3 \\ 0 & 1 & \frac{1}{2} & \frac{7}{2} \\ 0 & -\underline{5} & \frac{5}{2} & -\frac{5}{2} \end{array}$$

The process continues with the remaining 2×2 matrix to yield ultimately the required upper triangular matrix.

$$\begin{array}{rrrrr} & \underline{4} & 4 & -3 & 3 \\ & 0 & -\underline{5} & \frac{5}{2} & -\frac{5}{2} \\ \frac{1}{5} & 0 & 1 & \frac{1}{2} & \frac{7}{2} \\ \hline & \underline{4} & 4 & -3 & 3 \\ & 0 & -\underline{5} & \frac{5}{2} & -\frac{5}{2} \\ & 0 & 0 & \underline{1} & 3 \\ \hline \end{array}$$

and then the obvious back-substitution gives the values:

$$x_3 = 3$$
$$x_2 = 2$$
$$x_1 = 1$$

This process can be easily written as an algorithm which uses the ability of most scientific programming languages to employ subscripted variables or arrays. The algorithm will be presented in terms of the elements of **A**, that is a_{ij} and these elements are easily converted to a specific language. There are in general $n - 1$ major steps denoted by the index variable r.

```
FOR r = 1 TO n − 1 DO
        w = abs(a_rr)
        l = r
        FOR i = r + 1 TO n DO
                IF abs(a_ir) > w THEN
                        w = abs(a_ir)
                        l = i
                ENDIF
        ENDDO
Comment: At the end of this loop l is the row
        position of the element of largest modulus,
        or the pivotal element.
        IF l ≠ n THEN
                FOR i = r TO n DO
                        w = a_li
                        a_li = a_ri
                        a_ri = w
                ENDDO
                w = b_l
                b_l = b_r
                b_r = w
        ENDIF
Comment: the elements of rows r and l have now been
        interchanged, including those in the vector b.
        FOR i = r + 1 TO n DO
                w = −a_ir/a_rr
                FOR j = r TO n DO
                        a_ij = a_ij + w * a_rj
                ENDDO
                b_i = b_i + w * b_r
        ENDDO
Comment: The multiplying factor is stored in w and the
        elimination completed
ENDDO
Comment: This completes the major steps
x_n = b_n/a_nn
FOR i = n − 1 DOWNTO 1 DO
        w = b_i
        FOR j = r + 1 TO n DO
                w = w − x_j * aij
        ENDDO
        x_i = w/a_ii
ENDDO
Comment: This completes the back-substitution
```

A count of the number of multiplications and divisions shows that it is of order $n^3/3$ giving a measure of the significant workload as additions and subtractions are so rapidly executed compared with multiplications and divisions that they may be ignored.

The process of elimination may also be carried out so that elements both above and below the diagonal are eliminated, and this process is often called Gauss–Jordan elimination. The resulting matrix is then diagonal and so requires no back-substitution to complete the computation of the **x** vector. The layout for this process then has the form:

$$
\begin{array}{ccccc}
 & 2 & 3 & -1 & 5 \\
 & 4 & 4 & -3 & 3 \\
 & 2 & -3 & 1 & -1 \\
\hline
 & \underline{4} & 4 & -3 & 3 \\
-\frac{1}{2} & 2 & 3 & -1 & 5 \\
-\frac{1}{2} & 2 & -3 & 1 & -1 \\
\hline
 & \underline{4} & 4 & -3 & 3 \\
 & 0 & 1 & \frac{1}{2} & \frac{7}{2} \\
 & 0 & -\underline{5} & \frac{5}{2} & -\frac{5}{2} \\
\hline
\end{array}
$$

The process at this stage is identical with the earlier Gauss elimination but now elimination of elements above the diagonal will be included.

$$
\begin{array}{ccccc}
\frac{4}{5} & \underline{4} & 4 & -3 & 3 \\
 & 0 & -\underline{5} & \frac{5}{2} & -\frac{5}{2} \\
\frac{1}{5} & 0 & 1 & \frac{1}{2} & \frac{7}{2} \\
\hline
1 & \underline{4} & 0 & -1 & 1 \\
-\frac{5}{2} & 0 & -\underline{5} & \frac{5}{2} & -\frac{5}{2} \\
 & 0 & 0 & \underline{1} & 3 \\
\hline
 & 4 & 0 & 0 & 4 \\
 & 0 & -5 & 0 & -10 \\
 & 0 & 0 & 1 & 3 \\
\hline
\end{array}
$$

and then the values:

$$x_3 = 3$$
$$x_2 = 2$$
$$x_1 = 1$$

follow immediately.

It is clear that the process of Gaussian elimination can be used for evaluating determinants. The process preserves the value of the determinant of **A** except for a sign change for every row interchange. Hence for the matrix **A** in the example, two interchanges take place leaving the overall sign still positive. The determinant is then just the product of the diagonal elements of the upper triangular matrix which immediately yields the value -20. There is no gain in using Gauss–Jordan elimination here as the upper triangular form is all that is needed.

Each part of the process is an elementary row operation which is algebraically equivalent to a pre-multiplication by an elementary matrix. Such a matrix can be determined by applying the row operation in question to the identity matrix. Hence the elementary matrix

$$\begin{pmatrix} 0 & 0 & 1 \\ 0 & 1 & 0 \\ 1 & 0 & 0 \end{pmatrix}$$

is obtained by exchanging rows 1 and 3 of the identity and simple experiment demonstrates that premultiplying any 3×3 matrix by this elementary matrix will yield a new matrix with its rows 1 and 3 interchanged. Hence the process of Gaussian elimination has the algebraic form:

$$\mathbf{N}_m \cdots \mathbf{N}_1 \mathbf{A} = \mathbf{U} \tag{3.2.1}$$

and Gauss–Jordan the form:

$$\mathbf{N}_m \cdots \mathbf{N}_1 \mathbf{A} = \mathbf{D} \tag{3.2.2}$$

where there are m elementary operations resulting in either an upper triangular matrix **U** or a diagonal matrix **D** in the two cases respectively.

There is a further related application of these operations which is to determine the inverse of a matrix, and has been hinted at earlier. Suppose a set of elementary row operations is used to reduce a matrix **A** to a unit matrix, then

$$\mathbf{NA} = \mathbf{I} \tag{3.2.3}$$

and hence post multiplying by $\mathbf{A}^{-1}$ gives

$$\mathbf{NI} = \mathbf{A}^{-1} \tag{3.2.4}$$

and it is clear that the same operations applied to the unit matrix yield the inverse of **A**. Aligning **A** and **I** side by side, and applying row operations to both **A** and

I will result in a scheme similar to the Gaussian and Gauss–Jordan elimination examples given earlier. Hence using the same example matrix gives:

	2	3	−1	1	0	0
	4	4	−3	0	1	0
	2	−3	1	0	0	1
—	—	—	—	—	—	—
	$\underline{4}$	4	−3	0	1	0
$-\frac{1}{2}$	2	3	−1	1	0	0
$-\frac{1}{2}$	2	−3	1	0	0	1
—	—	—	—	—	—	—
	$\underline{4}$	4	−3	0	1	0
	0	1	$\frac{1}{2}$	1	$-\frac{1}{2}$	0
	0	$-\underline{5}$	$\frac{5}{2}$	0	$-\frac{1}{2}$	1
—	—	—	—	—	—	—
$\frac{4}{5}$	$\underline{4}$	4	−3	0	1	0
	0	$-\underline{5}$	$\frac{5}{2}$	0	$-\frac{1}{2}$	1
$\frac{1}{5}$	0	1	$\frac{1}{2}$	1	$-\frac{1}{2}$	0
—	—	—	—	—	—	—
1	$\underline{4}$	0	−1	0	$\frac{3}{5}$	$\frac{4}{5}$
$-\frac{5}{2}$	0	$-\underline{5}$	$\frac{5}{2}$	0	$-\frac{1}{2}$	1
	0	0	$\underline{1}$	1	$-\frac{3}{5}$	$\frac{1}{5}$
—	—	—	—	—	—	—
	4	0	0	1	0	1
	0	−5	0	$-\frac{5}{2}$	1	$\frac{1}{2}$
	0	0	1	1	$-\frac{3}{5}$	$\frac{1}{5}$
—	—	—	—	—	—	—
	1	0	0	$\frac{1}{4}$	0	$\frac{1}{4}$
	0	1	0	$\frac{1}{2}$	$-\frac{1}{5}$	$-\frac{1}{10}$
	0	0	1	1	$-\frac{3}{5}$	$\frac{1}{5}$

to yield the inverse matrix

$$\begin{pmatrix} \frac{1}{4} & 0 & \frac{1}{4} \\ \frac{1}{2} & -\frac{1}{5} & -\frac{1}{10} \\ 1 & -\frac{3}{5} & \frac{1}{5} \end{pmatrix}$$

The Gauss elimination algorithm can therefore be used for a number of purposes with a few minor modifications. In the form specified in the text it is used to solve a set of linear algebraic equations. The reduced upper triangular form yields an effective method to evaluate determinants with a reasonable level of arithmetic work, compared with say multiplying the determinant out recursively. If the right-hand side **b** is omitted and a test put in for zero pivots, then the point in the routine at which this occurs gives the rank of the matrix. This point is given by the number of completed major steps denoted by the index variable r. With a simple modification to allow elimination above and below the diagonal the routine can be adjusted to carry out Gauss–Jordan elimination. The back substitution is then unrequired and is replaced by simple divisions to extract the **x** values from the resulting diagonal matrix. By extending the array declarations to have double width, the inverse of a matrix can be found, the loops now extending to $2n$ rather than n along the rows. The process of using elementary row operations will recur throughout the linear algebra work, and is therefore of fundamental importance.

3.3 Direct triangular decomposition

There is an alternative to the elimination processes used in the previous paragraph which is based on expressing the matrix **A** as a product of a lower triangular matrix and an upper triangular matrix in the form

$$\mathbf{A} = \mathbf{LU} \tag{3.3.1}$$

Once such a decomposition is found, the solution of a set of linear algebraic equations becomes a simple matter of forward and backward substitution, as

$$\mathbf{Ax} = \mathbf{LUx} = \mathbf{b} \tag{3.3.2}$$

and introducing the vector **y** as

$$\mathbf{Ux} = \mathbf{y} \tag{3.3.3}$$

leaves

$$\mathbf{Ly} = \mathbf{b} \tag{3.3.4}$$

Equation (3.3.4) has a lower triangular matrix, and so y_1 is immediately b_1/l_{11}, and then forward substitution yields y_2 and so on to y_n. With **y** found, equation (3.3.3) is just the usual back-substitution for the result vector **x** as **U** is upper triangular.

The decomposition is found by multiplying out a typical **L** matrix with a typical **U** matrix and comparing element for element with the original matrix **A**. This gives a series of relations for the elements of **L** and **U**. Hence

$$\begin{bmatrix} l_{11} & 0 & \cdots & \cdots & \cdots & 0 \\ l_{21} & l_{22} & & & & \\ \vdots & & \ddots & & & \\ l_{i1} & \cdots & \cdots & l_{ii} & & \\ \vdots & & & & \ddots & \\ l_{n1} & \cdots & \cdots & \cdots & \cdots & l_{nn} \end{bmatrix} \begin{bmatrix} u_{11} & u_{21} & \cdots & u_{j1} & \cdots & u_{n1} \\ 0 & u_{22} & & \vdots & & \vdots \\ & & \ddots & \vdots & & \vdots \\ & & & u_{jj} & \cdots & u_{nj} \\ & & & & \ddots & \vdots \\ & & & & & u_{nn} \end{bmatrix}$$

$$= \begin{bmatrix} u_{11}l_{11} & \cdots & u_{j1}l_{11} & \cdots & u_{n1}l_{11} \\ u_{11}l_{21} & \cdots & u_{j1}l_{21} + u_{j2}l_{22} & & \vdots \\ u_{11}l_{31} & \cdots & u_{j1}l_{31} + u_{j2}l_{32} + u_{j3}l_{33} & & \vdots \\ \vdots & & \vdots & & \vdots \\ \vdots & & u_{j1}l_{j1} + u_{j2}l_{j2} + \cdots + u_{jj}l_{jj} & & \vdots \\ \vdots & & \vdots & & \vdots \\ u_{11}l_{n1} & \cdots & u_{j1}l_{n1} + u_{j2}l_{n2} + \cdots + u_{jj}l_{nj} & \cdots & u_{n1}l_{n1} + \cdots + u_{nn}l_{nn} \end{bmatrix} \tag{3.3.5}$$

For a typical column of **A** it can be seen that the number of terms in each element increases by one until the last term in the sum reaches u_{jj}. As all the u's beyond this are zero, the number of terms in each element now remains constant. On the increasing part of the column, the comparison of elements allows the computation of u values, whereas in the constant column width area the comparison yields further l values. By working down each column in turn the complete set of u's and l's can be found. The result is the following algorithm:

$u_{ii} = 1 \qquad i = 1, n$
$l_{i1} = a_{i1} \qquad i = 1, n$
FOR $j = 2, n$ DO
  FOR $k = 1, j-1$ DO
    $$u_{jk} = \left(a_{kj} - \sum_{l=1}^{k-1} u_{jl}l_{kl} \right) / l_{kk}$$
  ENDDO
  $$l_{jj} = \left(a_{jj} - \sum_{l=1}^{j-1} u_{jl}l_{jl} \right) / u_{jj}$$
  FOR $k = j+1, n$ DO
    $$l_{kj} = \left(a_{kj} - \sum_{l=1}^{j-1} u_{jl}l_{kl} \right) / u_{kj}$$
  ENDDO
ENDDO

This algorithm can be demonstrated by solving the linear equations $\mathbf{Ax} = \mathbf{b}$ with

$$\mathbf{A} = \begin{bmatrix} 1 & 2 & 2 \\ 1 & 3 & 4 \\ -1 & -3 & -5 \end{bmatrix} \tag{3.3.6}$$

and $\mathbf{b} = (1, 1, 1)^T$. The decomposition gives

$$\begin{aligned} 2 + u_{22} &= 3 & u_{22} &= 1 \\ -2 + l_{32}u_{22} &= -3 & l_{32} &= -1 \\ 2 + u_{23} &= 4 & u_{23} &= 2 \\ -2 + l_{32}u_{23} + u_{33} &= -5 & u_{33} &= -1 \end{aligned} \tag{3.3.7}$$

with the other elements being immediate. $\mathbf{L}$ is taken to be unit lower. Hence the decomposition is

$$\mathbf{A} = \begin{bmatrix} 1 & 0 & 0 \\ 1 & 1 & 0 \\ -1 & -1 & 1 \end{bmatrix} \begin{bmatrix} 1 & 2 & 2 \\ 0 & 1 & 2 \\ 0 & 0 & -1 \end{bmatrix} \tag{3.3.8}$$

and the vector $\mathbf{y}$ in the forward substitution satisfies

$$\begin{bmatrix} 1 & 0 & 0 \\ 1 & 1 & 0 \\ -1 & -1 & 1 \end{bmatrix} \begin{bmatrix} y_1 \\ y_2 \\ y_3 \end{bmatrix} = \begin{bmatrix} 1 \\ 1 \\ 1 \end{bmatrix} \tag{3.3.9}$$

to give $y_1 = 1$, $y_2 = 0$ and $y_3 = 2$. Finally the back-substitution is given by

$$\begin{bmatrix} 1 & 2 & 2 \\ 0 & 1 & 2 \\ 0 & 0 & -1 \end{bmatrix} \begin{bmatrix} x_1 \\ x_2 \\ x_3 \end{bmatrix} = \begin{bmatrix} 1 \\ 0 \\ 2 \end{bmatrix} \tag{3.3.10}$$

which gives $x_3 = -2$, $x_2 = 4$ and $x_1 = -3$.

As the decomposition of a matrix $\mathbf{A}$ into a product $\mathbf{LU}$ is unique once one of either $\mathbf{L}$ or $\mathbf{U}$ is chosen to be a unit triangle, then similar conditions for failure can arise in triangular decomposition as in Gaussian elimination. Hence the issue of pivoting needs to be considered in the triangular decomposition to avoid the possibility of zero elements for l_{kk}, as these elements are used as divisors in the first k loop of the algorithm. To achieve pivoting, it is necessary to compute the elements which will arise below the current diagonal position, that is, the elements from which an element of maximum modulus was chosen in Gaussian elimination. This pivotal row can then be interchanged with the leading row. The final result will be a decomposition of a matrix which is a permutation of the original matrix. These computed elements can be stored in an extra column, say p_r, and laid out as in the display

$$\begin{bmatrix} u_{11} & u_{12} & u_{13} & u_{14} & u_{15} & b'_1 & p_1 \\ l_{21} & u_{22} & u_{23} & u_{24} & u_{25} & b'_2 & p_2 \\ l_{31} & l_{32} & a_{33} & a_{34} & a_{35} & b_3 & p_3 \\ l_{41} & l_{42} & a_{43} & a_{44} & a_{45} & b_4 & p_4 \\ l_{51} & l_{52} & a_{53} & a_{54} & a_{55} & b_5 & p_5 \end{bmatrix} \tag{3.3.11}$$

where the bottom 3×3 matrix is still to be processed. It is also important now to include the right-hand side $\mathbf{b}$ as this will be permuted and require computation. Hence there are n major steps of which the rth is:

FOR $i = r, n$ DO

$$p_i = a_{ir} - \sum_{j=1}^{r-1} l_{ij} u_{jr}$$

ENDDO

LET $|p_k|$ be the $\max_{j=r,n} |p_j|$

Interchange the whole of rows r and k

Overwrite p_r on a_{rr} to give u_{rr}

FOR $i = r + 1, n$ DO

Compute $l_{ir} = p_i / u_{rr}$ and overwrite on a_{ir}

ENDDO

FOR $i = r + 1, n$ DO

Compute $u_{ri} = a_{ri} - \sum_{j=1}^{r-1} l_{rj} u_{ji}$

and overwrite on a_{ri}

ENDDO

Compute $b'_r = b_r - \sum_{j=1}^{r-1} l_{rj} b'_j$

and overwrite on b_r

This version of the algorithm applied to the example

$$\begin{bmatrix} 1 & 4 & 2 \\ 1 & 3 & 4 \\ -2 & -3 & -5 \end{bmatrix} \tag{3.3.12}$$

with $\mathbf{b} = (9, 1, 4)^T$ gives:

$$\begin{array}{rrrrr} 1 & 4 & 2 & 9 & 1 \\ 1 & 3 & 4 & 1 & 1 \\ -2 & -3 & -5 & 4 & -2 \\ \hline -2 & -3 & -5 & 4 & -2 \\ 1 & 3 & 4 & 1 & 1 \\ 1 & 4 & 2 & 9 & 1 \\ \hline \end{array}$$

with the p's appearing in the last column as in (3.3.11), and the first interchange having been completed. The new elements in column one and row one are now computed according to the algorithm to give

$$\begin{array}{rrrrr} -2 & -3 & -5 & 4 & \\ -\frac{1}{2} & 3 & 4 & 1 & \frac{3}{2} \\ -\frac{1}{2} & 4 & 2 & 9 & \frac{5}{2} \\ \hline \end{array}$$

The new p values follow from

$$p_2 = a_{22} - l_{21}u_{12} = \tfrac{3}{2}$$

and

$$p_3 = a_{32} - l_{31}u_{12} = \tfrac{5}{2}$$

and appear again as the right-hand column. An interchange of rows two and three is required to give:

$$\begin{array}{ccccc} -2 & -3 & -5 & 4 & \\ -\frac{1}{2} & 4 & 2 & 9 & \frac{5}{2} \\ -\frac{1}{2} & 3 & 4 & 1 & \frac{3}{2} \\ \hline \end{array}$$

Then the formulae

$$\begin{aligned} u_{23} &= a_{23} - l_{21}u_{13} = 2 - \tfrac{5}{2} = -\tfrac{1}{2} \\ b_2' &= b_2 - l_{21}p_1 = 9 - \left(-\tfrac{1}{2}\right)4 \\ p_3 &= a_{33} - l_{31}u_{13} - l_{32}u_{23} = 4 - 5\left(\tfrac{1}{2}\right) - \tfrac{3}{2}\left(-\tfrac{1}{2}\right) = \tfrac{9}{5} \end{aligned}$$

give

$$\begin{array}{ccccc} -2 & -3 & -5 & 4 & \\ -\frac{1}{2} & \frac{5}{2} & -\frac{1}{2} & 11 & \\ -\frac{1}{2} & \frac{3}{5} & 4 & 1 & \frac{9}{5} \\ \hline \end{array}$$

with the final step yielding

$$\begin{array}{cccc} -2 & -3 & -5 & 4 \\ -\frac{1}{2} & \frac{5}{2} & -\frac{1}{2} & 11 \\ -\frac{1}{2} & \frac{3}{5} & \frac{9}{5} & -\frac{18}{5} \\ \hline \end{array}$$

Hence the final decomposed matrices are

$$\begin{bmatrix} 1 & 0 & 0 \\ -\frac{1}{2} & 1 & 0 \\ -\frac{1}{2} & \frac{3}{5} & 1 \end{bmatrix} \begin{bmatrix} -2 & -3 & -5 \\ 0 & \frac{5}{2} & -\frac{1}{2} \\ 0 & 0 & \frac{9}{5} \end{bmatrix} \tag{3.3.13}$$

which expands to give

$$\begin{bmatrix} -2 & -3 & -5 \\ 1 & 4 & 2 \\ 1 & 3 & 4 \end{bmatrix} \tag{3.3.14}$$

which is indeed a permutation of the original matrix. Moreover the forward substitution has been combined into the algorithm for the **b** values and so to solve the matrix all that is required is to back-substitute using the **U** values and the final right-hand side. This yields $x_3 = -2$, $x_2 = 4$ and $x_1 = -3$ correctly.

One disadvantage of straight Gaussian elimination is that in general no advantage can be taken of the symmetry of **A**. The process is such that any initial symmetry is soon destroyed. In the case of direct triangular decomposition, this is not the case and there is a special version of the algorithm in the symmetric case. This algorithm is called Cholesky decomposition and takes the form analogous with (3.3.1) of

$$\mathbf{A} = \mathbf{L}\mathbf{L}^T \tag{3.3.15}$$

and a similar process of multiplying out a typical decomposition gives the algorithm

$$\begin{aligned}
&l_{11} = \sqrt{a_{11}} \\
&l_{i1} = a_{1i}/l_{11} \qquad i = 2, n \\
&\text{FOR } j = 2, n \text{ DO} \\
&\qquad l_{jj} = \sqrt{a_{jj} - \sum_{k=1}^{j-1} l_{jk}^2} \\
&\qquad l_{ij} = \left(a_{ji} - \sum_{k=1}^{j-1} l_{jk} l_{ik} \right) / l_{jj} \\
&\text{ENDDO}
\end{aligned}$$

with clear economies based on symmetry. The problem with this algorithm is that unless **A** is also positive definite the steps with square roots can involve complex arithmetic. In the forward and backward elimination, steps which follow the complex numbers cancel out to yield the correct real solution to the original set of equations.

As an example of this process, consider the reduction of the matrix

$$\mathbf{A} = \begin{bmatrix} 4 & 2 & -1 \\ 2 & 3 & 1 \\ -1 & 1 & 2 \end{bmatrix}$$

to a product of lower triangular matrices and the resulting solution of the set of equations $\mathbf{Ax} = (1, 1, 1)^T$.

The Cholesky expansion into lower triangular matrices gives:

$$\mathbf{A} = \begin{bmatrix} 4 & 2 & -1 \\ 2 & 3 & 1 \\ -1 & 1 & 2 \end{bmatrix} = \begin{bmatrix} l_{11} & 0 & 0 \\ l_{21} & l_{22} & 0 \\ l_{31} & l_{32} & l_{33} \end{bmatrix} \begin{bmatrix} l_{11} & l_{21} & l_{31} \\ 0 & l_{22} & l_{32} \\ 0 & 0 & l_{33} \end{bmatrix}$$

$$= \begin{bmatrix} l_{11}^2 & l_{11}l_{21} & l_{11}l_{31} \\ l_{11}l_{21} & l_{21}^2 + l_{22}^2 & l_{21}l_{31} + l_{22}l_{32} \\ l_{11}l_{31} & l_{21}l_{31} + l_{32}l_{22} & l_{31}^2 + l_{32}^2 + l_{33}^2 \end{bmatrix} \tag{3.3.16}$$

to give the equations

$$
\begin{aligned}
l_{11} &= 2 \\
l_{21} &= 2/l_{11} = 1 \\
l_{31} &= -1/l_{11} = -\tfrac{1}{2} \\
l_{22}^2 &= 3 - 1 = 2; \qquad l_{22} = \sqrt{2} \\
l_{32} &= (1 - l_{21}l_{31})/l_{22} = \left(1 + \tfrac{1}{2}\right)\sqrt{2} = 3\sqrt{2}/4 = 1.060\,66 \\
l_{33}^2 &= 2 - \tfrac{1}{4} - 18/16 = 10/16; \quad l_{33} = \sqrt{10}/4 = 0.790\,569
\end{aligned}
\tag{3.3.17}
$$

Now to solve the linear equations $\mathbf{Ax} = (1, 1, 1)^T$ the forward substitution gives

$$
\begin{bmatrix} 2 & 0 & 0 \\ 1 & \sqrt{2} & 0 \\ -\frac{1}{2} & \frac{3\sqrt{2}}{4} & \frac{\sqrt{10}}{4} \end{bmatrix}
\begin{bmatrix} y_1 \\ y_2 \\ y_3 \end{bmatrix} = \begin{bmatrix} 1 \\ 1 \\ 1 \end{bmatrix}
\tag{3.3.18}
$$

with

$$
\begin{aligned}
y_1 &= \tfrac{1}{2} \\
y_2 &= \left(1 - \tfrac{1}{2}\right)/\sqrt{2} = \tfrac{1}{4}\sqrt{2} \\
y_3 &= \left(1 + \frac{1}{4} - \frac{6}{16}\right)\frac{4}{\sqrt{10}} = \frac{7\sqrt{10}}{20}
\end{aligned}
\tag{3.3.19}
$$

to leave the backward substitution

$$
\begin{bmatrix} 2 & 1 & -\frac{1}{2} \\ 0 & \sqrt{2} & \frac{3\sqrt{2}}{4} \\ 0 & 0 & \frac{\sqrt{10}}{4} \end{bmatrix}
\begin{bmatrix} x_1 \\ x_2 \\ x_3 \end{bmatrix} = \left[\tfrac{1}{2} \quad \tfrac{\sqrt{4}}{4} \quad \tfrac{7\sqrt{10}}{20}\right]
\tag{3.3.20}
$$

with the final solution $x_3 = 7/5$, $x_2 = -4/5$ and $x_1 = 1$, which check in the original equations.

These decomposition algorithms will be used in Chapter 4 in one of the methods for finding the eigenvalues of a non-symmetric matrix.

3.4 Condition factor for linear equations

In the last paragraph, some methods for the solution of linear algebraic equations have been covered. With elimination methods, there is no truncation error as the process is exact, and simply simulates the normal algebraic processes that would be employed in a hand computation. However, rounding errors are another issue. Because of the finite word length of real computers, all the elements of a matrix will carry a machine precision error, and the question arises as to how this error propagates through the solution of the equations. In general terms there are two sources of such error. The original problem can be very sensitive to the input errors, yielding far larger solution errors than the input errors would seem to merit. This is called inherent instability and is independent of the method used to solve the problem. In other circumstances, the error magnification is entirely introduced by the method used in the solution of the problem. Such methods can be avoided, and only stable methods employed. With inherent instability the only solution is to employ high precision arithmetic to cover the loss of accuracy. It is therefore crucial to know the level of such instability and in this paragraph an attempt to measure this effect for linear equations is made.

The method of perturbations is employed and the analysis is easiest attempted in two parts. First consider the effect of perturbing the right-hand side $\mathbf{b}$ to $\mathbf{b}+\mathbf{k}$, with the result that the solution vector $\mathbf{x}$ becomes $\mathbf{x}+\mathbf{h}$. Then the relations:

$$\mathbf{Ax} = \mathbf{b} \tag{3.4.1}$$

and

$$\mathbf{A}(\mathbf{x}+\mathbf{h}) = \mathbf{b}+\mathbf{k} \tag{3.4.2}$$

hold to yield

$$\mathbf{Ah} = \mathbf{k} \tag{3.4.3}$$

Hence

$$\mathbf{h} = \mathbf{A}^{-1}\mathbf{k} \tag{3.4.4}$$

and taking norms gives:

$$||\mathbf{h}|| \leq ||\mathbf{A}^{-1}||\ ||\mathbf{k}|| \tag{3.4.5}$$

On the other hand, taking norms of equation (3.4.1) gives

$$||\mathbf{x}|| \geq ||\mathbf{A}||^{-1}\ ||\mathbf{b}|| \tag{3.4.6}$$

and hence the relative error in the resulting vector can be expressed as

$$\frac{||\mathbf{h}||}{||\mathbf{x}||} \leq ||\mathbf{A}||\ ||\mathbf{A}^{-1}||\ \frac{||\mathbf{k}||}{||\mathbf{b}||} \tag{3.4.7}$$

and so the crucial multiplying factor is $||\mathbf{A}||\ ||\mathbf{A}^{-1}||$. In the mathematical introduction it is seen that the 'tightest' practical norm was norm-2, and this leads to the conventional definition of the condition factor $\kappa(\mathbf{A})$ as

$$\kappa(\mathbf{A}) = ||\mathbf{A}||_2\, ||\mathbf{A}^{-1}||_2. \tag{3.4.8}$$

There is no doubt that this factor can be very large, though never smaller than unity. Before pursuing this further, let us consider what happens if $\mathbf{A}$ is perturbed by say $\mathbf{F}$.

In this case the equation for the perturbed problem is

$$(\mathbf{A}+\mathbf{F})(\mathbf{x}+\mathbf{h})=\mathbf{b} \tag{3.4.9}$$

and now

$$(\mathbf{A}+\mathbf{F})\mathbf{h}=-\mathbf{F}\mathbf{x} \tag{3.4.10}$$

If $\mathbf{A}+\mathbf{F}$ is expressed as

$$\mathbf{A}+\mathbf{F}=\mathbf{A}(\mathbf{I}+\mathbf{A}^{-1}\mathbf{F}) \tag{3.4.11}$$

and if $||\mathbf{A}^{-1}\mathbf{F}||<1$ then $\mathbf{A}+\mathbf{F}$ is non-singular and $\mathbf{h}$ can be expressed as

$$\mathbf{h}=-(\mathbf{I}+\mathbf{A}^{-1}\mathbf{F})^{-1}\mathbf{A}^{-1}\mathbf{F}\mathbf{x} \tag{3.4.12}$$

Taking norms then leads to

$$||\mathbf{h}||\leq\frac{||\mathbf{A}^{-1}\mathbf{F}||\ ||\mathbf{x}||}{1-||\mathbf{A}^{-1}\mathbf{F}||}\leq\frac{||\mathbf{A}^{-1}||\ ||\mathbf{F}||\ ||\mathbf{x}||}{1-||\mathbf{A}^{-1}||\ ||\mathbf{F}||} \tag{3.4.13}$$

and the relative error becomes:

$$\frac{||\mathbf{h}||}{||\mathbf{x}||}\leq||\mathbf{A}||\ ||\mathbf{A}^{-1}||\frac{||\mathbf{F}||}{||\mathbf{A}||}\Bigg/\left(1-||\mathbf{A}||\ ||\mathbf{A}^{-1}||\frac{||\mathbf{F}||}{||\mathbf{A}||}\right) \tag{3.4.14}$$

which again displays the crucial multiplying factor $||\mathbf{A}||\ ||\mathbf{A}^{-1}||$, and emphasizes the definition in equation (3.4.8) of $\kappa(\mathbf{A})$.

The spectral condition number or condition factor $\kappa(\mathbf{A})$ has the simple property that

$$\kappa(c\mathbf{A})=\kappa(\mathbf{A}) \tag{3.4.15}$$

demonstrating that multiplying sets of equations by constants has no effect on the condition factor. It is also true that

$$\kappa(\mathbf{A})=||\mathbf{A}||_2\,||\mathbf{A}^{-1}||_2\geq||\mathbf{A}\mathbf{A}^{-1}||_2=1 \tag{3.4.16}$$

where equality occurs if $\mathbf{A}=c\mathbf{R}$ where $\mathbf{R}$ is unitary.

Recalling the definition of singular values σ_i from equation (3.1.15), the definition of the condition factor $\kappa(\mathbf{A})$ can be reformulated in the equivalent form

$$\kappa(\mathbf{A})=\frac{\sigma_1}{\sigma_n} \tag{3.4.17}$$

where σ_1 is the largest singular value and σ_n is the smallest. It then follows that if $\mathbf{A}$ is real and symmetric

$$\kappa(\mathbf{A})=\frac{\lambda_1}{\lambda_n} \tag{3.4.18}$$

where the λ's are the eigenvalues of $\mathbf{A}$.

From this analysis it is clear that care is needed in the solution of a set of linear algebraic equations to guard against the possibility of a large condition factor. It is satisfying that the level of difficulty can be very easily determined, and the loss of accuracy forecast. The computation of the eigenvalues involved in this calculation will be treated in Chapter 4. In very badly behaved cases multi-length arithmetic is required. Such arithmetic is available as software packages in some libraries, but is inevitably quite slow.

An excellent illustration of the level of ill-conditioning that can arise is given by the Vandermonde matrices. These arise in the generation of orthogonal polynomials in Chapter 5, Section 2, and are associated the Gaussian quadrature methods in Chapter 6, Section 3. At this stage, no knowledge of these chapters is required. The matrices form a set of families and have the form

$$\begin{bmatrix} c_0 & c_1 & c_2 & \cdots & c_{n-1} \\ c_1 & c_2 & c_3 & \cdots & c_n \\ \vdots & \vdots & \vdots & & \vdots \\ c_{n-1} & c_n & c_{n+1} & \cdots & c_{2n-2} \end{bmatrix} \tag{3.4.19}$$

for various sequences c_i. These sequences are defined by the orthogonal polynomials required, and some of the commonly used ones with their associated polynomial names are given in the next equations:

$$\begin{aligned} c_i &= \int_{-1}^{1} x^i \,\mathrm{d}x \\ &= \begin{cases} 0 & i \text{ odd} \\ \frac{2}{i+1} & i \text{ even} \end{cases} \quad \text{Legendre} \end{aligned} \tag{3.4.20}$$

$$\begin{aligned} c_i &= \int_{0}^{\infty} e^{-x} x^i \,\mathrm{d}x \\ &= i! \qquad \text{Laguerre} \end{aligned} \tag{3.4.21}$$

$$\begin{aligned} c_i &= \int_{-\infty}^{\infty} e^{-x^2} x^i \,\mathrm{d}x \\ &= \begin{cases} 0 & i \text{ odd} \\ \frac{(i-1)!!\sqrt{\pi}}{2^{i/2}} & i \text{ even} \end{cases} \quad \text{Hermite} \end{aligned} \tag{3.4.22}$$

$$\begin{aligned} c_i &= \int_{-1}^{1} \frac{x^i}{\sqrt{(1-x^2)}} \,\mathrm{d}x \\ &= \begin{cases} 0 & i \text{ odd} \\ \frac{(i-1)!!\pi}{i!!} & i \text{ even} \end{cases} \quad \text{Chebyshev} \end{aligned} \tag{3.4.23}$$

where the notation $i!!$ is defined as $2 \cdot 4 \cdot 6 .\ldots i$ for i even and $2 \cdot 4 \cdot 6) .\ldots i-1$ for i odd. The condition factors for these four series of matrices for increasing order are shown in Table 3.1.

Table 3.1

n	Legendre	Laguerre	Hermite	Chebyshev
2	3.0	6.9	2.0	2.0
3	14.1	188.7	3.8	13.0
4	67.6	1.4(4)	12.9	64.8
5	3.58(2)	1.8(6)	67.8	3.9(2)
6	1.9(3)	3.5(8)	4.84(2)	2.0(3)
7	1.02(4)	9.5(10)	4.2(3)	1.2(4)
8	5.5(4)	3.4(13)	4.2(4)	6.6(4)

It can be seen in the Laguerre case that the level of ill-conditioning rises extremely quickly with multi-length arithmetic being necessary on most computers with matrices of order eight where 13 figures will be lost. Even the more gentle cases result in a loss of four figures for order eight. These values also rise inexorably to take the value 2.35(13) at $n = 24$ for the Legendre example. In practice, these polynomials are computed to high orders such as 512 in quadrature, and special methods have been devised to circumnavigate the issue. The exercises contain some classic examples of ill-conditioned matrices, and in the way of these things, ill-conditioned matrices seem to arise quite commonly in applications. The important point is to always be aware that a linear equation solving programme, even using as direct an approach as Gaussian elimination, can lead to incorrect results. There is no truncation error involved now, the issue is entirely one of rounding.

3.5 Iterative solution of linear equations

In Section 3.2 one particular approach to the solution of linear algebraic equations was considered. This was an exact approach based on analytic elimination. An alternative method with some considerable advantages under certain circumstances is to consider an iterative solution. It was demonstrated in Chapter 1 how a single non-linear equation could be solved by direct iteration. To generate a method, the equation was rearranged in the form

$$x = F(x) \tag{3.5.1}$$

and then by adding subscripts in the form

$$x^{(r+1)} = F(x^{(r)}) \tag{3.5.2}$$

an iteration is obtained which will certainly converge to the required root if it converges at all. The conditions for convergence are then a matter of analysis. The same approach is possible for linear equations. Suppose the specific example

$$\begin{aligned} 4x_1 + 2x_2 + x_3 &= 1 \\ x_1 + 7x_2 + x_3 &= 4 \\ x_1 + x_2 + 20x_3 &= 7 \end{aligned} \tag{3.5.3}$$

with solution $x_1 = -0.102, x_2 = 0.539, x_3 = 0.328$ is considered. Then as the diagonal elements of this matrix are somewhat larger than the off-diagonal elements a very rough and ready estimate of the solution is obtained from the diagonal elements alone. This gives $x_1 = 0.25$, $x_2 = 0.57$, $x_3 = 0.35$. Hence the possibility of rearranging the equation to isolate the diagonal elements on the left-hand side of the equations and applying an iterative process akin to the above, arises. In the specific case the resulting equations are

$$\begin{aligned} 4x_1^{(r+1)} &= 1 - 2x_2^{(r)} - x_3^{(r)} \\ 7x_2^{(r+1)} &= 4 - x_1^{(r)} - x_3^{(r)} \\ 20x_3^{(r+1)} &= 7 - x_1^{(r)} - x_2^{(r)} \end{aligned} \tag{3.5.4}$$

and beginning with $\mathbf{x}^{(r)} = \mathbf{0}$ gives the iterates shown in Table 3.2.

It is clear that convergence to the true solution is occurring, and hence the question of the circumstances under which such convergence takes place needs to be addressed. The method is called Jacobi's method.

It is well-known from the theory of linear algebra that the vector iteration

$$\mathbf{e}^{(r+1)} = \mathbf{C}\,\mathbf{e}^{(r)} + \mathbf{c} \tag{3.5.5}$$

converges if and only if no eigenvalue of the defining matrix $\mathbf{C}$ has modulus greater than or equal to unity. In the case when the diagonalization theorem can be applied, the diagonalization transformation reduces the set of equations in (3.5.5) to a decoupled set with diagonal elements equal to the above eigenvalues, say λ_i. Hence as $\lambda_i^r \to 0$ as $r \to \infty$, the vectors $\mathbf{e}^{(r)} \to 0$ to yield the above result.

Hence if the iterative process defined in the example (3.5.4) can be expressed in the form of equation (3.5.5) then there is an immediate convergence condition, namely that the spectral radius (defined in (3.1.28)) of $\mathbf{C}$ should be less than unity.

Following this line of argument, $\mathbf{A}$ is split additively into the three parts

$$\mathbf{A} = \mathbf{L} + \mathbf{D} + \mathbf{U} \tag{3.5.6}$$

where $\mathbf{L}$ is lower triangular, $\mathbf{D}$ is diagonal and $\mathbf{U}$ is upper triangular. Then the separation of the diagonal elements to the left-hand side can be represented in

Table 3.2

r	x_1	x_2	x_3
0	0.0	0.0	0.0
1	0.25	0.57	0.35
2	−0.123	0.486	0.309
3	−0.070	0.545	0.332
4	−0.105	0.534	0.326

general terms to give the algebraic form of Jacobi's method as

$$\mathbf{D}\mathbf{x}^{(r+1)} = \mathbf{b} - (\mathbf{L} + \mathbf{U})\mathbf{x}^{(r)} \tag{3.5.7}$$

with the convergence criterion that convergence occurs if and only if no eigenvalue of $\mathbf{C}$ has modulus greater than or equal to unity where

$$\mathbf{C} = -\mathbf{D}^{-1}(\mathbf{L} + \mathbf{U}). \tag{3.5.8}$$

Because of (3.1.26) a necessary but not sufficient condition for convergence is that

$$||\mathbf{C}|| < 1 \tag{3.5.9}$$

for any matrix norm. The application of $|| \; ||_1$ or $|| \; ||_\infty$ yields an easily applicable test. If for all rows or for all columns, the sum of the moduli of the off-diagonal elements is less than the modulus of the corresponding diagonal element then

$$||\mathbf{D}^{-1}(\mathbf{L} + \mathbf{U})||_{1/\infty} < 1 \tag{3.5.10}$$

and a necessary condition for convergence has been obtained. A matrix satisfying this condition is said to be *diagonally dominant*. Clearly there will be matrices that are not diagonally dominant for which convergence occurs, and the full test using the eigenvalues of $\mathbf{C}$ is then needed to resolve the convergence issue.

A further side effect of the convergence condition is that the rate of convergence is determined by the magnitude of the eigenvalue of largest modulus, say λ_l, as it will be λ_l^r which goes to zero the slowest. Hence the number of iterations to get, say, m figure accuracy is the value of r for which

$$\lambda_l^r = 10^{-m}. \tag{3.5.11}$$

The next stage in the development is to look for modifications which will enhance the convergence rate and possibly broaden the types of matrices for which convergence occurs. One very simple idea is to re-consider the example in (3.5.4). Once a new estimate for x_1 is found in the iteration for the first equation, there is no obvious reason why this supposedly improved value should not immediately be used in the second and third equations, rather than updating at the end of a complete iteration. This modified method is the Gauss–Seidel method and has the form

$$\begin{aligned} 4x_1^{(r+1)} &= 1 - 2x_2^{(r)} - x_3^{(r)} \\ 7x_2^{(r+1)} &= 4 - x_1^{(r+1)} - x_3^{(r)} \\ 20x_3^{(r+1)} &= 7 - x_1^{(r+1)} - x_2^{(r+1)} \end{aligned} \tag{3.5.12}$$

and beginning with $\mathbf{x}^{(r)} = \mathbf{0}$ gives the iterates shown in Table 3.3.

The convergence appears to be more rapid than the original Jacobi method, and indeed the convergence criterion will now depend on the matrix

$$\mathbf{C} = -(\mathbf{D} + \mathbf{L})^{-1}\mathbf{U} \tag{3.5.13}$$

Table 3.3

r	x_1	x_2	x_3
0	0.0	0.0	0.0
1	0.25	0.57	0.35
2	−0.123	0.539	0.329
4	−0.102	0.539	0.328

with the necessary condition that convergence occurs if

$$||(\mathbf{D}+\mathbf{L})^{-1}\mathbf{U}|| < 1 \tag{3.5.14}$$

for some matrix norm.

For the Gauss–Seidel method, convergence can also be established if $\mathbf{A}$ is symmetric and positive definite. By the symmetry

$$(\mathbf{D}+\mathbf{L})^{-1}\mathbf{U} = (\mathbf{D}+\mathbf{L})^{-1}\mathbf{L}^T \tag{3.5.15}$$

and letting λ and $\mathbf{y}$ be a typical eigenvalue and eigenvector respectively then

$$(\mathbf{D}+\mathbf{L})^{-1}\mathbf{L}^T\mathbf{y} = \lambda\mathbf{y} \tag{3.5.16}$$

or

$$\mathbf{L}^T\mathbf{y} = \lambda(\mathbf{D}+\mathbf{L})\mathbf{y}. \tag{3.5.17}$$

The positive definiteness ensures that $\mathbf{D}+\mathbf{L}$ is non-singular, but $\mathbf{y}$ and λ may well be complex. Hence the relations

$$\bar{\mathbf{y}}^T\mathbf{L}^T\mathbf{y} = \lambda\bar{\mathbf{y}}^T(\mathbf{D}+\mathbf{L})\mathbf{y} = p + \mathrm{i}q \tag{3.5.18}$$

hold, and, transposing the equations, remembering each term in the equality is scalar, gives

$$\bar{\mathbf{y}}^T\mathbf{L}\mathbf{y} = \mathbf{y}^T\mathbf{L}^T\bar{\mathbf{y}} = \overline{\bar{\mathbf{y}}^T\mathbf{L}^T\mathbf{y}} = p - \mathrm{i}q \tag{3.5.19}$$

where p and q are the respective real and imaginary parts. Hence

$$\bar{\mathbf{y}}^T\mathbf{A}\mathbf{y} = \bar{\mathbf{y}}^T(\mathbf{L}+\mathbf{D}+\mathbf{L}^T)\mathbf{y} = 2p + \bar{\mathbf{y}}^T\mathbf{D}\mathbf{y} \tag{3.5.20}$$

which is positive since $\mathbf{A}$ is positive definite. Hence

$$\lambda = \frac{p + \mathrm{i}q}{\bar{\mathbf{y}}^T\mathbf{D}\mathbf{y} + p - \mathrm{i}q} \tag{3.5.21}$$

and

$$|\lambda|^2 = \frac{p^2 + q^2}{(\bar{\mathbf{y}}^T\mathbf{D}\mathbf{y} + p)^2 + q^2} \tag{3.5.22}$$

giving two cases to consider. As $\bar{\mathbf{y}}^T\mathbf{D}\mathbf{y} > 0$ since $\mathbf{D}$ is positive definite, then if $p > 0$ trivially $|\lambda| < 1$. On the other hand, if $p < 0$, then equation (3.5.20) gives

$\bar{\mathbf{y}}^T\mathbf{D}\mathbf{y} > 2|p|$ and again $|\lambda| < 1$, so establishing that no eigenvalue of the requisite $\mathbf{C}$ matrix has modulus greater than or equal to unity, and therefore convergence.

It might be expected from all this that Gauss–Seidel's method always converges more rapidly than Jacobi's method. Inevitably a counterexample exists to disprove the conjecture. The matrix

$$\mathbf{A} = \begin{pmatrix} 1 & 0 & 1 \\ -1 & 1 & 0 \\ 1 & 2 & -3 \end{pmatrix} \tag{3.5.23}$$

has a $\mathbf{C}$ matrix for Jacobi's method which is

$$\mathbf{C}_J = \begin{pmatrix} 0 & 0 & 1 \\ -1 & 0 & 0 \\ -\frac{1}{3} & -\frac{2}{3} & 0 \end{pmatrix} \tag{3.5.24}$$

and for Gauss–Seidel's method

$$\mathbf{C}_{GS} = \begin{pmatrix} 0 & 0 & 1 \\ 0 & 0 & 1 \\ 0 & 0 & 1 \end{pmatrix} \tag{3.5.25}$$

For $\mathbf{C}_J$ the eigenvalues are $\lambda_1 = 0.748$ and $\lambda_{2,3} = -0.374 \pm \mathrm{i}\,0.868$ with slow convergence, whereas the eigenvalues for $\mathbf{C}_{GS}$ are $\{1, 0, 0\}$ which diverges. However, it has been shown that excluding such pathological examples as these which are very close to divergence, that more rapid convergence can be expected from the Gauss–Seidel method.

Such convergence is illustrated in the following simple example with matrix $\mathbf{A}$ given by:

$$\mathbf{A} = \begin{bmatrix} 3 & 1 & -1 \\ 2 & 8 & -1 \\ 1 & 1 & -4 \end{bmatrix} \tag{3.5.26}$$

In the Jacobi iteration, the convergence matrix is $\mathbf{C}_J$ which has the value

$$\begin{aligned} \mathbf{C}_J &= \mathbf{D}^{-1}(\mathbf{L}+\mathbf{U}) \\ &= \begin{bmatrix} \frac{1}{3} & 0 & 0 \\ 0 & \frac{1}{8} & 0 \\ 0 & 0 & -\frac{1}{4} \end{bmatrix} \begin{bmatrix} 0 & 1 & -1 \\ 2 & 0 & -1 \\ 1 & 1 & 0 \end{bmatrix} = \begin{bmatrix} 0 & \frac{1}{3} & -\frac{1}{3} \\ \frac{1}{4} & 0 & -\frac{1}{8} \\ -\frac{1}{4} & -\frac{1}{4} & 0 \end{bmatrix} \end{aligned} \tag{3.5.27}$$

Similarly in the Gauss–Seidel case the requirement is to evaluate $\mathbf{C}_G = (\mathbf{D} + \mathbf{L})^{-1}\mathbf{U}$ which gives

$$\mathbf{C}_G = \begin{bmatrix} 3 & 0 & 0 \\ 2 & 8 & 0 \\ 1 & 1 & -4 \end{bmatrix}^{-1} \begin{bmatrix} 0 & 1 & -1 \\ 0 & 0 & -1 \\ 0 & 0 & 0 \end{bmatrix}$$

$$= \frac{1}{96}\begin{bmatrix} 32 & 0 & 0 \\ -8 & 12 & 0 \\ 6 & 3 & -24 \end{bmatrix}\begin{bmatrix} 0 & 1 & -1 \\ 0 & 0 & -1 \\ 0 & 0 & 0 \end{bmatrix}$$

$$= \frac{1}{96}\begin{bmatrix} 0 & 32 & -32 \\ 0 & -8 & -4 \\ 0 & 6 & -9 \end{bmatrix} \tag{3.5.28}$$

Hence the eigenvalues of $\mathbf{C}_J$ satisfy the characteristic equation

$$\begin{vmatrix} -\lambda & \frac{1}{3} & -\frac{1}{3} \\ \frac{1}{4} & -\lambda & \frac{1}{8} \\ -\frac{1}{4} & -\frac{1}{4} & -\lambda \end{vmatrix} = -\lambda^3 + \tfrac{19}{96}\lambda + \tfrac{1}{32} = 0 \tag{3.5.29}$$

and the corresponding eigenvalues of $\mathbf{C}_G$ satisfy

$$\begin{vmatrix} -\lambda & \frac{32}{96} & -\frac{32}{96} \\ 0 & -\frac{8}{96} - \lambda & -\frac{4}{96} \\ 0 & \frac{6}{96} & -\frac{9}{96} - \lambda \end{vmatrix} = -\lambda\left[\lambda^2 + \tfrac{17}{96}\lambda + \tfrac{48}{962}\right] = 0 \tag{3.5.30}$$

Using a non-linear equation technique from Chapter 1 such as Newton's method yields the roots $-0.195\,8534$, $-0.313\,3478$ and $0.509\,2024$ for the eigenvalues of $\mathbf{C}_J$ and the quadratic in the Gauss–Seidel case gives the roots $-0.139\,837$ and $-0.037\,245$, together with the trivial root 0. Hence the Gauss–Seidel method is converging at a rate commiserate with $(0.139\,837)^m$ at the mth step, giving roughly one extra decimal place per iteration. On the other hand, the method of Jacobi has an equivalent factor $(0.509\,2024)^m$ requiring approximately three iterations per figure.

It will be seen in Chapter 9 on partial differential equations, that the principle method of solution is to reduce the problem to an equivalent one in linear algebra by some form of discretization. One natural outcome of this process is matrices of large order with only a limited number of non-zero elements in each row. Such a matrix is called *sparse* and if an elimination process is applied to such a matrix, then fill-in occurs as the process continues, resulting in a reduction in the maximum order of matrix which can be handled. Obviously the iterative methods outlined above do not fill-in and are admirably suited to solving matrix problems with their origins in partial differential equations. Moreover the nature of the approximations tends to yield matrices which lean towards being diagonally dominant so yielding acceptable rates of convergence. Cases arise when this diagonal dominance is not available and elimination methods which minimize fill-in have been set up for such cases.

Because of the significance of iterative methods in the solution of partial differential equations (see Chapter 9), considerable effort has been put into improving the convergence rates. One way of achieving this end, and providing some control over the situation, is the method of successive over-relaxation which is commonly shortened to the SOR method. The principle can be used on any iteration. Suppose the kth iteration of a process results in the value $\bar{\mathbf{x}}^{(k)}$ where the bar indicates that this will not be the accepted value in the usual way. Instead the accepted value is given by

$$\mathbf{x}^{(k)} = \mathbf{x}^{(k-1)} + \omega[\bar{\mathbf{x}}^{(k)} - \mathbf{x}^{(k-1)}] \tag{3.5.31}$$

where ω is a factor called the relaxation factor. Clearly $\omega = 1$ gives back the original un-modified method. The full iteration has the form

$$(\mathbf{I} + \omega\mathbf{D}^{-1}\mathbf{L})\mathbf{x}^{(k+1)} = (1 - \omega)\mathbf{x}^{(k)} + \omega[\mathbf{D}^{-1}\mathbf{b} - \mathbf{D}^{-1}\mathbf{U}\mathbf{x}^{(k)}] \tag{3.5.32}$$

with the convergence matrix $\mathbf{C}$ having the value

$$\mathbf{C} = (\mathbf{I} + \omega\mathbf{D}^{-1}\mathbf{L})^{-1}[(1 - \omega)\mathbf{I} - \omega\mathbf{D}^{-1}\mathbf{U}]. \tag{3.5.33}$$

The practical implementation will then be

$$x_i^{k+1} = (1 - \omega)x_i^k + \frac{\omega}{a_{ii}}\left[b_i - \sum_{j=1}^{i-1} a_{ij}x_j^{(k+1)} - \sum_{j=i+1}^{n} a_{ij}x_j^{(k)}\right] \tag{3.5.34}$$

The object is now to minimize the spectral radius $\rho(\mathbf{C})$ by surreptitious choice of ω, and so enhance the rate of convergence. This process is non-trivial and is described in detail in Ortega (1972) but some of the main results can be quoted here. The Ostrowski–Reich Theorem says that if $\mathbf{A}$ is positive definite and $0 < \omega < 2$ then the SOR method converges for any initial $\mathbf{x}^{(0)}$. In the specific case when $\mathbf{A}$ is also tri-diagonal then

$$\rho(\mathbf{C}_{GS}) = \rho^2(\mathbf{C}_J) < 1 \tag{3.5.35}$$

and the optimum value for ω is given by

$$\omega = \frac{2}{1 + \sqrt{1 - \rho(\mathbf{C}_J)}} \tag{3.5.36}$$

The result in equation (3.5.35) has more general application and demonstrates that the Gauss–Seidel process converges at twice the rate of the Jacobi process, except in pathological cases referred to earlier.

As an example, the solution of

$$\begin{aligned} 2x_1 + 2x_2 &= 1 \\ x_1 + 2x_2 + x_3 &= 4 \\ x_2 + 2x_3 &= 2 \end{aligned} \tag{3.5.37}$$

with $\omega = 1$ to give the original Gauss–Seidel method converges to seven correct figures in 50 iterations. With $\omega = 1.3$, 40 iterations is sufficient to yield 13 figure accuracy. Even with a small change of ω to 1.25 the convergence is to 13 figures in 55 iterations, which shows considerable sensitivity to the ω values.

There are other devices which can be used to enhance the rates of convergence of these methods which depend on how the equations are ordered in relation to the underlying grid for the partial differential equation. Such a process is the ADI or alternating direction implicit method which will be pursued in Section 9.5 when partial differential equations are considered in detail.

3.6 Sparse systems

A matrix is sparse if the number of non-zero elements is small. It will be seen in Chapters 7 and 9 that boundary value problems in ordinary differential equations and both the finite difference and finite element approaches to partial differential equations, lead to sparse matrices. For grid points not near to boundaries, a simple finite difference approximation to Laplace's equation

$$\frac{\partial^2 \phi}{\partial x^2} + \frac{\partial^2 \phi}{\partial y^2} = 0 \tag{3.6.1}$$

gives

$$\frac{\phi_{i+1,j} - 2\phi_{i,j} + \phi_{i-1,j}}{h^2} + \frac{\phi_{i,j+1} - 2\phi_{i,j} + \phi_{i,j-1}}{h^2} = 0 \tag{3.6.2}$$

where the finite difference scheme has a step size of h and the value of $\phi(x, y)$ at the point (ih, jh) is $\phi_{i,j}$. In order to obtain a conventional representation for the set of equations (3.6.2) with i and j ranging over the space of interest, the ϕ's need to be re-indexed as a one-subscript vector. This would commonly be achieved by indexing row by row or column by column. Hence if row i had a total of r_i elements then the element $\phi_{i,j}$ would be re-indexed as $\phi[\sum_{s=1}^{i-1} r_s + j]$. In this way, as i and j vary over the space of interest, (3.6.2) represents the defining linear equations with only five elements in each row. By cancelling the term h^2 and collecting terms, the diagonal element has coefficient four and the other elements coefficient unity. Therefore this set of equations has a natural diagonal dominance which will make the iterative solutions of the previous paragraph one approach to a solution. Only the non-zero elements need to be stored and a number of ways of organizing such storage arise in the literature.

With the re-indexing given above, the five elements are not contiguous, three elements including the diagonal being so, and the other two being separated as these arise from the original previous row and original following row respectively. The storage schemes need to reflect this behaviour, and generally the simpler the scheme is made, the more wasteful it is of storage, though the more efficient it can be of access in the solving algorithm. The extreme would be to store the full matrix with normal access to the indexed elements. An economy on this is to store elements which fall into a fixed band centred on the diagonal say from b_l to b_u where these limits are chosen to include all non-zero elements. Some zero elements will be included here, namely those between extreme non-zeros, and if any particular row has a somewhat larger band width this will cause the other rows to have zeros stored beyond the extreme non-zeros. On the other hand the elements can be accessed by a simple integer addition which depends on the band width limits.

The complete elimination of stored zero values can be achieved if in parallel with the stored element, a corresponding integer array element is set up with the true row position for each element. Then in the example the ϕ's would be stored in an $n \times 5$ matrix and an integer array of the same size would contain the actual row position of the corresponding ϕ. Hence in a programme, instead of accessing an element with $a_{i,j}$, it is necessary to use $a_{i,ind(i,j)}$ where a is the array of coefficients and *ind* contains the true row positions. The result is very efficient storage but increased overheads in the accessibility of the elements. This is called the coordinate scheme.

Along the same lines is the possibility of using list processing ideas to generate a linked list in which a structure (or in pascal a record) is set up containing a current element and the addresses of its neighbours. Hence progress can be made from one element to another within the list. This approach has such high overheads that it is not normally employed.

There are some useful compromises which balance the above requirements. The full matrix can for example be broken down into a set of submatrices each of which is held as a full matrix, with some zeros now being stored. In addition a list of indices is needed to determine just where in the original matrix, the elements (1, 1) of each submatrix occur. Hence at the expense of extra storage some reduction in access costs is achieved. This is commonly known as clique storage.

The choice of storage method is also affected by the processing hardware. Even with a conventional machine, the balance between storage and accessibility of array elements can vary. If a vector machine is used, storage as a collection of vectors may prove best despite some storage of zeros. Hence the use of an index array to collect a vector into packed form becomes important. This process is called the gather operation, and its inverse is a scatter operation. The vector advantages can then be applied after the gather operation, and the results can be restored after processing with a scatter operation.

Once a storage system is established, the iterative methods of the previous paragraph are straightforward to implement. Often the equations being solved do not have the simple form shown in (3.6.2). The use of general coordinate systems can cause a loss of the diagonal dominance, as can the equations representing the boundary conditions. Convergence of iterative schemes can then be slow. It is tempting, therefore, to return to direct methods such as Gaussian elimination. The problem now is that as the process defined earlier is performed, new non-zero elements will be generated to give a fill-in effect. Two things need to be taken into account. Firstly the process needs modification to minimize the fill-in. For example partial pivoting as used in Section 3.2 will soon cause considerable fill-in with the associated storage costs. Secondly it is necessary to allow some increase in storage as the algorithm progresses. Either some dynamic process is required, or sufficient leeway has to be built in initially to allow the unavoidable fill-in to take place.

The pivotal strategy often used is that of Markowitz (1957). If the number of entries in row i is r_i and in column j is c_j then the process chooses a pivot to minimize

$$(r_i - 1)(c_j - 1) \tag{3.6.3}$$

and threshold pivoting is used in which the chosen pivot is taken from the set which has the property that

$$|a_{kk}| \leq u \max_i |a_{ik}| \tag{3.6.4}$$

where u limits the increase in size of the new elements and yet allows some freedom to reduce fill-in. The tuning parameter u can be chosen as small as 0.1 without apparent problems.

A compromise between direct and iterative methods can be very effective in sparse problems. Instead of the iteration being applied to each row, a block iterative method is employed. The matrix in question is split into blocks of submatrices of the form:

$$\begin{bmatrix} \mathbf{A}_{11} & \mathbf{A}_{12} & \mathbf{0} & \cdots \\ \mathbf{A}_{21} & \mathbf{A}_{22} & \mathbf{A}_{23} & \cdots \\ \vdots & & & \vdots \\ \cdots & \mathbf{0} & \mathbf{A}_{n,n-1} & \mathbf{A}_{nn} \end{bmatrix} \tag{3.6.5}$$

where the diagonal submatrices are square, but the off-diagonal sub-matrices can be rectangular to account for irregularity in the band structure. As in the normal iterative methods, an initial solution vector is used and then the iterations:

$$\mathbf{A}_{rr}\mathbf{x}_r^{k+1} = b_r - \mathbf{A}_{r,r-1}\mathbf{x}'^k_r - \mathbf{A}_{r,r+1}\mathbf{x}''^k_r \tag{3.6.6}$$

are computed where $\mathbf{x}_r$ is the set of $\mathbf{x}$ values pertaining to the block $\mathbf{A}_{rr}$, and $\mathbf{x}'_r$ and $\mathbf{x}''_r$ are the remaining elements of the vector $\mathbf{x}$ either side of the diagonal block. Hence a 'block's worth' of $\mathbf{x}$ elements is updated at each iteration and conventional sparse methods can be used for the direct solution in each block. Clearly if the blocks are reduced to a single element then the original iterative method is obtained and at the other extreme if only one block is used then the conventional sparse problem for a direct solution is recovered.

Partitioning is also employed in problems which are termed *huge*. Here the hardware limitations are being reached and so direct methods are employed on part of the system. One such approach is to partition into a 2×2 block of the form

$$\begin{bmatrix} \mathbf{A}_{11} & \mathbf{A}_{12} \\ \mathbf{A}_{21} & \mathbf{A}_{22} \end{bmatrix} \tag{3.6.7}$$

and look for a factorization of the form

$$\begin{bmatrix} \mathbf{L}_{11} & \\ \mathbf{L}_{21} & \mathbf{L}_{22} \end{bmatrix} \begin{bmatrix} \mathbf{U}_{11} & \mathbf{U}_{12} \\ & \mathbf{U}_{22} \end{bmatrix} \tag{3.6.8}$$

The object is to have special forms for block matrices which simplify the resulting solution of the partial problem. One such form is for $\mathbf{A}_{11}$ to be block triangular, so requiring only forward substitutions.

Another rather different approach to *huge* systems is to solve

$$\mathbf{Ax} = \mathbf{b} \tag{3.6.9}$$

by finding a matrix $\Delta\mathbf{A}$ for which the solution of

$$(\mathbf{A} + \Delta\mathbf{A})\mathbf{x} = \mathbf{b} \tag{3.6.10}$$

is easy, possibly because the new matrix has a banded structure. Perturbation methods are then used to recover the solution to the original matrix.

The application of vectorization and parallel methods adds a further layer to the approach for large systems which is currently an active area in research. For further reading in these areas the reader is referred to Duff, Erisman and Reid(1986) and Duff(1986).

The package MAPLE has the capability of defining vectors and matrices, and has routines to evaluate determinants and solve sets of linear algebraic equations. Such a package would be used for the day-to-day solution of small to moderate problems, leaving vector and parallel machines for the large problems discussed above.

Further reading

Wilkinson J. H. (1965) *Algebraic Eigenvalue Problem*, Oxford University Press, Oxford.
Fox L. (1968) *Introduction to Numerical Linear Algebra*, Oxford University Press, Oxford.
Duff I. S., Erisman A. M. and Reid J. K. (1986) *Direct Methods for Sparse Matrices*, Oxford University Press, Oxford.

Exercises

(3.1) Given the matrix

$$\mathbf{A} = \begin{pmatrix} 0.9 & 0.1 \\ 0.3 & 0.8 \end{pmatrix}$$

determine the values of the matrix norms $||\mathbf{A}||_\infty$, $||\mathbf{A}||_1$, $||\mathbf{A}||_E$ and $||\mathbf{A}||_2$. Verify the inequality

$$||\mathbf{A}||_2 \le ||\mathbf{A}||_E \le n^{1/2}||\mathbf{A}||_2$$

(3.2) Repeat the above procedure for the matrix

$$\mathbf{A} = \begin{pmatrix} 0.1 & 0.0 & 1.0 \\ 0.2 & 0.3 & 0.0 \\ 0.2 & 0.1 & 0.4 \end{pmatrix}$$

(3.3) Find $||\mathbf{A}||_2$ for the matrix

$$\begin{bmatrix} \frac{931}{625} & \frac{408}{625} & \frac{36}{125} \\ \frac{408}{625} & \frac{1169}{625} & \frac{48}{125} \\ \frac{36}{125} & \frac{48}{125} & \frac{66}{25} \end{bmatrix}$$

[Hint: Exact arithmetic will yield a polynomial with simple factors!].

(3.4) Calculate the spectral condition number with respect to inversion, $\kappa(\mathbf{A})$, for the matrices in exercises (3.1) and (3.2).

(3.5) Calculate the spectral condition number of the matrix

$$\mathbf{A} = \begin{pmatrix} 5 & 7 & 6 & 5 \\ 7 & 10 & 8 & 7 \\ 6 & 8 & 10 & 9 \\ 5 & 7 & 9 & 10 \end{pmatrix}$$

What is the expected loss of accuracy in the computation of the inverse of this matrix? Demonstrate this by using Gauss elimination to find the inverse.

(3.6) Use Cholesky decomposition to solve the linear equations

$$\begin{pmatrix} 5 & 7 & 6 & 5 \\ 7 & 10 & 8 & 7 \\ 6 & 8 & 10 & 9 \\ 5 & 7 & 9 & 10 \end{pmatrix} \begin{pmatrix} x_1 \\ x_2 \\ x_3 \\ x_4 \end{pmatrix} = \begin{pmatrix} 23 \\ 32 \\ 33 \\ 31 \end{pmatrix}$$

(3.7) Find the condition factor with respect to inversion for the system

$$x_1 + x_2 = 2$$

$$x_1 + 1.01x_2 = 2.01$$

and solve these equations using a fixed level of accuracy. Compare the solution with that of the equations

$$x_1 + x_2 = 2$$

$$1.001x_1 + x_2 = 2.01$$

in the light of the condition factor obtained.

(3.8) Consider the two sets of equations:

$$7x_1 + 8x_2 + 9x_3 = 24$$

$$8x_1 + 9x_2 + 10x_3 = 27$$

$$9x_1 + 10x_2 + 8x_3 = 27$$

and

$$\begin{aligned} 7x_1 + 8x_2 + 9x_3 &= 24.1 \\ 8x_1 + 9x_2 + 10x_3 &= 26.9 \\ 9x_1 + 10x_2 + 8x_3 &= 26.9 \end{aligned}$$

in the same way as for exercise (3.7).

(3.9) Investigate the condition factor of the matrix

$$\begin{bmatrix} 100 & 1 & 2 \\ 1 & 1 & -1 \\ 2 & -1 & \alpha \end{bmatrix}$$

for varying α. In particular find the value of α which gives an infinite factor, find out how the condition factor varies for large α, and investigate the smallest value the factor can take.

(3.10) Sketch the curve of $\kappa(\mathbf{A})$ against c for the matrix

$$\begin{bmatrix} 1 & 1 & 2 \\ 1 & 2 & 6 \\ 1 & 6 & c \end{bmatrix}$$

considering in particular the situation for large c, and where $\kappa(\mathbf{A})$ is singular.

(3.11) The row operation of interchanging two rows of a matrix is equivalent to pre-multiplying that matrix by another elementary matrix $\mathbf{N}$. Find $\mathbf{N}$ such that rows i and j, $i \neq j$, are interchanged in an $n \times n$ matrix. Establish that if elementary operations are applied to a matrix $\mathbf{A}$ which reduce it to the unit matrix $\mathbf{I}$, then the same operations applied to a unit matrix yield $\mathbf{A}^{-1}$. Hence find the inverse of

$$\begin{bmatrix} 1 & 2 & -1 \\ 0 & 1 & 1 \\ 2 & 1 & 1 \end{bmatrix}$$

(3.12) Apply direct triangular decomposition to the matrix

$$\begin{bmatrix} 1 & 2 & 2 \\ 1 & 3 & 4 \\ -1 & -3 & -5 \end{bmatrix}$$

to solve the set of linear algebraic equations with the right-hand side $\mathbf{b} = (1, 1, 1)^T$.

(3.13) Apply Cholesky decomposition to solve the linear equations

$$\begin{bmatrix} 1 & 1 & 2 \\ 1 & 2 & 6 \\ 2 & 6 & 24 \end{bmatrix} \mathbf{x} = \begin{bmatrix} 6 \\ 24 \\ 120 \end{bmatrix}$$

(3.14) By considering the algorithm for Gauss elimination, calculate the major computational effort by generating a formula for the number of multiplications and divisions required for the solution of n simultaneous equations. Repeat the procedure for Cholesky decomposition in the symmetric case, so demonstrating the gain in the use of the latter approach with symmetry present.

(3.15) Establish the rate of convergence for Jacobi's method on the matrix

$$\begin{bmatrix} 2 & 1 & 1 \\ -1 & 3 & 1 \\ 0 & -1 & 4 \end{bmatrix}$$

and estimate how many iterations are needed to obtain three-figure accuracy, seven-figure accuracy and 14-figure accuracy. Estimate the number of iterations for machine accuracy on your particular computer and test this estimate in practice.

(3.16) Use Jacobi's iterative method to solve the linear equations

$$\begin{bmatrix} 100 & 1 & 0.05 \\ 1 & 2 & 0.01 \\ 0.05 & 0.01 & 0.1 \end{bmatrix} \begin{bmatrix} x_1 \\ x_2 \\ x_3 \end{bmatrix} = \begin{bmatrix} 1 \\ 1 \\ 1 \end{bmatrix}$$

and explain how the ill-conditioning in the system manifests itself in the process.

(3.17) Use Gauss elimination to solve the set of linear equations

$$\begin{bmatrix} 9 & -2 & 1 \\ 4 & 5 & -2 \\ 1 & -3 & -5 \end{bmatrix} \begin{bmatrix} x_1 \\ x_2 \\ x_3 \end{bmatrix} = \begin{bmatrix} 17 \\ -9 \\ 4 \end{bmatrix}$$

and establish the condition factor for the system. Solve the system iteratively using Gauss–Seidel's method. Find the convergence rate for the matrix and confirm that this rate is achieved in practice.

(3.18) Use Jacobi's method and Gauss–Seidel's method to solve iteratively the system of equations

$$\begin{bmatrix} -2 & 2 & 0 & 0 \\ 1 & -3 & 1 & 0 \\ 0 & 1 & -6 & 1 \\ 0 & 0 & 1 & -11 \end{bmatrix} \begin{bmatrix} y_1 \\ y_2 \\ y_3 \\ y_4 \end{bmatrix} = \begin{bmatrix} -1 \\ -1 \\ -1 \\ -1 \end{bmatrix}$$

and confirm that the rates of convergence agree with the theoretical expectations.

(3.19) Solve the system of equations

$$0.732x - 5.421y + 1.013z = 4.256$$

$$3.491x + 2.203y + 0.782z = -7.113$$

$$0.961x - 1.523y + 4.265z = 3.727$$

using Gauss–Seidel's method, and confirm the rate of convergence obtained.

(3.20) Determine conclusively whether the Gauss–Seidel iteration converges on the matrices:

$$\text{(i)} \qquad \begin{bmatrix} 1 & 1 & 2 \\ 0 & 3 & 3 \\ 4 & 0 & 2 \end{bmatrix}$$

and

$$\text{(ii)} \qquad \begin{bmatrix} 1 & 0.1 & 0.1 \\ 0 & 3 & 0.2 \\ 0.1 & 0 & 2 \end{bmatrix}$$

(3.21) Show that the matrix

$$\mathbf{A} = \begin{bmatrix} 8 & -1 & 0 \\ -1 & 6 & -1 \\ 0 & -1 & 4 \end{bmatrix}$$

is symmetric and positive definite. Hence solve the linear equations $\mathbf{Ax} = (1, 1, 1)^T$ by Gauss–Seidel's method with $\mathbf{x}^{(0)} = (1/8, 1/6, 1/4)^T$. Confirm the convergence rate using the convergence matrix $\mathbf{C}$.

(3.22) Solve the equations

$$\begin{bmatrix} 4 & -1 & 0 & 0 \\ -1 & 4 & -1 & 0 \\ 0 & -1 & 4 & -1 \\ 0 & 0 & -1 & 4 \end{bmatrix} \begin{bmatrix} x_1 \\ x_2 \\ x_3 \\ x_4 \end{bmatrix} = \begin{bmatrix} 1 \\ 1 \\ 1 \\ 1 \end{bmatrix}$$

using the SOR method with a range of values of the relaxation factor ω. Investigate the optimum value of ω in this case.

4

Algebraic Eigenvalue Problem

4.1 Introduction

It soon becomes apparent in the early study of linear equations that not all sets of linear algebraic equations have unique solutions. If the set of equations has a zero right-hand side, then a very important special case arises. Hence one solution of the set of equations

$$\mathbf{Ax} = \lambda \mathbf{x} \tag{4.1.1}$$

for a real $n \times n$ matrix $\mathbf{A}$ is always $\mathbf{x} = \mathbf{0}$. However there is a set of values for the parameter λ for which non-trivial solutions for the vector $\mathbf{x}$ exist. These non-trivial solutions are termed eigenvectors and the corresponding values of λ are termed eigenvalues. A study of linear algebra (see Finkbeiner, 1960) soon establishes that the number of such eigenvalues is equal to the order of the matrix, and there is now a clear numerical problem to compute eigenvalues of matrices. The traditional linear algebra approach results in the equation

$$\det(\mathbf{A} - \mathbf{I}\lambda) = 0 \tag{4.1.2}$$

to give a polynomial of degree n for the n eigenvalues, some or all of which may be complex.

Eigenvalue problems arise in all branches of applied mathematics and seem to be a very fundamental part of the structure of the universe. The solutions of the Schrödinger equation of quantum mechanics are eigensolutions, the various eigenvalues representing the energy levels of the resulting orbitals. In a simpler vein, the solutions of homogeneous ordinary differential equations such as

$$\frac{d^2y}{dx^2} + \lambda^2 y = 0 \tag{4.1.3}$$

with homogeneous boundary conditions such as $y(0) = y(\pi) = 0$, are the non-trivial solutions for certain values of λ, namely $\lambda = n$ with n an integer. Again there is always the trivial solution $y = 0$. In this case there is an infinity of eigenvalues, which in a vibration problem represents the harmonics in the frequency pattern. The differential equation problem will be treated in more detail in Chapter 7 and later with integral equations in Chapter 8. In effect, any discretization of the differential equation problem results in an equivalent algebraic eigenvalue problem.

Eigenvalue problems arise in other ways, for example in space dynamics when the principal axes of inertia are required for a space vehicle, or in elasticity or rheology when principle axes are required for the stress matrix, or the viscosity matrix.

Eigenvalue problems are also important in a less direct manner in numerical applications. For example to discover the condition factor in the solution of a set of linear algebraic equations involves finding the ratio of the largest to the smallest singular values of the underlying matrix, as derived in Chapter 3. Equally the eigenvalue problem is involved when establishing the stiffness of ordinary differential equation problems (see Chapter 7).

One simple-minded method for the computation of eigenvalues would be to expand out the determinant of equation (4.1.2) and then solve the polynomial by Newton's method to obtain the roots. Consider the two stages of this endeavour from a practical point of view. In order to expand out even a quite low order, determinant programming techniques become necessary. There is no problem in writing a recursive routine which multiplies out a determinant by say its first column, and calls itself recursively for each sub-determinant of order $n - 1$. When the recursion arrives at a single-element determinant its value is immediate and the recursion can be unwound to yield the value of the determinant. The elements of the determinant are not real numbers but polynomials, the original diagonal elements being $a_{ii} - \lambda$. Hence a package for polynomial arithmetic is required, which is straightforward to write but quite expensive to run. A count of the number of operations needed is easy to make, being $n\times$ the count for an $(n - 1) \times (n - 1)$ determinant. Hence the total number of polynomial operations is $n!$, which is a very expensive exercise for even low-order problems. Nevertheless, the process is feasible and the result is the characteristic equation, or polynomial, whose roots are the required eigenvalues.

The final stage is to find the roots of the polynomial. We have seen in Chapter 1 that the real roots could be found by Newton's method if they were real and distinct, but for multiple roots and complex roots the process looks less certain. Bairstow's method from Chapter 1 answers some of these criticisms, but there is also a stability problem in that in general the roots of a polynomial are very sensitive to its coefficients, and so rounding errors imposed on the coefficients from the computation of say the first root or pair roots could lead to catastrophic loss of accuracy in the second stage, after deflation or dividing out. With multiple roots there would be a difficulty in knowing whether all the n eigenvalues were found or whether some were indeed repeated.

It is easy to see why such sensitivity arises. Consider a polynomial $f(z)$ with a root $z = \lambda_i$ so that

$$f(z) = (z - \lambda_i)h(z) = \sum_{r=0}^{n} a_r z^r \tag{4.1.4}$$

Consider perturbing one of the coefficients a_r, say the kth, to give the perturbed polynomial

$$g(z) = f(z) + \delta a_k\, z^k \tag{4.1.5}$$

Let the new root be $z = \lambda_i + \delta\lambda_i$, then

$$0 = \delta\lambda_i\, h(\lambda_i + \delta\lambda_i) + \delta a_k\, (\lambda_i + \delta\lambda_i)^k \tag{4.1.6}$$

But differentiating (4.1.4) gives

$$f'(z) = h(z) + (z - \lambda_i)h'(z) \tag{4.1.7}$$

and expanding (4.1.6) using Taylor's Theorem to first-order terms gives

$$0 = \delta\lambda_i\, [h(\lambda_i + \delta\lambda_i\, h'(\lambda_i) + \cdots] + \delta a_k\, (\lambda_i^k + k\delta\lambda_i\, \lambda_i^{k-1} + \cdots) \tag{4.1.8}$$

with the result that

$$\delta\lambda_i = -\frac{\delta a_k\, \lambda_i^k}{h(\lambda_i)} \tag{4.1.9}$$

Using (4.1.7) gives the equivalent results:

$$\delta\lambda_i = -\frac{\delta a_k\, \lambda_i^k}{f'(\lambda_i)} = -\frac{\delta a_k\, \lambda_i^k}{\prod\limits_{j\neq i} (\lambda_i - \lambda_j)} \tag{4.1.10}$$

Hence large perturbations can arise in two ways. Either λ_i^k becomes very large, or alternatively $f'(\lambda_i)$ becomes very small. Wilkinson(1965) gives a number of examples to illustrate these effects and even for a polynomial with linearly distributed roots satisfying $\lambda_r = 21 - r$ with $r = 1, \ldots, 20$ severe problems can arise. Using the results above, $i = 5$ and $k = 15$ gives a perturbation

$$\begin{aligned} |\delta\lambda_i| &\sim \varepsilon |a_k| \frac{(21 - i)^k}{(20 - i)!(i - 1)!} \\ &\sim \varepsilon 10^{10} \frac{16^{15}}{15!4!} = \varepsilon(0.38)10^{15} \end{aligned} \tag{4.1.11}$$

In a similar way, geometrically distributed roots can cause problems as $\lambda_i - \lambda_j$ can now be very small.

As a method, therefore, this simple approach has its problems, and it is the object of this chapter to develop alternative methods which in general work directly from the original matrix in order to generate the required eigenvalues. The chapter is split into three major parts. The first part covers what may be loosely described as elementary methods. For some problems these will be sufficient in their own right, but all will have shortcomings. In the second part of the chapter attention is restricted to symmetric matrices which are guaranteed to have real eigenvalues. A range of methods is found for this particular group of matrices culminating in a robust method which is applicable to large problems. In the third part, the more difficult problem of a general matrix is considered with extensions of the symmetric case being used as far as possible. Again shortcomings are encountered, and these are overcome in the so-called QR process, and its relative, the LR method.

4.2 The stability of eigenvector and eigenvalue problems

Before treating specific methods for the numerical computation of eigenvalues and eigenvectors, a more theoretical aspect of the work will be considered. As in the previous chapter, the question of the accuracy of computed values will be considered from the point of view of whether rounding errors are magnified in the solution process. A perturbation approach is used to analyse the problem and this approach requires a few mathematical preliminaries.

Firstly, define the quantities s_i as

$$s_i = \mathbf{y}_i^T \mathbf{x}_i \tag{4.2.1}$$

where $\mathbf{x}_i$ are the usual eigenvectors or right-hand eigenvectors, and $\mathbf{y}_i$ are the left-hand eigenvectors defined by

$$\mathbf{y}_i^T \mathbf{A} = \lambda_i \mathbf{y}_i^T. \tag{4.2.2}$$

These two sets of eigenvectors satisfy an orthogonality condition according to:

Theorem

$$\mathbf{x}_i^T \mathbf{y}_j = 0 \quad \text{if} \quad \lambda_i \neq \lambda_j \tag{4.2.3}$$

Proof

From the definitions of the right- and left-hand eigenvectors

$$\mathbf{x}_i^T \mathbf{A}^T = \lambda_i \mathbf{x}_i^T \tag{4.2.4}$$

and

$$\mathbf{A}^T \mathbf{y}_j = \lambda_j \mathbf{y}_j. \tag{4.2.5}$$

Post-multiply (4.2.4) by $\mathbf{y}_j$ and pre-multiply (4.2.5) by $\mathbf{x}_i^T$ to give the two equations

$$\mathbf{x}_i^T \mathbf{A}^T \mathbf{y}_j = \lambda_i \mathbf{x}_i^T \mathbf{y_j}$$
$$\mathbf{x}_i^T \mathbf{A}^T \mathbf{y}_j = \lambda_j \mathbf{x}_i^T \mathbf{y}_j$$

and subtracting gives

$$0 = (\lambda_i - \lambda_j)\mathbf{x}_i^T \mathbf{y}_j \tag{4.2.6}$$

Hence if $\lambda_i \neq \lambda_j$ then $\mathbf{x}_i^T \mathbf{y}_j = 0$.

In order to determine the sensitivity of the eigenvalues and eigenvectors of the matrix $\mathbf{A}$, consider the perturbed problem

$$(\mathbf{A} + \varepsilon\mathbf{B})\tilde{\mathbf{x}}_1(\varepsilon) = \tilde{\lambda}_1(\varepsilon)\tilde{\mathbf{x}}_1(\varepsilon) \tag{4.2.7}$$

The object will be to discover how far the perturbed eigenvalue, $\tilde{\lambda}_1$ differs from λ_1 and how far the perturbed eigenvector $\tilde{\mathbf{x}}_1$ differs from $\mathbf{x}_1$. Assume that the elements of **A** and **B** are normalized so that $|a_{ij}| < 1$ and $|b_{ij}| < 1$, and that $\mathbf{y}_i$ and $\mathbf{x}_i$ are normalized so that their second vector norms are unity. Further define B_{ij} as $B_{ij} = \mathbf{y}_i^T \mathbf{B} \mathbf{x}_j$ then

$$|B_{ij}| = |\mathbf{y}_i^T(\mathbf{B}\mathbf{x}_j)| \leq ||\mathbf{y}_i||_2||\mathbf{B}\mathbf{x}_j||_2$$
$$\leq ||\mathbf{B}||_2||\mathbf{y}_i||_2||\mathbf{x}_j||_2 \leq n \qquad (4.2.8)$$

Now $\tilde{\lambda}_1$ can be written as

$$\tilde{\lambda}_1 = \lambda_1 + \varepsilon k_1 + \varepsilon^2 k^2 + \cdots \qquad (4.2.9)$$

and

$$\tilde{\mathbf{x}}_1 = \mathbf{x}_1 + \varepsilon \mathbf{z}_1 + \varepsilon^2 \mathbf{z}_2 + \cdots \qquad (4.2.10)$$

But each of the perturbation vectors **z** can itself be expressed as a linear combination of the basis vectors made up of the eigenvectors themselves to give

$$\mathbf{z}_i = \sum_{j=1}^{n} s_{ij}\mathbf{x}_j \qquad (4.2.11)$$

Hence $\tilde{\mathbf{x}}_1$ rearranges to the form

$$\tilde{\mathbf{x}}_1(\varepsilon) = \mathbf{x}_1 + \varepsilon \sum_{j=1}^{n} s_{j1}\mathbf{x}_j + \varepsilon^2 \sum_{j=1}^{n} s_{j2}\mathbf{x}_j + \cdots$$
$$= (1 + \varepsilon s_{11} + \varepsilon^2 s_{12} + \cdots)\mathbf{x}_1 + (\varepsilon s_{21} + \varepsilon^2 s_{22} + \cdots)\mathbf{x}_2 + \cdots$$
$$+ (\varepsilon s_{n1} + \varepsilon^2 s_{n2} + \cdots)\mathbf{x}_n \qquad (4.2.12)$$

As eigenvectors are only determined to a multiplicative constant, division by the coefficient of $\mathbf{x}_1$ gives the normalized form

$$\tilde{\mathbf{x}}_1(\varepsilon) = \mathbf{x}_1 + (\varepsilon t_{21} + \varepsilon^2 t_{22} + \cdots)\mathbf{x}_2 + \cdots + (\varepsilon t_{n1} + \varepsilon^2 t_{n2} + \cdots)\mathbf{x}_n \qquad (4.2.13)$$

where the constants t_{ij} could be found from the s_{ij} if necessary.

Substitute these perturbations into (4.2.7), and the following relations are obtained by equating the coefficients of ε on both sides. The zeroth-order term gives correctly the original eigenvalue problem. Hence

$$\mathbf{A}\left(\sum_{i=2}^{n} t_{i1}\mathbf{x}_i\right) + \mathbf{B}\mathbf{x}_1 = \lambda_1\left(\sum_{i=2}^{n} t_{i1}\mathbf{x}_i\right) + k_1\mathbf{x}_1 \qquad (4.2.14)$$

or, using $\mathbf{A}\mathbf{x}_i = \lambda_i\mathbf{x}_i$,

$$\sum_{i=2}^{n} t_{i1}(\lambda_i - \lambda_1)\mathbf{x}_i + \mathbf{B}\mathbf{x}_1 = k_1\mathbf{x}_1. \qquad (4.2.15)$$

The two dominant perturbations, k_1 and the t_{i1} can now be found by simple algebra. Pre-multiply (4.2.15) by $\mathbf{y}_1^T$ and use the orthogonality result (4.2.3) to get

$$k_1 = \frac{\mathbf{y}_1^T \mathbf{B} \mathbf{x}_1}{\mathbf{y}_1^T \mathbf{x}_1} = \frac{B_{11}}{s_1} \tag{4.2.16}$$

and therefore the bound

$$|k_1| \leq \frac{n}{|s_1|} \tag{4.2.17}$$

For the quantities t_{i1}, pre-multiply (4.2.15) by $\mathbf{y}_i^T$ to obtain

$$(\lambda_i - \lambda_1) t_{i1} s_i + B_{i1} = 0 \qquad i = 2, 3, \ldots, n \tag{4.2.18}$$

to yield the first-order perturbation

$$\varepsilon \left[\frac{B_{21}\mathbf{x}_2}{(\lambda_1 - \lambda_2)s_2} + \frac{B_{31}\mathbf{x}_2}{(\lambda_1 - \lambda_3)s_3} + \cdots + \frac{B_{n1}\mathbf{x}_2}{(\lambda_1 - \lambda_n)s_n} \right] \tag{4.2.19}$$

From relation (4.2.17) it is clear that the eigenvalue problem for λ_i is well-conditioned as long as the corresponding quantity s_i is not small. Equally the eigenvector problem for $\mathbf{x}_1$ is well-conditioned as long as none of the s_i, $i = 2, 3, \ldots, n$ are small and as long as no eigenvalue λ_i, $i = 2, 3, \ldots, n$ is close to λ_1. These conditions are very akin to the condition factor $\kappa(\mathbf{A})$ used in Chapter 3 for the linear equation problem. These formulae can be demonstrated using hand computations for a simple example, namely the matrix

$$\begin{bmatrix} 4 & 3 \\ 1 & 6 \end{bmatrix} \tag{4.2.20}$$

which has the characteristic polynomial $\lambda^2 - 10\lambda + 21 = 0$ with eigenvalues $\lambda_1 = 7$ and $\lambda_2 = 3$. The right and left eigenvectors then satisfy

$$\begin{aligned} 4x_1 + 3x_2 &= 7x_1 \\ x_1 + 6x_2 &= 7x_2 \end{aligned} \tag{4.2.21}$$

to give $\mathbf{x}_1 = (1, 1)/\sqrt{2}$,

$$\begin{aligned} 4x_1 + 3x_2 &= 3x_1 \\ x_1 + 6x_2 &= 3x_2 \end{aligned} \tag{4.2.22}$$

to give $\mathbf{x}_2 = (3, -1)/\sqrt{10}$,

$$\begin{aligned} 4y_1 + y_2 &= 7y_1 \\ 3y_1 + 6y_2 &= 7y_2 \end{aligned} \tag{4.2.23}$$

to give $\mathbf{y}_1 = (1, 3)/\sqrt{10}$, and

$$\begin{aligned} 4y_1 + y_2 &= 3y_1 \\ 3y_1 + 6y_2 &= 3y_2 \end{aligned} \tag{4.2.24}$$

to give $\mathbf{y}_2 = (1, -1)/\sqrt{2}$. Hence

$$s_1 = \mathbf{y}_1^T \mathbf{x}_1 = 2/\sqrt{5} \tag{4.2.25}$$

and

$$s_2 = \mathbf{y}_2^T \mathbf{x}_2 = 2/\sqrt{5} \tag{4.2.26}$$

Neither s_1 nor s_2 is small implying from (4.2.17) that the eigenvalue problem is well-conditioned. Moreover, no pair of eigenvalues is close together and hence all the eigenvector problems are well-conditioned from (4.2.19).

With these theoretical considerations complete, attention now turns to the problem of how eigenvalues may be found without incurring stability problems.

4.3 Simple iterative methods

The first of these methods is called direct iteration and is simply defined by

$$\mathbf{y}^{(p)} = \mathbf{A}\mathbf{y}^{(p-1)} \tag{4.3.1}$$

with $\mathbf{y}^{(0)}$ arbitrary. If this process is iterated as it stands, and if the elements of $\mathbf{A}$ are say greater than unity, then in general the resulting $\mathbf{y}$ vector will have ever increasing components. Hence at each stage the vector $\mathbf{y}$ is normalized by dividing by $||\mathbf{y}||_\infty$, that is the element of largest modulus in the vector $\mathbf{y}$ (see Chapter 3, equation (3.1.2)). The outcome can be seen algebraically as

$$\mathbf{y}^{(p)} = \mathbf{A}\mathbf{y}^{(p-1)} = \mathbf{A}^2\mathbf{y}^{(p-2)} = \cdots = \mathbf{A}^p\mathbf{y}^{(0)} \tag{4.3.2}$$

and any vector in a vector space can be represented as a linear combination of basis vectors, including the set of eigenvectors, yielding

$$\mathbf{y}^{(p)} = \mathbf{A}^p\mathbf{y}^{(0)} = \sum_{r=1}^{n} \alpha_r \lambda_r^p \mathbf{x}_r \tag{4.3.3}$$

as the eigenvectors satisfy

$$\begin{aligned} \mathbf{A}\mathbf{x}_r &= \lambda_r \mathbf{x}_r \\ \mathbf{A}^2\mathbf{x}_r &= \lambda_r^2 \mathbf{x}_r \\ &\vdots \end{aligned} \tag{4.3.4}$$

If the eigenvalue of largest modulus is denoted by λ_1 and is real and distinct, then

$$\mathbf{y}^{(p)} = \lambda_1^p \left\{ \alpha_1 \mathbf{x}_1 + \sum_{r=2}^{n} \alpha_r \left(\frac{\lambda_r}{\lambda_1} \right)^p \mathbf{x}_r \right\} \tag{4.3.5}$$

and hence as $p \to \infty$

$$\mathbf{y}^{(p)} \to \lambda_1^p \alpha_1 \mathbf{x}_1. \tag{4.3.6}$$

The vector $\mathbf{y}^{(p)}$ therefore tends to the eigenvector $\mathbf{x}_1$ corresponding to the eigenvalue λ_1 of largest modulus. Remembering that eigenvectors are only determined to a multiplicative constant (they still satisfy (4.3.4)), the normalization at each step simply yields a normalized eigenvector. Not only does the process yield an eigenvector but the ratio

$$\frac{||\mathbf{y}^{(p)}||_\infty}{||\mathbf{y}^{(p-1)}||_\infty} \to \lambda_1 \tag{4.3.7}$$

which means that the normalization factor employed at each step tends to the eigenvalue λ_1. It is therefore easy to generate a simple scheme to find the values of λ_1 and $\mathbf{x}_1$. Consider the matrix

$$\mathbf{A} = \begin{pmatrix} 1 & 1 & 3 \\ 1 & -2 & 1 \\ 3 & 1 & 3 \end{pmatrix} \tag{4.3.8}$$

then the scheme in Table 4.1 implements the direct iterative process. The scaled vector is denoted by $\mathbf{z}^{(p)}$ and the normalization factor by k_p.

Hence the eigenvalue is converging towards 5.44 with the corresponding eigenvalue being {0.73, 0.23, 1.0}. What can be said about this process in terms of convergence rates and reliability? The convergence rate follows from (4.3.5) where the dominant term which has to tend to zero is the term λ_2/λ_1 assuming the eigenvalues are ordered according to modulus. It is the rate at which this ratio tends to zero when raised to a power which determines the convergence rate for the algorithm. Hence rapid convergence occurs if the two eigenvalues of largest modulus are well separated in modulus. Hence if λ_2 is a half of λ_1 then the rate of convergence is like $(\frac{1}{2})^p$ which is just over three iterations per decimal place. If all the

Table 4.1

p		$\mathbf{y}^{(p)}$			$\mathbf{z}^{(p)}$		k_p
0	1.0	1.0	1.0	1.0	1.0	1.0	
1	5.0	0.0	7.0	0.714	0.0	1.0	7.0
2	3.714	1.714	5.143	0.722	0.333	1.0	5.143
3	4.054	1.056	5.50	0.737	0.192	1.0	5.50
4	3.929	1.351	5.404	0.727	0.250	1.0	5.404
5	3.976	1.227	5.431	0.732	0.226	1.0	5.431
6	3.973	1.284	5.422	0.730	0.236	1.0	5.422

eigenvalues were complex, considerable problems would arise. Two equal largest modulus eigenvalues could be resolved if these were separate from the third largest modulus because then convergence would result in

$$\mathbf{y}^{(p)} \to \lambda_1^p \alpha_1 \mathbf{x}_1 + \lambda_1^p \alpha_2 \mathbf{x}_2 \tag{4.3.9}$$

and considering the three iterates $\mathbf{y}^{(p)}$, $\mathbf{y}^{(p+1)}$ and $\mathbf{y}^{(p+2)}$ gives sufficient equations to compute the common λ_1. Unfortunately this is a rather spurious exercise as it will not be known whether the required eigenvalues are multiple, nor to what degree. Slow convergence can occur with closely spaced eigenvalues, so switching to the multiple case if convergence is not forthcoming may be premature, and just what level of multiplicity is to be tried is another issue. Hence direct iteration is an effective method in some circumstances when the practical problem can guarantee the required dominance of one eigenvalue.

One improvement to the process of direct iteration is to modify the original matrix. The result is the modified direct iteration method defined by

$$\mathbf{y}^{(p)} = (\mathbf{A} - q\mathbf{I})\mathbf{y}^{(p-1)} \tag{4.3.10}$$

where q is called the modifier. Corresponding to equation (4.3.5) two possibilities arise

$$\mathbf{y}^{(p)} = (\lambda_1 - q)\left\{\alpha_1 \mathbf{x}_1 + \alpha_2 \left(\frac{\lambda_2 - q}{\lambda_1 - q}\right)^p \mathbf{x}_2 + \cdots + \alpha_n \left(\frac{\lambda_n - q}{\lambda_1 - q}\right)^p \mathbf{x}_n\right\} \tag{4.3.11}$$

or if $(\lambda_n - q)$ is the largest of these factors in magnitude then

$$\mathbf{y}^{(p)} = (\lambda_n - q)\left\{\alpha_n \mathbf{x}_n + \alpha_1 \left(\frac{\lambda_1 - q}{\lambda_n - q}\right)^p \mathbf{x}_1 + \cdots + \alpha_{n-1} \left(\frac{\lambda_{n-1} - q}{\lambda_n - q}\right)^p \mathbf{x}_{n-1}\right\} \tag{4.3.12}$$

Now there is a degree of freedom through the choice of the modifier q as to the convergence rate. In the first case the factor

$$\left|\frac{\lambda_2 - q}{\lambda_1 - q}\right| \tag{4.3.13}$$

has to be minimized and in the second case it is the factor

$$\left|\frac{\lambda_{n-1} - q}{\lambda_n - q}\right| \tag{4.3.14}$$

which needs consideration. Again the problem arises that if the eigenvalues were known then the optimum modifiers could be used namely $q = \frac{1}{2}(\lambda_2 + \lambda_n)$ in the first case and $q = \frac{1}{2}(\lambda_1 + \lambda_{n-1})$ in the second. But this begs the question to some extent. However, if q is set to an already computed λ_1 then convergence will be to $\lambda_n - q$, from which the eigenvalue of smallest modulus can be found. The combined processes of direct and modified direct iteration will yield the eigenvalues of largest and smallest magnitude as long as these eigenvalues are separated in modulus from their neighbours.

There is yet a third variation of the process called inverse iteration which is defined by

$$\mathbf{y}^{(p)} = (\mathbf{A} - q\mathbf{I})^{-1}\mathbf{y}^{(p-1)} \tag{4.3.15}$$

and equivalent to equation (4.3.5) is the relation

$$\mathbf{y}^{(p)} = \sum_{r=1}^{n} \alpha_r(\lambda_r - q)^{-p}\mathbf{x}_r \tag{4.3.16}$$

Convergence will now be towards the dominant term in the summation of equation (4.3.16). This term will be the one with λ_r nearest to the modifier q. If the modifier is exactly, or even to machine accuracy, equal to an eigenvalue, what happens? Is the convergence immediate in some sense to the corresponding eigenvector? The answer is effectively, yes. The method will be implemented as a set of linear algebraic equations for each iteration, either solved by Gauss elimination at each stage, or handled by first finding the inverse matrix and using this inverse repeatedly as only the right-hand sides vary from step to step. The linear equations have the form

$$\mathbf{y}^{(p)}(\mathbf{A} - \mathbf{I}q) = \mathbf{y}^{(p-1)} \tag{4.3.17}$$

and if q is an eigenvalue then the equations will be singular, and have rank $n - 1$. Gauss elimination will result in the final step giving a pivot of zero thus blocking the usual process of back-substitution. But the eigenvector required is only determined to a multiplicative constant. The *ratio* of the solution vector components is determined only by the first $n - 1$ equations which are perfectly straightforward. The way forward is to set $x_n = 1$ say and use the first $n - 1$ equations to determine the now-scaled solution vector. The right-hand sides are ignored as these are finite and can be ignored compared with the unscaled x values which are infinite due to the final zero pivot. The result is immediate convergence to the eigenvector corresponding to λ_r. It is correct that the right-hand side should play no part as the result is independent of the initial value $\mathbf{y}^{(0)}$. This method is commonly used to find eigenvectors given an eigenvalue from another method.

In fact the equations obtained in this limiting case are precisely the homogeneous equations which define the eigenvector problem in equation (4.1.1), demonstrating the validity of the process. It is when the modifier is just close, but not equal to an eigenvalue that the iterative process is required.

Continuation with the iteration in Table 4.1 gives the eigenvector $\{0.731, 0.233, 1.0\}$ and eigenvalue 5.425. This eigenvalue will now be employed in an example to compute the eigenvector by inverse iteration. The process of Gauss elimination is used as in Section 3.2, to give the layout:

	−4.425	1	3	1
0.226	1	−7.425	1	1
0.678	3	1	−2.425	1
	−4.425	1	3	1
	0	−7.199	1.678	1.226
0.2331	0	1.678	−0.3911	1.678
	−4.425	1	3	1
	0	−7.199	1.678	1.226
	0	0	-0.47×10^{-7}	1.9638

The pivot at the final step becomes zero to machine accuracy as expected, as seven figure arithmetic was employed in the steps and only four figures displayed. Treating x_3 as unity and using the first two equations with their right-hand sides ignored gives the eigenvector $\{0.730\,639\,81, 0.233\,082\,74, 1\}$ which is in agreement with the results of direct iteration on the same problem in Table 4.1. In fact there is no need to carry a right-hand column at all if the modifier is equal to an eigenvalue to machine accuracy. Obviously if the modifier is not close to an eigenvalue, then repeated iterations of the above type are required until convergence is reached. However this algorithm is most commonly used in the above manner to compute an eigenvector from a given eigenvalue. In this form, it has already been seen that the problem reduces to no more than the solution of the original homogeneous linear equations (4.1.1) for the vector $\mathbf{x}$ with a value of λ known.

So far three methods have been derived. Direct iteration will yield the eigenvalue of maximum modulus, modified direct iteration with the modifier equal to λ_1 yields the eigenvalue of smallest modulus and inverse iteration yields any eigenvector corresponding to a given eigenvalue. The final process which allows all eigenvalues to be computed in theory is the process of deflation. This process yields a matrix of order $n - 1$ which has all the eigenvalues of the original order n matrix $\mathbf{A}$ except λ_1 say. Direct iteration applied to this new deflated matrix will then yield the eigenvalue of largest modulus of this matrix which is the eigenvalue of second largest modulus of the original matrix. By repeated deflation, eigenvalues which have been computed can be eliminated and the next largest found until finally only a simple 2×2 matrix is left. Its eigenvalues can be found as the roots of a quadratic directly from its characteristic equation.

Suppose that an eigenvalue λ_1 is found and satisfies the problem

$$\mathbf{A}\mathbf{x}_1 = \lambda_1 \mathbf{x}_1 \tag{4.3.18}$$

Suppose further that there exists a non-singular matrix $\mathbf{H}$ such that

$$\mathbf{H}\mathbf{x}_1 = k_1 \mathbf{e}_1 \tag{4.3.19}$$

where

$$\mathbf{e}_1 = \begin{bmatrix} 1 \\ 0 \\ 0 \\ \vdots \end{bmatrix} \tag{4.3.20}$$

Then using equation (4.3.18) gives

$$\mathbf{HA}(\mathbf{H}^{-1}\mathbf{H})\mathbf{x}_1 = \lambda_1 \mathbf{H}\mathbf{x}_1 \tag{4.3.21}$$

or

$$\mathbf{HAH}^{-1}\mathbf{e}_1 = \lambda_1 \mathbf{e}_1 \tag{4.3.22}$$

and therefore the first column of the matrix $\mathbf{HAH}^{-1}$ is $\{\lambda_1, 0, 0, \ldots\}^T$ to give

$$\mathbf{HAH}^{-1} = \begin{bmatrix} \lambda_1 & \mathbf{b}_1^T \\ 0 & \mathbf{B} \end{bmatrix} \tag{4.3.23}$$

and as the transformation is a similarity transformation, eigenvalues are preserved and so the eigenvalues of the $(n-1) \times (n-1)$ matrix $\mathbf{B}$ are $\lambda_2, \ldots, \lambda_n$. Equation (4.3.19) gives sufficient information to construct the matrix $\mathbf{H}$. In Gaussian elimination in Section 3.2, zeros were placed below the leading element in successive columns by elementary row operations. These could also be considered as pre-multiplications by a sequence of elementary matrices (3.2.1). Treating the vector $\mathbf{x}_1$ as such a column will yield the requisite elementary operations to reduce $\mathbf{x}_1$ to $k_1\mathbf{e}_1$. These same operations applied to $\mathbf{A}$ will give $\mathbf{HA}$. Applying the equivalent column operations further post-multiplies the matrix by $\mathbf{H}^{-1}$, yielding the completely deflated matrix $\mathbf{HAH}^{-1} = \mathbf{B}$. There is a further simplification in that the final column operation will involve adding a multiple of column 1 to column i for $i = 2, n$. But column 1 will have the form indicated in (4.3.23) and will leave the elements of $\mathbf{B}$ unaffected. Hence this step can be omitted. The resulting deflation algorithm then takes on the simple following form.

(i) Assume $\mathbf{x}_1$ is normalized with maximum element having modulus unity.
(ii) Let the element of maximum modulus in $\mathbf{x}_1$ be in row l.
(iii) Interchange rows 1 and l of $\mathbf{x}_1$, calling the new vector $\mathbf{y}$.
(iv) For $i = 2$ to n
  Replace row i by row $i - (y_i \times$ row 1)

This process is well illustrated by an example with the calculation laid out in the manner of a Gaussian elimination but with the first column being that of the vector $\mathbf{x}_1$. Hence when row operations are applied to $\mathbf{x}$ in the usual way, the corresponding operations can be applied to $\mathbf{A}$ with the equivalent column operations where relevant, to generate $\mathbf{HAH}^{-1}$ as required.

0.731	1	1	3
0.233	1	−2	1
1.000	3	1	3
1.000	3	1	3
0.233	1	−2	1
0.731	1	1	3
1.000	3	1	3
0.233	1	−2	1
0.731	3	1	1
1.000	?	?	?
0.000	?	−2.233	0.3008
0.000	?	0.2694	−1.192

The first step interchanges rows 1 and 3 to put the element of largest modulus in $\mathbf{x}_1$ into row 1. The second step is the corresponding column interchange, and the third and final step carries out the elimination according to step (iv) of the algorithm, only the required elements of the new matrix **B** being computed. Hence the required second-order deflated matrix is

$$\mathbf{B} = \begin{bmatrix} -2.233 & 0.3008 \\ 0.2694 & -1.192 \end{bmatrix} \tag{4.3.24}$$

It can be confirmed that the computational process is in agreement with the algebraic derivation as the operations performed accumulate to yield the transformation matrix

$$\mathbf{H} = \begin{bmatrix} 0 & 0 & 1 \\ 0 & 1 & -0.233 \\ 1 & 0 & -0.731 \end{bmatrix} \tag{4.3.25}$$

and direct multiplying out gives

$$\mathbf{HAH}^{-1} = \begin{bmatrix} 5.426 & 1 & 3 \\ 0 & -2.233 & 0.301 \\ 0 & 0.269 & -1.192 \end{bmatrix} \tag{4.3.26}$$

to confirm the matrix **B** obtained by row operations.

The sequence of methods derived and discussed in this paragraph gives a partial answer as to how eigenvalues and eigenvectors may be computed. The use of direct iteration for the eigenvalue of largest modulus, followed by inverse iteration for the corresponding eigenvector, will allow deflation to be used and the process repeated until only a 2×2 matrix is left. It is also clear that convergence difficulties can occur and there are clear problems with pairs of complex eigenvalues. Attention will now be confined to symmetric matrices in which all the eigenvalues are real.

4.4 Eigenvalues of symmetric matrices

For symmetric matrices the eigenvalues are real, and the computational difficulties are much reduced. A very different approach to the problem than the iterative one used in the previous paragraph, is to ask whether it is possible to construct a similarity transformation which gives a new matrix whose eigenvalues are easy to determine? One such matrix would be the diagonal matrix whose eigenvalues are the diagonal elements themselves. The construction would conceivably consist not of one transformation but of a sequence of transformations which, stage by stage, moulds the matrix into the required form. To make the algorithm reasonably fast the transformation matrices used have to be fairly simple, and they probably need at least one free parameter to allow tuning to the particular operation required. It is also inconvenient to have to compute the inverse of the transformation matrices at every step, so a transformation matrix which is orthogonal and therefore has a trivial inverse is to be encouraged. The first of the methods to be considered in this light is the method of Jacobi (1846). The transformation matrices used are the familiar rotation matrices used for rotating axes in co-ordinate geometry. These matrices are denoted by $\mathbf{R}$ and are defined by

$$\begin{aligned} R_{pp} = R_{qq} &= \cos\theta \qquad & R_{pq} &= -R_{qp} = \sin\theta \\ R_{ii} &= 1 \text{ for } i \neq p, q \qquad & R_{ij} &= 0 \text{ otherwise} \end{aligned} \tag{4.4.1}$$

and clearly these matrices are orthogonal in that $\mathbf{R}^{-1} = \mathbf{R}^T$. Moreover there is a free parameter in the form of θ, not to mention a choice of the plane of rotation defined by p and q. When a general matrix is pre-multiplied by a rotation matrix of the above type then elements in rows p and q are altered. All other rows remain unchanged. Equally a post-multiplication affects only columns p and q. The following equations give the values of the altered rows and columns after the pre- and post-multiplication in the form

$$\mathbf{A}_k = \mathbf{R}_k \mathbf{A}_{k-1} \mathbf{R}_k^T \tag{4.4.2}$$

with the new elements denoted by

$$a_{ip}^{(k)} = a_{ip}^{(k-1)} \cos\theta + a_{iq}^{(k-1)} \sin\theta = a_{pi}^{(k)} \qquad i \neq p, q \tag{4.4.3}$$

$$a_{iq}^{(k)} = -a_{ip}^{(k-1)} \sin\theta + a_{iq}^{(k-1)} \cos\theta = a_{qi}^{(k)} \qquad i \neq p, q \tag{4.4.4}$$

$$a_{pp}^{(k)} = a_{pp}^{(k-1)} \cos^2\theta + 2a_{pq}^{(k-1)} \cos\theta \sin\theta + a_{qq}^{(k-1)} \sin^2\theta \tag{4.4.5}$$

$$a_{qq}^{(k)} = a_{pp}^{(k-1)} \sin^2\theta - 2a_{pq}^{(k-1)} \cos\theta \sin\theta + a_{qq}^{(k-1)} \cos^2\theta \tag{4.4.6}$$

$$a_{pq}^{(k)} = (a_{qq}^{(k-1)} - a_{pp}^{(k-1)})\cos\theta\sin\theta + a_{pq}^{(k-1)}(\cos^2\theta - \sin^2\theta) = a_{qp}^{(k)} \quad (4.4.7)$$

Having chosen a particular off-diagonal element to be made zero, this position is used to fix the values of p and q. Then equation (4.4.7) can be used to specify the value of θ which will ensure that $a_{pq}^{(k)} = a_{qp}^{(k)} = 0$ namely

$$\tan 2\theta = \frac{2a_{pq}^{(k-1)}}{a_{pp}^{(k-1)} - a_{qq}^{(k-1)}} \quad (4.4.8)$$

with $|\theta| \leq \pi/4$. Computational care is needed if $a_{pp}^{(k-1)} = a_{qq}^{(k-1)}$ which would cause overflow in equation (4.4.8), but yields the value $\theta = \pi/4$.

Hence the Jacobi algorithm consists of first choosing the element of largest modulus from the off-diagonal elements. The values of p and q are then fixed. The angle θ is then computed using equation (4.4.8) and the similarity transformation is completed with the update elements (4.4.3) through to (4.4.6). The process is then repeated. It is clear that as the transformation at any step is going to change all the elements in rows and columns p and q, then any of the zeros which have been generated can be lost in later steps when the same row or column is reaffected. Hence the process is not a finite one in that each element only needs to be made zero once. The process is iterative and elements are made zero in one step, later steps destroy this zero, and eventually the element is made zero again, and so on. Hence it is a matter of proof to establish that the process converges, and such a proof should also answer the question as to how rapidly convergence occurs.

Theorem

The Jacobi scheme defined above is such that

$$\mathbf{A}_k \to \text{diag}(\lambda_i) \quad \text{as} \quad k \to \infty \quad (4.4.9)$$

Proof

Let $\mathbf{A}_k = \text{diag}(a_{ii}^{(k)}) + \mathbf{E}_k$ where $\mathbf{E}_k$ is the matrix of off-diagonal elements. As the Jacobi process is a sequence of similarity transformations then it is necessary to prove that $||\mathbf{E}_k||_E \to \mathbf{0}$ as $k \to \infty$. The update equations (4.4.3) to (4.4.7) give

$$\sum_{i \neq p,q} [(a_{ip}^{(k)})^2 + (a_{iq}^{(k)})^2] = \sum_{i \neq p,q} [(a_{ip}^{(k-1)})^2 + (a_{iq}^{(k-1)})^2] \quad (4.4.10)$$

so that the sum of the squares of the off-diagonal elements excepting elements (p, q) and (q, p) is constant. But these are the elements which are made zero in the iterative process. Hence

$$||\mathbf{E}_k||_E^2 = ||\mathbf{E}_{k-1}||_E^2 - 2(a_{pq}^{(k-1)})^2 \quad (4.4.11)$$

But $a_{pq}^{(k-1)}$ is the element of $\mathbf{E}_{k-1}$ of largest modulus and there are $n^2 - n$ off-diagonal elements to give:

$$||\mathbf{E}_k||_E^2 \le \left[1 - \frac{2}{n^2 - n}\right] ||\mathbf{E}_{k-1}||_E^2 \le \left[1 - \frac{2}{n^2 - n}\right]^k ||\mathbf{E}_0||_E^2 \tag{4.4.12}$$

Hence $||\mathbf{E_k}||_E^2 \to \mathbf{0}$ as $k \to \infty$ as required.

As a simple example of this process consider the matrix

$$\mathbf{A}_0 = \begin{bmatrix} 0.6532 & 0.2165 & 0.0031 \\ & 0.4105 & 0.0052 \\ & & 0.2132 \end{bmatrix} \tag{4.4.13}$$

where only the upper triangular elements have been shown to take advantage of the symmetry. Then a rotation in the (1, 2) plane gives

$$\tan 2\theta = \frac{2 \times 0.2165}{0.6532 - 0.4105}$$

with the resulting values $\cos\theta = 0.8628$ and $\sin\theta = 0.5055$. The updated matrix is

$$\mathbf{A}_1 = \begin{bmatrix} 0.7801 & 0 & 0.0053 \\ & 0.2836 & 0.0029 \\ & & 0.2132 \end{bmatrix} \tag{4.4.14}$$

and now a rotation in the plane (1, 3) is used with

$$\tan 2\theta = \frac{2 \times 0.0053}{0.7801 - 0.2132}$$

to give $\cos\theta = 1$, $\sin\theta = 0.0094$ and

$$\mathbf{A}_2 = \begin{bmatrix} 0.7801 & 0 & 0 \\ & 0.2836 & 0.0029 \\ & & 0.2132 \end{bmatrix} \tag{4.4.15}$$

The final transformation is in the (2, 3) plane with

$$\tan 2\theta = \frac{2 \times 0.0029}{0.2836 - 0.2132}$$

and $\cos\theta = 0.9991$ and $\sin\theta = 0.0411$ which gives

$$\mathbf{A}_3 = \begin{bmatrix} 0.7801 & 0 & 0 \\ & 0.2837 & 0 \\ & & 0.2131 \end{bmatrix} \tag{4.4.16}$$

The resulting eigenvalues appear on the diagonal. It is important to note that in this example the off-diagonal elements are quite small to start with, so to the accuracy used, the zeros generated in one step are not affected in later steps. This will not be the case in general. Hence in $\mathbf{A}_2$ the transformation in the $(1, 3)$ plane would affect the previously found zero in position $(1, 2)$ but this is not significant to working accuracy because θ is close to zero.

There are some variations on strategy which can be used with this method. In the above approach the element with the largest off-diagonal modulus is made zero at each stage. Hence the 'worst' element is knocked out at each stage. Some attempt to reduce the searching required to determine this largest off-diagonal element in modulus occurs in the following alternatives.

The first is called the serial method. Here the elements are made zero in order row by row to complete a series of sweeps. In a sweep, every off-diagonal element is zeroed at least once. Convergence will be less rapid, inevitably.

The second is called the threshold method. Here sweeps are completed as in the previous case, but elements are only made zero if their modulus is greater than some current threshold. Once the current threshold is reached by all elements, it is itself reduced another stage and the process continued. A typical set of thresholds would be 10^{-1}, 10^{-2}, 10^{-4}, 10^{-8} and so on.

In order to assess the effectiveness of Jacobi's method, it is necessary to look at the convergence rate predicted by equation (4.4.12) of the proof. The rate of convergence depends on how rapidly

$$\left[1 - \frac{2}{n^2 - n}\right]^k \tag{4.4.17}$$

goes to zero as $k \to \infty$. The larger the order n of the matrix, the slower is the rate of convergence. Also the larger the value of n, the larger the workload to complete one sweep. Indeed a function count establishes that one sweep involves the order of $2n^3$ multiplications and divisions as well as $n^2/2$ square roots. Hence the method generates all the eigenvalues of a symmetric matrix, and the accumulated product of the rotation matrices will give the corresponding eigenvectors. But for moderate to large n, convergence will be very slow. The method could be described as robust but slow for larger problems. The methods which will be described next attempt to get round this problem of speed by removing the iterative nature of Jacobi's method. There is the inevitable cost to pay that the resulting matrix will turn out not to be the convenient diagonal form, and some further work will be needed to finally extract the eigenvalues themselves.

The first of the methods to be considered here is due to Givens (1954). Like Jacobi's the method is based on the use of rotation matrices to introduce zeros into a matrix but preserve eigenvalues. However Givens uses a different strategy which does not cause the destruction of zeros once they are found. The price to be paid for this refinement is that the matrix will only be reduced to tri-diagonal form and further work is needed to extract the eigenvalues from this form. The Givens strategy depends on the fact that the transformation defined by (4.4.2) and the update rules (4.4.3) to (4.4.7) causes alteration of all elements in rows and columns p and q. There is no reason that it should be the element (p, q) which is made zero by a rotation in the (p, q) plane. Givens suggests that a zero is placed in position (p, q) using a transformation in the plane $(p, q + 1)$. In other words the rotation is one column ahead of the column in which the zero is being placed. Hence as zeros are placed in this column, the transformations never affect the zeros previously found. Not only this, as the process moves to the next column of zeros, the elements in the previous column will be changed, but these will be combinations of elements with trigonometric functions, and all the elements will be zero, so yielding zero. Hence the whole process becomes finite and no iterative activity is required.

The snag is that to put a zero in say position (2, 1) requires a transformation in (2, 2) which does not exist. This is true for any sub- or super-diagonal element. Hence the resulting matrix will be tri-diagonal. The values of θ follow from equation (4.4.4) in which the element $(p, q + 1)$ is made zero. The following algorithm is then immediate:

There are $n - 2$ major steps of which the rth places zeros in column r and involves:

for $i = r + 2$ to n do

(i) Compute $x = (a_{r,r+1}^2 + a_{ri}^2)^{1/2}$

(ii) $\cos\theta = a_{r,r+1}/x$ and $\sin\theta = a_{ri}/x$
if $x = 0$ then $\cos\theta = 1$ and $\sin\theta = 0$.

(iii) for $j = r + 1$ to n do
Compute $a_{j,r+1}\cos\theta + a_{ji}\sin\theta$ to give w_1
Compute $-a_{j,r+1}\sin\theta + a_{ji}\cos\theta$ to give w_2
Replace $a_{j,r+1}$ by w_1 and a_{ji} by w_2

(iv) for $j = r + 1$ to n do
Compute $a_{r+1,j}\cos\theta + a_{ij}\sin\theta$ to give w_1
Compute $-a_{r+1,j}\sin\theta + a_{ij}\cos\theta$ to give w_2
Replace $a_{r+1,j}$ by w_1 and a_{ij} by w_2.

This process uses only $\frac{4}{3}n^3$ multiplications and divisions with $\frac{1}{2}n^2$ square roots, with the obvious gain in speed over even one sweep of Jacobi's method.

To demonstrate one step of this algorithm, consider the partially processed matrix

$$\mathbf{A} = \begin{bmatrix} 3 & 1 & 0 & 0 \\ 1 & -2 & 0.1 & 0.1 \\ 0 & 0.1 & 4 & 0.1 \\ 0 & 0.1 & 0.1 & -5 \end{bmatrix} \tag{4.4.18}$$

The final zero is now required in position (4, 2) and therefore a rotation is required in plane (4, 3). This transformation will now be applied from first principles to illustrate the formulae in the algorithm. Hence the transformation gives:

$$\begin{bmatrix} 1 & 0 & 0 & 0 \\ 0 & 1 & 0 & 0 \\ 0 & 0 & c & -s \\ 0 & 0 & s & c \end{bmatrix} \begin{bmatrix} 3 & 1 & 0 & 0 \\ 1 & -2 & 0.1 & 0.1 \\ 0 & 0.1 & 4 & 0.1 \\ 0 & 0.1 & 0.1 & -5 \end{bmatrix} \begin{bmatrix} 1 & 0 & 0 & 0 \\ 0 & 1 & 0 & 0 \\ 0 & 0 & c & s \\ 0 & 0 & -s & c \end{bmatrix} \tag{4.4.19}$$

where c represents $\cos\theta$, and s, $\sin\theta$. Multiplying out gives

$$\begin{bmatrix} 3 & 1 & 0 & 0 \\ 1 & -2 & 0.1c-0.1s & 0.1s+0.1c \\ 0 & 0.1c-0.1s & 4c^2-0.2sc-5s^2 & 0.1c^2+9sc-0.1s^2 \\ 0 & 0.1s+0.1c & 0.1c^2+9sc-0.1s^2 & -5c^2+0.2sc+4s^2 \end{bmatrix} \tag{4.4.20}$$

and so the element $0.1s + 0.1c$ is made zero if $s = 1/\sqrt{2}$ and $c = -1/\sqrt{2}$ to give the final tri-diagonal matrix

$$\begin{bmatrix} 3 & 1 & 0 & 0 \\ 1 & -2 & -1/5\sqrt{2} & 0 \\ 0 & -1/5\sqrt{2} & -2/5 & -9/2 \\ 0 & 0 & -9/2 & -3/5 \end{bmatrix} \tag{4.4.21}$$

It can be confirmed that the eigenvalues of both **A** and the above tri-diagonal matrix are correctly {3.192 17, −2.190 82, 4.003 15, −5.004 51}.

It was at one stage thought that this process due to Givens would be the most efficient technique for the reduction of a symmetric matrix to tri-diagonal form. However a further technique was developed which halves the operation count to $\frac{2}{3}n^3$ multiplications and divisions. This is Householder's method (1958), and uses elementary Hermitian matrices rather than rotation matrices for the similarity transformation. These matrices are defined by

$$\mathbf{P}_r = \mathbf{I} - 2\mathbf{w}\mathbf{w}^T \tag{4.4.22}$$

with the vectors $\mathbf{w}$ satisfying $\mathbf{w}^T\mathbf{w} = 1$. It is easy to show that $\mathbf{P}_r = \mathbf{P}_r^T = \mathbf{P}_r^{-1}$.

In general the leading elements of the vector $\mathbf{w}$ are zero and it is convenient to write the vector of non-zero elements as $\mathbf{v}$ so that

$$\mathbf{P}_r = \left[\begin{array}{c:c} \mathbf{I} & \mathbf{0} \\ \hdashline \mathbf{0} & \mathbf{I} - 2\mathbf{v}_r\mathbf{v}_r^T \end{array}\right] \tag{4.4.23}$$

The idea in Householder's method is to fix the elements of the vector $\mathbf{v}_r$ so that all elements below the sub-diagonal of the current column are made zero, and this is illustrated in block form as

$$\mathbf{A}_r = \mathbf{P}_r\mathbf{A}_{r-1}\mathbf{P}_r = \left[\begin{array}{c:c:c} \multicolumn{2}{c:}{\mathbf{C}_{r-1}} & \mathbf{0} \\ \cdots & \cdots & \mathbf{c}_r^T \\ \hdashline \mathbf{0} & \mathbf{c}_r & \mathbf{Q}_r\mathbf{B}_{r-1}\mathbf{Q}_r \end{array}\right] \tag{4.4.24}$$

where

$$\mathbf{A}_{r-1} = \left[\begin{array}{ccc:c} & \mathbf{C}_{r-1} & & \mathbf{0} \\ & & & \mathbf{b}_{r-1}^T \\ \hdashline \mathbf{0} & \vdots & \mathbf{b}_{r-1} & \mathbf{B}_{r-1} \end{array}\right] \tag{4.4.25}$$

with $\mathbf{Q}_r = \mathbf{I} - 2\mathbf{v}_r\mathbf{v}_r^T$ and $\mathbf{c}_r = \mathbf{Q}_r\mathbf{b}_{r-1}$. The object is now to force the vector $\mathbf{c}_r$ to have null elements excepting the first. This results in a little sub-problem whose solution can be applied to each step in the above process as required. The sub-problem is that given a vector $\mathbf{x}$, the matrix $\mathbf{P}$ is required which forces the reduction

$$\mathbf{Px} = k\mathbf{e}_1 \tag{4.4.26}$$

Taking second norms of both sides gives $k = \pm S$ with $S^2 = x_1^2 + \cdots + x_n^2$. Comparing like elements then gives the equations

$$x_1 - 2w_1(\mathbf{w}^T\mathbf{x}) = \pm S \tag{4.4.27}$$

$$x_i - 2w_i(\mathbf{w}^T\mathbf{x}) = 0 \qquad i = 2, n \tag{4.4.28}$$

and then writing $K = \mathbf{w}^T\mathbf{x}$, squaring, adding and dividing by two gives

$$2K^2 = s^2 \mp x_1S \tag{4.4.29}$$

$$\mathbf{P} = \mathbf{I} - \frac{\mathbf{uu}^T}{2K^2} \tag{4.4.30}$$

where $\mathbf{u}^T = (x_1 \mp S, x_2, \ldots, x_n)$. Hence the elements of the vector $\mathbf{u}$ and therefore those of the vector $\mathbf{v}$ are known. The result for the complete matrix is then the following set of rules for the rth step of Householder's process. There will be again $n - 2$ such steps, of which the rth is:

$$\begin{aligned} u_{ir} &= 0 \qquad i = 1, 2, \ldots, r \\ S_r &= \left(\sum_{i=r+1}^{n} a_{ri}^2\right)^{1/2} \\ u_{ir} &= a_{ri} \qquad i = r+2, \ldots, n \\ u_{r+1,r} &= a_{r,r+1} \mp S_r \\ 2K_r^2 &= S_r^2 \mp a_{r,r+1}S_r \end{aligned} \tag{4.4.31}$$

and

$$\mathbf{P}_r = \mathbf{I} - \frac{\mathbf{u}_r\mathbf{u}_r^T}{2K_r^2} \tag{4.4.32}$$

The sign ambiguity in the formula is resolved by a choice which avoids cancellation, so that if $a_{r,r+1}$ is positive, a positive sign is used for S_r, and the opposite for negative values.

The final step of multiplying out the product $\mathbf{P}_r\mathbf{A}_{r-1}\mathbf{P}_r$ is efficiently performed by using symmetry to advantage. Firstly define the vector $\mathbf{p}_r$ by

$$\mathbf{p}_r = \frac{\mathbf{A}_{r-1}\mathbf{u}_r}{2K_r^2} \tag{4.4.33}$$

and the vector $\mathbf{q}_r$ by

$$\mathbf{q}_r = \mathbf{p}_r - \tfrac{1}{2}\mathbf{u}_r(\mathbf{u}_r^T\mathbf{p}_r/2K_r^2) \tag{4.4.34}$$

then

$$\mathbf{A}_r = \mathbf{A}_{r-1} - \mathbf{u}_r\mathbf{q}_r^T - \mathbf{q}_r\mathbf{u}_r^T \tag{4.4.35}$$

and the multiplying out process has been reduced to a sequence of inner products.

To demonstrate Householder's method, consider the first step in the reduction of the matrix

$$\begin{bmatrix} 4 & 1 & 2 & 3 \\ 1 & 5 & 1 & 1 \\ 2 & 1 & 2 & 3 \\ 3 & 1 & 3 & 1 \end{bmatrix} \tag{4.4.36}$$

to tri-diagonal form. Then the sub-problem is to reduce the vector $(1, 2, 3)$ to $(k, 0, 0)$ so that

$$\mathbf{P}\begin{bmatrix} 1 \\ 2 \\ 3 \end{bmatrix} = \begin{bmatrix} k \\ 0 \\ 0 \end{bmatrix} \tag{4.4.37}$$

Then $k = \pm S = \pm\sqrt{14}$ and equations (4.4.27) and (4.4.28) give

$$\begin{aligned} 2Kw_1 &= 1 \mp \sqrt{14} \\ 2Kw_2 &= 2 \\ 2Kw_3 &= 3 \end{aligned} \tag{4.4.38}$$

which square, add and divide by two to give $2K^2 = 14 \mp \sqrt{14}$ with $\mathbf{u}^T = (1 \mp \sqrt{14}, 2, 3)$. Hence the resulting matrix $\mathbf{P}$ is

$$\begin{bmatrix} 1 & 0 & 0 & 0 \\ 0 & 1 - \frac{(1+\sqrt{14})^2}{14+\sqrt{14}} & \frac{2(1+\sqrt{14})}{14+\sqrt{14}} & \frac{3(1+\sqrt{14})}{14+\sqrt{14}} \\ 0 & \frac{2(1+\sqrt{14})}{14+\sqrt{14}} & 1 - \frac{4}{14+\sqrt{14}} & \frac{6}{14+\sqrt{14}} \\ 0 & \frac{3(1+\sqrt{14})}{14+\sqrt{14}} & \frac{6}{14+\sqrt{14}} & 1 - \frac{9}{14+\sqrt{14}} \end{bmatrix} \tag{4.4.39}$$

The multiplying out can then be completed using the formula from equation (4.4.34).

The methods of Givens and Householder give efficient ways of reducing a symmetric matrix to tri-diagonal form. It can be shown that the resulting tri-diagonal forms only differ in certain elements which have opposite signs, but still result in the same characteristic equation. The problem remains as to how to determine the eigenvalues of the reduced tri-diagonal form. This problem is solved by using Sturm Sequences which are now discussed. Some of the underlying theorems are assumed in this work.

The resulting tri-diagonal matrix is assumed to have the form

$$\mathbf{C} = \begin{bmatrix} \alpha_1 & \beta_2 & & & & \\ \beta_2 & \alpha_2 & \beta_3 & & & \\ & \beta_3 & \alpha_3 & & & \\ & & & \ddots & & \\ & & & & & \beta_n \\ & & & & \beta_n & \alpha_n \end{bmatrix} \tag{4.4.40}$$

If in this matrix any of the β_i are zero, then clearly the matrix will split into two separate tridiagonal matrices which together have the eigenvalues of the original (just look at the characteristic equations). Hence it is reasonable to assume that none of the β_i are zero, and this has a theoretical implication later. The leading principle minors of $\det(\mathbf{C} - \mu\mathbf{I})$ of order r can be denoted by $p_r(\mu)$ and are given by

$$\begin{aligned} p_0(\mu) &= 0, \quad \text{say} \\ p_1(\mu) &= \alpha_1 - \mu \\ p_i(\mu) &= (\alpha_i - \mu)p_{i-1}(\mu) - \beta_i^2 p_{i-2}(\mu) \quad i = 2, n \end{aligned} \tag{4.4.41}$$

The p_i's are easily computed from the matrix $\mathbf{C}$ and in particular the zeros of $p_n(\mu)$ are the eigenvalues of $\mathbf{C}$. There is a Separation theorem which states that the eigenvalues of the leading principal sub-matrix of order r of $\mathbf{C}$, say $\mathbf{C}_r$, strictly separate the eigenvalues of $\mathbf{C}_{r+1}$ if no β_i is zero, Wilkinson(1965). The method for computing the eigenvalues depends on the very important Sturm Sequence Property which is defined by the following theorem.

Theorem

Let $p_0(\mu), \ldots, p_n(\mu)$ be found for some μ. Then $s(\mu)$, the number of agreements in sign of consecutive members of this sequence, is the number of eigenvalues of $\mathbf{C}$ which are strictly greater than μ. If $p_r(\mu) = 0$ then the convention is to take its sign to be opposite to that of $p_{r-1}(\mu)$.

Proof

Assume that the number of agreements of sign, s, in the sequence $p_0, \ldots, p_r$ is the number of eigenvalues of $\mathbf{C}_r > \mu$. Consideration will then be given to the relationship between the eigenvalues $x_1, \ldots, x_r$ of $\mathbf{C}_r$ and the eigenvalues $y_1, \ldots, y_{r+1}$ of $\mathbf{C}_{r+1}$ to see how many extra agreements occur under the various possibilities. The eigenvectors $\mathbf{x}_1, \ldots, \mathbf{x}_r$ can be ordered in relation to μ to give

$$x_1 > x_2 > \cdots > x_s > \mu \geq x_{s+1} > x_{s+2} > x_r \tag{4.4.42}$$

No eigenvalues are coincident since the super-diagonal elements are non-zero. The separation theorem says that the eigenvalues of $\mathbf{C}_{r+1}$ strictly separate those of $\mathbf{C}_r$ and hence

$$y_1 > x_1 > y_2 > x_2 > \cdots > y_s > x_s > y_{s+1} > x_{s+1} > \cdots > y_r > x_r > y_{r+1} \tag{4.4.43}$$

Comparing (4.4.42) and (4.4.43) it is clear that μ must lie between x_s and x_{s+1}, and may be either side of y_{s+1} or identical with it. Hence either s or $s+1$ eigenvalues of $\mathbf{C}_{r+1}$ are $> \mu$ and three possibilities arise.

Firstly $p_r(\mu)$ and $p_{r+1}(\mu)$ can be written as

$$p_r(\mu) = \prod_{i=1}^{r} (x_i - \mu) \tag{4.4.44}$$

$$p_{r+1}(\mu) = \prod_{i=1}^{r+1} (y_i - \mu) \tag{4.4.45}$$

Now if $\mu \neq x_i$ or y_i for any i, then if $y_{s+1} > \mu$ then $p_r(\mu)$ and $p_{r+1}(\mu)$ have the same sign and correctly there is one more agreement, as there is one extra eigenvalue greater than μ.

If on the other hand $y_{s+1} < \mu$ then $p_r(\mu)$ and $p_{r+1}(\mu)$ have different signs and so correctly there remain s eigenvalues strictly greater than μ.

The case remaining is when $x_{s+1} = \mu$ when the sign associated with $p_r(\mu)$ is opposite to that of $p_{r-1}(\mu)$. But zeros of p_{r-1}, p_r and p_{r+1} are strictly separated. Hence p_{r+1} must have exactly one extra zero greater than μ and one less than μ. Hence p_{r+1} is of opposite sign to p_{r-1} and therefore of the same sign as p_r. Hence correctly there is one more zero $> \mu$, so completing the proof.

The final stage of the eigenvalue finding process for a symmetric matrix is to use the Sturm sequence property to extract the eigenvalues from the tri-diagonal matrices which the processes of Givens and Householder yield. This step uses the very basic bisection method discussed in Chapter 1 as an algorithm.

The process allows the kth eigenvalue to be located independently of the others to any degree of accuracy in a predeterminable number of steps. We know from the study of norms in Chapter 3, (3.1.26) that

$$\lambda \leq ||\mathbf{A}|| \tag{4.4.46}$$

and so the set of eigenvalues can be bounded above and below by

$$\max_i \{|\beta_i| + |\alpha_i| + |\beta_{i+1}|\} \tag{4.4.47}$$

using the definition of either $|| \ ||_1$ or $|| \ ||_\infty$. Hence let $b_0 > a_0$ be such that $\lambda_k \in (a_0, b_0)$ and hence $s(a_0) \geq k$ and $s(b_0) < k$. Further let c_1 be the point of bisection $(a_0 + b_0)/2$, then $s(c_1)$ will allow the decision to made as to which half-interval λ lies in.

$$\text{if } s(c_1) \leq k \text{ then } \lambda_k \in (c_1, b_0)$$

$$\text{if } s(c_r) < k \text{ then } \lambda_k \in (a_0, c_1)$$

In this way the length of the containing interval can be reduced until the required accuracy is obtained. After m steps the interval length will be $(b_0 - a_0)/2^m$, and hence m is chosen sufficiently large to force this interval length to be less than the required accuracy. Almost one extra decimal figure of accuracy is obtained for every three iterations.

The process can be illustrated in its early stages using the tri-diagonal result from the example on Givens' method in matrix (4.4.21), namely

$$\begin{bmatrix} 3 & 1 & 0 & 0 \\ 1 & -2 & -1/5\sqrt{2} & 0 \\ 0 & -1/5\sqrt{2} & -2/5 & -9/2 \\ 0 & 0 & -9/2 & -3/5 \end{bmatrix} \tag{4.4.48}$$

For this matrix $||\mathbf{C}||_1 = 5.1$ so that $\lambda \in (-5.1, 5.1)$. The Sturm sequences are

$$\begin{aligned} p_0(\mu) &= 1 \\ p_1(\mu) &= 3 - \mu \\ p_2(\mu) &= (-2 - \mu)(3 - \mu) - 1 = \mu^2 - \mu - 7 \\ p_3(\mu) &= (-\tfrac{2}{5} - \mu)(\mu^2 - \mu - 7) - \tfrac{1}{50}(3 - \mu) \\ &= -\mu^3 + 0.6\mu^2 + 7.42\mu + 2.74 \\ p_4(\mu) &= \mu^4 - \frac{2803}{100}\mu^2 + \frac{6529}{500}\mu + \frac{70053}{500} \\ &= \mu^4 - 28.03\mu^2 + 13.058\mu + 140.106 \end{aligned} \tag{4.4.49}$$

Some values of $s(\mu)$ are given in Table 4.2, from which it is clear that $\lambda_1 \in (-4, -5.1)$, $\lambda_2 \in (-2, -3)$, $\lambda_3 \in (3, 4)$ and $\lambda_4 \in (4, 5)$. Each interval is then bisected and the evaluation of the $s(\mu)$ continued to yield the eigenvalues $-5.004\,51$, $-2.190\,82$, $3.192\,17$ and $4.003\,15$.

In summary, the eigenvalues of a symmetric matrix can be found by Jacobi's method and a reduction to diagonal form as long as the matrix is not too big in order. For large matrices the arithmetic load and the rate of convergence become prohibitive, and the more economical and non-iterative methods of Givens and Householder come into their own. Now the reduction is limited to a tri-diagonal

Table 4.2

$s(\mu)$	p_0	p_1	p_2	p_3	p_4	s
$s(0)$	+	+	−	+	+	2
$s(2)$	+	+	−	+	+	2
$s(-2)$	+	+	−	+	+	2
$s(3)$	+	0	−	+	+	2
$s(-3)$	+	+	+	+	−	3
$s(4)$	+	−	+	−	−	1
$s(-4)$	+	+	+	+	−	3

form and the eigenvalues of this form are elegantly and robustly found using Sturm sequences and the bisection method. It is the property that the eigenvalues are real which makes these approaches feasible. In the non-symmetric case, complex eigenvalues are possible and the processes become more difficult.

4.5 Eigenvalues of a non-symmetric matrix

In the non-symmetric case, only limited progress can be made using transformations such as Givens and Householder in a manner analogous to the symmetric problems of the previous paragraph. Clearly only the elements below the sub-diagonal will be made zero and hence the reduced matrices will be in upper Hessenberg form. The eigenvalues will be less easy to find from these forms than from the tri-diagonal matrices of the symmetric case. The algorithm for Givens method is slightly different from that in the symmetric case namely:
There are $n-2$ major steps of which the rth places zeros in column r and involves:

for $i = r+2$ to n do
- (i) Compute $x = (a_{r+1,r}^2 + a_{ir}^2)^{1/2}$
- (ii) $\cos\theta = a_{r+1,r}/x$ and $\sin\theta = a_{ir}/x$
  if $x = 0$ then $\cos\theta = 1$ and $\sin\theta = 0$.
- (iii) for $j = r+1$ to n do
  Compute $a_{r+1,j}\cos\theta + a_{ij}\sin\theta$ to give w_1
  Compute $-a_{r+1,j}\sin\theta + a_{ji}\cos\theta$ to give w_2
  Replace $a_{r+1,j}$ by w_1 and a_{ij} by w_2
- (iv) for $j = 1$ to n do
  Compute $a_{j,r+1}\cos\theta + a_{ji}\sin\theta$ to give w_1
  Compute $-a_{j,r+1}\sin\theta + a_{ji}\cos\theta$ to give w_2
  Replace $a_{j,r+1}$ by w_1 and a_{ji} by w_2.

Other than symmetry considerations the difference lies in step (iv) where the loop now runs from 1 to n to account for the non-zero elements in the upper right-hand half of the matrix.

Similarly Householder's method can be employed with the following algorithm. There will be again $n - 2$ such steps, with

$$\begin{aligned} u_{ir} &= 0 \qquad i = 1, 2, \ldots, r \\ S_r &= \left(\sum_{i=r+1}^{n} a_{ir}^2 \right)^{1/2} \\ u_{ir} &= a_{ir} \qquad i = r+2, \ldots, n \\ u_{r+1,r} &= a_{r+1,r} \mp S_r \\ 2K_r^2 &= S_r^2 \mp a_{r+1,r} S_r \end{aligned} \tag{4.5.1}$$

and

$$\mathbf{P}_r = \mathbf{I} - \frac{\mathbf{u}_r \mathbf{u}_r^T}{2K_r^2} \tag{4.5.2}$$

The sign ambiguity in the formula is again resolved by a choice which avoids cancellation, so that if $a_{r,r+1}$ is positive, a positive sign is used for S_r, and the opposite for negative values.

As an example of the Givens process, consider the matrix

$$\begin{bmatrix} 5 & -2 & 1 \\ 1 & 1 & 3 \\ 1 & 4 & -1 \end{bmatrix} \tag{4.5.3}$$

then only the element (3, 1) requires transforming to reduce this matrix to upper Hessenberg form. The transformation is

$$\begin{bmatrix} 1 & 0 & 0 \\ 0 & c & s \\ 0 & -s & c \end{bmatrix} \begin{bmatrix} 5 & -2 & 1 \\ 1 & 1 & 3 \\ 1 & 4 & -1 \end{bmatrix} \begin{bmatrix} 1 & 0 & 0 \\ 0 & c & -s \\ 0 & s & c \end{bmatrix} \tag{4.5.4}$$

which multiplies out to give

$$\begin{bmatrix} 5 & -2c + s & 2s + c \\ c + s & c^2 + 7cs - s^2 & -4s^2 - 2sc + 3c^2 \\ -s + c & 4c^2 - 2cs - 3s^2 & s^2 - 7cs - c^2 \end{bmatrix} \tag{4.5.5}$$

where $c = \cos\theta$ and $s = \sin\theta$, and hence the choice of θ to force element (3, 1) to zero is that $s = c$ or $s = c = 1/\sqrt{2}$. The resulting upper Hessenberg form is

$$\begin{bmatrix} 5 & -1/\sqrt{2} & 3/\sqrt{2} \\ 2/\sqrt{2} & 7/2 & -3/2 \\ 0 & -1/2 & -7/2 \end{bmatrix} \tag{4.5.6}$$

which has the same characteristic equation namely

$$-\lambda^3 + 5\lambda^2 + 12\lambda - 70 = 0 \tag{4.5.7}$$

as the original matrix.

As an example of the Householder process in the non-symmetric case, consider the matrix

$$\begin{bmatrix} 4 & 1 & 3 \\ 2 & -1 & 4 \\ 6 & 1 & 3 \end{bmatrix} \tag{4.5.8}$$

then in terms of elementary Hermitian matrices the transformation is now

$$\begin{bmatrix} 1 & 0 & \\ 0 & 1-2w_1^2 & -2w_1w_2 \\ 0 & -2w_1w_2 & 1-2w_2^2 \end{bmatrix} \begin{bmatrix} 4 & 1 & 3 \\ 2 & -1 & 4 \\ 6 & 1 & 3 \end{bmatrix} \begin{bmatrix} 1 & 0 & \\ 0 & 1-2w_1^2 & -2w_1w_2 \\ 0 & -2w_1w_2 & 1-2w_2^2 \end{bmatrix} \tag{4.5.9}$$

On multiplying out the element in position (3, 1) becomes

$$-4w_1w_2 + 6 - 12w_2^2 = 0 \tag{4.5.10}$$

which together with the condition $w_1^2 + w_2^2 = 1$ gives two equations for w_1 and w_2 which satisfy

$$40w^4 - 40w^2 + 9 = 0 \tag{4.5.11}$$

yielding

$$\begin{aligned} w_1^2 &= 1/2 \pm \sqrt{10}/20 \\ w_2^2 &= 1/2 \mp \sqrt{10}/20 \end{aligned} \tag{4.5.12}$$

and $w_1w_2 = \mp 3\sqrt{10}/20$. Hence the transformation matrix $\mathbf{P}$ is

$$\begin{bmatrix} 1 & 0 & 0 \\ 0 & \mp\sqrt{10}/10 & \pm 3\sqrt{10}/10 \\ 0 & \pm 3\sqrt{10}/10 & \pm\sqrt{10}/10 \end{bmatrix} \tag{4.5.13}$$

in agreement with the formulae of the algorithm.

There is no Sturm sequence equivalent in the non-symmetric case, so the next stage is to find the eigenvalues directly from the upper Hessenberg forms which have been found. If the value of $\det(\mathbf{A} - \mathbf{I}z)$ could be efficiently found for a given z with $\mathbf{A}$ upper Hessenberg, then setting the determinant to zero would give a non-linear equation, and a method akin to those in Chapter 1 could be used. Clearly care would be needed with complex roots and equal roots, but in principle, if the determinant above can be evaluated for a given z then the required roots are determinable. If in addition the derivative of the determinant with respect to z can be found then a Newton method would become feasible. Hyman's method (1957) gives just such an algorithm.

Let $\mathbf{P} = \mathbf{A} - z\mathbf{I}$ and let the columns of $\mathbf{P}$ be $(\mathbf{p}_1, \ldots, \mathbf{p}_n)$. The principle of the method is very easy. Because $\mathbf{A}$ is upper Hessenberg, the last two columns can be used to place a zero into element (n, n) by subtracting a suitable multiple of the last but one column from the last. Similarly column $n-2$ can be used to place a zero in the last column in position $n-1, n$, and the zero below element $(n-1, n-2)$

will ensure that the previously zeroed element remains zero. Clearly the process can be repeated until the last column has only its first element non-zero. It is then trivial to multiply out the resulting matrix by the last column, as the bottom left-hand sub-matrix is diagonal (because the whole of **A** is upper Hessenberg). Hence the determinant of **P** is found for a given z. Suppose therefore that the multiplying factors for the elimination are $(x_1, \ldots, x_{n-1})$ then we require

$$x_1\mathbf{p}_1 + x_2\mathbf{p}_2 + \cdots + x_{n-1}\mathbf{p}_{n-1} + \mathbf{p}_n = k(z)\mathbf{e}_1 \tag{4.5.14}$$

with $\mathbf{e}_1$ being the usual vector $(1, 0, 0, \ldots, 0)^T$. Then the matrix **Q** obtained from **P** by these elementary column operations has the same determinant as **P** and expanding by the last column gives

$$\det(\mathbf{A} - z\mathbf{I}) = \det\mathbf{P} + \det\mathbf{Q} = \pm k(z)(a_{21}a_{32}\cdots a_{n,n-1}) \tag{4.5.15}$$

Renaming the crucial sub-diagonal elements as $a_{i+1,i} = b_{i+1}$ and equating elements in (4.5.14) beginning with the nth, gives the recurrence relations

$$b_r x_{r-1} + (a_{rr} - z)x_r + a_{r,r+1}x_{r+1} + \cdots + a_{rn}x_n = 0 \tag{4.5.16}$$

for $r = n, \ldots, 2$ and $x_n = 1$ for completeness.

Comparing the first elements gives

$$(a_{11} - z)x_1 + a_{12}x_2 + \cdots + a_{1n}x_n = k(z) \tag{4.5.17}$$

As $k(z)$ differs from the required determinant by only a constant it is only necessary to find the zeros of $k(z)$ itself. In order to apply a Newton method to this problem it is necessary to have the derivatives $k'(z)$ and realizing that the recurrence relations in (4.5.16) have x_i's which depend on z, means that these relations can be differentiated with respect to z to give

$$b_r x'_{r-1} + (a_{rr} - z)x'_r + a_{r,r+1}x'_{r+1} + \cdots + a_{rn}x'_n - x_r = 0 \tag{4.5.18}$$

with

$$(a_{11} - z)x'_1 + a_{12}x'_2 + \cdots + a_{1n}x'_n - x_1 = k'(z) \tag{4.5.19}$$

and now $x'_n = 0$.

Finally an iteration of the form

$$z^{m+1} = z^{(m)} - \frac{k(z^{(m)})}{k'(z^{(m)})} \tag{4.5.20}$$

is used to recover the eigenvalues.

The processes involved in this method can be seen by considering the matrix

$$\begin{bmatrix} 8 & 1 & -2 \\ 3 & -1 & 4 \\ 0 & 2 & -4 \end{bmatrix} \tag{4.5.21}$$

which is in upper Hessenberg form. The process will be carried out for the starting value $z^{(0)} = 8$. Either the recurrence relations of equations (4.5.16) and (4.5.17) can be used, or from first principles by considering the determinant of the matrix:

$$\mathbf{A} - 8\mathbf{I} = \begin{bmatrix} 0 & 1 & -2 \\ 3 & -9 & 4 \\ 0 & 2 & -12 \end{bmatrix} \tag{4.5.22}$$

To reduce the final column to $(k, 0, 0)^T$, requires from the last row that the second column be multiplied by six and added to the last to yield:

$$\begin{bmatrix} 0 & 1 & 4 \\ 3 & -9 & -50 \\ 0 & 2 & 0 \end{bmatrix} \tag{4.5.23}$$

and then moving backwards to the second row, the first column needs to be multiplied by 50/3 and added to the last to give the values

$$x_1 = 50/3, \quad x_2 = 6, \quad x_3 = 1 \text{ and } k = 4 \tag{4.5.24}$$

For the derivative, k' it is necessary to use the recurrence relations derived in equations (4.5.18) and (4.5.19) which give

$$2x_2' - 12x_3' - 1 = 0 \tag{4.5.25}$$

But $x_3' = 0$, and hence $x_2' = \frac{1}{2}$. The final relation then gives $x_2' - 50/3 = k' = -97/6$. Then one application of Newton's method in the form of equation (4.5.20) gives

$$z^{(1)} = z^{(0)} - \frac{k(z^{(0)})}{k'(z^{(0)})} = 8 + \frac{24}{97} = 8.25 \tag{4.5.26}$$

This demonstrates good convergence, the true eigenvalue being 8.2353. Successive iterations along the above lines will rapidly yield high accuracy.

It is when equal roots and complex roots arise that this process will prove less than satisfactory. A somewhat different approach is needed to overcome these shortcomings and this is provided by the **LR** and **QR** processes which form the final algorithms of this chapter.

4.6 QR and LR algorithms

These algorithms are different in nature from the methods treated so far, but do rely on similarity transformations. The principle of the methods is easy to see, but both the refinements necessary to speed convergence and the proof of convergence are more involved. In Chapter 3 the process of direct triangular decomposition was employed to solve a set of linear algebraic equations. The underlying decomposition is

$$\mathbf{A} = \mathbf{LR} \tag{4.6.1}$$

where **L** is a lower triangular matrix and **R** is an upper triangular matrix, and was developed by Rutishauser (1958). To achieve a unique decomposition it is conventional to make **L** have unit elements on the diagonal. It is from this decomposition that the **LR** algorithm takes its name. The **LR** algorithm consists of the iterations

$$\mathbf{A}_s = \mathbf{L}_s\mathbf{R}_s \tag{4.6.2}$$

$$\mathbf{A}_{s+1} = \mathbf{R}_s\mathbf{L}_s \tag{4.6.3}$$

in which the next iterated **A** matrix is obtained by multiplying the **L** and **R** factors in reverse order. By continuing this process sufficiently often, the iterated matrices will begin to take up a special form. In general all elements below the diagonal will begin to converge to zero with some possible exceptions on the sub-diagonal. These non-converging subdiagonal elements can be considered to form the left-hand bottom element of a (2×2) sub-matrix on the diagonal. The eigenvalues of these sub-matrices will follow as the solution of the quadratic characteristic equations and will yield any complex pairs of eigenvalues in a simple manner. Where the sub-diagonal elements go to zero to machine accuracy, then the corresponding diagonal elements are the required eigenvalues, as the **LR** process preserves eigenvalues. Hence all the eigenvalues can be found, with the complications of equal roots and complex roots being neatly absorbed into the diagonal quadratic factors.

The process as described is very slow to converge, and the decomposition being used suffers from occasional accidental failure. In the eigenvalue problem it is possible that at some stage of the iteration the iterated matrix $\mathbf{A}_s$ could become singular, and then the decomposition will fail to exist. Various restarting procedures can be used to get round the problem, but a more satisfactory solution is to use the alternative **QR** decomposition, by Francis (1961, 1962). In this decomposition, **Q** is a unitary matrix and **R** is upper triangular. Such a decomposition can be generated using pre-multiplied rotation matrices to place zeros below the diagonal in the given matrix **A**. This process is very like the Givens method of Section 4.4 but involves only pre-multiplication by the rotation matrices, the algorithm now being as follows:

There are $n - 1$ major steps of which the rth places zeros in column r and involves:

for $i = r + 1$ to n do
- (i) Compute $x = (a_{r,r}^2 + a_{ir}^2)^{1/2}$
- (ii) $\cos\theta = a_{r,r}/x$ and $\sin\theta = a_{ir}/x$
  if $x = 0$ then $\cos\theta = 1$ and $\sin\theta = 0$.
- (iii) For $j = r + 1$ to n do
  Compute $a_{rj}\cos\theta + a_{ij}\sin\theta$ to give w_1
  Compute $-a_{rj}\sin\theta + a_{ij}\cos\theta$ to give w_2
  Replace a_{rj} by w_1 and a_{ij} by w_2

The result is

$$\mathbf{R_N} \cdots \mathbf{R}_1\mathbf{A} = \mathbf{R} \tag{4.6.4}$$

where $\mathbf{R}_i$ are the rotation matrices. Their product gives $\mathbf{Q}^T$ and hence the decomposition is defined.

Equivalently the elementary Hermitian matrices of the Householder process may be used to the same end. Hence

$$\mathbf{P}_r = \begin{bmatrix} \mathbf{I} & \vdots & \mathbf{0} \\ \cdots & \cdots & \cdots \\ \mathbf{0} & \vdots & \mathbf{I} - 2\mathbf{v}_r\mathbf{v}_r^T \\ & \vdots & \end{bmatrix} \tag{4.6.5}$$

and

$$\mathbf{A}_r = \begin{bmatrix} \mathbf{U}_r & \vdots & \mathbf{V}_r \\ \cdots & \cdots & \cdots \\ \mathbf{0} & \vdots & \mathbf{W}_{n-r} \end{bmatrix} \tag{4.6.6}$$

to give the transformed matrix

$$\mathbf{A}_{r+1} = \mathbf{P}_r\mathbf{A}_r = \begin{bmatrix} \mathbf{U}_r & \vdots & \mathbf{V}_r \\ \cdots & \cdots & \cdots \\ \mathbf{0} & \vdots & (\mathbf{I} - 2\mathbf{v}\mathbf{v}^T)\mathbf{W}_{n-r} \end{bmatrix} \tag{4.6.7}$$

where the matrix $\mathbf{U}_r$ is upper triangular of order r.

The result for the complete matrix is then the following set of rules for the rth step of Householder's process. There will be $n - 1$ such steps with

$$\begin{aligned} u_i &= a_{r+i,r+1} \qquad i + 2, \ldots, n - r \\ S^2 &= \sum_{i=1}^{n-r} a_{r+i,r+1}^2 \\ u_1 &= a_{r+1,r+1} \mp S_r \\ 2K^2 &= S_r^2 \mp a_{r+1,r+1}S_r \end{aligned} \tag{4.6.8}$$

and the modified matrix is found from

$$(\mathbf{I} - 2\mathbf{v}\mathbf{v}^T)\mathbf{W}_{n-r} = (\mathbf{I} - \mathbf{u}\mathbf{u}^T/2K^2)\mathbf{W}_{n-r} \tag{4.6.9}$$

As in the method based on rotation matrices, the product of the matrices $\mathbf{P}_r$ gives the required Hermitian matrix $\mathbf{Q}^T$.

Hence with this decomposition the **QR** algorithm is defined by

$$\mathbf{A}_s = \mathbf{Q}_s\mathbf{R}_s \tag{4.6.10}$$

$$\mathbf{A}_{s+1} = \mathbf{R}_s\mathbf{Q}_s \tag{4.6.11}$$

The full proof of convergence can be found in Wilkinson (1965), but if the eigenvalues are real and distinct a simpler proof can be made based on the diagonalization theorem of linear algebra. Parts of this proof are illustrative as to how the rate of convergence can be improved, and practical methods of making the process viable are discussed.

Consider therefore the iteration with

$$\mathbf{A}_s = \mathbf{Q}_s\mathbf{R}_s \tag{4.6.12}$$

and

$$\mathbf{R}_s\mathbf{Q}_s = \mathbf{A}_{s+1} = \mathbf{Q}_s^H\mathbf{A}_s\mathbf{Q}_s = \mathbf{Q}_s^H\mathbf{Q}_s\mathbf{R}_s\mathbf{Q}_s \tag{4.6.13}$$

It is now clear that the process is indeed a similarity transformation, and hence the proof reduces to showing that the form obtained is ultimately upper triangular in the case of real distinct eigenvalues. Hence

$$\mathbf{A}_{s+1} = \mathbf{Q}_s^H\mathbf{Q}_{s-1}^H\mathbf{A}_{s-1}\mathbf{Q}_{s-1}\mathbf{Q}_s = \mathbf{Q}_s^H\cdots\mathbf{Q}_1^H\mathbf{A}_1\mathbf{Q}_1\cdots\mathbf{Q}_s \tag{4.6.14}$$

from which it follows that

$$\mathbf{Q}_1\mathbf{Q}_2\cdots\mathbf{Q}_s\mathbf{A}_{s+1} = \mathbf{A}_1\mathbf{Q}_1\cdots\mathbf{Q}_s \tag{4.6.15}$$

Define

$$\mathbf{P}_s = \mathbf{Q}_1\cdots\mathbf{Q}_s \tag{4.6.16}$$

and

$$\mathbf{U}_s = \mathbf{R}_s\cdots\mathbf{R}_1 \tag{4.6.17}$$

then

$$\begin{aligned}\mathbf{P}_s\mathbf{U}_s &= \mathbf{Q}_1\cdots\mathbf{Q}_{s-1}(\mathbf{Q}_s\mathbf{R}_s)\mathbf{R}_{s-1}\cdots\mathbf{R}_1\\ &= \mathbf{Q}_1\cdots\mathbf{Q}_{s-1}\mathbf{A}_s\mathbf{R}_{s-1}\cdots\mathbf{R}_1\\ &= \mathbf{A}_1\mathbf{Q}_1\cdots\mathbf{Q}_{s-1}\mathbf{R}_{s-1}\cdots\mathbf{R}_1\\ &= \mathbf{A}_1\mathbf{P}_{s-1}\mathbf{U}_{s-1}\end{aligned} \tag{4.6.18}$$

to yield iteratively

$$\mathbf{P}_s\mathbf{U}_s = \mathbf{A}_1^s \tag{4.6.19}$$

Now make the assumption that for $\mathbf{A}_1$

$$|\lambda_1| > |\lambda_2| > \cdots > |\lambda_n| \tag{4.6.20}$$

The diagonalization theorem then ensures the existence of a non-singular matrix $\mathbf{X}$ such that

$$\mathbf{A}_1^s = \mathbf{X}\,\mathrm{diag}(\lambda_i^s)\mathbf{X}^{-1} = \mathbf{X}\mathbf{D}^s\mathbf{Y}\text{ , say} \tag{4.6.21}$$

Let the decompositions of the matrices **X** and **Y** be $\mathbf{X} = \mathbf{QR}$ and $\mathbf{Y} = \mathbf{LU}$ where **Q** is unitary, **R** and **U** are upper triangular, and **L** is a unit lower triangular matrix. Then

$$\mathbf{A}_1^s = \mathbf{QRD}^s\mathbf{LU} = \mathbf{QR}(\mathbf{D}^s\mathbf{LD}^{-s})\mathbf{D}^s\mathbf{U}. \tag{4.6.22}$$

But $\mathbf{D}^s\mathbf{LD}^{-s}$ is a unit lower triangular matrix and its (i, j)th element is

$$l_{ij}\left(\frac{\lambda_i}{\lambda_j}\right)^s \quad \text{for } i > j \tag{4.6.23}$$

Therefore

$$\mathbf{D}^s\mathbf{LD}^{-s} = \mathbf{I} + \mathbf{E}_s \tag{4.6.24}$$

where $\mathbf{E}_s \to \mathbf{0}$ as $s \to \infty$. Hence the matrix $\mathbf{A}_1^s$ can be expressed as

$$\begin{aligned} \mathbf{A}_1^s &= \mathbf{QR}(\mathbf{I} + \mathbf{E}_s)\mathbf{D}^s\mathbf{U} \\ &= \mathbf{Q}(\mathbf{I} + \mathbf{RE}_s\mathbf{R}^{-1})\mathbf{RD}^s\mathbf{U} \\ &= \mathbf{Q}(\mathbf{I} + \mathbf{F}_s)\mathbf{RD}^s\mathbf{U} \end{aligned} \tag{4.6.25}$$

where $\mathbf{F}_s \to \mathbf{0}$ as $s \to \infty$.

Let $\mathbf{I} + \mathbf{F}_s = \tilde{\mathbf{Q}}_s\tilde{\mathbf{R}}_s$ with $\tilde{\mathbf{Q}}_s$ being unitary and $\tilde{\mathbf{R}}_s$ being upper triangular. Then as $\mathbf{F}_s \to \mathbf{0}$ as $s \to \infty$, $\tilde{\mathbf{Q}}_s$ and $\tilde{\mathbf{R}}_s$ both tend to the unit matrix. Hence

$$\mathbf{A}_1^s = (\mathbf{Q}\tilde{\mathbf{Q}}_s)(\tilde{\mathbf{R}}_s\mathbf{RD}^s\mathbf{U}) \tag{4.6.26}$$

and as this factorization is essentially unique (apart from certain signs), it follows that

$$\mathbf{P}_s = \mathbf{Q}\tilde{\mathbf{Q}}_s\hat{\mathbf{D}} \tag{4.6.27}$$

$$\to \mathbf{Q}\hat{\mathbf{D}} \tag{4.6.28}$$

and the matrix $\hat{\mathbf{D}}$ is diagonal and unitary and can be found if it is imposed that all the **R**'s have positive diagonal elements. Hence

$$\begin{aligned} \mathbf{A}_s &= \mathbf{P}_{s-1}^{-1}\mathbf{A}_1\mathbf{P}_{s-1} \\ &\to \mathbf{Q}_s^{-1}\mathbf{A}_1\mathbf{Q}_s \\ &= \mathbf{RX}^{-1}\mathbf{A}_1\mathbf{XR}^{-1} \\ &= \mathbf{R}(\operatorname{diag}\lambda_i)\mathbf{R}^{-1} \end{aligned} \tag{4.6.29}$$

where the matrix $\hat{\mathbf{D}}$ has been absorbed, so giving the required form.

Despite theoretical convergence being obtained, in practice the rate of convergence can be very slow. A number of devices are used to improve this situation. The most obvious is always to reduce the given general matrix to upper Hessenberg form using either Givens' or Householder's methods in the first instance. The **QR** algorithm is then only involved in reducing to zero elements on the sub-diagonal. The process does indeed preserve the upper Hessenberg form, so there is no question of 'fill-in'. In addition, if a pre-reduction to upper Hessenberg form has been completed, then the triangularization in the **QR** process involves only the removal of one (sub-diagonal) element in every column. Hence the gains of the Householder process on a full matrix are lost, and Given's process becomes the most efficient.

A second device is to introduce shifts of origin so that the algorithm becomes

$$\mathbf{A}_s - k_s\mathbf{I} = \mathbf{Q}_s\mathbf{R}_s \tag{4.6.30}$$

$$\mathbf{R}_s\mathbf{Q}_s + k_s\mathbf{I} = \mathbf{A}_{s+1} \tag{4.6.31}$$

for a suitable choice of k_s. Various strategies are possible now. A simple one is to choose k_s to encourage rapid convergence at the bottom right-hand end of the matrix. Converged parts of the matrix can then be deflated off so reducing the matrix size. If the bottom right-hand (2×2) matrix has eigenvalues $p_s \pm iq_s$, then in the complex case take $k_s = p_s$ and in the real case take

$$k_s = \begin{cases} p_s & \text{if } |p_s - a_{nn}^s| < |q_s - a_{nn}^2| \\ q_s & \text{otherwise} \end{cases} \tag{4.6.32}$$

This will force the ratio λ_i/λ_j in the proof of convergence to become more rapidly small and so encourage locally more rapid convergence.

The third device is to recognize when deflation can be carried out at the earliest stage. If any sub-diagonal element becomes zero to machine accuracy then clearly the problem splits into two submatrices with the same totality of eigenvalues as the original matrix. Indeed if two consecutive sub-diagonal elements have a product which is zero to machine accuracy, a similar split can be carried out. Each submatrix is then processed in turn. Care is needed in the programming to keep track of the various sub-matrices as they arise. This can be done by having a separate 'marker' vector with an element corresponding to each diagonal element of the complete matrix. A code can be set into the 'marker' vector to mark points of deflation, say with a zero, elements awaiting processing, say with a one, and the currently processed matrix with a two. As deflation occurs this 'marker' vector is updated, and can be consulted when part of a matrix is complete and another submatrix is sought for further processing.

By its nature, this process is best executed by programme, and two forms of the algorithm are reported in the following examples. The first example is a simple-minded application of the **QR** method with no refinements, except for initial reduction to upper Hessenberg form. This form is

$$\begin{bmatrix} 0.32 & 0.754\,717 & 0.503\,401 & 0.223\,17 & 0.27 \\ 0.53 & 1.2 & 0.287\,097 & 0.503\,675 & 0.25 \\ 0.0 & 0.974\,151 & 0.472\,367 & 0.291\,502 & 0.574\,528 \\ 0.0 & 0.0 & 0.794\,857 & -0.327\,882 & 0.466\,5233 \\ 0.0 & 0.0 & 0.0 & 0.377\,939 & -0.244\,485 \end{bmatrix} \tag{4.6.33}$$

and by the tenth step the matrix has the form

$$\begin{bmatrix} 1.944\,94 & 0.492\,783 & -0.294\,952 & 0.809\,262 & -0.654\,465 \\ 0.130\,287(-4) & -0.356\,997 & 0.448\,784 & 0.278\,429 & -0.072\,6476 \\ 0.0 & -0.859\,804(-2) & 0.0651\,788 & -0.108\,134 & 0.354\,092 \\ 0.0 & 0.0 & -0.803\,560 & -0.100\,920 & 0.374\,735 \\ 0.0 & 0.0 & 0.0 & -0.349\,436(-4) & -0.132\,201 \end{bmatrix} \tag{4.6.34}$$

At this stage partial convergence is apparent. The preservation of upper Hessenberg form is noted, and the sub-diagonal elements are beginning to reduce to zero, in particular elements (2,1) and (5,4). Element (3,2) is reducing far more slowly and element (4,3) is not reducing. In fact this element is oscillating between 0.1 and -0.8 in successive steps. A complex pair might be anticipated here. After 101 iterations the picture had changed a little to give:

$$\begin{bmatrix} 1.944\,94 & -0.435\,953 & -0.221\,147 & -0.312\,293 & 0.654\,372 \\ 0.402\,501(-73) & -0.196\,209 & -0.324\,805 & -0.017\,0956 & 0.141\,793 \\ 0.0 & 0.077\,6635 & -0.485\,852 & -0.779\,693 & -0.441\,028 \\ 0.0 & 0.0 & 0.110\,747(-7) & 0.289\,458 & -0.238\,004 \\ 0.0 & 0.0 & 0.0 & 0.896\,825(-19) & -0.132\,339 \end{bmatrix} \tag{4.6.35}$$

From this matrix it is clear that three real eigenvalues have emerged, namely 1.944 94 in the position (1,1), 0.289 458 in (4,4) and $-0.132\,339$ in (5,5). A 2×2 matrix is then left to yield a pair of eigenvalues as the solution of the corresponding quadratic characteristic equation:

$$\det \begin{vmatrix} -0.196\,209 - \lambda & -0.324\,805 \\ 0.077\,6635 & -0.485\,852 - \lambda \end{vmatrix} = 0 \tag{4.6.36}$$

or

$$\lambda^2 + 0.682\,061\lambda + 0.120\,554 = 0 \tag{4.6.37}$$

with roots $-0.341\,031 \pm i0.065\,2089$.

The shortcomings of the unrefined process are very evident. Certainly deflation could have been employed to isolate the first eigenvalue at an early stage, such as by the 21st step when the element (2,1) has reached $-0.23(-13)$. The fifth eigenvalue could also be deflated, as by the 36th step the element (5,4) is reduced to $0.84(-13)$. Presumably had the first eigenvalue been taken out at the 21st step, then the fifth would have gone more rapidly than without this earlier deflation. Only a 3×3 matrix is then left for further reduction.

Contrast this rate of convergence with the refined algorithm with shifts and deflation. Now after only 11 steps the matrix has the form

$$\begin{bmatrix} 1.944\,94 & -0.270\,048(-2) & 0.609\,526(-1) & -0.329\,473 & 0.115\,515 \\ 0.512\,263(-9) & 0.289\,458 & -0.177\,175 & -0.248\,834 & -0.184\,499 \\ 0.0 & -0.302\,373(-6) & -0.132\,339 & 0.375\,375 & -0.215\,863 \\ 0.0 & 0.0 & 0.166\,053(-9) & -0.603\,304 & -0.316\,231 \\ 0.0 & 0.0 & 0.0 & 0.230\,969 & -0.787\,571(-1) \end{bmatrix} \tag{4.6.38}$$

which gives an isolated real eigenvalue in position (1,1) as before and two 2×2 sub-matrices which yield the real pair of eigenvalues 0.289 458 and −0.132 339 once again, from the topmost of the submatrices and the complex pair $-0.341\,031 \pm i0.065\,2089$ from the bottom sub-matrix. It is likely that convergence of the upper sub-matrix would have continued until the sub-diagonal element became zero to working accuracy, but once a deflation into 2×2 matrices has occurred it is worth solving these problems directly.

In summary, the non-symmetric matrix poses a more difficult eigenvalue problem. An initial reduction to upper Hessenberg form by either Givens' or Householder's process is an effective starting point. It then depends on the type of eigenvalues being expected, how many are needed, and what the order of the matrix is as to what is the best way forward. The Hyman process will yield real eigenvalues with little problem as long as they are not multiple roots. Where multiple roots and complex roots are expected, the use of the **QR** algorithm, despite its complexity, would seem justified.

Further reading

Finkbeiner D. T. (1960) *Introduction to Matrices and Linear Transformations*, W. H. Freeman, San Francisco.
Fox L. (1968) *Introduction to Numerical Linear Algebra*, Oxford University Press, Oxford.
Wilkinson J. H. (1965) *Algebraic Eigenvalue Problem*, Oxford University Press, Oxford.

Exercises

(4.1) Find the dominant eigenvalue and the corresponding eigenvector of the matrix

$$\begin{bmatrix} 9 & 10 & 8 \\ 10 & 5 & -1 \\ 8 & -1 & 3 \end{bmatrix}$$

using direct iteration. Now modify the method so that you obtain the smallest eigenvalue and the corresponding eigenvector.

(4.2) Find the dominant and the smallest eigenvalues and the corresponding eigenvectors of the following moment of inertia matrix which results from considerations of a three-dimensional rigid body:

$$mr^2 \begin{bmatrix} 10.000 & 0.134 & -0.866 \\ 0.134 & 6.500 & -1.000 \\ -0.866 & -1.000 & 7.500 \end{bmatrix}$$

Deflate using elementary row and column operations to obtain the middle eigenvalue and corresponding eigenvector.

(4.3) Find the three eigenvalues and eigenvectors of the following matrix given that the eigenvalues are real and well-spaced:

$$\begin{bmatrix} 9.2706 & 7.3412 & 7.5529 \\ -17.718 & -11.035 & -18.988 \\ 6.3529 & -3.2941 & 9.7647 \end{bmatrix}$$

(4.4) Find the quantities s_i for the matrices in exercises (4.1) and (4.2) and discuss the condition of both the eigenvalue and the eigenvector problem.

(4.5) Apply the process of direct iteration to the matrix

$$\begin{bmatrix} 10 & 1 & -1 \\ 2 & 1 & 2 \\ 1 & -1 & 1 \end{bmatrix}$$

for the eigenvalue of largest modulus. Using *approximate* deflation, estimate the moduli of the other two eigenvalues, and hence deduce the rate of convergence of the initial direct iteration. Confirm agreement between this rate and the actual rate obtained in practice.

(4.6) Use Jacobi's method to find the eigenvalues and eigenvectors of the matrix

$$\begin{bmatrix} 2 & 1 & 6 \\ 1 & 3 & 2 \\ 6 & 2 & 4 \end{bmatrix}$$

(4.7) Use Jacobi's method to solve exercise (4.2) and so take advantage of symmetry.

(4.8) Use Jacobi's method to find the eigenvalues and eigenvectors of the matrix

$$\begin{bmatrix} 15.010 & 0.000 & 0.0082 \\ 0.000 & 15.01 & -0.0058 \\ 0.0082 & -0.0058 & 15.01 \end{bmatrix}$$

Would direct iteration have worked in this case ?

(4.9) Show that in one sweep of Jacobi's method the number of multiplications and divisions is of order $2n^3$, the number of additions and subtractions is of order n^3 and the number of square roots is of order n^2. Find precise values for these counts. Note that the formula for $\tan 2\theta$ in terms of $\tan\theta$ has been used to obtain a quadratic for $\tan\theta$, from which the required values of $\sin\theta$ and $\cos\theta$ can then be computed without finding θ explicitly. Is this process better than using inverse tangents to find θ explicitly?

(4.10) Given that 1.0, −1.2, 4.0 and 7.3 are eigenvalues of the matrix

$$\begin{bmatrix} 3.6571 & 3.1253 & -4.0418 & 1.9648 \\ -4.2286 & -6.4703 & 4.0901 & -5.1978 \\ -2.3439 & -1.9978 & 3.9659 & -2.4813 \\ 5.4857 & 10.688 & -3.0626 & 9.9473 \end{bmatrix}$$

find the corresponding eigenvectors. If these given eigenvalues are in fact only approximate, can the problem still be solved? Try the set 0.8, −1.4, 4.6, 10.0.

(4.11) Analyse the level of ill-conditioning, if any, for each eigenvalue and eigenvector of the matrix

$$\begin{bmatrix} 1.003 & 0.996 & 0.998 \\ 0.003 & 2.996 & 1.998 \\ -0.003 & -2.496 & -1.498 \end{bmatrix}$$

giving the expected loss of figures in each case.

(4.12) Reduce the eigenvalue problem

$$\mathbf{Ax} = \lambda \mathbf{Bx}$$

to conventional form. Hence find the eigenvalues and eigenvectors for the problem

$$\begin{bmatrix} -70 & 2 & 1 \\ -65 & 7 & 1 \\ 14 & -1 & -1 \end{bmatrix} \mathbf{x} = \lambda \begin{bmatrix} -15/2 & -7/2 & -3/2 \\ -7 & -2 & -3 \\ 3/2 & 1/2 & 1/2 \end{bmatrix} \mathbf{x}$$

(4.13) Given that $\lambda = -5$ is an eigenvalue of the matrix

$$\begin{bmatrix} 74 & 29 & 21 \\ -132 & -53 & -36 \\ -87 & -33 & -26 \end{bmatrix}$$

deflate the matrix to yield a 2×2 matrix with the same eigenvalues except for $\lambda = 5$. Hence find the second and third eigenvalues.

(4.14) Show that the matrix

$$\begin{bmatrix} 9 & 10 & 8 \\ 10 & 5 & -1 \\ 8 & -1 & 3 \end{bmatrix}$$

has the characteristic equation

$$\lambda^3 - 17\lambda^2 - 78\lambda + 654 = 0$$

and hence find the eigenvalues of the matrix.

Use Givens' method to tri-diagonalize the matrix, find the new characteristic equation and confirm that the eigenvalues are in agreement with the first computation.

Repeat using Householder's method, in particular noticing the difference in the two tri-diagonalizations.

Finally show independently that

$$|\lambda_i| \le 27 \qquad i = 1, 2, 3$$

and hence use the Sturm sequence property and bisection to find the eigenvalues of the tri-diagonal forms.

(4.15) By reduction to tri-diagonal form and by the use of Sturm sequences show that the condition number with respect to inversion (see Chapter 3) for the problem $\mathbf{Ax} = \mathbf{b}$ with

$$\mathbf{A} = \begin{bmatrix} 10 & 7 & 8 & 7 \\ 7 & 5 & 6 & 5 \\ 8 & 6 & 10 & 9 \\ 7 & 5 & 9 & 10 \end{bmatrix}$$

is $\kappa(\mathbf{A}) = 2984.1$.

(4.16) Consider Givens reduction of a real $n \times n$ matrix to upper Hessenberg form. Let the element a_{ir} be the current element being made zero. Show that the relevant values for the rotation angle θ are given by

$$\cos\theta = a_{r+1,r}/x \quad \text{and} \quad \sin\theta = a_{ir}/x$$

with

$$x = (a_{r+1,r}^2 + a_{ir}^2)^{1/2}$$

Show that the new element in position $(r + 1, r)$ has the value x. Hence apply the method from first principles to the matrix

$$\begin{bmatrix} 1 & -1 & 2 \\ 3 & -1 & 4 \\ 2 & -1 & 8 \end{bmatrix}$$

confirming that a_{21} is indeed x.

(4.17) Establish for the Givens reduction to upper Hessenberg form of a real $n \times n$ matrix, that the number of multiplications and divisions is of order $\frac{10}{3}n^3$ and for Householder's method the equivalent result is $\frac{5}{3}n^3$.

(4.18) Reduce the matrix

$$\begin{bmatrix} 9 & 10 & 7 \\ 10 & -5 & -1 \\ 8 & -1 & 2 \end{bmatrix}$$

to upper Hessenberg form using both Givens' and Householder's methods. Compare the resulting forms. Hence use Hyman's method to find the eigenvalues.

(4.19) Apply the **LR** algorithm to find the eigenvalues of the matrix

$$\begin{bmatrix} 4 & 3 & 2 \\ 5 & 1 & 6 \\ 0 & 1 & 3 \end{bmatrix}$$

(4.20) Apply the **QR** algorithm to the matrix in exercise (4.19).

(4.21) Use Householder's process to reduce the matrix

$$\begin{bmatrix} 17/2 & -5/3 & 49/6 \\ 15/2 & -1/3 & 77/6 \\ 3/2 & 5/3 & 11/6 \end{bmatrix}$$

to upper Hessenberg form. Apply the **QR** process to this reduced form. Complete one step only and comment on the result given that the eigenvalues are 10, 3 and -2.

5

Approximation Theory

5.1 Introduction

Approximation theory is concerned with the representation of generally commonly used functions in a simple easy to compute form. The most obvious example is the function $\sin x$ as used in almost any computer library as a standard function. Remembering the high frequency of use of this function, it is of great economic concern to reduce the workload in evaluating the function as much as possible. The analytic representations of $\sin x$, though representing the function for all x, are not necessarily computationally practicable. Hence the truncated Taylor series

$$\sin x = x - \frac{x^3}{3!} + \frac{x^5}{5!} - \cdots + (-1)^{m+1} \frac{x^{2m-1}}{(2m-1)!} \tag{5.1.1}$$

used in Chapter 1 (see equation (1.1.5)) is not suitable for other than very small x because of stability considerations. Analytically the infinite series converges for all x. The error in a Taylor series representation also grows as the distance from the point of expansion increases. For computational purposes it is better to have the error much more uniformly distributed across the interval of interest. Representations as polynomials and as rational functions of polynomials are commonly employed computationally, and such forms can have very high accuracy. Typically an error of less than 2×10^{-9} is obtained over the interval $0 \leq x \leq \pi/2$ for the approximation

$$\sin x/x = 1 + a_2x^2 + a_4x^4 + a_6x^6 + a_8x^8 + a_{10}x^{10} \tag{5.1.2}$$

with

$$\begin{aligned} a_2 &= -0.166\,666\,6664 \qquad & a_8 &= 0.000\,002\,7526 \\ a_4 &= 0.008\,333\,3315 \qquad & a_{10} &= -0.000\,000\,0239 \\ a_6 &= -0.000\,198\,4090 \end{aligned} \tag{5.1.3}$$

A whole host of mathematical and computational questions now arise. How are approximations like this found in practice? How can the error be sensibly measured? What is the ultimate accuracy a polynomial can deliver? If the polynomial has a limited maximum degree, what is the best approximation and how does it manifest itself? These questions all have useful and effective answers. A secondary line of thought then follows. If highly accurate approximations are available, can these be used to advantage in other branches of numerical work where functions are represented in a different but simple manner, to allow such processes as numerical differentiation and integration to be applied? This chapter consists of a series of theorems which answer the above questions. Once a theorem is proved its manifestations in practice can be pursued.

The object in this chapter will be to find an approximation to a given function $f(x)$ which belongs to $C[a, b]$, the set of continuous functions on the interval $[a, b]$. If the approximation is a polynomial of degree n, it will be represented by $p_n(x)$, with the associated error

$$E(x) = p_n(x) - f(x) \tag{5.1.4}$$

Some means is required to measure this error over the interval, and the usual concept of a norm is introduced. Define $||f||$ to satisfy the axioms

$$\begin{aligned}
&\text{(i)} \quad && ||f|| > 0 \qquad (||f|| = 0 \text{ iff } f = 0) \\
&\text{(ii)} \quad && ||cf|| = |c| \, ||f|| \qquad \text{for any scalar } c \\
&\text{(iii)} \quad && ||f_1 + f_2|| \le ||f_1|| + ||f_2||
\end{aligned} \tag{5.1.5}$$

As in the linear algebra of Chapter 3, the main concern here is with norms which allow computational evaluation. Two such norms are commonly employed. The first is the root mean square norm, often abbreviated to RMS, and defined by

$$||E||_{RMS} = \sqrt{\frac{\int_a^b E^2(x)\,\mathrm{d}x}{b-a}} \tag{5.1.6}$$

and the second is called the uniform norm and is defined by

$$||E||_U = \max_{a \le x \le b} |E(x)| \tag{5.1.7}$$

The root mean square norm effectively forms an average of the errors over the interval of interest. There is a computational danger here. If a function is very smooth except for a sharp narrow peak, like the Dirac delta function, then the RMS norm will give a very small error measure despite the presence of a large error over a short range. The uniform norm will yield a large error as the narrow peak will dominate the norm. It is not acceptable to have an approximation which is very accurate except at a particular point. The measure needs to *guarantee* that the error is never above some required threshold, often machine accuracy in the particular situation. Hence the most far-reaching results are based on the uniform norm, but the RMS norm can also yield some useful approximations. One such approximation is that based on orthogonal polynomials which is pursued in the next paragraph.

5.2 Orthogonal polynomials

If a set of polynomials can be found which have the property that

$$\int_a^b w(x)\phi_r(x)\phi_s(x)\,dx = \delta_{rs} \tag{5.2.1}$$

then the set of polynomials is said to be orthonormal on the interval $[a, b]$ with respect to the weight function $w(x)$. If the integral in equation (5.2.1) is not unity when $r = s$, but some other constant, then the set is called orthogonal. It is easy to normalize an otherwise orthogonal set dividing $\phi_r(x)$ by the normalization integral

$$\int_a^b w(x)\phi_r^2(x)\,dx \tag{5.2.2}$$

Once an orthogonalization rule is established, approximations of the form

$$f(x) = \sum_{i=0}^{\infty} a_i\phi_i(x) \tag{5.2.3}$$

become simple to find as the coefficients a_i follow by multiplying both sides of (5.2.3) by $w(x)\phi_r(x)$ and integrating from a to b, to give

$$\begin{aligned}\int_a^b w(x)f(x)\phi_r(x)\,dx &= \sum_{i=0}^{\infty} a_i \int_a^b w(x)\phi_r(x)\phi_i(x)\,dx \\ &= a_r \int_a^b w(x)\phi_r^2\,dx \end{aligned} \tag{5.2.4}$$

and hence the usual generalized Fourier coefficient

$$a_r = \frac{\displaystyle\int_a^b w(x)f(x)\phi_r(x)\,dx}{\displaystyle\int_a^b w(x)\phi_r^2(x)\,dx} \tag{5.2.5}$$

Just what a generalized Fourier expansion means in terms of norms and errors can be seen by considering the problem from a different angle. Suppose the RMS error for the expansion in (5.2.3) is given by

$$S = \int_a^b w(x)\left[f(x) - \sum_{r=0}^{n} a_r\phi_r(x)\right]^2 dx \tag{5.2.6}$$

then the question can be asked as to what values the coefficients a_r need to take to minimize S. The necessary conditions are that

$$\frac{\partial S}{\partial a_s} = 2\int_a^b w(x)\left[f(x) - \sum_{r=0}^{n} a_r\phi_r(x)\right][-\phi_s(x)]\,dx = 0 \tag{5.2.7}$$

which yields

$$a_s = \int_a^b w(x)f(x)\phi_s(x)\,\mathrm{d}x \tag{5.2.8}$$

which is the usual Fourier coefficient. Hence the generalized Fourier expansion is the best approximation in the RMS sense.

Hence given a weight function $w(x)$ and an interval $[a, b]$ it would be useful to be able to generate the resulting set of orthogonal polynomials. Two methods will be discussed to achieve this object. The first method is known as Rodriguez' method and is an analytic technique. Because of this, it is restricted to problems for which the solutions are analytically viable, which are commonly called the classical orthogonal polynomials. These same polynomials also arise in the solution of various partial differential equations using the technique of splitting the variable and therefore they are familiar in applied mathematics.

In Rodriguez' method, the orthogonality condition, (5.2.1) is replaced by the equivalent form

$$\int_a^b w(x)\phi_r(x)q_{r-1}(x)\,\mathrm{d}x = 0 \tag{5.2.9}$$

in which $q_{r-1}(x)$ is an arbitrary polynomial of degree $r-1$ or less. Now set

$$w(x)\phi_r(x) = \frac{\mathrm{d}^r U_r(x)}{\mathrm{d}x^r} \tag{5.2.10}$$

so that the orthogonality condition (5.2.9) becomes

$$\int_a^b U_r^{(r)}(x)q_{r-1}(x)\,\mathrm{d}x = 0 \tag{5.2.11}$$

Integrating by parts r times then yields:

$$[U_r^{(r-1)}q_{r-1} - U_r^{(r-2)}q'_{r-1} + U_r^{(r-3)}q''_{r-1} - \cdots + (-1)^{r-1}U_r q_{r-1}^{(r-1)}]_a^b = 0 \tag{5.2.12}$$

as $q_{r-1}^{(r)} = 0$. But

$$\phi_r(x) = \frac{1}{w(x)}\frac{\mathrm{d}^r U_r(x)}{\mathrm{d}x^r} = \text{poly of degree } r \tag{5.2.13}$$

and therefore the required orthogonal polynomials can be generated as the solution of the differential equation

$$\frac{\mathrm{d}^{r+1}}{\mathrm{d}x^{r+1}}\left[\frac{1}{w(x)}\frac{\mathrm{d}^r U_r(x)}{\mathrm{d}x^r}\right] = 0 \tag{5.2.14}$$

with the boundary conditions from (5.2.12), namely

$$U_r(a) = U'_r(a) = U''_r(a) = \cdots = U_r^{(r-1)}(a) = 0$$
$$U_r(b) = U'_r(b) = U''_r(b) = \cdots = U_r^{(r-1)}(b) = 0 \tag{5.2.15}$$

The solution of this differential equation for general r is not trivial and solutions can only be found in certain cases where simplicity allows the form of the solution to become apparent. Such an example is the common case of $w(x) = 1$ with the interval $[-1, 1]$. For this example the differential equation is

$$\frac{d^{2r+1}U_r}{dx^{2r+1}} = 0 \tag{5.2.16}$$

with boundary conditions

$$U_r(\pm 1) = U_r'(\pm 1) = \cdots = U_r^{(r-1)}(\pm 1) = 0 \tag{5.2.17}$$

with the obvious solution

$$U_r(x) = C_r(x^2 - 1)^r \tag{5.2.18}$$

The Rodriguez form of the orthogonal polynomials is then

$$\phi_r(x) = C_r \frac{d^r}{dx^r}(x^2 - 1)^r \tag{5.2.19}$$

with the classical coefficients

$$C_r = \frac{1}{2^r r!} \tag{5.2.20}$$

though this choice of C_r does not normalize the polynomials, but arises from the associated applied mathematical considerations. These polynomials are the classic Legendre polynomials which arise in the axi-symmetric solution of Laplace's partial differential equation in spherical polar coordinates. From (5.2.19) the first few polynomials are

$$\begin{aligned}
P_0(x) &= 1 \\
P_1(x) &= x \\
P_2(x) &= \tfrac{1}{2}(3x^2 - 1) \\
P_3(x) &= \tfrac{1}{2}(5x^3 - 3x)
\end{aligned} \tag{5.2.21}$$

A simple example can be set up using the orthonormalized Legendre polynomials

$$\phi_n(x) = \sqrt{n + \frac{1}{2}}\, \frac{1}{2^n\, n!} \left[\frac{d^n}{dx^n}(x^2 - 1)^n \right] \tag{5.2.22}$$

which gives

$$\begin{aligned}
\phi_0(x) &= \frac{1}{\sqrt{2}} \\
\phi_1(x) &= x\sqrt{\frac{3}{2}} \\
\phi_2(x) &= (3x^2 - 1)\sqrt{\frac{5}{8}}
\end{aligned} \tag{5.2.23}$$

Hence to expand x^4 as a series of normalized Legendre functions in the form

$$x^4 = a_0\phi_0 + a_1\phi_1 + a_2\phi_2 \tag{5.2.24}$$

gives the generalized Fourier coefficients

$$\begin{aligned}
a_0 &= \int_{-1}^{1} \frac{x^4}{\sqrt{2}}\,\mathrm{d}x = \frac{\sqrt{2}}{5} \\
a_1 &= \int_{-1}^{1} \sqrt{\frac{3}{2}}x^5\,\mathrm{d}x = 0 \\
a_2 &= \int_{-1}^{1} x^4(3x^2-1)\sqrt{\frac{5}{8}}\,\mathrm{d}x = \frac{8}{35}\sqrt{\frac{5}{2}}
\end{aligned} \tag{5.2.25}$$

and the resulting expansion is then

$$x^4 \simeq \frac{6x^2}{7} - \frac{3}{35} \tag{5.2.26}$$

There is a more direct method of obtaining orthogonal polynomials given the weight function $w(x)$ and the interval $[a, b]$ which lends itself to a computational approach. Let the required polynomials have the form

$$P_n(x) = x^n + a_{n,n-1}x^{n-1} + \cdots + a_{n,1}x + a_{n,0} \tag{5.2.27}$$

then the coefficients $a_{n,r}$ need to be chosen to satisfy

$$\int_a^b w(x)P_n(x)P_m(x)\,\mathrm{d}x = 0 \qquad m \neq n \tag{5.2.28}$$

or equivalently

$$\int_a^b w(x)x^m P_n(x)\,\mathrm{d}x = 0 \qquad 0 \leq m < n \tag{5.2.29}$$

Substitute for $P_n(x)$ from (5.2.27) into (5.2.29) to give successively for increasing m,

$$\begin{aligned}
c_0a_{n,0} + c_1a_{n,1} + \cdots + c_{n-1}a_{n,n-1} &= -c_n \\
c_1a_{n,0} + c_2a_{n,1} + \cdots + c_na_{n,n-1} &= -c_{n+1} \\
\vdots \qquad\qquad & \qquad \vdots \\
c_{n-1}a_{n,0} + c_na_{n,1} + \cdots + c_{2n-2}a_{n,n-1} &= -c_{2n-1}
\end{aligned} \tag{5.2.30}$$

with the integrals c_m defined by

$$c_m = \int_a^b x^m w(x)\,\mathrm{d}x \tag{5.2.31}$$

which are called the monomial integrals.

The solution of this set of linear equations will give the coefficients of the required orthogonal polynomial and appears to be a straightforward linear algebra problem. However, the matrix involved is known as the Vandermonde matrix and is notoriously ill-conditioned, as seen in Section 3.4. As a result, high precision arithmetic is required for any but the smallest values of n. Some of the classical orthogonal polynomials are listed in Table 5.1, and their properties are documented in Abramowich and Stegun (1965).

For reference purposes a few of the low-order polynomials are listed below:

$$
\begin{aligned}
&P_0 = 1 \qquad P_1 = x \qquad P_2 = \tfrac{1}{2}(3x^2 - 1) \\
&P_3 = \tfrac{1}{2}(5x^3 - 3x) \qquad P_4 = \tfrac{1}{8}(35x^4 - 30x^2 + 3) \\
&T_0 = 1 \qquad T_1 = x \qquad T_2 = 2x^2 - 1 \\
&T_3 = 4x^3 - 3x \qquad T_4 = 8x^4 - 8x^2 + 1 \\
&L_0 = 1 \qquad L_1 = 1 - x \qquad L_2 = \tfrac{1}{2}(x^2 - 4x + 2) \\
&L_3 = \tfrac{1}{6}(-x^3 + 9x^2 - 18x + 6) \\
&L_4 = \tfrac{1}{24}(x^4 - 16x^3 + 72x^2 - 96x + 24) \\
&H_0 = 1 \qquad H_1 = 2x \qquad H_2 = 4x^2 - 2 \\
&H_3 = 8x^3 - 12x \qquad H_4 = 16x^4 - 48x^2 + 12 \\
&U_0 = 1 \qquad U_1 = 2x \qquad U_2 = 4x^2 - 1 \\
&U_3 = 8x^3 - 4x \qquad U_4 = 16x^4 - 12x^2 + 1
\end{aligned}
\tag{5.2.32}
$$

These polynomials will assume considerable importance in Chapter 6 on quadrature in which the Gaussian rules have abscissae which are the zeros of the orthogonal polynomials for the given integration range and weight function.

Orthogonal polynomials exhibit many properties which include a recurrence relation. Hence if $\phi_0(x), \phi_1(x), \ldots$ are polynomials with their leading coefficient set to unity, and if these polynomials are orthogonal with respect to the inner product

Table 5.1

Interval	$w(x)$	Symbol	Name
$[-1, 1]$	1	$P_n(x)$	Legendre
$[-1, 1]$	$(1 - x^2)^{-1/2}$	$T_n(x)$	First kind Chebyshev
$[0, \infty)$	e^{-x}	$L_n(x)$	Laguerre
$(-\infty, \infty)$	e^{-x^2}	$H_n(x)$	Hermite
$[-1, 1]$	$(1 - x^2)^{1/2}$	$U_n(x)$	Second kind Chebyshev

(ϕ_n, ϕ_m) defined by

$$\int_a^b w(x)\phi_n(x)\phi_m(x)\,dx = (\phi_n, \phi_m) \tag{5.2.33}$$

then

$$\phi_{n+1}(x) = (x - a_{n,n})\phi_n(x) - a_{n,n-1}\phi_{n-1}(x) - \cdots - a_{n,0}\phi_0(x)$$
$$n = 0, 1, 2, \ldots \tag{5.2.34}$$

where $\phi_0(x) = 1$ and

$$a_{n,s} = (x\phi_n, \phi_s)/(\phi_s, \phi_s) \qquad s = 0, 1, \ldots, n \tag{5.2.35}$$

If in addition the inner product satisfies the extra condition that $(x\phi_n, \phi_m) = (\phi_n, x\phi_m)$, which is true of the inner product defined by equation (5.2.33), then the recurrence relation (5.2.34) reduces to a three-term relation

$$\phi_{n+1} = (x - \alpha_n)\phi_n(x) - \beta_n\phi_{n-1}(x), \qquad n = 0, 1, \ldots \tag{5.2.36}$$

with $\phi_{-1}(x) = 0$ and

$$\alpha_n = \frac{(x\phi_n, \phi_n)}{(\phi_n, \phi_n)}, \qquad n = 0, 1, \ldots \tag{5.2.37}$$

and

$$\beta_n = \frac{(x\phi_n, \phi_{n-1})}{(\phi_{n-1}, \phi_{n-1})} = \frac{(\phi_n, \phi_n)}{(\phi_{n-1}, \phi_{n-1})}, \qquad n = 1, 2, \ldots \tag{5.2.38}$$

These recurrence relations can be used to compute values of the polynomials though care is required to avoid instability in this process. For the specific polynomials listed above, the relevant three-term recurrence relations are

$$(n+1)P_{n+1} = (2n+1)xP_n - nP_{n-1} \tag{5.2.39}$$

$$(n+1)L_{n+1} = [(2n+1) - x]L_n - nL_{n-1} \tag{5.2.40}$$

$$H_{n+1} = 2xH_n - 2nH_{n-1} \tag{5.2.41}$$

$$U_{n+1} = 2xU_n - U_{n-1} \tag{5.2.42}$$

Of somewhat special importance are the Chebyshev polynomials, $T_n(x)$, and these are now presented in more detail. At first sight there seems little to recommend these particular polynomials for special attention. These functions are defined by

$$T_n(x) = \cos(r \arccos(x)) = \cos r\theta \tag{5.2.43}$$

where $x = \cos\theta$, and are the orthogonal polynomials for the interval $[-1, 1]$ with respect to the weight function $w(x) = 1/\sqrt{1-x^2}$, and as such satisfy the orthogonality condition

$$\int_{-1}^{1} \frac{T_r(x)T_s(x)}{\sqrt{1-x^2}}\,dx = \begin{cases} 0 & r \neq s \\ \frac{\pi}{2} & r = s \neq 0 \\ \pi & r = s = 0 \end{cases} \tag{5.2.44}$$

The particular three-term recurrence relation satisfied by the Chebyshev polynomials is

$$T_{r+1}(x) = 2xT_r(x) - T_{r-1}(x) \tag{5.2.45}$$

and the differential equation

$$(1-x^2)y'' - xy' + n^2y = 0. \tag{5.2.46}$$

is satisfied by $y = T_n(x)$.

The importance of these functions in approximation theory will only become clear after the following two fundamental theorems have been stated and proved. These theorems will answer many of the questions posed in the introduction to the chapter. Orthogonal polynomials and the associated root mean square norm will now be left, with orthogonal polynomials being revisited in the quadrature work of Chapter 6. The use of the uniform norm and its implications is considered next. A very different theory arises in this case.

5.3 Weierstrass's theorem

One of the questions posed in the introduction was how accurate can a polynomial approximation be? It is not immediately obvious that given a sufficiently large degree polynomial that arbitrary accuracy can be achieved if $f(x)$ is continuous on $[a, b]$. Weierstrass's theorem establishes this result.

Theorem

For any continuous function $f(x)$ on the interval $[a, b]$ and for any $\varepsilon > 0$, $\exists$ a polynomial $P(x)$ s.t. $|f(x) - P(x)| < \varepsilon\ \forall\ x \in [a, b]$.

Proof

Consider $f(x)$ defined on $[0, 1]$ in the first instance. Define the nth. Bernstein polynomial for $f(x)$ as

$$B_n(f; x) = \sum_{k=0}^{n} f\left(\frac{k}{n}\right)\binom{n}{k} x^k(1-x)^{n-k} \tag{5.3.1}$$

These Bernstein polynomials have the properties that $B_n(f; 0) = f(0)$ and $B_n(f; 1) = f(1)$. The main proof depends on a lemma, namely

Lemma

Let $f(x)$ be bounded on [0, 1] then

$$\lim_{n\to\infty} B_n(f; x) = f(x) \tag{5.3.2}$$

for any point $x \in [0, 1]$ at which f is continuous. In the case when $f \in C[0, 1]$ then the lemma holds uniformly in [0, 1].

Proof

Note first that the inequality

$$\left|\frac{k}{n} - x\right| \geq \delta \tag{5.3.3}$$

implies that

$$\frac{1}{\delta^2}\left(\frac{k}{n} - x\right)^2 \geq 1 \tag{5.3.4}$$

and hence

$$\begin{aligned}
\sum_{|k/n-x|\geq\delta} \binom{n}{k} x^k(1-x)^{n-k} &\leq \frac{1}{\delta^2} \sum_{|k/n-x|\geq\delta} \left(\frac{k}{n} - x\right)^2 \binom{n}{k} x^k(1-x)^{n-k} \\
&\leq \frac{1}{\delta^2} \sum_{k=0}^{n} \left(\frac{k}{n} - x\right)^2 \binom{n}{k} x^k(1-x)^{n-k} \\
&= \frac{nx(1-x)}{\delta^2 n^2} \leq \frac{1}{4n\delta^2}
\end{aligned} \tag{5.3.5}$$

In order to establish

$$\sum_{k=0}^{n} \left(\frac{k}{n} - x\right)^2 \binom{n}{k} x^k(1-x)^{n-k} = \frac{nx(1-x)}{\delta^2 n^2} \tag{5.3.6}$$

the reader is referred to the exercises. The assumption that $f(x)$ is bounded in [0, 1] implies that $\exists$ an m s.t. $|f(x)| \leq M$ and for any two values $\alpha, \beta \in [0, 1]$ then $|f(\alpha) - f(\beta)| \leq 2M$. In addition, if x is a point of continuity of f then given an $\varepsilon > 0$, $\exists\ \delta$ s.t. $|f(x) - f(y)| < \varepsilon$ whenever $|y - x| < \delta$. Applying these results to the inequality established in (5.3.5) gives

$$
\begin{aligned}
|f(x) - B_n(f;x)| &\le \sum_{|k/n-x|<\delta} \left| f(x) - f\left(\frac{k}{n}\right)\right| \binom{n}{k} x^k(1-x)^{n-k} \\
&+ \sum_{|k/n-x|\ge\delta} \left| f(x) - f\left(\frac{k}{n}\right)\right| \binom{n}{k} x^k(1-x)^{n-k} \\
&\le \varepsilon \sum_{|k/n-x|<\delta} \binom{n}{k} x^k(1-x)^{n-k} + 2M \sum_{|k/n-x|\ge\delta} \binom{n}{k} x^k(1-x)^{n-k} \\
&\le \varepsilon \sum_{k=0}^{n} \binom{n}{k} x^k(1-x)^{n-k} + \frac{2M}{4n\delta^2} \\
&= \varepsilon + \frac{M}{2n\delta^2}
\end{aligned}
\tag{5.3.7}
$$

and hence for n sufficiently large, $|f(x) - B_n(f;x)| \le 2\varepsilon$ and since ε is arbitrary the lemma is proved.

If $f(x) \in C[0,1]$ then given an $\varepsilon > 0$, $\exists\ \delta$ s.t. $|f(x) - f(y)| < \varepsilon\ \forall\ x, y \in [0,1]$ which satisfies $|x - y| < \delta$ and therefore (5.3.6) holds uniformly in $[0,1]$. Therefore for $f(x) \in C[0,1]$ then given $\varepsilon > 0$ there is a sufficiently large n such that

$$|f(x) - B_n(f;x)| \le \varepsilon \qquad 0 \le x \le 1 \tag{5.3.8}$$

The full theorem follows using the linear transformation

$$y = \frac{x-a}{b-a} \tag{5.3.9}$$

to convert the interval $[0,1]$ to $[a,b]$.

The Bernstein polynomials used in this constructive proof can be found for a concrete example. Consider $B_3(e^{-x}; 3)$, which according to the definition has the form

$$
\begin{aligned}
B_3 &= e^0(1-x)^3 + 3e^{-1/3}x(1-x)^2 + 3e^{-2/3}x^2(1-x) + e^{-1}x^3 \\
&= x^3(e^{-1} - 3e^{-2/3} + 3e^{-1/3} - 1) + x^2(3e^{-2/3} - 6e^{-1/3} + 3) \\
&\quad + x(3e^{-1/3} - 3) + 1 \\
&= -0.022\,777\,984x^3 + 0.241\,063\,4934x^2 - 0.850\,406\,0682x + 1
\end{aligned}
\tag{5.3.10}
$$

As a comparison, $B_3(0.5) = 0.632\,2156$ whereas the exact value is $e^{-1/2} = 0.606\,5306$ and Taylor's series up to and including cubic terms gives

$$e^{-1/2} \sim 1 - \frac{1}{2} + \frac{\left(\frac{1}{2}\right)^2}{2!} - \frac{\left(\frac{1}{2}\right)^3}{3!} = 0.604\,1667 \tag{5.3.11}$$

Hence the Bernstein polynomials do not give particularly accurate approximations, but in this proof the emphasis is on the degree being sufficiently large.

This theorem answers the most fundamental question of approximation theory, and shows that a polynomial approximation can reach arbitrary accuracy given the use of sufficiently high a degree. In the next paragraph the second most important question is answered. If the degree of the polynomial approximation is fixed at some value n, then what choice of coefficients yields the best approximation in the practical uniform norm sense? Such best approximations will guarantee a required accuracy throughout their range.

5.4 Oscillation theorem

The Oscillation theorem characterizes the best polynomial approximation $p_n(x)$ of degree n to a continuous function $f(x)$ in the uniform norm sense. If the error function $\varepsilon(x)$ is defined as $\varepsilon(x) = f(x) - p_n(x)$ then the uniform norm is defined as

$$\|\varepsilon(x)\| = \max_{a \le x \le b} |\varepsilon(x)| \tag{5.4.1}$$

The object is to minimize this error by the choice of polynomial of given degree n, to yield the Oscillation theorem in the form:

Theorem

Let $f(x) \in \mathcal{C}[a, b]$ and $p_n(x)$ be the best uniform approximant to $f(x)$ of degree n. Let

$$E_n = \max_{a \le x \le b} |f(x) - p_n(x)|$$

and

$$\varepsilon(x) = f(x) - p_n(x).$$

Then there are at least $n - 2$ points $a \le x_1 < x_2 < \cdots < x_{n+2} \le b$ where $\varepsilon(x)$ assumes the values $\pm E_n$ and with alternating signs such that

$$\varepsilon(x_i) = \pm E_n \qquad i = 1, \ldots, n + 2$$

and

$$\varepsilon(x_i) = -\varepsilon(x_{i+1}) \qquad i = 1, \ldots, n + 1$$

Proof

By the uniform continuity of $\varepsilon(x)$, an ε can be selected such that $|x_1 - x_2| < \varepsilon$ implies that $|\varepsilon(x_1) - \varepsilon(x_2)| \le E_n/2$ for any x_1 and x_2. Hence divide the interval

$[a, b]$ into consecutive closed intervals of width $\leq \varepsilon$, and call those intervals on which $|\varepsilon(x)|$ assumes a maximum, $I_1, \ldots, I_m$. Since $\varepsilon(x)$ can vary at most $E_n/2$ in any of these intervals then

$$\text{either } \varepsilon(x) > E_n/2 \text{ or } \varepsilon(x) < -E_n/2 \tag{5.4.2}$$

Let $u_1, \ldots, u_m (= \pm 1)$ be the sign of $\varepsilon(x)$ in these intervals. Then the object of the proof is to show that there are at least $(n + 1)$ sign changes in this sequence. This will be done by showing that if there were fewer than $(n + 1)$ sign changes then there would exist a polynomial whose E_n was less than that of the best approximation $p(x)$ which would be a contradiction.

If all the u's were of the same sign then a better approximation could be simply obtained by adding a constant to $p(x)$ to establish the contradiction in this case.

Suppose all the intervals in which consecutive I's have the same sign are grouped together to give:

1st group	$I^{(1)} \ldots I^{(j_1)}$	1st sign
2nd group	$I^{(j_1+1)} \ldots I^{(j_2)}$	2nd sign
$\vdots$		$\vdots$
kth group	$I^{(j_{k-1}+1)} \ldots I^{(j_k)}$	kth sign

where $I^{(j_l)} = I_{j_l}$. There are $k - 1$ changes in sign, so the assumption is that $k - 1 < n + 1$ or $k < n + 2$.

Now consider two consecutive intervals $I^{(j_1)}$ and $I^{(j_1+1)}$. These cannot be adjacent, for $\varepsilon(x)$ is zero in neither and yet takes opposite signs in each. Hence $\exists\, x_1$ s.t.

$$I^{(j_1)} < x_1 < I^{(j_1+1)} \tag{5.4.3}$$

and similarly

$$I^{(j_{k-1})} < x_{k-1} < I^{(j_{k-1}+1)} \tag{5.4.4}$$

Now form the polynomial

$$q(x) = (x_1 - x) \cdots (x_{k-1} - x) \tag{5.4.5}$$

which vanishes only at the x_i. Since the points x_i are between the intervals, $q(x)$ must have constant sign on each interval and indeed in the first group $q(x)$ is positive, in the second it is negative and so on. Hence by selecting $u = \pm 1$ appropriately then $uq(x)$ will coincide with $\varepsilon(x)$ in sign over all the intervals $I_1 \cdots I_m$. The assertion is now that for η sufficiently small, the polynomial $p(x) + \eta u q(x)$ is a better approximation to $f(x)$ than $p(x)$ so forcing the required contradiction. This assertion will now be proved to complete the step.

Let J be the closure of $[a, b] - I_1 - I_2 - \cdots - I_m$. Then write $E'_n = \max_{x \in J} |\varepsilon(x)|$. Since the maxima of $\varepsilon(x)$ are assumed only on the I's then $E'_n < E_n$.

Choosing η to satisfy

$$0 < \eta < \frac{(E_n - E'_n)}{2 \max\limits_{a \le x \le b} |q(x)|} \tag{5.4.6}$$

then the assertion is that

$$\max_{a \le x \le b} |f(x) - p(x) - \eta q(x)| = \max_{a \le x \le b} |\varepsilon(x) - \eta q(x)| < E_n \tag{5.4.7}$$

which would yield a better approximation than $p(x)$.

On I_n, (5.4.6) gives

$$0 < \eta q(x) < \frac{q(x)(E_n - E'_n)}{2 \max\limits_{a \le x \le b} |q(x)|} \le \frac{E_n}{2} < \varepsilon(x) \tag{5.4.8}$$

Therefore $\varepsilon(x) > \eta q(x)$ and

$$\begin{aligned} |\varepsilon(x) - \eta q(x)| &= \varepsilon(x) - \eta q(x) \\ &\le E_n - \min_{x \in I_n} \eta q(x) = E_n - \text{a positive quantity} < E_n \end{aligned} \tag{5.4.9}$$

A similar result will hold for I_{n+1} and hence for J, using (5.4.6) gives

$$\begin{aligned} \max_{x \in J} |\varepsilon(x) - \eta q(x)| &\le \max_{x \in J} |\varepsilon(x)| + \eta \max_{x \in J} |q(x)| \\ &< E'_n + \frac{\max\limits_{x \in J} |q(x)|}{\max\limits_{a \le x \le b} |q(x)|} \frac{(E_n - E'_n)}{2} \\ &< E'_n + E_n - E'_n = E_n \end{aligned} \tag{5.4.10}$$

which proves the assertion and establishes the proof.

The converse theorem is easier to prove and can be stated as follows.

Theorem

Let $f(x) \in C[a, b]$. Given a polynomial $q(x)$ of degree n, that is $q(x) \in \mathcal{P}_n$, with

$$\max_{a \le x \le b} |f(x) - q(x)| = \delta \tag{5.4.11}$$

then if $\exists\ n + 2$ points $a \le x_1 < x_2 < \cdots < x_{n+2} \le b$ such that $f(x_i) - q(x_i) = \pm\delta$ for $i = 1, 2, \ldots, n + 2$ in an alternating fashion, then $\delta = E_n(f)$ and $q(x)$ is the best uniform approximation to f in $\mathcal{P}_n$.

Proof

By definition $E_n(f) \leq \delta$, hence assume that $E_n(f) < \delta$. Let $p(x)$ be the best uniform approximation. Then

$$q(x) - p(x_i) = q(x_i) - f(x_i) - (p(x_i) - f(x_i)) \tag{5.4.12}$$

and since

$$\max_{a \leq x \leq b} |f(x) - p(x)| = E_n(f) < \delta \tag{5.4.13}$$

letting

$$\tau_i = p(x_i) - f(x_i) \tag{5.4.14}$$

gives

$$|\tau_i| < \delta \tag{5.4.15}$$

$$\text{and} \qquad q(x_i) - p(x_i) = \pm\delta - \tau_i \tag{5.4.16}$$

Hence the function $q(x) - p(x) \in \mathcal{P}_n$ and has $n + 2$ points of alternation. Hence it has $n + 1$ zeros and therefore must be itself identically zero. Hence $q(x) \equiv p(x)$.

The implication of these theorems is considerable. Firstly, the best approximation in the uniform norm sense is simply recognized in that the error will have $n + 2$ equal and opposite maxima and minima. This is often called the oscillation property or the minimax property. Conversely, such an approximation can be set up by forcing the coefficients of the approximating polynomial to yield the minimax conditions. Hence to find the best quadratic approximation to $\arcsin x$ on the interval (0, 1), set

$$\varepsilon(x) = a + bx + cx^2 - \arcsin x \tag{5.4.17}$$

and the minimax conditions are

$$h = -a \tag{5.4.18}$$

$$-h = \arcsin x_1 - a - bx_1 - cx_1^2 \tag{5.4.19}$$

$$h = \arcsin x_2 - a - bx_2 - cx_2^2 \tag{5.4.20}$$

$$-h = \pi/2 - a - b - c \tag{5.4.21}$$

where the monotonic nature of $\arcsin x$ implies that two of the points of alternation are the end points and the other two are taken as x_1 and x_2. Also at x_1 and x_2 the error function must have a turning point and this yields two more equations as

$$\varepsilon'(x) = \frac{1}{\sqrt{1 - x^2}} - b - 2cx \tag{5.4.22}$$

which when set to zero at x_1 and x_2 gives

$$b + 2cx_1 = \frac{1}{\sqrt{1 - x_1^2}} = y_1 \tag{5.4.23}$$

$$b + 2cx_2 = \frac{1}{\sqrt{1 - x_2^2}} = y_2 \tag{5.4.24}$$

and there are six equations for the unknowns a, b, c and h which appear linearly, and x_1 and x_2 which appear non-linearly. Eliminating the linear variables gives the pair of non-linear equations

$$(y_1x_2 - y_2x_1) - (y_1 - y_2)(x_1 - 1)/2 = \frac{(x_2 - x_1)}{(x_1 - 1)}\left(\arcsin x_1 - \frac{\pi}{2}\right) \tag{5.4.25}$$

$$(y_1x_2 - y_2x_1) - (y_1 - y_2)x_2/2 = \frac{(x_2 - x_1)}{x_2}\arcsin x_2 \tag{5.4.26}$$

These can be solved numerically using one of the methods of Chapter 1 for sets of non-linear equations and the result is

$$p(x) = 0.095\,37 - 4.436 \times 10^{-9}x + 1.3801x^2 \tag{5.4.27}$$

which bears comparison with the Taylor expansion of the same function. Indeed at $x = 1$ the Taylor error is 0.404 whereas the minimax error is $h = 0.095\,37$ showing the gain in accuracy. In both cases the maximum error in the interval has been used.

It is clearly not a trivial matter to obtain best approximations of other than very low degree. In general there will be a residue of n simultaneous non-linear equations to be solved. These can be set up computationally by using Gauss elimination to eliminate the linear variables for every evaluation of the non-linear equations, as the non-linear variables will have current values obtained from the non-linear equation algorithm. Therefore normal elimination processes can be used to get the linear variables and hence evaluate the non-linear equations for use in say a multi-dimensional Newton as in Section 1.6.

A useful practical possibility is whether approximations can be found which, though not exactly best approximations, are close enough to this ideal and yet are easier to find. It is here that the importance of the Chebyshev polynomials becomes apparent, as $T_{n+1}(x)$ has $n + 2$ equal and opposite maxima and minima as is required by the Oscillation theorem. Hence if a function is expanded as a series of Chebyshev functions in the form

$$f(x) = \sum_{r=0}^{n} a_rT_r(x) + a_{n+1}T_{n+1}(x) + \cdots \tag{5.4.28}$$

then the error term in truncating the series after the term $r = n$, is dominated by the $T_{n+1}(x)$ term with the oscillation property. If this were the only part of the error term, then the finite sum to n terms would indeed be the best approximation. However, if the terms a_r fall off quite quickly then the resulting finite series, though not exactly a best approximation will be near to this ideal, and can be used with advantage. Hence there has been considerable effort to use Chebyshev expansions in the solution of many numerical problems. One such use has already been seen in the differentiation method of Section 2.6. Another will arise in the next chapter on quadrature when the method of Clenshaw and Curtis is considered.

There is also a direct application of this work in approximation theory called Economizing series and this process illustrates the power of the Chebyshev expansion very well. In Appendix D is shown a list of Chebyshev polynomials and their expansions in powers of x, together with the inverse table of powers of x expressed as series of Chebyshev polynomials. There are indeed simple relations between the various coefficients which render them easy to programme if necessary. Economization of series involves the simple but tedious procedure of starting with a given series expansion of $f(x)$, often a Taylor expansion. This expansion is re-written in terms of Chebyshev polynomials from the tables in the Appendix. The resulting series of Chebyshev polynomials is now truncated at some suitable point and the error incurred will then have an approximate minimax behaviour. The truncated series can then be re-expressed in powers of x to give an 'almost' best approximation. It certainly turns out that the shorter series has almost the accuracy of the longer original series, with a clear saving in effort on evaluation. This process is best illustrated by an example.

Consider the economization of the series for $\tan x$ given by

$$\tan x = x + \frac{x^3}{3} + \frac{2x^5}{15} + \frac{17x^7}{315} + \cdots \tag{5.4.29}$$

The resulting conversion to a series of Chebyshev polynomials using the formulae of Appendix D is tabulated in Table 5.2.

Truncating after the column for T_3 gives the result

$$\tan x \sim \left(1 + \frac{1}{4} + \frac{1}{12} + \frac{17}{9.64}\right) T_1 + \left(\frac{1}{12} + \frac{1}{24} + \frac{17}{15.64}\right) T_3$$

$$= \frac{785}{576} T_1 + \frac{137}{960} T_3 \tag{5.4.30}$$

Table 5.2

	T_0	T_1	T_2	T_3	T_4	T_5	T_6	T_7
x		1						
$\frac{x^3}{3}$		$\frac{1}{4}$		$\frac{1}{12}$				
$\frac{2x^5}{15}$		$\frac{5}{8}\frac{2}{15}$		$\frac{5}{16}\frac{2}{15}$		$\frac{2}{15.16}$		
$\frac{17x^7}{315}$		$\frac{17}{315}\frac{35}{64}$		$\frac{17}{315}\frac{21}{64}$		$\frac{17}{315}\frac{7}{64}$		$\frac{17}{315}\frac{1}{64}$

Conversion back to powers of x then yields the economized series

$$\tan x \sim \frac{785}{576}x + \frac{137}{960}(4x^3 - 3x)$$
$$= \frac{2692}{2880}x + \frac{137}{240}x^3 \tag{5.4.31}$$

It is instructive to compare this economized form with the original Taylor expansion, at say $x = 1$. The true value is 1.5574, whereas the original four terms of the Taylor expansion gives 1.5206. Taking only two terms of the Taylor expansion gives 1.3333, compared with the equivalent two-term economized series with a value of 1.5206. Hence the economized series retains much of its original accuracy despite having been truncated. This arises as the truncated terms almost show the minimax property.

Attention will now be given to alternatives to the direct use of the Oscillation theorem and the resulting solution of sets of non-linear equations to obtain accurate best approximations.

5.5 Exchange algorithm

The exchange algorithm is an iterative process designed to generate best uniform norm approximations for a given function and interval. The algorithm forces the minimax property to hold in a stage by stage approach. The algorithm has the following form.

(1) Choose $n + 2$ points in $[a, b]$, $x_i \quad i = 1, n + 2$.

(2) Fit a polynomial $P_n(x)$ to these points such that the errors at these points are alternately $\pm h$.
This process is called levelling, and there are $n + 2$ equations for the unknowns h and the $n + 1$ coefficients of P_n.

(3) Find the maximum error H in the entire interval, say at $x = x_{max}$.
If $|h| = H$ then the process is complete.
If not continue with:

(4) Reject one of the $n + 2$ points and exchange it for x_{max} such that the signs still alternate.
This is the exchange aspect of the algorithm.

(5) Repeat the process with the updated $n + 2$ points.

The convergence of this process requires proof, and to this end the steps of the algorithm require more precise mathematical representation. Suppose that after the kth step the current polynomial is P^k with points $x_1^k < \cdots < x_{n+2}^k$ such that

$$|f(x_i^k) - P^k(x_i^k)| = |h_k| > 0 \qquad i = 1, \ldots n + 2 \tag{5.5.1}$$

and such that $f(x_i^k) - P^k(x_i^k)$ is alternately $+|h_k|$ and $-|h_k|$. Let x_{max} of step (3) of the algorithm be more succinctly represented by y with the property that

$$\max_{a \le x \le b} |f(x) - P^k(x)| = |f(y) - P^k(y)| = M_k \tag{5.5.2}$$

Then if y is one of the x_i, $p^k(x)$ is the desired polynomial by the oscillation theorem. If not, then the new set of x's is formed, namely $x_1^{k+1} < \cdots < x_{n+2}^{k+1}$, consisting of all the x_i^k except one which is replaced with y in such a way as to maintain the alternation of signs.

The levelling process then consists of finding a new polynomial $P^{k+1}(x)$ and h_{k+1} such that

$$|f(x_i^{k+1}) - P^{k+1}(x_i^{k+1})| = |h_{k+1}| \qquad i = 1, n+2 \tag{5.5.3}$$

with $f(x_i^{k+1}) - P^{k+1}(x_i^{k+1})$ alternately positive and negative. This process can be achieved by a suitable modification to Lagrange interpolation. Let $Q(x)$ be a polynomial of degree $n+1$ which takes the values r_i at the points $x = x_i^{k+1}$. Then Lagrange's formula gives immediately

$$Q(x) = \sum_{i=1}^{n+2} r_i \prod_{\substack{j=1 \\ j \ne 1}}^{n+2} \frac{x - x_j^{k+1}}{x_i^{k+1} - x_j^{k+1}} \tag{5.5.4}$$

This polynomial is one degree greater than the required polynomial. A condition that the coefficient of x^{n+1} of $Q(x)$ should be zero leads to a choice of r_i which gives the required polynomial P^{k+1}. Hence let

$$\lambda_i = \prod_{\substack{j=1 \\ j \ne 1}}^{n+2} (x_i - x_j)^{-1} \tag{5.5.5}$$

dropping the superscripts $k+1$. The coefficient of x^{n+1} in $Q(x)$ is

$$\sum_{i=1}^{n+2} \lambda_i r_i \tag{5.5.6}$$

and hence $Q(x)$ is of degree n iff

$$\sum_{i=1}^{n+2} Q(x_i)\lambda_i = 0 \tag{5.5.7}$$

But $|\lambda_i| > 0$, and the sign of λ_i is different from the sign of λ_{i+1} for each i. Hence defining

$$\text{signum}(x) = \begin{cases} +1 & \text{if } x \ge 0 \\ -1 & \text{if } x < 0 \end{cases} \tag{5.5.8}$$

then the step is completed if h_{k+1} and P^{k+1} can be found such that

$$P^{k+1}(x_i) = f(x_i) \pm \mathrm{signum}(\lambda_i)|h_{k+1}| \tag{5.5.9}$$

Put $Q = P^{k+1}$ in (5.5.7) to give

$$\sum_{i=1}^{n+2} f(x_i)\lambda_i = \mp|h_{k+1}| \sum_{i=1}^{n+2} |\lambda_i| \tag{5.5.10}$$

and as $P^k(x)$ is itself a polynomial of degree n then

$$\begin{aligned}\sum_{i=1}^{n+2} f(x_i)\lambda_i &= \sum_{i=1}^{n+2}(f(x_i) - P^k(x_i))\lambda_i \\ &= \mp \sum_{i=1}^{n+2} |e_i\lambda_i|\end{aligned} \tag{5.5.11}$$

where e_i is either $\pm h_k$ or $\pm M_k$ by the construction of x^{k+1}. Therefore

$$|h_{k+1}| = \frac{\sum_{i=1}^{n+2} |e_i\lambda_i|}{\sum_{i=1}^{n+2} |\lambda_i|} \neq 0 \tag{5.5.12}$$

and

$$P^{k+1}(x) = \sum_{i=1}^{n+2}[f(x_i) \pm \mathrm{signum}(\lambda_i)|h_{k+1}|] \prod_{\substack{j=1 \\ j\neq 1}}^{n+2} \frac{x - x_j^{k+1}}{x_i^{k+1} - x_j^{k+1}} \tag{5.5.13}$$

which is the required polynomial $P^{k+1}(x)$ at the $k+1$th step.

The proof of convergence depends on a Lemma usually accredited to De la Vallée-Poussin.

Lemma

Let there be points $a \leq x_1 < x_2 < \cdots < x_n \leq b$, and positive numbers λ_i, $i = 1, \ldots, N$ where $N = n+2$ and a polynomial $P \in \mathcal{P}_n$ such that

$$f(x_i) - P(x_i) = (-1)^{i+m}\lambda_i \tag{5.5.14}$$

for some fixed integer m equal to 0 or 1. Then

$$\min_{Q\in\mathcal{P}_n} ||f - Q|| \geq \min(\lambda_1, \ldots \lambda_N) \tag{5.5.15}$$

Proof

Suppose that for some $Q \in \mathcal{P}_n$ that

$$||f - Q|| < \min(\lambda_1, \ldots \lambda_n) \tag{5.5.16}$$

Let $\Delta(x) = Q(x) - P(x) \equiv [f - P] - [f - Q]$ then

$$\begin{aligned} |\Delta(x_i)| &\leq |f(x_i) - P(x_i)| - |f(x_i) - Q(x_i)| \\ &= \lambda_i - |f(x_i) - Q(x_i)| > 0 \end{aligned} \tag{5.5.17}$$

and therefore $\operatorname{sign} \Delta(x_i) = (-1)^{i+m}$. Hence $\Delta(x)$ changes sign $N - 1 = n + 1$ times, and hence $P - Q$ is a polynomial of degree n vanishing $n + 1$ times. Hence $P \equiv Q$ to force a contradiction and hence establish the lemma.

The proof of convergence of the exchange algorithm is now fairly direct. Equation (5.5.12) gives

$$\begin{aligned} |h_{k+1}| &= \frac{\sum_{i=1}^{n+2} |\lambda_i||e_i|}{\sum_{i=1}^{n+2} |\lambda_i|} \\ &> \min_i |e_i| > |h_k| \end{aligned} \tag{5.5.18}$$

and hence using $|h_k|$ for λ_k in the lemma gives

$$|h_{k+1}| < |h_k| < \cdots < E < M_k \tag{5.5.19}$$

and also there exists a θ_k satisfying $0 < \theta_k < 1$ such that

$$|h_{k+1}| = (1 - \theta_k)|h_k| + \theta_k M_k \tag{5.5.20}$$

It can be proved that the x_i^k are separated, and hence $|x_i^k - x_j^k|$ are bounded away from zero. Further the $|\lambda_i|$ are bounded to give a θ such that $\theta_k > \theta > 0$ and

$$\begin{aligned} |h_{k+1}| - |h_k| &= \theta_k(M_k - |h_k|) > \theta[M_k - |h_k|] \\ &> \theta[E - |h_k|] \end{aligned} \tag{5.5.21}$$

and hence

$$E - |h_{k+1}| < (1 - \theta)(E - |h_k|) \tag{5.5.22}$$

Thus

$$E - |h_{k+1}| < q^{k+1}[E - |h_0|] \tag{5.5.23}$$

where $0 < q < 1$ and hence $\lim_{k\to\infty} |h_k| = E$ as required.

The oscillation theorem and the exchange algorithm are now compared on a simple example, that is finding the best approximation in the uniform norm sense to $\sqrt{x}$ on [0, 10]. Consider first the direct application of the oscillation theorem along the same lines as the example in Section 5.4.

The error function is given by

$$\varepsilon(x) = f(x) - ax^2 - bx - c \tag{5.5.24}$$

with $f(x) = \sqrt{x}$, and the derivative $\varepsilon'(x)$ by

$$\varepsilon'(x) = f'(x) - 2ax - b \tag{5.5.25}$$

The conditions of the oscillation theorem then insist that $\varepsilon(x)$ has four equal and opposite maxima and minima of magnitude say, h, at the points 0, x_1, x_2 and 10. Hence the defining equations are

$$\begin{aligned} h &= f(0) - c \\ -h &= f(x_1) - ax_1^2 - bx_1 - c \\ h &= f(x_2) - ax_2^2 - bx_2 - c \\ -h &= f(10) - 100a - 10b - c \qquad (5.5.26) \\ f'(x_1) &= 2ax_1 + b \\ f'(x_2) &= 2ax_2 + b \end{aligned}$$

the last two equations defining the internal turning points. The linear variables can be eliminated to yield

$$a = \frac{1}{4}\frac{(x_2^{-1/2} - x_1^{-1/2})}{(x_2 - x_1)} \tag{5.5.27}$$

$$b = \frac{1}{2}\frac{(x_2 x_1^{-1/2} - x_1 x_2^{1/2})}{(x_2 - x_1)} \tag{5.5.28}$$

which yield the two non-linear equations for x_1 and x_2 in the form

$$F_1 = 0 = x_2^{1/2} - ax_2^2 - bx_2 \tag{5.5.29}$$

$$F_2 = 0 = x_1^{1/2} - 10^{1/2} - ax_1^2 - bx_1 + 100a + 10b \tag{5.5.30}$$

with a and b given by (5.5.27) and (5.5.28) to make for easy coding. The pair of equations are then solved numerically using the methods of Chapter 1 to give $x_1 = 0.809\,013\,24$ and $x_2 = 6.038\,5572$, and the quadratic approximation is then

$$\sqrt{x} \sim -0.033\,695\,37x^2 + 0.610\,414\,26x + 0.213\,836\,06 \tag{5.5.31}$$

The alternative of using the exchange algorithm yields the following steps. Suppose the starting points are the set {0, 3, 6, 10}, then the levelling polynomial of degree two which misses the curve $\sqrt{x}$ by $\pm h$ at these points is

$$0.122\,55 + 0.583\,045\,58x - 0.029\,132\,881x^2 \tag{5.5.32}$$

and the corresponding h is $-0.122\,555$. This approximation has maximum errors at the points 0.885 0563 and 6.688 5498 where the errors are $-0.325\,011$ and 0.132 75 respectively. The former has the largest error in absolute terms and is therefore exchanged with the original point at $x = 3$ to maintain the alternating maxima and minima.

Hence at step two, the set of points is $\{0, 0.885\,0563, 6, 10\}$ and the new levelling polynomial is

$$0.213\,491\,55 + 0.610\,326\,54x - 0.033\,679\,709x^2 \tag{5.5.33}$$

with $h = -0.213\,491\,55$. The second stage turning points are at 0.809 239 14 and 6.040 5651 with errors $-0.214\,241\,29$ and 0.213 533 07 respectively. Hence the exchange is with the elements 0.885 0563 and 0.809 239 14 to give the set for the third step as $\{0, 0.809\,239\,14, 6, 10\}$. Note at this early stage that the oscillation property is beginning to manifest itself with errors around ± 0.2.

The levelling polynomial is now

$$0.213\,830\,79 + 0.610\,428\,32x - 0.033\,691\,671x^2 \tag{5.5.34}$$

with $h = -0.213\,830\,79$, turning points 0.808 969 77 and 6.038 5245, and respective errors $-0.213\,8308$ and 0.213 868 26. Now it is the turn of the six to be exchanged to give the set $\{0, 0.809\,239\,14, 6.038\,5245, 10\}$ with the fourth levelling polynomial

$$0.213\,836\,06 + 0.610\,414\,26x - 0.033\,695\,37x^2 \tag{5.5.35}$$

and $h = -0.213\,836\,06$ which is the result obtained by the direct application of the oscillation theorem.

The computational aspects of this process need some care especially the problem of finding the set of maxima and minima in the errors after levelling. This is the old difficulty of finding *all* the turning points and can sometimes prove tricky if two such points are close together. Any iterative process used for this section of the code will require initial points, and it will be non-trivial to choose these to ensure the process converges to both of the close values. Nevertheless this process proves effective in generating best approximations in the uniform norm sense.

The theory can be extended beyond polynomials to rational approximations of the form

$$R_N(x) = \frac{a_0 + a_1x + \cdots + a_nx^n}{1 + b_1x + \cdots + b_mx^m} \tag{5.5.36}$$

with $N = n + m$. The coefficient b_0 has been set at zero to eliminate the possibility of an arbitrary divisor. If a situation arises in which $b_0 = 0$ then a different constant in the approximation needs to be normalized. In the polynomial case with degree n, there are $n + 1$ free constants or coefficients and the oscillation theorem expects $n+2$ equal and opposite maxima and minima. In the rational case the number of free constants is $n+m+1$, and now the number of equal and opposite maxima and minima is $n+m+2$. Hence the defining equations can be set up as for the polynomial case, and the result is a set of partially non-linear equations. Now, however, not only do the turning points x_i appear non-linearly but also the coefficients b_i.

As an example of this extended process, consider the rational approximation of $\sin x$ on $[0, \pi/10]$ by

$$\sin x \sim \frac{ax^2+b}{1+cx} \tag{5.5.37}$$

The defining equations are, by analogy with the polynomial case,

$$\begin{aligned}
b &= h \\
\frac{ax_1^2+b}{1+cx_1} - \sin x_1 &= -h \\
\frac{ax_2^2+b}{1+cx_2} - \sin x_2 &= h \\
\frac{a\pi^2/10^2+b}{1+c\pi/10} - \sin \pi/10 &= -h \\
\frac{2ax_1(1+cx_1)-c(ax_1^2+b)}{(1+cx_1)^2} - \cos x_1 &= 0 \\
\frac{2ax_2(1+cx_2)-c(ax_2^2+b)}{(1+cx_2)^2} - \cos x_2 &= 0
\end{aligned} \tag{5.5.38}$$

where h is the error at each turning point and x_1 and x_2 are the two turning points. Hence there are six equations in six unknowns, but these have a higher proportion of non-linear variables than for the polynomial case. Elimination yields a convenient form for computation with

$$b = \frac{(1+cx_1)x_2^2 \sin x_1 - (1+cx_2)x_1^2 \sin x_2}{(2+cx_1)x_2^2 - cx_2x_1^2} \tag{5.5.39}$$

and

$$a = \frac{(1+cx_1)cx_2 \sin x_1 - (1+cx_2)(2+cx_1)\sin x_2}{cx_2x_1^2 - (2+cx_1)x_2^2} \tag{5.5.40}$$

allowing the three non-linear equations

$$F_1(x_1, x_2, c) = \frac{a(\pi/10)^2+b}{1+c\pi/10} - \sin(\pi/10) + b \tag{5.5.41}$$

$$F_2(x_1, x_2, c) = \frac{2ax_1(1+cx_1)-c(ax_1^2+b)}{(1+cx_1)^2} - \cos x_1 \tag{5.5.42}$$

$$F_3(x_1, x_2, c) = \frac{2ax_2(1+cx_2) - c(ax_2^2 + b)}{(1+cx_2)^2} - \cos x_2 \tag{5.5.43}$$

to be coded simply. These are now solved numerically by the methods of Chapter 1 to yield the required best rational approximation.

A final issue arises in approximations of both the polynomial and rational type. The shorter the interval of approximation, the lower can the degree of the approximation be for the same accuracy. There are numerous devices for interval reduction which are best illustrated using specific examples. Hence to approximate $\ln y$ over $(1, \infty)$, only values on $(1, e)$ are required, as

$$\ln y = \ln\left(e^n \frac{y}{e^n}\right) = n + \ln\left(\frac{y}{e^n}\right) \tag{5.5.44}$$

and n is chosen to ensure $y/e^n \in (1, e)$. This process is no more than removing exact powers of e from the original argument. Using non-integer n allows further reduction to say $(1, e^{1/2})$, $(1, e^{1/3})$ or smaller intervals, as long as $\ln 1/2$ or $\ln 1/3$ are known. In a similar way for $\sqrt{x}$, exact squares can be extracted using

$$\sqrt{x} = \sqrt{n^2 y} = n\sqrt{y} \tag{5.5.45}$$

or indeed any known square root can replace n^2. Hence the interval is reduced to (0.1,1) for integer n or say, $(0.1, \sqrt{2})$ if $n^2 = 2$.

For the trigonometric functions, multi-angle formulae can be used, for example

$$\sin 5\theta = 5\sin\theta - 20\sin^3\theta + 16\sin^5\theta \tag{5.5.46}$$

which allows $\sin\theta$ to be approximated for $(0, \pi/20)$ and then (5.5.46) is used to recover the required sine value on the whole quarter cycle $(0, \pi/2)$. It may be thought that this type of process could be used instead of making accurate approximations. That is, the interval is reduced until say, only a simple linear approximant is sufficient to deliver the full required accuracy. The restriction on this line of thought is stability. For example, in the case of the sine example, the coefficients of the multi-angle formulae become very large and alternating. Hence there will be a loss of accuracy as the final result is of order unity, being $\sin x$. The loss of accuracy will follow the example in Chapter 1, and will be equal to the number of digits in the largest coefficient in modulus. Even with $\sin 5\theta$ the loss is around two digits, which may be acceptable.

Best uniform norm approximations yield very useful forms for practical use. The generation of a best approximation of this type is non-trivial, and many of those used most often are available in the many tables of Mathematical Functions such as Abramowitch and Stegun. By using a series of Chebyshev functions 'almost best approximations' become feasible with very little workload for a slight loss of accuracy. Such approximations appeared in Chapter 1, equations (1.2.12) and (1.2.18) and will be taken up again under the auspices of quadrature in Chapter 6.

These formulae have also been used in Chapter 2 to effect numerical differentiation in Section 2.6.

There is another aspect to approximation theory which has been ignored so far and will be treated in the final paragraph. This topic is that of Padé approximation which is a very large subject and which will be treated here only briefly.

5.6 Padé approximation

Only limited attention has been paid so far to rational approximations. Padé approximations are rational approximations which bear a close analogy with Taylor expansions for polynomials. As in the previous paragraph, define $R_N(x)$ as

$$R_N(x) = \frac{a_0 + a_1 x + \cdots + a_n x^n}{1 + b_1 x + \cdots + b_m x^m} \tag{5.6.1}$$

with $N = n + m$. To obtain a Padé expansion about $x = 0$ for the given function $f(x)$, agreement is forced at $x = 0$ between $f(x)$ and $R_N(x)$ together with their first N derivatives. Consider, therefore,

$$f(x) - R_N(x) = (c_0 + c_1 x + \cdots + C_N x^N) - \frac{a_0 + \cdots + a_n x^n}{1 + b_1 x + \cdots + b_m x^m} \tag{5.6.2}$$

where

$$c_i = \frac{f^{(i)}(0)}{i!} \tag{5.6.3}$$

that is, $f(x)$ is representated by the usual Tayor expansion about $x = 0$. Rearranging (5.6.2) gives

$$f(x) - R_N(x) = \frac{(c_0 + c_1 x + \cdots + C_N x^N)(1 + b_1 x + \cdots + b_m x^m) - a_0 - \cdots - a_n x^n}{1 + b_1 x + \cdots + b_m x^m} \tag{5.6.4}$$

and if $f(x) = R_N(x)$ at $x = 0$ then the numerator in (5.6.4) must be zero to give

$$c_0 = a_0 \tag{5.6.5}$$

For the first N derivatives of $f(x)$ and $R_N(x)$ to be equal at $x = 0$ then the coefficients of the powers of x from x to x^N are zero to give the equations:

$$\begin{aligned} b_1 c_0 + c_1 - a_1 &= 0 \\ b_2 c_0 + b_1 c_1 + c_2 - a_2 &= 0 \\ b_3 c_0 + b_2 c_1 + b_1 c_2 + c_3 - a_3 &= 0 \\ &\vdots \\ \sum_{i=1}^{\min(n,m)} b_i c_{n-i} + c_n - a_n &= 0 \end{aligned} \tag{5.6.6}$$

which yield the numerator coefficients once the denominator coefficients have been found from the remaining equations, which are solved as a set of simultaneous linear algebraic equations. These equations are

$$\begin{aligned} b_m c_{n-m+1} + b_{m-1} c_{n-m+2} + \cdots + c_{n+1} &= 0 \\ b_m c_{n-m+2} + b_{m-1} c_{n-m+3} + \cdots + c_{n+2} &= 0 \\ &\vdots \\ b_m c_n + b_{m-1} c_{n+1} + \cdots + c_{n+m} &= 0 \end{aligned} \tag{5.6.7}$$

In these equations, if $j < 0$ then for consistency define $c_j = 0$. These equations are called the Padé equations and the resulting approximation is the $[n/m]$ approximant. As $b_0 = 1$, equations (5.6.7) are often written in matrix form:

$$\begin{bmatrix} c_{n-m+1} & c_{n-m+2} & \cdots & c_n \\ c_{n-m+2} & c_{n-m+3} & \cdots & c_{n+1} \\ \vdots & & & \vdots \\ c_n & c_{n+1} & \cdots & c_{n+m-1} \end{bmatrix} \begin{bmatrix} b_m \\ b_{m-1} \\ \vdots \\ b_1 \end{bmatrix} = - \begin{bmatrix} c_{n+1} \\ c_{n+2} \\ \vdots \\ c_{n+m} \end{bmatrix} \tag{5.6.8}$$

It is common to consider a sequence of Padé approximants, and it is often the case that the region of convergence of these Padé sequences is larger than the original circle of convergence of the originating series. The approximants are usually laid out in a Padé table such as in Table 5.3.

Consider the example

$$\arctan x = x - \tfrac{1}{3}x^3 + \tfrac{1}{5}x^5 - \tfrac{1}{7}x^7 + \tfrac{1}{9}x^9 \tag{5.6.9}$$

with a Padé approximation [5/4] to give $R_9(x)$. Then the Padé equations are

$$\begin{aligned} a_0 &= 0 \\ a_1 &= 1 \\ a_2 &= b_1 \\ a_3 &= -\tfrac{1}{3} + b_2 \\ a_4 &= -\tfrac{1}{3}b_1 + b_3 \\ a_5 &= \tfrac{1}{5} - \tfrac{1}{3}b_2 + b_4 \end{aligned} \tag{5.6.10}$$

Table 5.3

n m	0	1	2	$\cdots$
0	[0/0]	[1/0]	[2/0]	$\cdots$
1	[0/1]	[1/1]	[2/1]	$\cdots$
2	[0/2]	[1/2]	[2/2]	$\cdots$
$\vdots$	$\vdots$	$\vdots$	$\vdots$	$\vdots$

which exhausts the supply of a's, and in addition

$$\begin{aligned} \tfrac{1}{5}b_1 - \tfrac{1}{3}b_3 &= 0 \\ -\tfrac{1}{7} + \tfrac{1}{5}b_2 - \tfrac{1}{3}b_4 &= 0 \\ -\tfrac{1}{7}b_1 + \tfrac{1}{5}b_3 &= 0 \\ \tfrac{1}{9} - \tfrac{1}{7}b_2 + \tfrac{1}{5}b_4 &= 0 \end{aligned} \tag{5.6.11}$$

These equations solve to give the Padé approximation

$$\arctan x = \frac{x + \frac{7}{9}x^3 + \frac{64}{945}x^5}{1 + \frac{10}{9}x^2 + \frac{5}{21}x^4} \tag{5.6.12}$$

In order to compare the various expansions, consider Table 5.4.

It can be seen that the Taylor expansion to x^9 is, as expected, very accurate for the smaller x values, but falls off quite badly at $x = 1.0$. By comparison, the Padé expansion holds its accuracy quite well even as far as $x = 1$. This type of approximation does not have the oscillation property in which the errors are uniformly spread across the interval rather than getting worse away from the point of expansion as in this case. As a further example, a Padé Table for the function e^x is shown in Table 5.5, each entry having been computed along the above lines.

Padé sequences can be employed in a huge range of applications which fall outside the scope of this book. For example if $x = 1$, then defining the Padé approximants for a function $f(x)$ as $[n_j/m_j](x)$ for $j = 1, 2, \ldots$ the useful relation

$$\sum_{i=0}^{\infty} c_i = \lim_{j\to\infty} [n_j/m_j](1) \tag{5.6.13}$$

Table 5.4

x	True	Padé	Taylor
0.2	0.197 40	0.1940	0.197 40
0.4	0.380 51	0.380 51	0.380 51
0.6	0.540 42	0.540 42	0.540 67
0.8	0.674 74	0.674 77	0.679 82
1.0	0.785 40	0.785 58	0.874 92

Table 5.5

n / m	0	1	2
0	$\frac{1}{1}$	$\frac{1+x}{1}$	$\frac{2+2x+x^2}{2}$
1	$\frac{1}{1-x}$	$\frac{2+x}{2-x}$	$\frac{6+4x+x^2}{6-2x}$
2	$\frac{2}{2-2z+z^2}$	$\frac{6+2x}{6-4x+x^2}$	$\frac{12+6x+x^2}{12-6x+x^2}$

follows. This relation can be used directly to sum series, and to accelerate sequences such as $\{A_n\}$ by the device of setting

$$c_0 = A_0$$
$$c_{n+1} = A_{n+1} - A_n \quad n = 0, 1, 2, \ldots \tag{5.6.14}$$

By considering the diagonal sequence of Padé approximants in the Padé Table, the sequence accelerator of Shanks (Chapter 1, equation (1.7.9)) can be reproduced. Another use of Padé approximants is in the computation of the position of singularities given just the power series expansion. Hence the series

$$1/(1 - z) = 1 + z + z^2 + z^3 + \cdots \tag{5.6.15}$$

would yield $z = 1$ as a singular point despite the innocence of the series expansion. This topic is linked with the QD or quotient-difference algorithm, and readers are referred to Baker and Graves-Morris (1981) for further reading.

Further reading

Baker G. A. and Graves-Morris P. (1981) *Padè Approximants*, Encyclopaedia of mathematics, **13** and **14**.
Fox L. and Parker I. (1968) *Chebyshev Polynomials in Numerical Analysis*, Oxford Press, London.
Hasting C. Jr. (1955) *Approximations for Digital Computers*, Princeton University Press, Princeton, New Jersey.
Lorentz G.G. (1966) *Approximation of Functions*, Holt, Rinehart and Winston, New York.
Wendroff B. (1966) *Theoretical Numerical Analysis*, Academic press, New York.

Exercises

(5.1) Use Rodriguez's method to determine the orthogonal polynomials for the interval $[0, \infty]$ with weight function e^{-x}. These are Laguerre's polynomials. Compare the results with (5.2.32).

(5.2) Use Rodriguez's method to obtain the orthogonal polynomials for the interval $[-\infty, \infty]$ with the weight function e^{-x^2} in the form

$$\phi_r(x) = c_r e^{x^2} \frac{\mathrm{d}^r}{\mathrm{d}x^r}(e^{-x^2})$$

where c_r is the usual multiplicative constant. These are Hermite polynomials.

(5.3) In the proof of Weierstrass's Theorem the result

$$\sum_{k=0}^{n} \left(\frac{k}{n} - x\right)^2 \binom{n}{k} x^k (1-x)^{n-k} = \frac{x(1-x)}{n}$$

is used. Prove this by using the binomial expansion

$$\sum_{k=0}^{n} \binom{n}{k} (ax)^k (1-x)^{n-k} = (1-x+ax)^n$$

By differentiating this result twice by a, a series for

$$\sum_{k=0}^{n} k \binom{n}{k} x^k (1-x)^{n-k}$$

and

$$\sum_{k=0}^{n} k^2 \binom{n}{k} x^k (1-x)^{n-k}$$

can be found to combine to give the required result.

(5.4) Expand $\arcsin x$ as a series of Chebyshev polynomials using orthogonality. Find the general term and prove that

$$\tfrac{1}{8}\pi = 1 + \tfrac{1}{9} + \tfrac{1}{25} + \cdots$$

(5.5) Use the power series expansion of $\arctan x$ up to and including powers of x^9 to obtain an economized series up to and including x^3. Compare the errors in this economized series, with the same number of terms of a Taylor series.

(5.6) Economize the Taylor series

$$J_0(z) = 1 - \frac{1}{4}z^2 + \frac{(\frac{1}{4}z^2)^2}{(2!)^2} - \frac{(\frac{1}{4}z^2)^3}{(3!)^2}$$

to a quadratic. Compare the values of $J_0(1)$ obtained from the series above, two terms of Taylor, the economized series and the true value.

(5.7) Use the oscillation theorem directly to find the best approximation of degree zero to $\sqrt{x}$ on the interval $[1, 2]$.

(5.8) Use the oscillation theorem to find the best approximations of degree one and two to $\sin x$ on the interval $[0, \pi/4]$.

(5.9) Use the extended oscillation theorem to find the best rational approximation

$$\frac{ax^2 + bx + c}{x + d}$$

to $x^{1/3}$ on the interval $[0, 10]$.

(5.10) Repeat the procedure of exercise (5.8) for $\ln(x)$ on the reduced interval $[1, e]$ for the approximation

$$\ln(x) \sim \frac{ax + bx^2}{1 + cx}$$

(5.11) Use the exchange algorithm to generate the best approximation of degree two to $\sin x$. It is advisable here to have to hand routines for solving linear algebraic equations for the levelling process, and a non-linear equation solver for the maxima and minima.

(5.12) Use the exchange algorithm for the problem in exercise (5.6). [It is useful to know that the final points of alternation are $x_1 = 0$, $x_2 = 0.014$, $x_3 = 0.558$, $x_4 = 4.18$ and $x_5 = 10$].

(5.13) Construct a Padé approximant to $\sin x$ with numerators and denominators of degree four. Find the maximum error and where it occurs on the interval $[0, \pi/4]$. How does this compare with a best uniform norm approximation?

(5.14) The Fresnel integral $C(x)$ is defined by the Taylor series

$$C(z) = \sum_{n=0}^{\infty} \frac{(-1)^n \left(\frac{\pi}{2}\right)^{2n}}{(2n!)(4n+1)} z^{4n+1}$$

Find a Padé approximation to $C(z)$ of the form

$$\sum_{i=0}^{5} a_i z^i \Big/ \sum_{j=0}^{4} b_j z^j$$

Given that $C(1) = 0.779\,8934$ compare the Taylor and Padé approximations.

6

Quadrature

6.1 Introduction

One of the fundamental problems of analysis is that of integration. It is soon apparent to students of mathematics that this analytic process leads to closed-form solutions in a very limited set of cases. Even when such a closed form exists such as

$$\int_0^\infty t^{z-1}e^{-t}\,\mathrm{d}t = \Gamma(z) \tag{6.1.1}$$

it is often the case that the integral representation is commonly used as the principle definition. Hence the problem has just been given a convenient name, in this case the Gamma function, and without actually evaluating the integral, many of the analytic properties such as series expansions, can be derived. A glance through Abramowitch and Stegun (1965) will show that a great number of the commonly used integrals fall into this category. Numerical quadrature is all about evaluating integrals of all sorts from a knowledge of the integrand alone.

It is very easy to derive a simple and effective method by considering the integral as the area under a curve. If the curve is taken as series of straight lines then the area under each of these lines is a trapezium. The situation is illustrated in Figure 6.1.

Hence a simple quadrature formula can be obtained in the form

$$\int_a^b f(x)\,\mathrm{d}x = \frac{h}{2}[f(a) + 2f(a+h) + \cdots + 2f(a+(n-1)h) + f(b)] + E \tag{6.1.2}$$

using the fact that the area of a typical trapezium is

$$\tfrac{1}{2}h[f(a+ih) + f(a+(i+1)h)] \tag{6.1.3}$$

Clearly this is an approximation in general and hence the error is represented by E. It is to be expected that the smaller that h is made, the less the error E will be. The relationship between the error E and the step h will be pursued in more detail later.

There is a close relationship between the last chapter on approximation theory and quadrature. If $f(x)$, the integrand, can be well-approximated by a simpler function, and if this function or series of functions is itself integrable analytically,

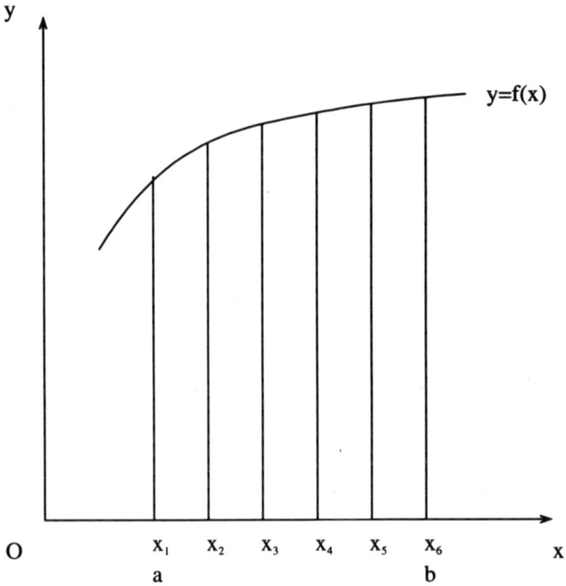

Figure 6.1

then the result is a quadrature rule. In the case of the trapezoidal formula above, the approximation is a piece-wise straight line fit and the analytic integration stage is trivial.

Hence consider the integral

$$\int_0^h f(x)\,\mathrm{d}x \tag{6.1.4}$$

which is typical of any finite integral by moving the point a to the origin and scaling x to give an upper limit h. Then fitting a straight line to pass through $(0, f(0))$ and $(h, f(h))$ gives

$$y = f_0 + \left(\frac{f_1 - f_0}{h}x\right) \tag{6.1.5}$$

where $f_0 = f(0)$ and $f_1 = f(h)$. Hence

$$\begin{aligned}\int_0^h f(x)\,\mathrm{d}x &= \int_0^h f_0 + \left(\frac{f_1 - f_0}{h}x\right)\,\mathrm{d}x\\ &\sim f_0 h + \tfrac{1}{2}(f_1 - f_0)h\\ &= \tfrac{1}{2}h(f_0 + f_1)\end{aligned} \tag{6.1.6}$$

to give the trapezoidal rule again.

An equivalent approach is to take Lagrange's interpolation formula of Section 2.2 for a two-point fit and integrate the approximation to give

$$f(x) \sim \frac{(x-h)}{(x_0-h)} f_0 + \frac{(x-x_0)}{(h-x_0)} f_1 + \frac{f''(\xi)}{3!}(x-x_0)(x-h) \tag{6.1.7}$$

with $x_0 = 0$. Integrating both sides yields both the trapezoidal quadrature rule and an error in the form

$$\int_0^h f(x)\,\mathrm{d}x = \frac{h}{2}(f_0 + f_1) + \frac{1}{3!}\int_0^h x(x-h)f''(\xi)\mathrm{d}x \tag{6.1.8}$$

Because of the constant sign of $x(x-h)$ in the interval of integration it is possible to use the integral mean value theorem to obtain the error E in the form

$$E = \frac{h^2}{6} f''(\eta) \tag{6.1.9}$$

where η is again an unknown point in the interval of integration.

There is another way to look at these rules. The trapezoidal rule fits a linear function to $f(x)$ and hence the general quadrature form

$$\int_0^h f(x)\,\mathrm{d}x \simeq W_1 f(0) + W_2 f(h) \tag{6.1.10}$$

is exact for $f(x) = 1$ and $f(x) = x$. This gives two linear algebraic equations for W_1 and W_2

$$W_1 + W_2 = h \tag{6.1.11}$$

$$W_2 h = h^2/2 \tag{6.1.12}$$

with the solution $W_1 = W_2 = h/2$ to yield the trapezoidal rule for the third time. This method is called the method of undetermined coefficients, and will be used extensively in the following work.

There is an obvious way to seek a more accurate rule. If a quadratic rather than a linear polynomial is fitted to $f(x)$ at the points $(0, f(0))$, $(h, f(h))$ and $(2h, f(2h))$ then the resulting quadratic can be integrated to give the rule

$$\int_0^{2h} f(x)\,\mathrm{d}x = \frac{h}{3}(f_0 + 4f_1 + f_2) + \frac{h^4}{90} f^{iv}(\xi) \tag{6.1.13}$$

which is known as Simpson's rule. The error term is not as straightforward to obtain as in the trapezoidal case, as the factor $x(x-h)(x-2h)$ which will occur under the integral, is no longer of constant sign. A slightly more involved calculation is needed and the result is the error in (6.1.13) which is one order higher than might be expected for a quadratic fit. This analysis will be considered later. The practical effect is that Simpson's rule, though designed to integrate a quadratic exactly will also integrate a cubic exactly. This idea of something for nothing has endeared Simpson's rule to many users, but the gain is very little compared with that which can be made by more advanced techniques. Just as for the trapezoidal rule, the method of undetermined coefficients can also be used to generate Simpson's rule.

Consider the interval now from 0 to $2h$, and the integral

$$\int_0^{2h} f(x)\mathrm{d}x \simeq W_1 f(0) + W_2 f(h) + W_3 f(2h) \tag{6.1.14}$$

Then the equations for the undetermined coefficients follow by making (6.1.14) exact for $f(x) = 1$, $f(x) = x$ and $f(x) = x^2$. This yields the equations

$$\begin{aligned} W_1 + W_2 + W_3 &= 2h \\ W_2 h + W_3(2h) &= 2h^2 \\ W_2 h^2 + W_3(2h)^2 &= 8h^3/3 \end{aligned} \tag{6.1.15}$$

with the solution $W_1 = W_3 = h/3$ and $W_2 = 4h/3$, which is again Simpson's rule.

The trapezoidal and Simpson's rules are usually generalized by subdivision, that is by summing repeated applications of the elementary rules over each of the sub-intervals, to give the quadrature formulae

$$\int_a^b f(x)\mathrm{d}x = \frac{h}{2}(f_0 + 2f_1 + \cdots + 2f_{n-1} + f_n) - \frac{nh^3}{12} f''(\xi) \tag{6.1.16}$$

and

$$\int_a^b f(x)\mathrm{d}x = \frac{h}{3}(f_0 + 4f_1 + 2f_2 + 4f_3 + \cdots + 4f_{n-1} + f_n) - \frac{nh^5}{180} f^{iv}(\xi). \tag{6.1.17}$$

These rules are referred to as the generalized trapezoidal rule and the generalized Simpson's rule respectively.

Based on the type of formulae which arise in these simple-minded derivations, and on the analytic definition of a Riemann integral it is reasonable to define a general quadrature rule to have the form

$$\int_a^b w(x) f(x)\,\mathrm{d}x = \sum_{i=1}^{N} W_i f(x_i) + E \tag{6.1.18}$$

where the W_i are called the weights, x_i the abscissae and $w(x)$ the weight function. Clearly the weights and abscissae depend on the interval $[a, b]$ and the weight function $w(x)$. The choice of weight function is used to remove some unsavoury but commonly occurring difficulty from the integrand, often a singularity of some sort. Quadrature rules can also be classified into open rules, which have abscissae which exclude the end-points, and closed rules which have abscissae which do include the end-points. Open rules are sometimes used to integrate functions with end-point singularities, as no function evaluation is then needed at the singular point.

It is now timely to consider extensions of the simple rules above by fitting a polynomial of arbitary degree to $f(x)$, or equivalently integrating the general Lagrange interpolation polynomial. This gives a set of rules called the Newton–Cotes formulae.

6.2 Newton–Cotes quadrature

The two lowest-order rules in this class have been derived above either by integrating Lagrange's interpolation formula or by using undetermined coefficients. Higher order rules using polynomials of general degree can also be obtained by either approach. The resulting quadrature rules form the set of Newton–Cotes rules and are tabulated in Table 6.1 up to the nine-point rule.

These rules can be computed in general as in Alaylioglu, Evans and Hyslop (1973), but examination of Table 2.1 shows that even at nine-points the coefficients for the weights are becoming alternating in sign. This heralds the onset of instability in the formulae, and a loss of accuracy equal to the number of digits in the largest coefficient can be expected due to subtractive cancellation of the type discussed in Chapter 1. As a result of this phenomenon, it is normal to use sub-division together with a repeatedly-used low-order rule to obtain convergence.

It is also possible to obtain a set of formulae in which only interior abscissae are used, so generating a set of open quadrature rules. Hence, either the Lagrangian fit is made on the interior points only, or the method of undetermined coefficients is employed with no terms involving $f(a)$ or $f(b)$ to yield rules of the form:

$$\int_0^{3h} f(x)\,\mathrm{d}x = \frac{3h}{2}(f_1 + f_2) + \frac{3h^3}{4} f''(\xi) \tag{6.2.1}$$

Table 6.1 Newton–Cotes quadrature rules

$\int_{x_1}^{x_n} f(x)\mathrm{d}x = kh\sum_{i=1}^{n} W_i f_i - Kh^{m+1} f^{(m)}(\xi)$
with $f_i = f(x_i)$

n	k	$W_i = W_{n-i+1},\quad i = 1, [(n+1)/2]$					m	K
2	$\frac{1}{2}$	1					2	$\frac{1}{12}$
3	$\frac{1}{3}$	1	4				4	$\frac{1}{90}$
4	$\frac{3}{8}$	1	3				4	$\frac{3}{80}$
5	$\frac{2}{45}$	7	32	12			6	$\frac{8}{945}$
6	$\frac{5}{288}$	19	75	50			6	$\frac{275}{12096}$
7	$\frac{1}{140}$	41	216	27	272		8	$\frac{9}{1400}$
8	$\frac{7}{17280}$	751	3577	1323	2989		8	$\frac{8183}{518400}$
9	$\frac{4}{14175}$	989	5888	−928	10496	−4540	10	$\frac{2368}{467775}$

$$\int_0^{4h} f(x)\,\mathrm{d}x = \frac{4h}{3}(2f_1 - f_2 + 2f_3) + \frac{14h^5}{45} f^{(iv)}(\xi) \tag{6.2.2}$$

$$\int_0^{5h} f(x)\,\mathrm{d}x = \frac{5h}{24}(11f_1 + f_2 + f_3 + 11f_4) + \frac{95h^5}{144} f^{(iv)}(\xi) \tag{6.2.3}$$

$$\int_0^{6h} f(x)\,\mathrm{d}x = \frac{3h}{10}(11f_1 - 14f_2 + 26f_3 - 14f_4 + 11f_5) + \frac{41h^7}{140} f^{(vi)}(\xi) \tag{6.2.4}$$

where $f_i = f(ih)$ and ξ is some point in the interval of integration. Again there are alternating signs with the associated instabilities that this will imply. The importance of these open formulae is in their use in algorithms for the numerical solution of ordinary differential equations. They can also be employed on integrals with end-point singularities in the manner indicated in the previous paragraph. The result is a crude technique for evaluating singular integrands.

The error estimate $E_n(x)$ in Lagrange's formula is integrated in the derivation of the Newton–Cotes rules, and results in the quadrature error

$$E_n = \frac{1}{(n+1)!} \int_0^n s(s-1)\cdots(s-n)F^{(n+1)}(\xi_1)\mathrm{d}s \tag{6.2.5}$$

with $\xi_1 \in (0, n)$ and the change of variable

$$x = a + \frac{(b-a)}{n}s$$

and

$$F(s) = f(x)$$

being employed. Steffensen (1927) demonstrated that (6.2.5) can be reduced to either

$$E_n = \frac{F^{(n+1)}(\xi_2)}{(n+1)!} \int_0^n s(s-1)\cdots(s-n)\mathrm{d}s \tag{6.2.6}$$

for n odd, or to

$$E_n = \frac{F^{(n+1)}(\xi_2)}{(n+2)!} \int_0^n \left(s - \frac{n}{2}\right) s(s-1)\cdots(s-n)\mathrm{d}s \tag{6.2.7}$$

for n even. Hence formulae with odd point numbers have an error which is one order higher than might be expected. As Simpson's rule is such a formula, and as it is low order so that any instabilities are absent, it has proved very popular in the past.

Asymptotic investigation of these error formulae yields another effect which has significance throughout quadrature. It is not unexpected that integrands with singularities cause difficulties to quadrature rules even if the abscissae are chosen to avoid the singularities themselves. It is more surprising that an integrand with no singularities on the real axis such as $1/(1+x^2)$ can be equally troublesome. Indeed it is the complex singularities at $\pm i$ which effect the convergence of the quadrature rule. If $f(x)$ has a Taylor expansion with radius of convergence R then using Stirling's approximation for factorial, the error estimate

$$|E_n| = O\left[\frac{1}{n^{3/2}(\ln n)^2}\left(\frac{b-a}{eR}\right)^n\right] \tag{6.2.8}$$

can be derived for a subdivision of length h. Hence if $R < (b-a)/e$ the error will ultimately increase for increasing n. Hence even singularities off the real axis can adversely effect the convergence of quadrature rules, and such integrands can be very difficult to spot if the function values arise as a table or from some computational exercise, so that the analytical form is not known.

Newton–Cotes formulae of low order have enjoyed much use as they are often the only rules learnt, being simple in nature. It is easy to see that rules with full numerical stability and higher point numbers would be far more effective in reducing the number of function evaluations needed to obtain a given accuracy, than the use of a low point number rule with subdivision. For example, the error in the generalized Simpson's rule is $nh^5 f^{(iv)}(\xi)/180$. With an effective doubling of point number to a five-point rule, which places intermediate points between each of the original points, the corresponding error from Table 6.1 is now $8(n/4)h^7 f^{(vi)}(\xi)/945$. Hence assuming no significant differences in the values of the two derivative terms, the error reduces from order h^5 to order h^7. On the other hand if the doubling of point numbers is achieved by a subdivision, the error is now $2n(h/2)^5 f^{(iv)}(\xi)/180$ and is only reduced by the factor 2^4. Hence for typical h values of say 0.1 or less the high point number strategy is far superior to the use of subdivision. Clearly stable high point number formulae are required to take advantage of this effect, whereas Newton–Cotes rules are forced into employing the less effective subdivision approach.

There is no reason why Newton–Cotes rules with weight functions $w(x)$ should not be used as long as high point numbers are avoided. Two such rules which are illustrative and useful are

$$\int_0^1 \ln(x) f(x)\,dx = \sum_{i=1}^{n} W_i f(x_i) + E \tag{6.2.9}$$

and

$$\int_a^b f(x) \begin{matrix}\sin\\ \cos\end{matrix}\, \omega x\,dx = \sum_{i=1}^{n} W_i f(x_i) + E \tag{6.2.10}$$

where the rules are easily generated if the monomial integrals

$$\int_0^1 x^n \ln(x)\,dx \quad \text{and} \quad \int_a^b x^n \begin{matrix}\sin\\ \cos\end{matrix}\, \omega x\,dx \tag{6.2.11}$$

are found analytically. Then the method of undetermined coefficients will generate a set of linear algebraic equations and produce two special families of rules. The logarithmically weighted rule appears in the exercises, and the trigonometrically weighted rules will be considered in more detail in Section 6.7, where this approach is one of the earliest methods used for oscillatory integrands.

There is one feature of these special rules which can make the Newton–Cotes approach worthwhile. If a problem arises in which a somewhat unusual weight function $w(x)$ occurs, and in which the direct integration of the full integrand is heavily hampered by the presence of $w(x)$, then being able to generate a rule with the weight function $w(x)$, may give a practical way to evaluate the integral fairly efficiently. This is particularly true in the oscillatory situation. Though some of the more refined rules which will be considered later have versions for differing weight functions, these rules are particularly hard to generate, and so it would be unlikely that it would be economic to attempt to set up a new rule for a non-standard weight function. A common dichotomy of quadrature arises. Should a high power method be used on the full integrand, or is it feasible to generate a Newton–Cotes rule and use subdivision on the much easier non-weight function part of the integrand?

Another device to improve the accuracy and yet retain Newton–Cotes formulae, due to Romberg (1955) is called Romberg quadrature. The idea here is to compute a sequence of trapezoidal approximations of the form

$$I = \int_a^b f(x)\,\mathrm{d}x = \frac{h_m}{2}\left[f(a) + f(b) + 2\sum_{j=1}^{n-1} f(x_j)\right] - \frac{b-a}{12}h_m^2 f''(\xi) \quad (6.2.12)$$

where $h_m = (b-a)/m$ and with m taking successively the values 1,2,4,8 to 2^{n-1} for some n. These approximations are denoted by the symbols $R_{1,1}, R_{2,1}, \ldots$ and are clearly related by

$$R_{k,1} = \tfrac{1}{2}\left[R_{k-1,1} + h_{k-1}\sum_{i=1}^{2^{k-2}} f\left(a + \left(i - \tfrac{1}{2}\right)\right) h_{k-1}\right]. \quad (6.2.13)$$

Given two successive trapezoidal estimates of the above type, it is possible to eliminate the h^2 error term using the equations

$$I = R_{k-1,1} - Ah^2 + O(h^4) \quad (6.2.14)$$

and

$$I = R_{k,1} - A(h/2)^2 + O(h^4) \quad (6.2.15)$$

to yield

$$I = \frac{4R_{k,1} - R_{k-1,1}}{3} + O(h^4) \quad (6.2.16)$$

in which the h subscripts have been dropped. Hence the general relation

$$R_{i,j} = \frac{4^{j-1}R_{i,j-1} - R_{i-1,j-1}}{4^{j-1} - 1} \tag{6.2.17}$$

gives a recurrence relation by which a sequence of integrals with increasing accuracy is generated. This set of estimates is often illustrated in the scheme

$$\begin{array}{lllll}
R_{1,1} & & & & \\
R_{2,1} & R_{2,2} & & & \\
R_{3,1} & R_{3,2} & R_{3,3} & & \\
\vdots & \vdots & & & \\
R_{n,1} & R_{n,2} & R_{n,3} & \cdots & R_{n,n}
\end{array}$$

which converges along the diagonal to the required integral and can be extended row by row as required by simply updating equation (6.2.13) and then extending the table along a further row using equation (6.2.17).

It has been demonstrated that stability requirements force the user of Newton–Cotes quadrature rules into repeated use of low-order rules, despite the general assertion that the best way to reduce truncation errors is to increase the point number of the rule. Were it possible to find a sequence of rules with guaranteed positive weights then the best strategy for approaching convergence by using increasingly high point numbers is validated. If at the same time, the truncation error can also be reduced then a powerful quadrature rule would have been found. Such a set of rules is derived in the next section.

6.3 Gaussian quadrature

To set the scene for Gaussian quadrature, a simple two-point formula is first derived. The general derivation then follows. Consider a formula of the form

$$\int_{-1}^{1} f(x)\,\mathrm{d}x = W_1 f(x_1) + W_2 f(x_2) + E \tag{6.3.1}$$

The Newton–Cotes approach is to fix the x_i as equally spaced, and use the two degrees of freedom in W_1 and W_2 to force exactness for $f(x) = 1$ and $f(x) = x$. But why throw away the two degrees of freedom which are represented by x_1 and x_2? The Gaussian approach also leaves the x's free and so the quadrature rule can be made exact for twice the range of powers of x. The potential dangers of this approach are considerable. There seems no obvious way to solve the resulting equations after using the method of undetermined coefficients. These equations will be non-linear. Even if some solution is found, would it not be remarkable if the resulting x_i were all real and lay in the interval of integration? Let us try the simple example first. Making (6.3.1) exact for $f(x) = 1,\ x,\ x^2,$ and x^3 gives the non-linear equations

$$
\begin{aligned}
W_1 + W_2 &= 2 \\
W_1x_1 + W_2x_2 &= 0 \\
W_1x_1^2 + W_2x_2^2 &= 2/3 \\
W_1x_1^3 + W_2x_2^3 &= 0
\end{aligned}
\tag{6.3.2}
$$

where the right-hand sides are the usual monomial integrals

$$\int_{-1}^{1} x^i \, \mathrm{d}x \tag{6.3.3}$$

defined in general in (5.2.31). It is of immediate note that the W's appear linearly and hence may be eliminated conventionally to give

$$
\begin{aligned}
W_2(x_1 - x_2) &= 2x_1 \\
W_2x_2(x_1 - x_2) &= -2/3 \\
W_2x_2^2(x_1 - x_2) &= 2x_1/3
\end{aligned}
\tag{6.3.4}
$$

and

$$-3x_1x_2 = 1 \tag{6.3.5}$$

$$x_1 + x_2 = 0 \tag{6.3.6}$$

It is worth remarking at this stage that the x's appear as the sum of the x's taken one at a time, namely $x_1 + x_2$ and the sum taken two at a time which is simply x_1x_2. The roots of a quadratic are related to the coefficients of a quadratic precisely by combinations of the x_i of this form. If x_1 and x_2 are roots of $x^2 + ax + b$ then

$$-a = x_1 + x_2 \tag{6.3.7}$$

$$b = x_1x_2 \tag{6.3.8}$$

This observation will become important in the general approach, but for the moment (6.3.5) and (6.3.6) give x_1 and x_2 as roots of $x^2 - \frac{1}{3} = 0$. Hence $x_1 = \sqrt{1/3}$ and $x_2 = -\sqrt{1/3}$, and returning to the original equations $W_1 = W_2 = 1$. Hence the remarkable has occurred. Not only do the x's turn out real but they also lie in the interval of integration. How does all this generalize?

For the general quadrature rule

$$\int_a^b f(x)\,\mathrm{d}x = \sum_{i=1}^{N} W_i f(x_i) + E \tag{6.3.9}$$

E can be made zero for

$$f(x) = 1, x, x^2, \ldots, x^{2n-1} \tag{6.3.10}$$

which gives the set of $2n$ non-linear equations:

$$
\begin{array}{lllll}
W_1+ & W_2+ & \cdots & +W_n & = c_0 \\
W_1x_1+ & W_2x_2+ & \cdots & +W_nx_n & = c_1 \\
W_1x_1^2+ & W_2x_2^2+ & \cdots & +W_nx_n^2 & = c_2 \\
\vdots & & & & \\
W_1x_1^{2n-1} & +W_2x_2^{2n-1} & +\cdots & +W_nx_n^{2n-1} & = c_{2n-1}
\end{array}
\tag{6.3.11}
$$

which consist of $2n$ unknowns of which the W_i's appear linearly. By taking any block of $(n+1)$ consecutive equations starting with the rth, the determinantal eliminant can be written down in the form

$$
\begin{vmatrix}
x_1^r & x_2^r & \cdots & \cdots & x_n^r & c_r \\
x_1^{r+1} & x_2^{r+1} & \cdots & \cdots & x_n^{r+1} & c_{r+1} \\
\vdots & & & & & \vdots \\
\vdots & & & & & \vdots \\
x_1^{r+n} & x_2^{r+n} & \cdots & \cdots & x_n^{n+r} & c_{n+r}
\end{vmatrix} = 0 \cdot
\tag{6.3.12}
$$

with r varying from 0 to $n-1$ and the c's defined in (5.2.31). If these determinants are expanded by their last columns, and if the factors $x_1^r x_2^r \cdots x_n^r$ are cancelled from the determinantal minors, the expansion below is formed.

$$
c_r x_1 x_2 \cdots x_n
\begin{vmatrix}
1 & 1 & \cdots & 1 \\
x_1 & x_2 & \cdots & x_n \\
\vdots & & & \vdots \\
x_1^{n-1} & x_2^{n-1} & \cdots & x_n^{n-1}
\end{vmatrix}
- c_{r+1}
\begin{vmatrix}
1 & 1 & \cdots & 1 \\
x_1^2 & x_2^2 & \cdots & x_n^2 \\
\vdots & & & \vdots \\
x_1^n & x_2^n & \cdots & x_n^n
\end{vmatrix}
+ \cdots
$$

$$
+(-1)^n c_{r+n}
\begin{vmatrix}
1 & 1 & \cdots & 1 \\
x_1 & x_2 & \cdots & x_n \\
\vdots & & & \vdots \\
x_1^{n-1} & x_2^{n-1} & \cdots & x_n^{n-1}
\end{vmatrix}
= 0
\tag{6.3.13}
$$

It is now clear that the expressions

$$(x_1 - x_2)(x_1 - x_3)\cdots(x_2 - x_3)\cdots(x_i - x_j)\cdots$$

for $i < j$ are all factors of these determinants which must expand to yield homogeneous functions of degree which depends on the power of the omitted row. Hence the final determinant with the $x_1^n \cdots x_n^n$ row missing must have a remaining factor which is linear, that is $x_1 + x_2 + \cdots + x_n$, after the above factors have been cancelled. The previous determinant must leave the sum of the products of the x's taken two at a time, and so on to end with the first determinant leaving just the product of the x's. Hence a direct comparison can be made with the coefficients of a polynomial

$$x^n + a_{n-1}x^{n-1} + \cdots + a_0 = 0 \tag{6.3.14}$$

in which a_{n-1} is minus the sum of the roots taken singly, down to a_0 which is $(-1)^n$ times the product of the roots. Hence as r varies from 0 to $n-1$ the set of n simultaneous linear algebraic equations arises for the coefficients $a_{n-1}, \ldots, a_0$ which define a polynomial with roots $x_1, \ldots, x_n$. These simultaneous equations can be written as

$$\begin{pmatrix} c_0 & c_1 & c_2 & \cdots & c_{n-1} \\ c_1 & c_2 & c_3 & \cdots & c_n \\ c_2 & c_3 & c_4 & \cdots & c_{n+1} \\ \vdots & & & & \vdots \\ c_{n-1} & c_n & c_{n+1} & \cdots & c_{2n-2} \end{pmatrix} \begin{pmatrix} a_0 \\ a_1 \\ a_2 \\ \vdots \\ a_{n-1} \end{pmatrix} = - \begin{pmatrix} c_n \\ c_{n+1} \\ c_{n+2} \\ \vdots \\ c_{2n-1} \end{pmatrix}. \tag{6.3.15}$$

These are precisely the equations for the generation of the set of orthogonal polynomials on the interval $[a, b]$ with respect to the weight function $w(x)$ as detailed in Chapter 5, the main matrix being of the Vandermonde type. This set of equations is notoriously ill-conditioned and high precision arithmetic is required to use this approach to generate Gaussian weights and abscissae. Practical ways of avoiding the ill-conditioning have been published by many authors including a sequence of papers by Gautschi (1968), and by Golub and Welsch (1969) and Sack and Donovan (1972), and are considered in Evans (1993). In these papers (5.2.31) is replaced by modified moments defined by

$$\int_a^b P_k(x)w(x)\,\mathrm{d}x \tag{6.3.16}$$

for a subtle choice of a possibly non-orthogonal polynomial $P_k(x)$. With modern computers it is arguable whether the use of multilength arithmetic on the direct equations is not in fact competitive with the methods used to circumvent the ill-conditioning.

Because of the computational expense of generating a new Gaussian formula, only commonly used combinations of interval and weight function are normally tabulated. Of these the most commonly used is the Gauss–Legendre rule with interval $[-1, 1]$ and weight function $w(x) = 1$. Any finite-range quadrature, say on the interval $[a, b]$, can be transformed using the linear transformation

$$x = \frac{(b-a)}{2}t + \frac{(b+a)}{2} \tag{6.3.17}$$

to this standard interval. Other commonly used combinations are shown in Table 6.2.

Many extensive tables of the weights and abscissae for these formulae are available. In particular, a range of orders can be found in Abramowitch and Stegun (1965), and some very extensive values in the more specialist book by Stroud and

Table 6.2

Interval	$w(x)$	Name
$[-1, 1]$	1	Gauss-Legendre
$[0, \infty]$	e^{-x}	Gauss-Laguerre
$[-\infty, \infty]$	e^{-x^2}	Gauss-Hermite
$[-1, 1]$	$1/\sqrt{(1-x^2)}$	Gauss-Chebyshev
$[0, 1]$	$\ln(x)$	Gauss-log

Secrest (1966). In addition computer mathematical libraries offer Gaussian quadrature rules, and have the necessary database with the coefficients. For reference purposes some low point number formulae are quoted in Appendix B.

The Gauss-Chebyshev rule stands out from the other rules in that its weights and abscissae can be expressed in closed analytic form, rather than as tables of high-precision real numbers. This yields the formula

$$\int_{-1}^{1} \frac{f(x)}{\sqrt{1-x^2}}\,\mathrm{d}x = \frac{\pi}{m}\sum_{k=1}^{m} f\left(\cos\left[\frac{(2i-1)\pi}{2m}\right]\right) \tag{6.3.18}$$

and it is of interest in that all the weights are equal. In fact equal weight quadrature rules are often called Chebyshev rules.

The approach used in the above derivation yields a computational method very directly, but gives little insight into the formulae and their errors. The more traditional derivation is that based on the interpolation formula of Hermite, which was covered in Section 2.3 and has the form:

$$y(x) = \sum_{k=1}^{m} h_k(x) f(x_k) + \sum_{k=1}^{m} \bar{h}_k(x) f'(x_k) + E(x) \tag{6.3.19}$$

where

$$h_i(x) = [1 - 2l_i'(x_i)(x - x_i)][l_i(x)]^2, \tag{6.3.20}$$

$$\bar{h}_i(x) = (x - x_i)[l_i(x)]^2 \tag{6.3.21}$$

and the error term is given by

$$E(x) = \frac{f^{(2m)}(\xi)}{(2m)!}[\pi(x)]^2 \tag{6.3.22}$$

By integrating both sides of this formula a Hermite quadrature rule is obtained which involves both values of the integrand $f(x)$ and its derivative $f'(x)$ at the abscissae. However, these abscissae are still free to be chosen. In particular they may be so chosen as to make zero the terms in the summation involving derivatives of $f(x)$. Hence the integrated Hermite formula is expressed as

$$\int_a^b w(x)f(x)\mathrm{d}x = \sum_{k=1}^{m} W_k f(x_k) + \sum_{k=1}^{m} \bar{H}_k f'(x_k) + E \tag{6.3.23}$$

where

$$W_i = \int_a^b w(x)h_i(x)\mathrm{d}x,$$

$$\text{and} \quad \bar{H}_i = \int_a^b w(x)\bar{h}_i(x)\mathrm{d}x$$

and the error term can be expressed in a number of equivalent forms

$$E = \int_a^b w(x)E(x)\mathrm{d}x = \int_a^b \frac{f^{(2m)}(\xi)w(x)[\pi(x)]^2}{(2m)!}\mathrm{d}x \tag{6.3.24}$$

$$= \frac{f^{(2m)}(\eta)}{(2m)!}\int_a^b w(x)[\pi(x)]^2\mathrm{d}x \tag{6.3.25}$$

if $w(x) \geq 0$ in $[a, b]$. In the usual way the values ξ and η lie in the interval of integration but are not known explicitly. The weight $\bar{H}_i$ can be expressed in the form

$$\bar{H}_i = \frac{1}{\pi'(x_i)}\int_a^b w(x)\pi(x)l_i(x)\mathrm{d}x \tag{6.3.26}$$

and it is clear now that $\bar{H}_i$ can be made to vanish if $\pi(x)$ is made orthogonal to $l_i(x)$ for $i = 1, m$ over $[a, b]$ with respect to the function $w(x)$. But each $l_i(x)$ is a polynomial of degree $m - 1$ and it is therefore sufficient that $\pi(x)$ be orthogonal to all polynomials of degree less than m over the interval $[a, b]$ w.r.t the function $w(x)$. The resulting $\pi(x)$ is completely defined by the choice of the abscissae $x_1, x_2, \ldots, x_m$ which are, therefore, the roots of the relevant orthogonal polynomials for the given interval and weight function which agrees with the previous algebraic approach. Unlike the previous derivation, we now have an error bound and indeed explicit formulae for the weights W_i.

The relative advantages and disadvantages of the formulae in this chapter will be pursued later, but one important property of the Gaussian formulae in general is that the weights W_i are positive as long as the weight function $w(x)$ is non-negative on $[a, b]$ and the formula is exact for all polynomials of degree $2n - 2$ or less.

To prove this result consider the integration formula

$$\int_a^b w(x)f(x)\mathrm{d}x = \sum_{i=1}^{n} W_i f(x_i) \tag{6.3.27}$$

which is exact for all polynomials of degree $2n - 2$ or less. Then if $\phi_n(x)$ is a member of the set of orthogonal polynomials for the interval $[a, b]$ and weight function $w(x)$, then the function

$$f(x) = \left[\frac{\phi_n(x)}{(x - x_i)}\right]^2 \tag{6.3.28}$$

is of degree $2n - 2$ and

$$f(x_k) = \begin{cases} 0, & \text{if } k \neq i \\ [\phi_n'(x_i)]^2, & \text{if } k = i. \end{cases} \tag{6.3.29}$$

Hence

$$\int_a^b w(x) \left[\frac{\phi_n(x)}{x - x_i}\right]^2 \mathrm{d}x = W_i [\phi_n'(x_i)]^2 \tag{6.3.30}$$

and

$$W_i = \int_a^b w(x) \left[\frac{\phi_n(x)}{(x - x_i)\phi_n'(x_i)}\right]^2 \mathrm{d}x \tag{6.3.31}$$

which establishes the result $W_i > 0$ as required.

This is a good point to take stock of the rules so far obtained, and introduce some further concepts.

One such point arises in the consideration of quadrature rules, and induces doubt as to the superiority of Gaussian rules. It is common to increase the point number of a quadrature rule in a successive manner as a means of obtaining the practical measure of the convergence of the rule. When a required number of figures agree consistently as the point number is increased, an accepted value for the integral can be quoted. One strategy is to begin with a two-point rule, and move on with four, eight and 16-point rules, etc, until convergence is observed. With a formula such as the Generalized Simpson's rule the four-point formula uses the abscissae of the two-point formula (and two new points). This is true for any doubling of the point number. If at the end of the process the last point number used was say 32, then only 32 function evaluations would have been used in total. Such a formula is said to be *progressive*. For a Gaussian formula, the abscissae for any point number are in general quite different from those for any other point number. Hence if convergence is confirmed at 32 points the total number of function evaluations is $2 + 4 + 8 + 16 + 32 = 62$. Hence there is scope for a progressive formula, which though less accurate than Gauss, could in practice be competitive, as there is effectively a factor of two to be gained in a practical progressive rule against the non-progressive alternative. Two such formulae are those of Clenshaw and Curtis (1960) and Patterson (1968). The first of these methods uses an expansion as a series of Chebyshev polynomials for the approximation of the integrand. The resulting abscissae are cosine weighted and hence progressive, and the precision is almost Gaussian as is demonstrated in an error analysis by O'Hara and Smith (1968). The rules due to Patterson (1968) have Gaussian formulae as their basis, but these formulae have been extended progressively by the addition of intermediate abscissae. Like Gaussian formulae these new points have to be computed at high precision and then stored. Clenshaw–Curtis has the property of being a self-contained algorithm (of remarkable simplicity). In this way, a gain in efficiency may be obtained by seeking a progressive rule with accuracy only slightly below that of the theoretically maximal order rules of Gauss. Both the rules of Clenshaw–Curtis and Patterson are considered in more detail in Sections 6.4 and 6.5 respectively.

A second consideration concerns the acceptance of convergence when two successive terms of a progressive rule agree. It is very easy to generate an example in which such a criterion is misleading. Suppose a progressive rule with point numbers from two through to 256 is being used. Then the counter example requires an integrand which is sufficiently smooth that it will be integrated to machine accuracy with say eight points, such as a low degree polynomial. A large sharp peak is now placed on the otherwise smooth curve so that it lies entirely between any pair of abscissae of, say, the 16-point rule. Hence the low-order formulae up to and including the 16-point rule will be unable to 'see' the peak and convergence will appear to be occurring with agreement of the eight and 16-point rules. If the progressive test is continued, the 32 point rule will be the first in which an abscissa is affected by the peak, and new different values will arise. As high degree formulae are used, eventually the peak will itself be fully integrated and true convergence will occur. The problem with such 'blackbox' convergence criteria is always that, if it is known how the process works, then it is usually easy to invent a possibly artificial example which defeats the rule. Software writers will usually claim that their practical convergence algorithms will perform correctly on reasonably smooth examples. If a user knows that his example is unusual, extra care should be taken.

Another term used to describe quadrature rules is *adaptive*. A rule is adaptive if it compensates for a difficult sub-range of an integrand by automatically increasing the number of quadrature points in the awkward region. Adaptive formulae are usually based on a standard underlying quadrature rule, often a progressive one. A simple adaptive rule could be set up by taking a given quadrature rule and attempting to use it to integrate the complete interval $[a, b]$, but with some fixed upper limit on the number of points allowed. An obvious approach is to use a progressive rule with a succession of points such as $2, 4, 8, 16 \ldots$ up to the set limit. If the values obtained from this sequence of quadratures exhibits practical convergence to the required accuracy, as observed by comparing successive elements of the quadrature sequence, then the result is accepted. However, if the sequence does not show convergence to a sufficient level, then the interval is bisected and a similar procedure is carried out on each subinterval. Clearly, any difficult region in the range $[a, b]$ will cause convergence difficulties in the specific subintervals covering that region. Further sub-division will occur until the concentration of quadrature points is sufficient to overcome the convergence problems. A recursive call to the underlying quadrature rule will achieve this effect with elegance.

This discussion leads naturally to the method of Clenshaw and Curtis (1960) based on the approximation theory idea that expansions in Chebyshev polynomials lead to 'almost best' approximations. The resulting rule may be less effective on powers of x, but the gain of progressivity can more than offset this deficiency.

6.4 Clenshaw–Curtis quadrature

Without loss of generality the integral

$$I = \int_{-1}^{1} f(t)\mathrm{d}t \tag{6.4.1}$$

is considered, as transformation (6.3.16) can be used to reduce the general interval to the required range $[-1, 1]$. Then in accordance with the earlier discussion, an approximation to $f(t)$ is made using Chebyshev polynomials in the form

$$f(t) = \sum_{r=0}^{N}{}'' a_r T_r(t) \tag{6.4.2}$$

where the $''$ indicates that the first and last terms of the series are to be halved. This expansion exists if $f(x)$ is continuous and of bounded variation. The a_r follow quite easily as there is a useful discrete orthogonalization condition from (1.2.13), namely

$$\sum_{s=0}^{N}{}'' \cos\frac{\pi is}{N} \cos\frac{\pi js}{N} = \begin{cases} 0, & \text{if } i \neq j \\ N, & \text{if } i = j = 0 \text{ or } N \\ \frac{N}{2}, & \text{if } i = j \neq 0 \text{ or } N \end{cases} \tag{6.4.3}$$

Hence if t is set according to

$$t_s = \cos\frac{\pi s}{N} \tag{6.4.4}$$

then multiplying both sides of (6.4.2) by $\cos(\pi sk/N)$ and summing over s with t set to t_s gives

$$\begin{aligned} \sum_{s=0}^{N}{}'' f(t_s)\cos\frac{\pi sk}{N} &= \sum_{s=0}^{N}{}''\sum_{r=0}^{N}{}'' a_r T_r(t_s)\cos\frac{\pi sk}{N} \\ &= \sum_{s=0}^{N}{}''\sum_{r=0}^{N}{}'' a_r \cos\frac{\pi rs}{N}\cos\frac{\pi sk}{N} \\ &= \frac{N}{2}a_k. \end{aligned} \tag{6.4.5}$$

Hence the fitting coefficients a_r are conveniently computed from the formula

$$a_r = \frac{2}{N}\sum_{s=0}^{N}{}'' f_s \cos\frac{\pi rs}{N} \tag{6.4.6}$$

with

$$f_s = f\left(\cos\frac{\pi s}{N}\right).$$

The second stage of generating a quadrature formula is to integrate the fitting approximation and this is also straightforward using the result

$$\int_{-1}^{1} T_r(t)\mathrm{d}t = \begin{cases} -2/r^2 - 1 & r \text{ even} \\ 0 & r \text{ odd} \end{cases} \tag{6.4.7}$$

to yield the complete formula

$$I_N = -\frac{4}{N}\sum_{s=0}^{N}{}'' f_s \sum_{r=0}^{N}{}'' \begin{cases} \dfrac{\cos\frac{\pi sr}{N}}{r^2-1} & r \text{ even} \\ 0 & r \text{ odd} \end{cases}. \tag{6.4.8}$$

Once it is realized that the abscissae are given by the argument of f_s in equation (6.4.6), then it is apparent that a Lagrangian approach could be used to generate the rule, that is by integrating Lagrange's interpolation formula which fits $f(x)$ on the interval $[-1, 1]$ with collocation at the above cosine weighted abscissae. Hence Clenshaw–Curtis quadrature rules are exact for polynomials of degree N or less. It will be shown in the error analysis to follow, that even though this precision is comparable to the Newton–Cotes rules, the actual accuracy is far greater than might be expected, and in practice approaches that of the Gaussian formulae themselves.

A more elegant form of the quadrature rule can be generated using the following simple recurrence due to Clenshaw (1957). The Chebyshev expansion (6.4.2) can be integrated in the form

$$\int_{-1}^{t} f(t)\mathrm{d}t = \frac{1}{2}b_0 + b_1T_1(t) + b_2T_2(t) + \cdots \tag{6.4.9}$$

where

$$b_r = \frac{a_{r-1} - a_{r+1}}{2r} \qquad r = 1, 2, \ldots \tag{6.4.10}$$

to yield a novel formula for the *indefinite* integral of $f(t)$. By putting $t = -1$ the value for b_0 is obtained according to

$$b_0 = 2b_1 - 2b_2 + 2b_3 - \cdots \tag{6.4.11}$$

and hence the convenient form of the quadrature rule is now

$$\begin{aligned}\int_{-1}^{1} f(t)\mathrm{d}t &= \frac{1}{2}b_0 + b_1 + b_2 + \cdots \\ &= 2(b_1 + b_3 + b_5 + \cdots)\end{aligned} \tag{6.4.12}$$

The errors in Clenshaw–Curtis quadrature formulae have been extensively studied by O'Hara and Smith (1968). Equation (6.4.2) is an approximate expansion of $f(t)$ in Chebyshev polynomials. The infinite, generalized Fourier expansion can also be written down to give

$$f(t) = \sum_{r=0}^{\infty}{}' A_r T_r(t) \tag{6.4.13}$$

which can be integrated to obtain an error estimate. Define E_N to be the true value of the integral I minus the value found using the N term series, namely I_N. The coefficients a_r and the coefficients A_r are related by substituting $f(t)$ from (6.4.13) into equation (6.4.6) so yielding

$$\begin{aligned} a_r &= \frac{2}{N}\sum_{s=0}^{N}{}'' \cos\frac{\pi rs}{N}\sum_{i=0}^{\infty}{}' A_i \cos\frac{\pi is}{N} \\ &= A_r + A_{2N-r} + A_{2N+r} + A_{4N-r} + A_{4N+r} + \cdots \end{aligned} \tag{6.4.14}$$

From (6.4.8) the two expansions for I and I_N can be written as

$$I_N = a_0 - \frac{2a_2}{1.3} - \frac{2a_4}{3.5} - \cdots - \frac{2a_{N-2}}{(N-3)(N-1)} - \frac{a_N}{(N-1)(N+1)} \tag{6.4.15}$$

and

$$I = A_0 - \frac{2A_2}{1.3} - \frac{2A_4}{3.5}\cdots \tag{6.4.16}$$

to give on subtraction

$$\begin{aligned} E_N &= 2A_{2N} - \frac{2}{1.3}(A_{2N-2} + A_{2N+2}) - \cdots - \frac{2}{(N-3)(N-1)}(A_{N+2} + A_{3N-2}) \\ &\quad + \sum_{r=1}^{N-1}\frac{2A_{N+2r}}{(N+2r-1)(N+2r+1)} \end{aligned} \tag{6.4.17}$$

On the basis that the coefficients A_r fall off reasonably quickly for large r, (an assumption valid for sufficiently smooth a function $f(x)$), terms higher than A_{3N} can be ignored. Also taking N to be even allows E_N to be recast as

$$\begin{aligned} E_N = &\sum_{r=0}^{N/2-1}{}' \frac{2A_{2N-2r}}{4r^2-1} + \sum_{r=0}^{N/2-1}{}' \frac{2A_{2N+2r}}{4r^2-1} \\ &- \sum_{r=1}^{N-1}\frac{2A_{N+2r}}{(N+2r+1)(N+2r-1)}. \end{aligned} \tag{6.4.18}$$

Further reduction can be obtained by combining terms, with cancellation arising particularly in the first and third summations, yielding

$$E_N = \frac{16.1.N}{(N^2-1^2)(N^2-3^2)}A_{N+2} + \frac{16.2.N}{(N^2-3^2)(N^2-5^2)}A_{N+4} + \cdots$$
$$+ \frac{16(N/2-1)N}{3(2N-1)(2N-3)}A_{2N-2} - \left(2+\frac{2}{4N^2-1}\right)A_{2N}$$
$$+ \left(\frac{2}{3} - \frac{2}{(2N+1)(2N+3)}\right)A_{2N+2} + \cdots \qquad (6.4.19)$$

In this form it is clear why the Clenshaw–Curtis quadrature rule exhibits higher accuracy than its degree would suggest. The dominant coefficients A_{N+2} and A_{N+4} have the smallest multiplying factors which are of order $1/N^3$, giving a uniformly small sum of terms. This effect is in accordance with the underlying approximation theory which is the original philosophy of the method.

Clenshaw–Curtis quadrature rules therefore exhibit some very competitive features. Like Gaussian formulae they are numerically stable to high degree, they have errors only slightly less good than an equivalent Gaussian rule, and more practically they are progressive. The quadrature rules are also completely self-contained requiring no tabulated coefficients, thus removing any restriction from the order of formula which may be employed. Related to this property is the ability that this algorithm has to be coded on a very small computer system such as might be used in mobile applications.

There is an *open* version of the Clenshaw–Curtis rule which does not have the end-points of the interval of integration as abscissae. This version of the Clenshaw–Curtis rule can be especially useful in dealing with singular integrands with the singularities at one or other end-point. The rule is derived using the alternative discrete orthogonalization formula

$$\sum_{s=1}^{N} \cos\frac{i(2s-1)\pi}{2N}\cos\frac{j(2s-1)\pi}{2N} = \begin{cases} 0, & \text{if } i \neq j \\ \frac{N}{2}, & \text{if } i = j \end{cases} \qquad (6.4.20)$$

rather than (6.4.3). The result is the open rule for the a_i's

$$a_i = \frac{2}{N}\sum_{s=1}^{N}\cos\frac{i(2s-1)\pi}{2N}\, f\left(\cos\frac{(2s-1)\pi}{2N}\right) \qquad (6.4.21)$$

which is analogous to (6.4.6).

Before leaving Clenshaw–Curtis rules, a brief mention should be made of yet a further advantageous property. The practical use of a progressive rule is usually based on a succession of quadratures each with double the number of points of the successor, the old points being re-used in the new quadrature. At the end of this process the required accuracy will be obtained for a specific point number rule, but the user will only realize this when a further point doubling has been completed and the two previous results compared. Hence, inevitably, twice the number of points are actually used than is strictly required. This last doubling of points could be completely ignored if a *reliable* error bound could be computed as part of the quadrature process. The error bounds of O'Hara and Smith (1968) make some headway in this direction but tests indicate that there is still a degree of unreliability in the process. Typical of their bounds are formulae such as

$$|E_N| < \max\{2|a_{N-4}|, 2|a_{N-2}|, |a_N|\} \tag{6.4.22}$$

in which an attempt is made to estimate the rate of fall-off of the coefficients A_r so that the calculated coefficients a_r can be employed to estimate the dominant terms in E_N. Such an estimate can be based on a range of particular a_r's, but there is always the risk of accidental zeros in the a's and hence the use of a set of highest order terms.

There is an alternative to Clenshaw–Curtis quadrature to obtain a progressive rule. This is the rule due to Patterson which is based on extending Gaussian quadratures.

6.5 Patterson's method

The major failing from a practical point of view of the Gaussian rules, lies in their non-progressive nature. The question then arises as to whether it is possible to generate some form of progressive rule from a Gaussian formula. Kronrod (1965) shows how an n-point Gaussian rule may be augmented by a further set of $n+1$ abscissae to yield a rule which will integrate exactly a polynomial of degree $3n+1$ for n even and $3n+2$ for n odd. This degree is equal to the number of degrees of freedom in the extended rule in that the extended rule has the extra $n+1$ weights and abscissae free, and in addition there are the n original Gaussian weights which are reset. It is only the original Gaussian abscissae which remain unchanged. Hence the total of $3n+2$ degrees of freedom is present for n odd and in the case of n even it is one less, as one of the abscissae is zero from the symmetry. The inevitable ill-conditioning in the generation of these extended rules enforces the use of very high precision arithmetic particularly in the higher degree rules which are the ones of particular interest, though Calio, Gautschi and Marchetti (1986) demonstrate how modified moments can be employed in a manner similar to that of Section 2.3. The use of Newton's method is investigated in Gautschi and Notaris (1988) as a means to generate extended formulae, and the general area of Kronrod quadrature is overviewed in Monegato (1976).

Patterson (1968) developed an alternative approach by considering augmenting an n point formula by p points. The polynomial $G_{n+p}(x)$ is introduced whose roots are the $n+p$ abscissae of the new formula. A general polynomial of degree $n+2p-1$ can be written in the form

$$F_{n+2p-1}(x) = Q_{n+p-1}(x) + G_{n+p}(x) \sum_{k=0}^{p-1} c_k x^k \tag{6.5.1}$$

where $Q_{n+p-1}(x)$ is a general polynomial of degree $n + p - 1$. By forcing the relation

$$\int_{-1}^{1} G_{n+p}(x) x^k \mathrm{d}x = 0 \qquad k = 0, 1, \ldots, p - 1, \tag{6.5.2}$$

to hold, the polynomial defined in (6.5.1) can be integrated exactly by an $n + p$ point formula. The result is a formula with $n + p$ points which has degree $n + 2p - 1$. A completely equivalent form for equation (6.5.2) is

$$\int_{-1}^{1} G_{n+p}(x) P_k(x) \mathrm{d}x = 0 \qquad k = 0, 1, \ldots, p - 1 \tag{6.5.3}$$

where $P_k(x)$ are the Legendre polynomials. By expressing $G_{n+p}(x)$ as a series of Legendre polynomials according to

$$G_{n+p}(x) = \sum_{t=0}^{n+p} t_i P_i(x) \tag{6.5.4}$$

and substituting into (6.5.3) yields

$$\sum_{i=0}^{n+p} t_i \int_{-1}^{1} P_i(x) P_k(x) \mathrm{d}x = 0, \quad k = 0, 1, \ldots, p - 1 \tag{6.5.5}$$

with the immediate implication that $t_i = 0$ for $i = 0, 1, \ldots, p - 1$ from the orthogonality of the Legendre polynomials on the interval $[-1, 1]$. Hence equation (6.5.4) reduces to

$$G_{n+p}(x) = \sum_{i=p}^{n+p} t_i P_i(x) \tag{6.5.6}$$

with a further simplification to

$$G_{n+p}(x) = \sum_{i=1}^{[n/2]+1} c_i P_{2i-2+p+q}(x), \tag{6.5.7}$$

with $q = n - 2[n/2]$, when account is taken of the symmetry of the abscissae. There is no loss in generality in choosing $c_{[n/2]+1}$ to be unity to leave the equations

$$\sum_{i=1}^{[n/2]} c_i P_{2i-2+p+q}(x_j) = -P_{n+p}(x_j), \qquad j = 1, 2, \ldots, [n/2]. \tag{6.5.8}$$

This set of linear algebraic equations may be solved for the coefficients $c_1, c_2, \ldots, c_{[n/2]}$ and then the extended abscissae follow as roots of the now known polynomial $G_{n+p}(x)$.

As pointed out by Morrow (1977), few such Patterson extension rules exist. However, the most commonly required rule based on the three-point Gauss–Legendre formulae with extensions to seven points, to 15 points and indeed through to 511 points does exist and is tabulated. This is a highly competitive rule and as a method appears to vie with Clenshaw–Curtis for efficient practical use. Some low point number weights and abscissae are given in Appendix C for this method.

A range of methods have been presented for general purpose quadrature, and the problem remains as to which to recommend in a given situation. To resolve this issue the four main methods, generalized Simpson, Gauss–Legendre, Clenshaw–Curtis and Patterson have been applied to a well-known set of test examples due to Kahaner (1971). These test integrals are defined in Table 6.3.

These integrals illustrate a range of practical problems. These include singular derivatives, which will imply that the error formulae are no longer bounded, cusps,

Table 6.3 The test integrals of Kahaner

n	$G_n(x)$	a	b	$\int_a^b G_n(x)\mathrm{d}x$
1	$\exp(x)$	0	1	1.718 281 828
2	$\begin{cases} 0, & 0 \le x < 0.3 \\ 1, & 0.3 \le x \le 1 \end{cases}$	0	1	0.700 000 000
3	$x^{1/2}$	0	1	0.666 666 667
4	$(23/25)\cosh x - \cos x$	-1	1	0.479 428 266
5	$1.0/(x^4 + x^2 + 0.9)$	-1	1	1.582 232 997
6	$x^{3/2}$	0	1	0.400 000 00
7	$x^{-1/2}$	0	1	2.000 000 00
8	$1/(1 + x^4)$	0	1	0.866 972 987
9	$2/(2 + \sin(10.0\pi x))$	0	1	1.154 700 538
10	$1/(1 + x)$	0	1	0.693 147 1806
11	$1/(1 + \exp(x))$	0	1	0.379 885 4930
12	$x/(\exp(x) - 1)$	0	1	0.777 504 634
13	$\sin(100.0\pi x)/\pi x$	0.1	1	0.009 098 645 257
14	$\sqrt{50}\exp(-50.0\pi x^2)$	0	10	0.500 000 000
15	$25\exp(-25x)$	0	10	1.000 000 00
16	$50/(\pi(1 + 2500x^2))$	0	10	0.499 363 381 07
17	$(\sin(50\pi x))^2/(50\pi x)^2$	0.01	1	0.112 139 308 27
18	$\cos(\cos(x) + 3\sin(x)$ $+2\cos(2x) + 3\cos(3x)$ $+3\sin(2x)))$	0	π	0.838 676 3426
19	$\ln(x)$	0	1	$-1.000\,000\,00$
20	$1/(x^2 + 1.005)$	-1	1	1.564 396 450
21	$(\mathrm{sech}(10(x - 0.2)))^2$ $+(\mathrm{sech}(100(x - 0.4)))^4$ $+(\mathrm{sech}(1000(x - 0.6)))^6$	0	1	0.210 802 7355

discontinuities, oscillatory behaviour and poles in the complex plane near to the real axis (which will also upset the error bounds). The four methods have been applied to these examples and the results are shown in Table 6.4. Two types of comparison have been made. Firstly, under the column marked A, the number of correct figures obtained with the 16-point rule is given. An F indicates that the method failed completely, for example function 7 is singular at the origin and Simpson's rule requires a function evaluation at this point, hence overflow occurs. The second comparison gives the number of quadrature points used to achieve 10-figure accuracy, with an f being used where this end was not achieved for a reasonable effort, such as over 500 quadrature points.

It is clear that the generalized Simpson's rule performs less well than any of the other three methods, and this rule should be used when either simplicity or low accuracy are required. There is less to choose between the three more powerful methods as the results stand, but it should be remembered that the Gauss–Legendre formulae are not progressive, so that the total number of function evaluations

Table 6.4 Comparison of general quadrature methods: A; Number of correct figures for 16 pt. rule, N; Number of points for 10 fig. accuracy, f; Failure to reach 10 fig. accuracy, F; Failure to run at all

G_n	Gen. Simpson		Gauss–Leg.		Patterson		Clenshaw	
	A	N	A	N	A	N	A	N
1	8	65	14	4	14	7	14	9
2	1	f	1	f	1	f	1	f
3	3	f	4	f	4	255	4	f
4	7	f	14	8	14	7	14	9
5	7	65	12	32	10	15	9	33
6	5	f	7	64	8	63	7	65
7	F	F	1	f	1	f	F	F
8	6	f	14	16	14	15	12	17
9	10	17	1	f	2	255	2	257
10	7	65	14	8	14	15	14	17
11	9	33	14	8	14	7	14	9
12	9	33	14	8	14	7	14	9
13	0	f	0	f	0	f	0	f
14	0	f	0	128	0	127	0	129
15	0	f	0	64	2	63	0	65
16	0	f	0	f	1	255	1	257
17	0	f	1	128	1	127	1	257
18	3	f	3	64	2	63	1	65
19	F	F	2	f	3	f	F	F
20	10	16	12	16	10	15	10	17
21	2	f	1	f	2	f	0	f

involved in a series of trial integrations with doubling point numbers will be approximately double the final point number indicated. The Patterson and Clenshaw rules perform very similarly, the main difference in these results being that Patterson's formulae are open and the standard Clenshaw–Curtis rule is closed. Hence integrals with end-point singularities fail in the Clenshaw case. In these cases, the Patterson accuracy is much reduced and the methods of the next section would be more appropriate. This is a case when the open Clenshaw formula (6.4.21) could have been used, but again the singularity will cause reduced accuracy.

6.6 Singular quadratures

The discussion so far has covered general quadrature methods. However, there are situations in which these standard methods will invariably fail, and special methods are required. Such problems arise in two important cases which will be the subjects of the next two sections. The case to be considered in this section is when the integrand has singularities in the range of integration, often at the end-points. It is common to sub-divide the range of integration so that only end-point singularities occur, and there is no loss of generality in considering only this case.

The methods for singular problems seem to fall into four general categories. The obvious, and probably most simple method, has been coined 'ignoring the singularity'. As the name implies, the process employs the direct use of a general purpose but open method. Hence if the singularity is taken to be at one of the end-points, then an open formula does not involve the evaluation of the integrand at its singular point. Clearly such an approach will involve a greater quadrature error than would be the case with a smooth function, but using a sufficiently high point number will eventually yield convergence. Gaussian rules and Patterson's method are both feasible routines for this process.

In some special cases, a Gaussian formula with the necessary weight function which will remove the singularity may be used. In this category can be included the singularity $1/\sqrt{1-x^2}$ on the interval $[-1, 1]$ which would use Gauss–Chebyshev quadrature, and $\ln x$ on $[0, 1]$ which would use Gauss–log quadrature. The obvious problem with such Gauss formulae arises for a non-standard singularity for which there is no standard pre-tabulated method. In general the expense of generating a special rule for the purpose is too great for a one-off exercise. However, it may be feasible to generate a weighted Newton–Cotes rule such as those in (6.2.9) and (6.2.10). As long as low-order formulae are used to avoid any instabilities, and as the equations for the relevant weights W_i are linear, then the approach is feasible. Higher accuracy can be obtained by some form of sub-division, possibly using the special rule only in the immediate vicinity of the singularity and reverting to Gauss–Legendre in the non-singular region.

The method usually referred to as 'removing the singularity' is based on subtracting out an analytically known integral which takes out the singularity to leave a conventional problem for a standard method. As a simple example consider

$$I = \int_0^1 \frac{\ln x}{1+x^2}\,\mathrm{d}x \tag{6.6.1}$$

whose singularity can be removed by the device

$$I = \int_0^1 \left(\frac{\ln x}{1+x^2} - \ln x\right)\mathrm{d}x + \int_0^1 \ln x\,\mathrm{d}x \tag{6.6.2}$$

to yield the two integrals

$$I = -\int_0^1 \frac{x^2 \ln x}{1+x^2}\,\mathrm{d}x + \int_0^1 \ln x\,\mathrm{d}x \tag{6.6.3}$$

of which the first is non-singular and the second can be obtained analytically. Clearly care is needed in evaluating the integrand in the first integral in any computer code which needs to avoid evaluating $x^2 \ln x$ at $x = 0$ as such, but needs to recognize that as $x \to 0$, $x^2 \ln x$ also tends to zero.

The third of the methods is called 'creeping up on the singularity', and as the phrase implies, involves evaluating integrals whose range of integration approaches the singular point. In this way a sequence of integrals can be generated, and for suitable sequences, accelerators of the type introduced in Section 1.7, can be employed to predict the required limit.

The fourth class of method uses transformations of the integrand by a subtle change of variable. If the choice can be arranged not only to expunge the singularity, but also to have zero end-point derivatives for the resulting integrand, then the enhanced trapezoidal rule based on the Euler–Maclaurin summation formula of Chapter 2 (2.1.61) becomes a viable proposition.

The first method to be considered in detail is in the spirit of the work of Chisholm, Genz and Rowlands (1972) in that sequences of quadrature approximations are generated and then subjected to acceleration procedures, primarily the ε-algorithm, as introduced in Chapter 1. The first method to be described here due to Evans, Hyslop and Morgan (1983), involves integration over increasingly small sub-intervals utilizing the process described as 'creeping up on the singularity'. Conventional general quadrature methods are used to compute the sequence of integrals, and if progressive methods such as Clenshaw–Curtis or Patterson are employed then the process yields accurate values rapidly.

Attention is confined to singular integrals of the form

$$S = \int_0^1 Q(x)\mathrm{d}x = \int_0^1 f(x)g(x)\mathrm{d}x \tag{6.6.4}$$

in which $g(x)$ is analytic and $f(x)$ is singular at $x = 0$. A typical case would be

$$f(x) = x^{-\alpha}(\ln x)^{\beta} \qquad (0 < \alpha < 1,\ \beta > 0) \tag{6.6.5}$$

Generalization is straightforward. For example, integrals over the range $[a, b]$ with an end-point singularity may be mapped onto $[0, 1]$ by a linear transformation.

Again, if an internal singularity occurs, sub-division at the singularity and subsequent linear transformations yield two integrals of the form (6.6.4).

The sequence $\{S_n\}$ which tends to the limit S as $n \to \infty$ is defined by

$$S_n = \int_{x_n}^{1} Q(x)\mathrm{d}x \tag{6.6.6}$$

where $x_n \to 0$ as $n \to \infty$. The successive terms in the sequence are generated according to

$$S_n = \sum_{i=1}^{n} T_i \tag{6.6.7}$$

where

$$T_i = \int_{x_i}^{x_{i-1}} Q(x)\mathrm{d}x \qquad (i \geq 1) \tag{6.6.8}$$

In this result $x_0 = 1$ and $x_n \to 0$ as $n \to \infty$.

Various possibilities are feasible for the monotonic decreasing sequence $\{x_n\}$ and the choice

$$x_n = \theta^n \qquad (0 < \theta < 1) \tag{6.6.9}$$

is found to be most effective in practice. The parameter θ may be chosen to optimize the process. Extensive numerical tests were carried out by Evans, Hyslop and Morgan (1983) to investigate the optimum choice of θ, with the result that the compromise value of $\theta = 0.2$ is suggested as reasonable for all the cases considered. To illustrate these methods the examples in Table 6.5 were considered.

These examples are chosen to demonstrate the high accuracy attainable for an economic number of function evaluations and also to demonstrate the effect of varying the parameter θ. I_1 is an integral which is 'nearly infinite', I_2 is an example considered by Cohen (1980) and I_3, I_4 and K_1 to K_9 exhibit some common singularities.

The results obtained are presented in Tables 6.6 and 6.7 where the total number of function evaluations required to attain a relative error of $E = 10^{-10}$ is depicted for various values of the parameter θ.

Note that, for the near pathological integral I_1, the successive terms T_i are very close to one another. Hence although it is possible to obtain the T_i values accurately using $N = 7$ or 15 points in the Patterson quadrature formula, the ε-algorithm is subject to severe cancellation when the differences of the T_i values are calculated. Consequently, in a single precision calculation to accuracy $p \simeq 11$ figures, it is possible to obtain I_1 to only about six significant figures when extrapolation to the limit is carried out. This fact is reflected by quoting the actual relative errors obtained in Table 6.5. Nevertheless, it is quite remarkable that such an accurate result is obtained, as predicted, from only the first three terms $\{S_1, S_2, S_3\}$ of the sequence, even for this extreme value of α. Many other methods failed completely on this integral.

Table 6.5 Preliminary test integrals

$$I_1 = \int_0^1 x^{-\alpha}\mathrm{d}x = 10^6 \qquad (\alpha = 1 - 10^6)$$

$$I_2 = \int_0^1 x^{0.95}e^x\mathrm{d}x = 1.020\,457\,359$$

$$I_3 = \int_0^1 (1+x^2)^{-1}(\ln x)^2\mathrm{d}x = 1.937\,892\,293$$

$$I_4 = \int_0^1 x^{1/2}e^{-x}(1+x)^{-1}\mathrm{d}x = 1.237\,643\,927$$

$$K_1 = \int_0^1 x^{1/2}\mathrm{d}x = 2/3$$

$$K_2 = \int_0^1 x^{-1/3}\mathrm{d}x = 3/2$$

$$K_3 = \int_0^1 x^{-2/3}\mathrm{d}x = 3$$

$$K_4 = \int_0^1 x^{7/2}\mathrm{d}x = 2/9$$

$$K_5 = \int_0^1 (\ln x)^2\mathrm{d}x = 2$$

$$K_6 = \int_0^1 (\ln x)^4\mathrm{d}x = 24$$

$$K_7 = \int_0^1 (1+x^2)^{-1}\mathrm{d}x = \pi/4$$

$$K_8 = \int_0^1 x^{-1/2}\ln x\mathrm{d}x = -4$$

$$K_9 = \int_0^1 (1-\ln x)^{-1}(-\ln x)^{-1/2}\mathrm{d}x = 1.343\,293\,422$$

Table 6.6 Number of function evaluations required to attain relative error $E = 1.0(-10)$

Integral	$\theta = 0.1$	$\theta = 0.2$	$\theta = 0.5$
I_1	45 4.3(−6)	21 4.4(−6)	21 2.3(−6)
I_2	32	27	40
I_3	104	111	91
I_4	105	127	84

The results for the other integrals show that accurate values are obtainable for moderate expenditure in function evaluations, especially for Cohen's integral I_2. The algorithm was implemented completely automatically in that the quadrature results were monitored to investigate convergence to a specified precision, as the number of points was increased.

Table 6.7

Integral	EHM($\theta = 0.2$)	
	$E(N_f = 35)$	$N_f(10^{-10})$
K_1	2.2(−8)	45
K_2	0.4(−7)	45
K_3	1.1(−6)	45
K_4	1.0(−10)	17
K_5	9.5(−4)	97
K_6	9.8(−2)	147
K_7	1.0(−10)	29
K_8	1.8(−6)	75
K_9	1.3(−5)	127

Examination of Table 6.6 indicates that the results are not particularly sensitive to the choice of the parameter θ. The 'extreme' values $\theta = 0.1$ and $\theta = 0.5$ may occasionally produce rapid convergence for a particular integral but it is considered safer, in general, to adopt the compromise value of $\theta = 0.2$. Similar tests were carried out on all the other test integrals with substantially the same conclusions, and these results are shown in Table 6.7.

The column $E(N_f = 35)$ gives the relative error E for 35 function evaluations, whereas the final column shows the number of function evaluations N_f for a relative error of 10^{-10}.

The second method to be covered in detail is the well-known technique of variable transformation of which two particular variants have proved highly successful. One approach is to use a simple polynomial transformation as employed by Evans, Forbes and Hyslop (1983). This process is equivalent to a standard analytic device for singular integrals. There is also the group of transformations which move the singular points of the integrand to infinity, thus converting the improper integral into a convergent infinite one for which the trapezoidal rule is particularly efficient. Such procedures have been proposed by many authors including Schwartz (1969a), Takahasi and Mori (1973) and Stenger (1973) whose work has been developed by Squire (1976, 1979). This group includes transformations into a finite interval which also employ the trapezoidal rule.

It is sufficient to consider singular integrals of the form

$$S = \int_0^1 f(x)\mathrm{d}x \tag{6.6.10}$$

where $f(x)$ and/or its derivatives are singular at one or both of the end-points, $x = 0$ and $x = 1$, of the normalized range.

In general the transformation

$$x = \psi(t) \tag{6.6.11}$$

is utilized which, for convenience, is assumed to map the interval (0, 1) onto itself. Various simple polynomial forms for $\psi(t)$ are tested for the transformed integral

$$S = \int_0^1 f(\psi(t))\dot{\psi}(t)\mathrm{d}t \tag{6.6.12}$$

It is the derivative $\dot{\psi}(t)$ which produces the alleviation of the singularity and it is of interest to investigate the effect of increasing the degree of the polynomial to achieve this end for various types of singularity.

The simplest change of variable, $x = t^n$ is, of course, ideal for alleviating singularities of the form $f(x) \sim x^{-\alpha}$. Indeed, the choice $n = (1 - \alpha)^{-1}$ will remove this type of singularity completely, but this approach is really only of value in analytic treatment of specific cases. The idea here is to use integral values of n which are sufficiently large to alleviate many commonly occurring singularities, examples including $x^{-\alpha}(1 - x)^{-\beta}$, $x^{-\alpha}(\ln x)^{\beta}$, $\ln \ln x^{-1}$, etc.

In cases where $f(x)$ is singular at $x = 1$ as well as at $x = 0$, the transformation used in Evans, Hyslop and Morgan (1983) is adopted. The singularity at $x = 1$ is mapped onto $x = 0$ and integral (6.6.10) is hence expressed in the form

$$S = \frac{1}{2}\int_0^1 [f(x/2) + f(1 - x/2)]\mathrm{d}x \tag{6.6.13}$$

where the integrand is now singular at $x = 0$ only and the transformation $x = t^n$ may now be applied directly.

It will be noted that care must be exercised in computing the second term, $f(1 - x/2)$, in the integrand as $x \to 0$. This point has been commented on by many authors and the commonly adopted remedy is to express $f(z)$ as $z \to 1$ in the form

$$f(z) = f(1 - \delta) \tag{6.6.14}$$

where the 'residual' δ as well as the variable z must be accurately available in the function routine. In the present context , $\delta = t^n/2$ and even though t may be of order 10^{-7} when implementing the highest order ($N = 511$) Patterson quadrature formula, underflow may be avoided, even for fairly large values of n. Consequently terms such as $(1 - x)^{-\beta}$ may be evaluated without overflow occurring since cancellation is avoided. Note also that, for certain 'composite' singularities such as $\ln \ln x^{-1}$, this device is not successful since $\ln \ln(1 - \delta)^{-1}$ cannot be evaluated without overflow under these extreme circumstances, unless arithmetic precision is increased prohibitively. It is necessary in these cases to use analytic expansions such as $\ln(\delta + \delta^2/2 + \cdots)$ in the function routines.

In Table 6.8 results are presented for tests carried out on the various integrals listed in Table 6.5. The number of quadrature points N required to obtain a relative error of 10^{-10} in each of the integrals is depicted for values of n ranging from $n = 1$ to $n = 8$. Patterson quadrature is utilized with $N = 3, 7, 15, 31, \ldots, 511$ points, the N-point formula integrating exactly a polynomial of degree $(3N + 1)/2$. (The corresponding Gaussian formula has polynomial precision $(2N - 1)$). For comparison purposes, the last column shows the number of points used in the acceleration procedure of Evans, Hyslop and Morgan (1983) described earlier.

Table 6.8 Number of Patterson Quadrature points (N) required to attain relative error $E = 1.0(-10)$ for various integrals using $x = t^n$ transformation

n	1	2	4	6	8	EHM
I_1	—	—	—	—	—	21*
I_2	63	15	15	15	31	27
I_3	—	127	31	31	31	111
I_4	—	15	15	15	15	127

*6-figure accuracy only
EHM — Evans, Hyslop and Morgan (1983)

The $n = 1$ column corresponds to ignoring the singularity and it will be observed that even the highest order Patterson formula fails in many cases to produce convergence to the required accuracy. This is, indeed, a measure of the difficulty of the integrals considered. Nonetheless, the $x = t^n$ transformation rapidly smooths the integrands and convergence is attained in all cases (except I_1) with a maximum of $N = 31$ quadrature grid-points. It is necessary to use values of n as large as $n = 6$ to achieve convergence in general for the types of singularity tested here. Calculations were carried out for larger values of n, and it was found that numerical instability due to round-off only began to appear for values approaching $n = 10$. There is slight evidence of this effect appearing at $n = 8$ for I_2 and for K_4, so that $n = 6$ may be a good practical limit.

An alternative approach is to employ a transformation which maps the singular points of the integrand to infinity. A *low*-order quadrature rule such as the trapezoidal rule then proves to be particularly effective, as the derivatives in the Euler–Maclaurin summation formula will fall off rapidly at infinity. The most general singular integral considered is of the form

$$S = \int_a^b f(x)\mathrm{d}x \tag{6.6.15}$$

where $f(x)$ may be singular at both end-points. Integrands possessing more singularities may be reduced to several integrands of this form by sub-division at the singularities. Generalizing the procedure of Squire (1979), the substitution

$$x = \tfrac{1}{2}(b + a) + \tfrac{1}{2}(b - a)\tanh g(t) \tag{6.6.16}$$

is proposed in which $g(t) \to \pm\infty$ as $t \to \pm\infty$. For example, the choice $g(t) = t$ gives the classical tanh transformation employed by Squire. Consequently

$$S = \frac{1}{2}(b - a)\int_{-\infty}^{\infty} f\left[\frac{1}{2}(b + a) + \frac{1}{2}(b - a)\tanh g(t)\right] \mathrm{sech}^2(g(t))\dot{g}(t)\mathrm{d}t \tag{6.6.17}$$

in which $\dot{g}(t)$ represents the derived function $\mathrm{d}g/\mathrm{d}t$. With a suitable choice of $g(t)$, the $\mathrm{sech}^2\, g(t)$ factor falls off extremely rapidly as $t \to \pm\infty$ and controls the behaviour of the transformed integrand even though the singular values of $f(a)$ and

$f(b)$ are being approached. For this reason, when the trapezoidal rule is applied by setting

$$t = jh \qquad j = 0, \pm 1, \pm 2, \ldots \tag{6.6.18}$$

the summation may be truncated at $j = \pm N$, say, yielding

$$S \simeq 2h(b-a) \sum_{j=-N}^{N} A_j f(x_j) \tag{6.6.19}$$

where

$$x_j = \tfrac{1}{2}(b+a) + \tfrac{1}{2}(b-a)(1-q_j)(1+q_j)^{-1}, \tag{6.6.20}$$

$$A_j = q_j(1+q_j)^{-2}\dot{g}(jh) \tag{6.6.21}$$

and

$$q_j = \exp[-2g(jh)] \tag{6.6.22}$$

As mentioned in (6.6.14) difficulty arises in evaluating $f(x_j)$ for the larger values of j. The reason is that cancellation effects in computing x_j may give rise to overflow as the singularities are approached. The function $f(x)$ should be programmed as a two-parameter function $f(x, \delta)$ in which the 'residual' δ, representing the distance from the singularity, should also be specified. Hence x_j in equation (6.6.20) should be expressed as

$$x_j = a + (b-a)q_j(1+q_j)^{-1} \tag{6.6.23}$$

in the vicinity of $x = a$ and as

$$x_j = b - (b-a)(1+q_j)^{-1} \tag{6.6.24}$$

near $x = b$, so providing accurate numerical values for δ. Some simplification is possible when attention is confined to *odd* functions of $g(t)$, suitable examples being

$$g(t) = t^{\nu} \qquad (\nu = 1, 3, 5, \ldots) \tag{6.6.25}$$

or

$$g(t) = c \sinh t. \tag{6.6.26}$$

In such circumstances, it is easily seen that (6.6.19) reduces to

$$S \simeq 2h(b-a)\left\{ A_0 f\left[\frac{1}{2}(b+a)\right] + \sum_{j=1}^{N} A_j[f(a+\delta_j) + f(b-\delta_j)] \right\} \tag{6.6.27}$$

in which

$$\delta_j = (b-a)q_j(1+q_j)^{-1} \tag{6.6.28}$$

Equation (6.6.27) is the form adopted for computational purposes, remembering that the functions are programmed as two-parameter procedures.

Table 6.9 Number of function evaluations required to attain relative error $E = 1.0(-10)$

Integral	EHM	EFH	$\tanh(c\sinh t)$
I_1	21	f	f
I_2	27	15	25
I_3	111	31	31
I_4	127	15	29
K_1	45	7	25
K_2	45	3	23
K_3	45	3	25
K_4	17	15	29
K_5	97	95	25
K_6	141	31	25
K_7	29	31	32
K_8	75	31	32
K_9	129	31	32

To conclude this section a comparison is made on the set of test integrals between the accelerator method of Evans, Hyslop and Morgan (EHM), the polynomial transformation of Evans Forbes and Hyslop (EFH) and the use of the transformation $x = \tanh(c\sinh t)$. The number of function evaluations for an accuracy of 10^{-10} is shown in each case in Table 6.9.

The spectacular success of EFH in certain cases (e.g. $K_1 - K_3$) is explained by the simplicity of the integrands which become trivial when $x = t^6$ is employed. In general the $\tanh(c\sinh t)$ transformation seems to perform very well, and is not particularly sensitive to the constant c, and values between $c = 1$ and $c = 5$ were used.

6.7 Oscillatory quadrature

The problem of evaluating oscillatory integrals arises in many applications, and, together with singular integrals, is one of the main classes of integral for which the general methods of the earlier paragraphs prove inadequate. For example, such integrals arise in the analysis of water waves, in diffraction problems, and in many quantum mechanical computations. The classical motivation for the study of such problems was the numerical inversion of integral transforms which occur in the solution of ordinary or partial differential equations.

The general problem is the evaluation of the integral

$$I = \int_a^b w(x)f(x)\mathrm{d}x \tag{6.7.1}$$

where $w(x)$ is the oscillatory part and $f(x)$ is smooth. For a finite range $[a, b]$ somewhat different methods are applicable than for an infinite range with b set at infinity. These two cases will be treated separately.

The oscillatory part $w(x)$ is most commonly $\sin \omega x$ or $\cos \omega x$ for some large ω, though Hankel transformations and cylindrical polar coordinates often result in Bessel function oscillations such as $J_0(\omega x)$, and some recent work concentrates on such non-trigonometric forms. This work also includes oscillatory parts of the form $\sin(\omega q(x))$ and $\cos(\omega q(x))$ for a smooth function $q(x)$ which is treated in Section 6.8. The infinite-range techniques cope with any of these cases, and indeed with difficult functions such as $\sin(1/x)$ for x small.

Again there is no loss of generality in considering integrals over the interval $[-1, 1]$ as the transformation

$$x = \frac{b-a}{2}y + \frac{b+a}{2} = my + c \tag{6.7.2}$$

reduces (6.7.1) to

$$m \int_{-1}^{1} w(my+c) f(my+c) \mathrm{d}y \tag{6.7.3}$$

and for $w(x) = \sin(\omega x)$ or $w(x) = \cos(\omega x)$ we have

$$w(my+c) = \begin{cases} \cos m\omega y \cos c\omega - \sin m\omega y \sin c\omega \\ \sin m\omega y \cos c\omega + \cos m\omega y \sin c\omega \end{cases} \tag{6.7.4}$$

to yield in each case two integrals of the standard form

$$K \int_{-1}^{1} \begin{matrix}\cos\\ \sin\end{matrix} (\Omega y) F(y) \mathrm{d}y \tag{6.7.5}$$

where K is either $\sin c\omega$ or $\cos c\omega$, $\Omega = \omega m$ and $F(y) = f(my+c)$.

The methods fall into two main categories: those which are oscillatory equivalents of Newton–Cotes rules and depend on fitting $f(x)$ by a relatively low degree polynomial and fixing the abscissae in an equally spaced fashion, and those which are equivalent to the Clenshaw–Curtis rule where the fit is with Chebyshev polynomials. The alternative use of sine and cosine weighted Gaussian formulae is of very limited application as a different set of expensive-to-calculate coefficients is required for each particular range and ω value, though these formulae are available in Gautschi (1970) and Piessens (1970a,b,c). Hence following this introduction, attention is first drawn to the equivalent of the Newton–Cotes or classical methods.

The earliest methods followed the same historical route as general quadrature rules with equally-spaced Newton–Cotes rules being used first. For integrals of the form

$$\int_{a}^{b} f(x) \begin{matrix}\sin\\ \cos\end{matrix} \omega x \mathrm{d}x \tag{6.7.6}$$

Filon (1928) produced a simple method based on fitting $f(x)$ to a quadratic and integrating monomials of the form $x^n \sin \omega x$ and $x^n \cos \omega x$ for $n = 0, 1$ and 2 to generate an equivalent rule to Simpson's but for an oscillatory weight. An equivalent of the trapezium rule was found by Clendenin (1966) and a rule with polynomials of degree five by Flinn (1960).

Filon's formula involves first subdividing the interval $[a, b]$ of integration into an even number of $2n$ equal parts in the same way as Simpson's rule, so that $b = a + 2nh$. This allows high accuracy to be obtained without having to use high-order formulae, which will, like the Newton–Cotes rules of the previous chapter, be unstable. The quadratic fit is then made to each sub-interval in the form

$$f(x) = A + B(x - x_r) + C(x - x_r)^2 \tag{6.7.7}$$

where

$$x_r = a + rh$$

and A, B and C are constants.

A little manipulation gives explicit formulae for A, B and C to fit $f(x)$ at the points x_{r-1}, x_r and x_{r+1}. It is then a simple matter of integrating the quadratic approximation to $f(x)$ on each sub-interval to obtain a quadrature rule of the form

$$\int_a^b f(x) \sin \omega x \mathrm{d}x = h[\alpha\{f(a)\cos \omega a - f(b)\cos \omega b\} + \beta S_{2r} + \gamma S_{2r-1}] \tag{6.7.8}$$

where

$$\alpha = 1/\theta + \cos\theta \sin\theta/\theta^2 - 2\sin^2\theta/\theta^3$$

$$\beta = 2[(1 + \cos^2\theta)/\theta^2 - 2\sin\theta\cos\theta/\theta^3]$$

$$\gamma = 4(\sin\theta/\theta^3 - \cos\theta/\theta^2)$$

and $\theta = \omega h$. The terms S_{2r} and S_{2r-1} are the two components of a generalized Simpson's rule for the same integral, namely

$$2S_{2r} = f(a) + 2\sum_{r=1}^{n-1} f(x_{2r}) + f(b) \tag{6.7.9}$$

and

$$S_{2r-1} = \sum_{r=1}^{n} f(x_{2r-1}) \tag{6.7.10}$$

Following the analogy with general quadrature methods, an equivalent approach to that of Clenshaw and Curtis is presented in the oscillatory case, as an improvement over the above Newton–Cotes technique. It is clear that if a fit to $f(x)$ is made based on a series of Chebyshev polynomials, then integrals of the form $\int_{-1}^{1} T_k(x) \sin \omega x \, \mathrm{d}x$ are required, and the problem is that no straightforward closed form seems to be available. Hence any method will require the evaluation of such integrals by a special technique.

If $f(x)$ is expanded as a Chebyshev series then

$$f(x) = \sum_{k=0}^{n}{}' a_k T_k(x) \tag{6.7.11}$$

where the prime indicates the first term of the series is halved. By using the equation

$$a_k = \frac{2}{\pi} \int_{-1}^{1} \frac{T_k(x) f(x) \mathrm{d}x}{(1-x^2)^{1/2}} \tag{6.7.12}$$

with a Gauss–Chebyshev rule yields a quadrature rule of the form

$$I = \int_{-1}^{1} f(x) \begin{matrix}\cos\\ \sin\end{matrix} \omega x \mathrm{d}x = \sum_{i=0}^{n}{}' a_i N_i(\omega) \tag{6.7.13}$$

provided integrals of the form

$$\mathrm{i}^k I_k(\omega) = \int_{-1}^{1} T_k(x) e^{\mathrm{i}\omega x} \mathrm{d}x = N_k(\omega) \tag{6.7.14}$$

can be found.

There is no closed form as a standard function for these integrals, and it is to evaluate (6.7.14) that all the effort has been directed.

Piessens and Poleunis (1971) used an infinite expansion of $e^{\mathrm{i}\omega x}$ in the form

$$e^{\mathrm{i}\omega x} = 2\sum_{j=0}^{\infty}{}' \mathrm{i}^j J_j(\omega) T_j(x) \tag{6.7.15}$$

to get an infinite series of Bessel functions for I_k as

$$\mathrm{i}^k I_k = -2\sum_{j=0}^{\infty}{}' \mathrm{i}^j J_j(\omega) \left[\frac{1}{(k+j)^2-1} + \frac{1}{(k-j)^2-1}\right] \tag{6.7.16}$$

The obvious shortcoming with this result is the dependence on ω, and indeed for increasing ω the computational effort increases.

Patterson (1976) evaluates (6.7.14) by expanding $T_j(x)$ as a series of Legendre functions and then the standard integral

$$\int_{-1}^{1} e^{\mathrm{i}\omega x} P_k(x)\, \mathrm{d}x = \mathrm{i}^k \left(\frac{2\pi}{\omega}\right)^{1/2} J_{k+1/2}(\omega) \tag{6.7.17}$$

can be used to complete the quadrature. Hence the integral reduces to

$$\int_{-1}^{1} f(x) e^{\mathrm{i}\omega x} \mathrm{d}x = \left(\frac{\pi}{2\omega}\right)^{1/2} \sum_{j=0}^{n}{}' a_j \sum_{k=0}^{j} \mathrm{i}^k (2k+1) R_{k,j} J_{k+1/2}(\omega) \tag{6.7.18}$$

with

$$R_{k,j} = \int_{-1}^{1} P_k(x)T_j(x)\mathrm{d}x \tag{6.7.19}$$

and

$$R_{k+2,j} = \left(\frac{j^2 - k^2}{j^2 - (k+3)^2}\right) R_{k,j} \tag{6.7.20}$$

giving a recurrence relation to find $R_{k,j}$. This process is completely stable and has the only disadvantage that an efficient Bessel function procedure is necessary for a good implementation.

An almost identical approach appeared in Littlewood and Zakian (1976), the resulting Bessel functions being found by a recurrence relation based on continued fractions. Hence it is clear that a number of closely related methods exist, some of which avoid the main issue, that is the evaluation of (6.7.14), whereas others rely on the existence of an analytic form for (6.7.14) if the Chebyshev functions are replaced by Legendre functions to find a route to a feasible algorithm. The cost is the evaluation of spherical Bessel functions, and the gain is an algorithm with no stability limit on point number. However in practice, point numbers as high as 64 or 128 are not commonly used, as it may well be more expedient to use adaptive sub-division to 'hit' the problem areas with a higher density of abscissae. This then opens the possibility of a simpler algorithm than the above, but one with some point number restriction.

A simpler method to evaluate (6.7.14) which avoids the Bessel function approaches is presented next, with some discussion on the stability of the process. The basis of the method due to Alaylioglu, Evans and Hyslop (1976) is the evaluation of $N_k(\omega)$, by picking out the coefficients, $D_{i,r}$, of x^r in $T_i(x)$ and then making use of the results

$$\int_{-1}^{1} x^r \cos\omega x \mathrm{d}x = \sum_{l=0}^{r} l!\binom{r}{l}\frac{x^{r-l}}{\omega^{l+1}} \sin\left(\omega x + \frac{1}{2}l\pi\right)\Bigg|_{-1}^{1} \tag{6.7.21}$$

$$\int_{-1}^{1} x^r \sin\omega x \mathrm{d}x = -\sum_{l=0}^{r} l!\binom{r}{l}\frac{x^{r-l}}{\omega^{l+1}} \cos\left(\omega x + \frac{1}{2}l\pi\right)\Bigg|_{-1}^{1} \tag{6.7.22}$$

(Gradshteyn and Ryzhik,1965).

The Chebyshev polynomials $T_i(x)$ are defined in Chapter 1 and have the coefficients, $D_{i,r}$, of x^r which can be easily calculated by means of the recurrence relation:

$$D_{i,r} = 2D_{i-1,r-1} - D_{i-2,r} \qquad i \geq 2, \quad r \leq i \tag{6.7.23}$$

yielding a 'Pascal' triangle which facilitates computation. The basic integrals $N_k(\omega)$ are then easily obtained. In practice these are usually separated into their real and imaginary parts for the separate calculation of the integrals involving $\cos\omega$ or $\sin\omega$ using, either (6.7.21) or (6.7.22) for even or odd r respectively.

Therefore to generate a quadrature rule, the Clenshaw–Curtis fitting of $f(x)$ to a series of Chebyshev polynomials $T_i(x)$ given by (6.4.2), the series with coefficients given by (6.4.6) is employed. Hence to evaluate the integral

$$I_c = \int_a^b f(x) \cos px \mathrm{d}x \tag{6.7.24}$$

a linear transformation is used to enable the result to be written in the form

$$I_c = \frac{1}{2}(b-a)\cos K \int_{-1}^{1} F(t)\cos \omega t \mathrm{d}t - \frac{1}{2}(b-a)\sin K \int_{-1}^{1} F(t) \sin \omega t \mathrm{d}t \tag{6.7.25}$$

where

$$K = \tfrac{1}{2}p(b+a) \tag{6.7.26}$$

$$\omega = \tfrac{1}{2}p(b-a) \tag{6.7.27}$$

and

$$F(t) = f\left[\tfrac{1}{2}(b+a) + \tfrac{1}{2}(b-a)t\right]. \tag{6.7.28}$$

Approximating $F(t)$ by the polynomial of degree n

$$F(t) \simeq \sum_{i=0}^{n}{}'' a_i T_i(t) \tag{6.7.29}$$

collocating at the $(n+1)$ points t_j where

$$t_j = \cos\frac{\pi j}{n} \qquad j = 0(1)n \tag{6.7.30}$$

yields

$$\begin{aligned} I_c \simeq {} & \frac{1}{2}(b-a)\cos K \sum_{i=0}^{n}{}'' a_i \sum_{r=0}^{i} D_{i,r} \int_{-1}^{1} t^r \cos \omega t \mathrm{d}t \\ & - \frac{1}{2}(b-a)\sin K \sum_{i=0}^{n}{}'' a_i \sum_{r=0}^{i} D_{i,r} \int_{-1}^{1} t^r \sin \omega t \mathrm{d}t \end{aligned} \tag{6.7.31}$$

where

$$a_i = \frac{2}{n}\sum_{j=0}^{n}{}'' F(t_j)\cos\frac{\pi j i}{n}. \tag{6.7.32}$$

Similarly

$$I_s = \int_a^b f(x) \sin px \mathrm{d}x \tag{6.7.33}$$

yields

$$I_s \simeq \frac{1}{2}(b-a)\sin K \sum_{i=0}^{n}{}'' a_i \sum_{r=0}^{i} D_{i,r} \int_{-1}^{1} t^r \cos \omega t \mathrm{d}t$$
$$+ \frac{1}{2}(b-a)\cos K \sum_{i=0}^{n}{}'' a_i \sum_{r=0}^{i} D_{i,r} \int_{-1}^{1} t^r \sin \omega t \mathrm{d}t \qquad (6.7.34)$$

The basic integrals required in (6.7.31) and (6.7.34) are supplied by (6.7.21) and (6.7.22). These methods are illustrated in the following examples taken from the test integrals of Piessens and Poleunis. For the purpose of illustration the integral

$$\int_0^{2\pi} x \cos x \sin px \mathrm{d}x = \begin{cases} -2\pi p(p^2-1)^{-1} & (p = 2, 3, 4, \ldots) \\ -\pi/2 & (p = 1) \end{cases} \qquad (6.7.35)$$

is considered. The numerical results are depicted in Table 6.10 for $p = 1, 2, 4, 16, 64$ and 256. To facilitate direct comparison, double precision calculations were carried out. The order of the formulae is increased until the errors are less than those obtained by Piessens and Poleunis for 30 function evaluations. This accuracy is achieved for all values of p for about 19 or 20 function evaluations, thus representing an improvement over the alternative approaches. This is due presumably, to a decrease in round-off errors generated by the present algorithm compared with the earlier Bessel-function series prescription for the evaluation of the basic integrals. The stability of the algorithms is tested by proceeding to larger values of n and it is possible to go to $n = 25$ even for $p = 1$ and still use the ω^{-1} series, (6.7.21) and (6.7.22).

The method of Patterson gives the same results as AEH, as the fitting is identical and only the method of evaluating (6.7.14) varies. As long as the method of Alayli-oglu, Evans and Hyslop (AEH) is kept within its stability range, both methods will deliver identical results.

Table 6.10 Comparison of the absolute errors in the numerical evaluation of $\int_0^{2\pi} x \cos x \sin px \mathrm{d}x$.

		Piessens and Poleunis		AEH	
p	Exact	Absolute error	Number of function evaluations	Absolute error	Number of function evaluations
1	−1.570 796 326 794 8966	5(−15)	30	4(−16)	19
2	−4.188 790 204 786 3910	9(−15)	30	6(−16)	19
4	−1.675 516 081 914 5564	3(−15)	30	1(−15)	20
16	−0.394 239 078 097 5427	1(−14)	30	5(−15)	20
64	−0.098 198 744 727 5930	3(−15)	30	2(−16)	20
256	−0.024 544 067 118 9132	1(−15)	30	2(−16)	19

There is a very recent method to evaluate (6.7.14) which avoids both the stability restrictions of AEH and the evaluation of series of Bessel functions due to Evans (1995). The suggestion here is to return to the three-term recurrence relation

$$-\frac{\omega I_{n-1}}{2(n-1)} + I_n + \frac{\omega I_{n+1}}{2(n+1)} = -\frac{2\sin\omega}{(n^2-1)} \qquad n \text{ odd} \tag{6.7.36}$$

$$\frac{\omega I_{n-1}}{2(n-1)} + I_n - \frac{\omega I_{n+1}}{2(n+1)} = -\frac{2\cos\omega}{(n^2-1)} \qquad n \text{ even} \tag{6.7.37}$$

This formula used in the forward direction is stable for $n < \omega$ and in the backward direction is stable for $n \gg \omega$. Hence the gap arises for $n > \omega$ and upwards to some vague limit defined by the symbol $\ll$. It is certainly feasible to write accurate formulae for I_n valid for $n \gg \omega$. A number of asymptotic formulae have been tried but the most economical was obtained as follows. Consider the equivalent integral

$$J = \int_0^{\pi} \sin\theta \cos n\theta e^{\mathrm{i}\omega\theta}\, \mathrm{d}\theta \tag{6.7.38}$$

or

$$J = \int_0^{\pi} g(\theta)\cos n\theta\, \mathrm{d}\theta \tag{6.7.39}$$

where

$$g(\theta) = \begin{cases} \sin\theta\cos(\omega\cos\theta) \\ \sin\theta\sin(\omega\cos\theta) \end{cases} \tag{6.7.40}$$

Then successive integration by parts will yield the asymptotic series

$$J = \sum_{i=0}^{\infty} (-1)^{[i/2]} \left[\frac{g^{(i)}(\theta) cs_i(n\theta)}{n^{i+1}}\right]_0^{\pi} \tag{6.7.41}$$

and

$$cs_i(n\theta) = \begin{cases} \sin n\theta & i \text{ even} \\ \cos n\theta & i \text{ odd} \end{cases} \tag{6.7.42}$$

with

$$g(\theta) = gc(\theta) = \sin\theta\cos(\omega\cos\theta) \tag{6.7.43}$$

or

$$g(\theta) = gs(\theta) = \sin\theta\sin(\omega\cos\theta) \tag{6.7.44}$$

The high derivatives of g follow using Leibnitz' rule which can be implemented by setting

$$gc(\theta) = \sin(\theta)vc(\theta) \tag{6.7.45}$$

$$gs(\theta) = \sin(\theta)vs(\theta) \tag{6.7.46}$$

with $vc(\theta) = \cos(\omega \cos\theta)$ and $vs(\theta) = \sin(\omega\cos\theta)$. Then

$$vc^{(m)}(\theta) = \omega\, gs^{(m-1)}(\theta) \tag{6.7.47}$$

and

$$vs^{(m)}(\theta) = -\omega\, gc^{(m-1)}(\theta) \tag{6.7.48}$$

The mth derivative of the product in equations (6.7.43) and (6.7.44) can then be evaluated using

$$D^m(uv) = vD^m(u) + mD^{n-1}(u) + \cdots + \frac{n(n-1)\cdots(n-k+1)}{k!} D^k(v)D^{n-k}(u) + \cdots \tag{6.7.49}$$

A recursive algorithm is now feasible based on the rules:

(i) Declare gc, vs and vc as FORWARD functions to be fully declared later (as in Pascal).

(ii) Define $u^{(i)}$ to be $\mathrm{d}^i/\mathrm{d}\theta^i(\sin(\theta))$ between the limits 0 and π. Hence $u^{(i)}$ is zero for even i, and for odd i is 1.0 if $[(i+1)/2]$ is even and -1 in the odd case. This is the u of equation (6.7.49).

(iii) $gs^{(i)}$ can now be coded using the series (6.7.49) based on (6.7.46) in terms of $vs^{(i)}$ and $u^{(i)}$.

(iv) Similarly $gc^{(i)}$ can be coded using $vc^{(i)}$ and $u^{(i)}$.

(v) Finally to complete the recursive definition, $vs^{(i)}$ is coded as zero for odd i, and $-\theta gc^{(m-1)}$ otherwise. The recursion will unwind down to a zero result and terminate.

(vi) Similarly $vc^{(i)}$ is 0 for odd i and $\theta gs^{(i-1)}$ otherwise.

The series (6.7.41) then forms the main routine to yield the required asymptotic series.

This formula is used only once to find a value for I_n for sufficiently large n. In practice, setting n equal to $[3\omega]$ has proved satisfactory.

This asymptotic result allows the three-term recurrence relations (6.7.36) and (6.7.37) to be used in a stable manner, in effect expressing I_n in terms of I_{n-1} and I_{n+1} to yield an implicit form, with n ranging from $[\omega]$ to $3[\omega]$. The resulting matrix is tri-diagonal and of order $N = 2[\omega]$, and the corresponding set of linear equations can be evaluated using a very restricted form of Gaussian elimination to place a zero in the sub-diagonal. Back-substitution is also then economized to yield the complete set of required I_n values. The storage requirement is the modest $3N$. In practice it is unusual to use even a 1000-point formula as is indicated by tabulated Gaussian and Patterson coefficients. The resulting quadrature rule is then simply coded using (6.7.13). Hence the rule can be used in the usual progressive manner with point numbers 2, 4, 8, ... with only evaluation of the set I_n for a given ω. This algorithm has been used without difficulty to order 500.

In summary, the algorithms of Patterson and the above method yield powerful methods analogous with the Clenshaw–Curtis rule which are applicable to highly oscillatory trigonometric integrands. This work leaves some further problems which are beyond the scope of this book but which are treated in Evans (1993). There is however, a further type of problem which can be treated here, namely the infinite-range oscillatory integrand.

The solution to this problem is somewhat different and is treated separately. Typically integrals of the form

$$\int_0^\infty f(x) \begin{matrix}\sin\\ \cos\end{matrix}\, \omega x \,\mathrm{d}x \tag{6.7.50}$$

are now being considered

The method, due to Alaylioglu, Evans and Hyslop (1973), involves applying the Shanks' acceleration techniques to the sequence of integrals generated by taking successive half-cycles of the oscillatory integrand. Hence the range of integration is sub-divided in accordance with the half-cycles of the integrand into sub-ranges $[a_n, a_{n+1}]$, $n = 0, 1, 2, \ldots$, where in the case of the integrals of the form (6.7.50), a_n is given by

$$a_n = n\pi/k \tag{6.7.51}$$

in the sin case. More general forms of the oscillatory function can clearly be tackled, for example, integrals of the form

$$\int_0^\infty f(x) \begin{matrix}\cos\\ \sin\end{matrix}\, kx^2 \mathrm{d}x \tag{6.7.52}$$

which arise in many applications are also treated here. For such integrals a_n is given by

$$a_n = (n\pi/k)^{1/2}. \tag{6.7.53}$$

Non-trigonometric oscillations such as those defined by Bessel functions are also possible as long as the zeros of the oscillatory function are known or can be found. A general purpose integrator is employed to carry out the integration over each half-cycle $[a_n, a_{n+1}]$. To minimize the contributions from truncation and round-off errors, the integrator is applied successively in each of the sub-intervals $[a_n, a_{n+1}]$ to yield the values

$$T_n = \int_{a_n}^{a_{n+1}} g(x)\mathrm{d}x \tag{6.7.54}$$

where $g(x)$ represents the appropriate integrand. For most of the examples which follow, no more than 24 Gauss–Legendre points are employed in the evaluation of the integrals over each cycle, though any of the general purpose integrators would have been applicable.

The actual integrals required are of course given by

$$A = \sum_{i=0}^{\infty} T_i \tag{6.7.55}$$

and an application of the Shanks' technique for accelerating convergence of the partial sums

$$A_n = \sum_{i=0}^{n} T_i. \tag{6.7.56}$$

is made. The process is best illustrated using examples.

The first example of the process concerns the evaluation of the integral

$$I_1(\alpha, k) = \int_0^{\infty} e^{-\alpha x} \sin kx \mathrm{d}x = k(\alpha^2 + k^2)^{-1}. \tag{6.7.57}$$

In this case the terms of the sequence $\{A_n\}$ are easily evaluated analytically yielding

$$A_n = k(\alpha^2 + k^2)^{-1}\{1 - (-1)^{n+1} \exp[-(n+1)\alpha\pi/k]\} \tag{6.7.58}$$

The geometric convergence of this sequence implies *immediate* convergence of the sequence $e_i\{A_n\}$ to the exact limiting value, that is $B_{1n} = k(\alpha^2 + k^2)^{-1}$ for all n.

As a practical example of this class of integral, consideration is given to

$$I_2 = \int_0^{\infty} x^{-1} \exp(-x/2) \sin x \mathrm{d}x = \arctan 2 = 1.107\,149 \cdots \tag{6.7.59}$$

The transformation e_1 was applied iteratively to the sequence $\{A_n\}$ to give the sequences

$$e_1\{A_n\}, e_1^2\{A_n\}, \ldots,$$

the first few members of which are shown in Table 6.11. For comparison purposes, Euler's transformation (see Knopp (1928)) is also applied to the terms T_n to produce the sequence of partial sums $\{E_n\}$.

The half-cycle contributions, T_n, are evaluated over $[n\pi, (n+1)\pi]$ using a 16-point Gauss–Legendre formula. It will be noticed that the Euler sequence $\{E_n\}$ converges less well than the original sequence $\{A_n\}$. In contrast $e_1\{A_n\}$

Table 6.11 Successive sequences for evaluating I_2

n	T_n	A_n	E_n	$e_1\{A_n\}$	${e_1}^2\{A_n\}$
0	1.148 148	1.148 148	0.574 074		
1	−0.045 820	1.102 328	0.849 656	1.107 254	
2	0.005 519	1.107 847	0.982 409	1.107 141	1.107 149
3	−0.000 809	1.107 038	1.046 562	1.107 150	
4	0.000 130	1.107 163	1.077 652		

converges extremely rapidly and $e_1{}^2\{A_n\}$ converges immediately to the limiting value. Further, the transformation e_2 could also be applied to $\{A_n\}$ and again yields the limiting value immediately, namely $B_{2,2} \simeq 1.107\,149$. A further example involving a different family of integrals is

$$I_3 = \int_{\pi}^{\infty} x^{-2} \sin x \mathrm{d}x = -\mathrm{ci}(\pi) \simeq -0.073\,668 \tag{6.7.60}$$

where ci is the cosine integral as defined by Gradshteyn and Ryzhik (1965). The results are demonstrated in Table 6.12.

The half-cycle contributions were evaluated over $[(n+1)\pi, (n+2)\pi]$ again using a 16-point Gauss–Legendre formula.

As a final example, the integral

$$I_4 = \int_0^{\infty} x^2 \sin 100x^2 \mathrm{d}x \tag{6.7.61}$$

which converges in the mean only (in the Abel sense) is considered. The exact result is

$$I_4 = (\pi/2)^{1/2}/4000 \simeq 3.133\,285 \times 10^{-4} \tag{6.7.62}$$

which is readily obtained by standard integration, using the integration factor $\exp(-\gamma x^2)$ as γ tends to zero. The numerical results are shown in Table 6.13.

The half-cycle contributions, T_n, are evaluated with $a_n = (n\pi/100)^{1/2}$ using twelve two-point Gauss–Legendre formulae. It is noticed in this case that the original sequence $\{A_n\}$ is divergent. The Euler transformation produces the sequence $\{E_n\}$ which is slowly convergent. The sequence $e_1\{A_n\}$ is also seen to be slowly convergent, but when the operator e_1 is used iteratively, the successive sequences $e_1{}^2\{A_n\}, e_1{}^3\{A_n\}, \ldots,$ are seen to be converging rapidly. The application of the operator e_2 to the original sequence $\{A_n\}$ produces the sequence

$$\{3.1268, 3.1354, 3.1322, 3.1337, 3.1329, \ldots \times 10^{-4}\}$$

and $e_2{}^2$ produces 3.1332×10^{-4} as its first term.

Table 6.12 Successive sequences for evaluating I_3

n	T_n	A_n	E_n	$e_1\{A_n\}$	$e_1{}^2\{A_n\}$	$e_1{}^3\{A_n\}$
0	−9.6230	−9.6230	−4.8115			
1	3.3180	−6.3049	−6.3877	−7.3496		
2	−1.6737	−7.9786	−7.5529	−7.3744	−7.3677	
3	1.0078	−6.9708	−7.4305	−7.3633	−7.3667	−7.3669
4	−0.6730	−7.6439	−7.3901	−7.3689	−7.3670	
5	0.4812	−7.1626	−7.3614	−7.3657		
6	−0.3612	−7.5238	−7.3651			
	$\times 10^{-2}$	$\times 10^{-2}$	$\times 10^{-2}$	$\times 10^{-2}$	$\times 10^{-2}$	$\times 10^{-2}$

Table 6.13 Successive sequences for evaluating I_4

n	T_n	A_n	E_n	$e_1\{A_n\}$	$e_1^2\{A_n\}$	$e_1^3\{A_n\}$	$e_1^4\{A_n\}$
0	1.2177	1.2177	6.0883				
1	−2.1650	−0.9474	3.7199	2.7356			
2	2.7998	1.8525	3.3292	3.3475	3.1236		
3	−3.3143	−1.4619	3.2090	2.9944	3.1366	3.1330	
4	3.7588	2.2970	3.1647	3.2324	3.1317	3.1332	3.1332
5	−4.1560	−1.8590	3.1468	3.0577	3.1340	3.1332	
6	4.5182	2.6593	3.1392	3.1931	3.1327		
7	−4.8535	−2.1943	3.1359	3.0842			
8	5.1671	2.9728	3.1344				
	$\times 10^{-3}$	$\times 10^{-3}$	$\times 10^{-4}$	$\times 10^{-4}$	$\times 10^{-4}$	$\times 10^{-4}$	$\times 10^{-4}$

In conclusion, it appears that the Shanks acceleration technique is a powerful tool for the evaluation of oscillatory integrals with an infinite range. This is particularly true when the oscillations are damped in a predominantly exponential manner or fall off, for example, as $1/x^p$ where p is sufficiently large (say $p \geq 1$). Also, for integrals which converge in the mean only, Shanks method proves successful, so giving a broadly applicable method for infinite-range oscillatory integrals.

6.8 Evaluating integrals with irregular oscillatory integrands

There are a number of rather different ways to approach finite-range irregular oscillatory integrals which have the general form

$$\int_a^b f(x) \begin{matrix}\sin\\ \cos\end{matrix}\, \omega q(x)\, \mathrm{d}x \tag{6.8.1}$$

The first of these methods is very easy to derive. The method is equivalent to that of Clendenin (1966) for regular oscillatory quadrature, or the trapezoidal rule for general purpose quadrature. Both $f(x)$ and $q(x)$ are approximated by linear forms and the resulting approximation is integrated analytically to yield a quadrature rule. Hence

$$I = \int_a^b f(x) \begin{matrix}\sin\\ \cos\end{matrix}\, \omega q(x)\, \mathrm{d}x \tag{6.8.2}$$

and sub-division gives

$$I = \sum_{i=1}^{N} \int_{a+(i-1)h}^{a+ih} f(x) \begin{matrix}\sin\\ \cos\end{matrix}\, \omega q(x)\, \mathrm{d}x$$

$$= \sum_{i=1}^{N} I_i \tag{6.8.3}$$

Then it is easy to establish

$$I_i = \int_{a+(i-1)h}^{a+ih} (mx+c) \begin{matrix}\sin\\ \cos\end{matrix} \omega(\alpha x+\beta)\,\mathrm{d}x \tag{6.8.4}$$

with the constants m, c, α and β taking the values

$$m = (f(a+ih) - f(a+(i-1)h))/h \tag{6.8.5}$$

$$c = f(a+ih) - m(a+ih) \tag{6.8.6}$$

$$\alpha = (q(a+ih) - q(a+(i-1)h))/h \tag{6.8.7}$$

$$\beta = q(a+ih) - \alpha(a+ih) \tag{6.8.8}$$

Expanding the sine and cosine in (6.8.4) gives the following form for the quadrature rule

$$\begin{aligned} I_i &= \int_{a+(i-1)h}^{a+ih} (mx+c) \left\{ \begin{matrix} \sin\omega\alpha x\cos\omega\beta + \cos\omega\alpha x\sin\omega\beta \\ \cos\omega\alpha x\cos\omega\beta - \sin\omega\alpha x\sin\omega\beta \end{matrix} \right. \mathrm{d}x \\ &= \left\{ \begin{matrix} \cos\omega\beta(mI_{1s}+cI_{0s}) + \sin\omega\beta(mI_{1c}+cI_{0c}) \\ \cos\omega\beta(mI_{1c}+cI_{0c}) - \sin\omega\beta(mI_{1s}+cI_{0s}) \end{matrix} \right|_{a+(i-1)h}^{a+ih} \end{aligned} \tag{6.8.9}$$

and the quantities I_{is} and I_{ic} are standard integrals

$$I_{is} = \int x^i \sin kx\,\mathrm{d}x \tag{6.8.10}$$

and

$$I_{ic} = \int x^i \cos kx\,\mathrm{d}x \tag{6.8.11}$$

It is also possible (Evans, 1994a) to continue to quadratic approximations to obtain a Filon or Simpson equivalent rule but this involves some tedious Fresnel integrals to yield the final quadrature rule, and a greater gain in accuracy can be obtained by other means. The error analysis of these methods shows that the oscillatory function $q(x)$ has to be approximated to an error which is better than $1/\omega$ for convergence. This condition automatically places a limit on the achievable error for convergence which depends on ω itself in the trapezoidal equivalent case and the cube root of ω in the Simpson equivalent case. Hence these methods become less successful for increasing ω, a state of affairs which is against the grain of thinking in oscillatory quadrature.

Another obvious device is to make the transformation $y = q(x)$ to yield the conventional oscillatory integral

$$\int_{q(a)}^{q(b)} \frac{f(x)}{q'(x)} \begin{matrix}\sin\\ \cos\end{matrix} \omega y\,\mathrm{d}y \tag{6.8.12}$$

with $x = q^{-1}(y)$. Again this method, due to Evans (1994b), requires $q'(x)$ and $q^{-1}(x)$, and again Newton's method is the basis for finding the inverse function, a linear interpolator being used to obtain the initial point. The resulting standard integral is then evaluated using a method such as that of Patterson (1976) or Alayli-oglu, Evans and Hyslop (1976). One of the major objections to this approach is that zeros of $q'(x)$ in the interval of integration will yield a singular quadrature problem where no such difficulty is present in the original. Where this occurs a standard Clenshaw–Curtis quadrature is made over a restricted range of the interval of integration of the original integral which is non-singular. This combination proves very effective especially if care is used in the Newton algorithm for the inverse function. A step reduction safeguard is introduced into the Newton algorithm which has the following form.

(i) Tabulate $q(x)$ for $x = a, a+h, \ldots,\ b = a + mh$ for a suitable m.
A quite small m such as 2 or 4 is quite sufficient.

(ii) Use a linear interpolation rule on these tabulated values to obtain x_0.
The idea is to put x_0 close to the final root to enhance the speed of the rootfinding.

(iii) Continue with steps (iv) and (v) until convergence.

(iv) Compute either $(q')^{-1}$ or q' depending on which avoids possible divisions by zero.
If $(q')^{-1}$ is computed and is $> 10^3$
or if q' is computed and is $< 10^{-3}$
then the routine *step* (see below) is called.

(v) Otherwise:
compute $x_1 = x_0 + (y - q(x_0)) * (q')^{-1}$
or $x_1 = x_0 + (y - q(x_0))/q'$
according as to whether $(q')^{-1}$ or q'
was computed in (iv).

If $x_1 < a$ then $x_1 = a$
else if $x_1 > b$ then $x_1 = b$.

The routine *step*
if $|x_1 - x_0| > maxstep$ then
if $x_1 > x_0$ take a step $+maxstep/2$
if $x_1 < x_0$ take a step $-maxstep/2$
if either process causes x_1 to lie beyond the end-points
$[a, b]$ then reset x_1 to the nearest end-point.
otherwise accept x_1.

The routine *step* takes a step *ee* in the direction of the zero, which can be determined by two sample points, unless this causes x_1 to lie beyond $[a, b]$ when x_1 is taken as the nearest end-point. A typical value for *maxstep* is $(b - a)/4$ and for *ee*, $(b - a)/20$.

These restrictions give the routine some robustness, especially in regions of high gradient which arise near to singular points, and when q' is very small. The device of using either $(q')^{-1}$ or q' avoids an artificial introduction of a division by zero. A set of test examples has been compiled from the examples employed by previous authors to demonstrate their methods. In addition some further examples have been included here to extend the effectiveness of the test examples in showing up weaknesses in the proposed methods. The method of comparison will need to consider not only the number of function evaluations of $f(x)$, but for the majority of methods, also the number of evaluations of $q(x)$ and $q'(x)$.

A very different approach is that of Levin (1982), which again handles varying oscillations of the form

$$I = \int_a^b f(x)e^{\mathrm{i}q(x)}\,\mathrm{d}x \tag{6.8.13}$$

The basis of the method is that if $f(x)$ were of the form

$$f(x) = \mathrm{i}q'(x)p(x) + p'(x) \tag{6.8.14}$$

then the integral in equation (6.8.13) could be evaluated directly to give

$$I = p(b)e^{\mathrm{i}q(b)} - p(a)e^{\mathrm{i}q(a)}. \tag{6.8.15}$$

Equation (6.8.14) is considered as a differential equation for the function $p(x)$ which is solved for $p(x)$ and used in equation (6.8.15). The general solution is

$$p(x) = e^{-\mathrm{i}q(x)}\left[\int_a^x f(t)e^{\mathrm{i}q(t)}\,\mathrm{d}t + c\right] \tag{6.8.16}$$

which means that in general the solution for $p(x)$ is as oscillatory as the original integrand, but as f and q' are slowly oscillatory there exists a slowly oscillatory solution p_0. Hence writing

$$p(x) = p_0(x) + ce^{-\mathrm{i}q(x)} \tag{6.8.17}$$

and using this form in (6.8.13) gives

$$I = p_0(b)e^{\mathrm{i}q(b)} - p_0(a)e^{\mathrm{i}q(a)} \tag{6.8.18}$$

and so $p_0(a)$ and $p_0(b)$ are all that are needed. To single out $p_0(x)$ from other highly oscillatory solutions, a representation is tried in which n-point collocation to the basis set $\{u_k(x)\}$ for $k = 1, n$ is set up to give

$$p_0(x) = \sum_{k=1}^{n} \alpha_k\, u_k(x) \tag{6.8.19}$$

This form is substituted into equation (6.8.14) and a set of linear algebraic equations is solved for the α's. Hence I can be easily evaluated from equation (6.8.18). Once again difficulties with large linear systems seem to arise for too great a value of n and the suggestion is to sub-divide once again.

Levin's method determines $p(x)$ explicitly, despite there being no well-defined boundary condition for the solution of (6.8.14). However, the exact relation in (6.8.14) can be used in a different way which avoids these problems. Instead of trying to find $p(x)$ explicitly, consider the relation

$$\int_a^b f(x)e^{\mathrm{i}\omega q(x)}\,\mathrm{d}x = \sum_{i=1}^{N} A_i f(x_i) + E_N \qquad (6.8.20)$$

with the free coefficients A_i chosen to make the formula exact for

$$f(x) = \mathrm{i}q'(x)p(x) + p'(x) \qquad (6.8.21)$$

with a suitable choice of $p(x)$. Now the left-hand sides can be evaluated directly to yield the equivalent of the monomial integrals for general quadrature rules. There are now a number of possible avenues to pursue.

The easiest approach is to take a 'Newton–Cotes' view and fix the abscissae $\{x_i\}$ as being equally spaced. This leaves the A_i to be computed from a set of linear equations whose form depends on the choice of functions used for $p(x)$. Again, the simplest choice seems to be $p(x) = 1, x, \ldots, x^{N-1}$. This is the exact function method (EF). The resulting formulae have been applied to the set of test integrals in Table 6.14 with a high degree of reliability and success.

Each of the five methods was applied to the test set and a count made of the number of evaluations made of the functions f, q and q' for a given accuracy. These counts appear as triples in the table in the above order. In cases where functions are combined together as in Gauss, then a single count is given of the number of integrand evaluations which will include an f and a q evaluation. Hence, for comparison, double this count is equivalent to the sum for the f and q counts in the other columns. In general an accuracy of 12 figures is achieved with these counts, but where this accuracy is not attained then the number of correct figures is shown in brackets after the function count. Where function counts are expressed as a product such as 28×9, subdivision has been used with 28 subdivisions of the

Table 6.14

f	Limits	Integrand	ω	value
f1	(0, 1)	$e^x \cos \omega x$	10	−0.178 899 602 8768
f2	(100, 200)	$(1 + \ln x)\cos(x \ln x)$	1.0	−1.774 298 974 906
f3	$(0, \beta)$	$\cos x \cos(\omega \cos x)/\beta$	40	0.050 194 456 10620
f4	(0, 1)	$\sin x \cos(\omega x(1 + x))$	500	4.598 593 978 40(−4)
f5	(0, 1)	$\cos x \cos(\omega\sqrt{1 - x^2})$	10.0	−3.961 556 279 8520(−1)
f6	(0, 1)	$\cos x \cos(\omega \sin x)$	10.0	8.468 680 691 183(−2)
f7	(0, 1)	$e^x \sin(\omega \cosh x)$	10.0	−2.556 593 290 493(−1)
f8	(0, 1)	$\sin^2 x \cos(\omega \tanh x)$	10.0	0.134 116 499 03305

Table 6.15

f	Gauss	Trap	Inverse	Levin	EF
f1	16	256(6)	(17, 19, 51)	(20, 2, 20)	2×8
f2	2×512	256(6)	(9, 31, 67)	(5, 2, 5)	2×2
f3	16	256(4)	(33, 130, 281) + 16	(80, 8, 80)(7)	8×12
f4	16×512	256(2)	(65, 262, 577)	(80, 8, 80)	8×12
f5	16	256(3)	16 + (33, 135, 291) + 8	fail	$32 \times 8 + 8(11)$
f6	16	256(5)	(3, 9, 17)	(14, 6, 14)	2×2
f7	16	256(5)	(65, 267, 587) + 16	(286, 38, 286)	8×12
f8	16	256(6)	(33, 129, 279)	(94, 14, 94)	2×12

nine-point rule. The results are shown in Table 6.15, where the entries consist of the three counts in the above order.

Where singularities are introduced at one or other end point by the transformation, these regions have been integrated by Clenshaw–Curtis on the original integral and the counts for this rule appear as being added to the normal triple for the irregular oscillatory integral. In the case of the last method, the number of function evaluations for f, q and q' is identical as long as values of q and q' are stored after being computed for the first row of the matrix defining the underlying linear equations. These stored values are then used repeatedly in the later rows. Hence only the count for f values appears in this last column. The apparent speed with which this method completes f2 and f6 is because these were chosen to have analytic values for comparison, and hence $p(x) = q'(x)$, so resulting in the numerical method being exact.

6.9 Multiple quadrature

The problem to be addressed with multi-dimensional integrals is to evaluate

$$I = \int_{a_n}^{b_n} \int_{a_{n-1}(x_n)}^{b_{n-1}(x_n)} \cdots \int_{a_1(x_1,\ldots,x_n)}^{b_1(x_1,\ldots,x_n)} f(x_1, x_2, \ldots, x_n)\mathrm{d}x_1\mathrm{d}x_2 \cdots \mathrm{d}x_n \qquad (6.9.1)$$

and a number of new difficulties arise by comparison with the one-dimensional case. Firstly the regions of integration do not now fall into the three linearly transformable canonical types of a single dimension, that is, a finite range, a semi-infinite range and a doubly infinite range. Hence families of formulae for all manner of different regions are available, all with somewhat more limited applicability than the three one-dimensional cases. Regions such as squares, cubes, n-cubes, circles, spheres and simplexes are commonly encountered. It is also apparent that each dimension will require some minimum number of integration points for suitable accuracy, and hence, even for quite small n, a large number of function evaluations can be expected. For higher dimensions this factor is often the overriding difficulty.

Any additional problems concerning the behaviour of the integrand, such as those covered in the previous chapters for single-dimensional integrals, will be greatly diversified in many dimensions. Hence subspaces of singularities can provide a vast spectrum of complications. Integrands with oscillatory parts can have 'overlapping' wave forms in different dimensions making any approach along the lines of the previous paragraphs quite unworkable.

The classical approach is to treat equation (6.9.1) as a recursive set of one-dimensional integrals to yield a product rule of the form

$$I \simeq \sum_{i_n=1}^{N_n} \cdots \sum_{i_1=1}^{N_1} W_{i_1} W_{1_2} \cdots W_{i_n} f(x_{i_1}, x_{i_2}, \ldots, x_{i_n}) \tag{6.9.2}$$

where the weights W_{i_j} and the abscissae x_{i_j} are chosen to be appropriate for the specific dimension to which they are applied. The number of function evaluations is immediately obvious with $N_1 N_2 \cdots N_n$ integration points which may be quite large. On the other hand, judicious choice of methods for each dimension can absorb some of the special properties, for example using Gauss–Laguerre on a semi-infinite range. With such an approach it is always possible to increase the number of points used to achieve higher accuracy in a practical convergence test, though clearly the actual number of points will rise by a factor of 2^n if each dimension is doubled. Although product rules are easy to use, relying on the extensive work from one dimension, there usually exists non-product rules which have an equivalent degree d to a given product rule but with fewer points. However in practice, the more expensive product rule yields far greater accuracy then the non-product alternatives which are often restricted to quite low-order rules. Not all regions of integration are feasible for the use of a product rule, but more complicated regions can often be handled by breaking them up into sub-regions and applying a product rule to each sub-region.

There are a great number of special formulae for specific regions such as a circle, a rectangle or a cube which are not product rules, but whose weights and abscissae are chosen to make the rule exact for some choice of $f(x_1, \ldots, x_n)$. Even the equivalent of the Newton–Cotes rules for two-dimensional integrals may not exist as the interpolating monomials such as $1, x, y, x^2, y^2, xy, \ldots$ cannot necessarily be formed into linear combinations to give prescribed values at given points. Equivalently the use of undetermined coefficients to force a rule to be exact for some suitable monomials yields a set of simultaneous equations in which symmetries can result in linear dependence occurring so that two different monomials lead to the same equation. Hence it is common for rules generated in this way to be 'one-off' rules rather than members of sequences which can be applied successively for increasing accuracy. These rules are beyond the scope of this book and are covered in the specialist text by Stroud (1971).

Tchakaloff (1957) showed that formulae of degree d exist which have at most $(n+d)!/(n!d!)$ integration points and in addition have the desirable properties that all the points lie in the region of integration and the weights are all positive. A number of constructive methods have been devised to generate such methods and include, Davis (1967, 1969), Wilson (1969) and Dennis and Goldstein (1969), though the automatic computation of weights and abscissae from these methods seems non-trivial (Stroud 1971, Chapter 3).

Typical of formulae in this class is the fifth degree, seven-point rule of Radon (1948) which is derived in detail in Stroud (1971) and has the form

$$\int_{-1}^{1}\int_{-1}^{1} f(x, y)\mathrm{d}x\mathrm{d}y = \tfrac{8}{7}f(0, 0) + \tfrac{20}{63}[f(0, t) + f(0, -t)]$$
$$+ \tfrac{5}{9}[f(r, s) + f(r, -s) + f(-r, s) + f(-r, -s)] \qquad (6.9.3)$$

$$\text{with} \quad r = \sqrt{\tfrac{3}{5}}, \quad s = \sqrt{\tfrac{1}{3}} \quad \text{and} \quad t = \sqrt{\tfrac{14}{15}}$$

An extension to this rule was attempted by Cohen and Gismalla (1985) by trying to obtain a quadrature rule with the form

$$Q(f) = \int_{-1}^{1}\int_{-1}^{1} f(x, y)\mathrm{d}x\mathrm{d}y \simeq Af(0, 0) + B[f(0, t) + f(0, -t)]$$
$$+ C[f(r, s) + f(r, -s) + f(-r, s) + f(-r, -s)] \qquad (6.9.4)$$

and forcing exactness for the monomials $1, x^2, y^2, x^2y^2$ and y^4 to give five equations for t, r, s, A and B in terms of C. The final value C is then found by minimizing the expression

$$S(C) = \sum_{\substack{m+n>2 \\ (m,n)=(2,0)}}\sum \left(\int_{-1}^{1}\int_{-1}^{1} x^{2m}y^{2n}\mathrm{d}x\mathrm{d}y - Q(x^{2m}y^{2n})\right)^2 \qquad (6.9.5)$$

to get an optimum formula. Rules of this type were studied further in Cohen and Gismalla (1985) in which integrals of symmetric two-dimensional functions were considered.

Yet another approach is to consider a quadrature rule defined by

$$\int_{R_s} f(\mathbf{x})\mathrm{d}\mathbf{x} \simeq |\mathrm{det}\mathbf{A}| \sum_{n\in Z^s} f(\mathbf{An}) \qquad (6.9.6)$$

where the real matrix $\mathbf{A}$ is non-singular with real elements, Z is the set of integers and R_s the space $(-\infty, \infty)^s$. Clearly $\mathbf{A}$ characterizes the method completely, and is chosen to yield good approximations for a given class of integrals. Hence this method is called the method of good matrices, and is closely related to the method of good lattice points(Korobov, 1963; Hlawka, 1962) which has the form

$$\int_{[0,1)^s} f(\mathbf{x})\mathrm{d}\mathbf{x} = N^{-1}\sum_{i=1}^{N} f(\{(n/N)\mathbf{g}\}) \qquad (6.9.7)$$

where n and N are integers and $\mathbf{g}$ is an integer vector or good lattice point. Typical of these rules is the two-dimensional rule

$$\tfrac{1}{2}\sqrt{3}h^2 \sum_{n_1=-\infty}^{\infty} \sum_{n_2=-\infty}^{\infty} f\left(h\left(n_1+\tfrac{1}{2}n_2\right), h(\tfrac{1}{2}\sqrt{3}n_2)\right) \tag{6.9.8}$$

Sugihara (1987) uses such methods with the tanh transformation of the singular quadrature section to obtain multiple quadrature rules. One by-product of this approach is that singularities on the boundaries of hypercubes can be encompassed in the method.

Two slightly more unusual methods which appear to have had little airing in the literature but would seem to merit mention here are those of Schwartz (1969b), and Gismalla, Jenkins and Cohen (1988). The latter is easy to outline. The idea is simply to expand any denominator as an infinite series, hence dealing with singularities in passing. The resultant series is then summed using one of Levin's transformations. Marked success was obtained for integrals such as

$$\int_0^{\pi}\int_0^{\pi}\int_0^{\pi} \frac{\mathrm{d}x\,\mathrm{d}y\,\mathrm{d}z}{1-\cos x\cos y\cos z} \tag{6.9.9}$$

The former method converts the problem into a variational form. Hence to find

$$I = \int_R \frac{\rho(\mathbf{x})\mathrm{d}\mathbf{x}}{W(\mathbf{x})} \tag{6.9.10}$$

over a given region R, the functional

$$J = 2\int_R \phi(\mathbf{x})\rho(\mathbf{x})\,\mathrm{d}\mathbf{x} - \int_R \phi(\mathbf{x})W(\mathbf{x})\phi(\mathbf{x})\rho(\mathbf{x})\,\mathrm{d}\mathbf{x} \tag{6.9.11}$$

is considered, and this functional has a stationary value when $J = I$. If $W(\mathbf{x})$ contains a zero, so forming the singular part of I, then the functional which needs to be evaluated is non-singular. The suggestion is to choose ϕ to be a linear combination of basis functions which will yield a simple matrix form for the required stationary point. The elements of this matrix will each involve evaluating integrals and the efficiency of the method will depend on the judicious choice of the basis set.

For fairly small dimensions n, certainly for the very common cases for which $n \leq 3$, the product rules have a great deal in their favour. Arguably the most interesting aspect of such rules, whose theory and practice is well-known from one-dimensional work, is the code required to implement them. As this is non-standard, and involves some delicate programming to achieve full generality, a programme in the common and portable language Pascal is given here. Recursion is used freely and this eliminates the use of Fortran, although many implementations of the Fortran language do allow recursion despite the lack of official support or documentation. The code is presented firstly with a suitable declarative heading and then a subprogramme to effect the product rule algorithm. The specific one-dimensional rules which are used are not specifically presented, but should present no problem to the user.

```
CONST dimm=6;
TYPE float=longreal;
vec=ARRAY[1..dimm] OF float;
ivec=ARRAY[1..dimm] OF integer;

FUNCTION genint(FUNCTION f(VAR xd:vec):float;
FUNCTION fl(it:integer;VAR xd vec):float;
FUNCTION fu(it:integer;VAR xd vec):float;
VAR meth,ord:ivec;VAR h vec;dim:integer):float;
VAR xd:vec;
curdim,max,it:integer;
a,b,prec:float;

FUNCTION fsp(x:float):float; FORWARD;

FUNCTION integ(it:integer;a,b,prec:float;max:integer):float;
BEGIN
        CASE it OF
                1:integ:=gaussleg(fsp,a,b,prec,max);
                2:integ:=curtclen(fsp,a,b,prec,max);
                3:integ:=patterson(fsp,a,b,prec,max);
                4:integ:=newcot(fsp,a,b,prec,max)
        END
END

FUNCTION fsp(x:float):float;
VAR w:float;
BEGIN
        xd[curdim]:=x;
        curdim:=curdim-1;
        IF curdim=0 THEN
                w:=f(xd);
        ELSE
        BEGIN
                it:=meth[curdim];prec:=h[curdim];max:=ord[curdim];
                a:=fl(curdim,xd);b:=fu(curdim,xd);
                w:=integ(it,a,b,prec,max)
        END;
        curdim:=curdim+1;
        fsp:=w
END;
```

```
BEGIN
        curdim:=dim;
        it:=meth[curdim];prec:=h[curdim];max:=ord[curdim];
        a:=fl(curdim,xd);b:=fu(curdim,xd);
        genint:=integ(it,a,b,prec,max)
END:
```

The principle of this programme is very straightforward in that the individual dimensions are so handled to allow the one-dimensional rules available to see a conventional problem, that is to be able to integrate the FUNCTION fsp from a lower limit a to an upper limit b. The choice of one-dimensional rule is made through the integer it, and each rule is coded to have a standard parameter set (though these parameters may be actually used in different ways). The FUNCTION fsp can either be another integral, which is the ELSE part of its definition, or if all the dimensions have been activated, that is if the current dimension, *curdim*, has been reduced from its initial value of n to zero, then fsp delivers the value of the multi-dimensional integrand itself. At this stage all the components of the argument *xd* will have been set.

In this example code, four standard methods have been made available, Gauss–Legendre (*gaussleg*), Clenshaw–Curtis (*curtclen*), Patterson(*patterson*) and Newton–Cotes (*newcot*), each with a standard parameter set, namely the function to be integrated (fsp), the limits of integration a and b and two further parameters which may be used in different ways. For the first three methods the parameter *prec* is the required precision, a sequence of rules being executed up to some upper limit *max*. For *curtclen* and *patterson* these sequences are coded progressively. The Newton–Cotes rules use *prec* for the step size and *max* for the actual point number to generate just one estimate of the integral. Clearly further parameters could be added, such as a boolean which would deliver FALSE if convergence was not attained, and so initiate a suitable error message.

As the recursion unwinds, a further component of the n-vector variable of integration *xd* is set in the first statement of the FUNCTION fsp. Not until all n components are fixed can $f(xd)$ be executed in the THEN part of this function. Similarly the limits a and b are themselves set by calls to the limiting functions fl and fu which may themselves be functions of some of the components of *xd* as in equation (6.1.1). Of course only the components which have been set can appear in these functions at any stage. This device allows full variable limit capability to be achieved.

Hence the specification of *genint* can now be defined. The routine delivers the multiple integral of equation (6.9.1), the function f being coded as the FUNCTION f(VAR *xd* : *vec*) : $float$. The limiting functions are the FUNCTIONS fl and fu, each of which has an integer parameter to indicate to which integral the limit belongs. This would normally be coded as a CASE statement. Hence $a_1(x_1, \ldots, x_n)$ is $fl(1, xd)$, and $b_2(x_2, \ldots, x_n)$ is $fu(2, xd)$. The ARRAY parameters *meth* and *ord* give the required methods(by number) for each dimension to correspond with the CASE statement in the FUNCTION *integ*, and the maximum point number to be used respectively. Similarly the ARRAY h is used to fix *prec* for each dimension. The variable *dim* is the dimension of the integral. It should be noted that the usual Pascal limitations require that a global constant upper limit to arrays is set and this is done in the time-honoured way with the CONST dimm.

For high dimensionality, the methods discussed so far become very expensive, and there is a need for a less computationally intensive method with a corresponding lower accuracy. A statistical or Monte Carlo approach becomes feasible in such a case and is based on sampling the height of the integrand. In one dimension it is easy to see how the method operates by considering the integral

$$I = \int_a^b f(x)\,\mathrm{d}x \tag{6.9.12}$$

and then forming the average of the sample

$$f_{av} = \frac{1}{n}\sum_{i=1}^{n} f(x_i). \tag{6.9.13}$$

But it might be expected that

$$f_{av} \simeq I/(b-a) \tag{6.9.14}$$

and hence

$$I \simeq \frac{b-a}{n}\sum_{i=1}^{n} f(x_i). \tag{6.9.15}$$

The various methods rely on the judicious choice of the sampling points $(x_1, x_2, \ldots, x_n)$ with a view to reducing the resulting error. It is a consequence of the Central Limit Theorem that the error in a quadrature like equation (6.9.15) is of order $1/\sqrt{n}$, and this analysis applies to all dimensions.

Before leaving this topic, it is worth noting that Mathematica contains two useful integration routines. The first generates the analytic values for conventional quadratures using the construction

$$\text{Integrate}[f(x), x]$$

and the second completes a numerical quadrature from a to b using

$$\text{NIntegrate}[f(x), \{x, a, b\}]$$

These packages give warning of difficulties which are beyond their scope, and hence are less liable to misuse than would otherwise be the case.

Further reading

Davis P. J. and Rabinowitz P. (1984) *Methods of Numerical Integration*, Academic Press, New York.
Evans G. A. (1993) *Practical Numerical Integration*, John Wiley and Sons, Chichester.
Stroud A. H. (1971) *Approximate Calculation of Multiple Integrals*, Prentice-Hall, Englewood Cliffs, New Jersey.
Stroud A. H. and Secrest D. (1966) *Gaussian Quadrature Formulae*, Prentice-Hall, Englewood Cliffs, New Jersey.

Exercises

(6.1) Derive the open Newton–Cotes formulae

$$\int_0^{3h} f(x)\,dx = \frac{3h}{2}(f_1 + f_2) + \frac{3h^3}{4} f''(\xi)$$

and

$$\int_0^{4h} f(x)\,dx = \frac{4h}{3}(2f_1 - f_2 + 2f_3) + \frac{14h^5}{45} f^{(iv)}(\xi)$$

using both a Lagrangian approach and the method of undetermined coefficients.

(6.2) Show that the n-point Newton–Cotes rule also integrates polynomials of degree $(n+1)$ exactly if n is even.

(6.3) Obtain the weights A_1, A_2 and A_3 and an estimate of the error E in the quadrature rule defined by

$$-\int_0^1 \ln(x) f(x)\,dx = A_1 f(0) + A_2 f(0.5) + A_3 f(1.0) + E$$

by making $E = 0$ for $f(x) = 1,\ x$ and x^2. Use this rule to compute

$$-\int_0^1 x^3 \ln(x)\,dx$$

comparing the result with a Gauss–Legendre three-point rule and the true solution.

(6.4) By making $E = 0$ for $f(x) = 1, x,$ and x^2 in the quadrature rule

$$\int_{-1}^1 \frac{f(x)}{\sqrt{1-x^2}}\,dx = Af(1) + Bf(0) + Cf(-1) + E$$

determine A, B, and C. Find an estimate of the error E in terms of derivatives of f at zero. Hence evaluate

$$\int_{-1}^1 \frac{\tan x^2}{\sqrt{1-x^2}}\,dx$$

and by using the transformation $x = \cos\theta$ to remove the singularities, apply Simpson's rule to the new integral and compare the two approaches.

(6.5) Prove that the weights A_1, A_2 and A_3 for the Gaussian quadrature formula

$$\int_{-1}^{1} \frac{f(x)\,\mathrm{d}x}{\sqrt{1-x^2}} = A_1 f(x_1) + A_2 f(x_2) + A_3 f(x_3) + E$$

are equal and find the abscissae x_1, x_2, and x_3, by forcing the error E to be zero for $f(x) = 1,\ x,\ x^2, \ldots, x^5$. Apply the formula to the integral

$$\int_{-1}^{1} \frac{\arccos(x)}{\sqrt{1-x^2}}\,\mathrm{d}x$$

using exact arithmetic. Compare the value with the true solution and explain the agreement.

(6.6) Evaluate the integral

$$\int_{-\infty}^{\infty} e^{-x^2} \cos x\,\mathrm{d}x$$

using Gauss–Hermite quadrature, and compare your results with the exact value which is $\sqrt{\pi}\exp(-1/4)$.

(6.7) Evaluate the integral

$$\int_{-1}^{1} \frac{\cos x}{\sqrt{1-x^2}}\,\mathrm{d}x$$

using Gauss–Chebyshev quadrature, and compare the results with the exact value of $\pi J_0(1)$.

(6.8) Derive the two-point Gauss quadrature rule of the form

$$\int_{0}^{\pi} f(x)\sin(mx)\,\mathrm{d}x = W_1 f(x_1) + W_2 f(x_2) + E$$

which is valid for positive integers m. Apply the rule to the integrals

$$\int_{0}^{\pi} e^{x} \sin mx\,\mathrm{d}x$$

for $m = 1$, 10 and 100. Compare the results with the true values.

(6.9) Derive the two point Gauss–log rule of the form

$$\int_{0}^{1} f(x)\ln x\,\mathrm{d}x = -W_1 f(x_1) - W_2 f(x_2)$$

to obtain the values

$$x_1 = \frac{15 - \sqrt{106}}{42} \qquad x_2 = \frac{15 + \sqrt{106}}{42}$$

$$W_1 = \frac{212 + 9\sqrt{106}}{424} \qquad W_2 = \frac{212 - 9\sqrt{106}}{424}$$

Hence evaluate

$$\int_0^1 e^x \ln x \, dx$$

and compare the result with an application of Gauss–Legendre with the singularity effectively ignored.

(6.10) Derive formulae of the general form

$$\int_{-1}^{1} f(x)\, dx = A_1 f(-1) + \sum_{k=2}^{n} A_k f(x_k) + E$$

by making the rule exact for $f(x) = 1,\ x,\ x^2,\ \ldots,\ x^{2k-2}$. Use $n = 2$ and $n = 3$, so generating the first two Radau formulae.

(6.11) Derive formulae of the general form

$$\int_{-1}^{1} f(x)\, dx = A_1 f(-1) + \sum_{k=2}^{n-1} A_k f(x_k) + A_n f(1) + E$$

by making the rule exact for $f(x) = 1,\ x,\ x^2,\ \ldots,\ x^{2k-3}$. Use $n = 2$ and $n = 3$, so generating the first two Lobatto formulae.

(6.12) Evaluate the integral

$$\int_{-3}^{3} \frac{dx}{1 + x^2}\, dx$$

using Gauss–Legendre quadrature and compare the results with the true value.

(6.13) Using Hermite's interpolation rule with $f(x)$ and $f'(x)$ collated at $x = -1, 0$ and 1, derive the quadrature rule

$$\int_{-1}^{1} f(x)\, dx = [7f(-1) + 16f(0) + 7f(1) + f'(-1) - f'(-1)]/15$$

and show the rule has degree 5.

(6.14) Establish the recurrence relation

$$c_i = i\left(\frac{\pi}{2}\right)^{i-1} - i(i-1)c_{i-2}$$

for the monomial integrals

$$c_i = \int_0^{\pi/2} x^i \sin x \, dx.$$

Hence find the three-point Gaussian quadrature rule for the interval $[0, \pi/2]$ and the weight function $\sin x$, computing both the abscissae and the weights. Apply the formula to the integral

$$\int_0^{\pi/2} e^x \sin x \, dx$$

and compare your result with the true value.

(6.15) Consider the generalized Fourier expansion

$$f(x) = \sum_{r=0}^{\infty} A_r T_r(x)$$

with coefficients

$$A_r = \int_{-1}^{1} \frac{f(x) T_r(x)}{\sqrt{1 - x^2}} \, dx$$

Apply the Gauss–Chebyshev quadrature rule to this integral to get an open version of the Clenshaw–Curtis rule. Further apply the transformation $x = \cos\theta$ and the trapezoidal rule to get a closed version of the Clenshaw–Curtis rule. How does this last closed rule differ from the standard version in (6.4.8)? Why is the trapezoidal rule so effective?

(6.16) Extend the two-point Gauss–Legendre rule to obtain a five-point Patterson rule.

(6.17) Apply the transformation $\theta = t^n$ for a suitable n to the integral:

$$\int_0^{\pi/2} \sin^{1/2} x \, dx$$

and investigate how the function count varies with n for a given accuracy.

(6.18) Evaluate the singular integral

$$\int_0^1 \frac{e^{-x^{1/2}}}{x^{1/2}} \, dx$$

by finding analytically the sequence

$$A_i = \int_{1/2^{i+1}}^{1} \frac{e^{-x^{1/2}}}{x^{1/2}} \, dx$$

for $i = 0, 1, \ldots, 4$ and then employing the ε-algorithm to accelerate the sequence. Compare the result with the true value.

(6.19) Evaluate the singular integral

$$\int_0^1 \frac{\sin x}{x^{1/2}}\,\mathrm{d}x$$

by using a sequence of integrals beginning with

$$I_1 = \int_{1/4}^1 \frac{\sin x}{x^{1/2}}\,\mathrm{d}x$$

and the Shanks' transformation e_1. Use any suitable numerical method for the evaluation of the sequence of non-singular integrals. Apply a polynomial transformation to the problem and compare the workload with the above acceleration technique.

(6.20) Apply a suitable method to evaluate the integrals

$$\int_0^1 \frac{\ln x}{\sqrt{1-x^2}}\,\mathrm{d}x = -\frac{\pi}{2}\ln 2$$

$$\int_0^1 \frac{\ln x}{(x(1-x^2)^2)^{1/3}}\,\mathrm{d}x = -\frac{1}{8}\left[\Gamma\left(\frac{1}{3}\right)\right]^3$$

$$\int_0^1 \frac{(\ln x)^3}{1-x}\,\mathrm{d}x = -\frac{\pi^4}{15}$$

(6.21) Apply the method of Filon to find the value of

$$\int_0^1 \frac{\sin 1000x}{\sqrt{1+x^2}}\,\mathrm{d}x$$

(6.22) Apply the method of Alaylioglu, Evans and Hyslop to the integral

$$\int_0^1 x\cos x \sin 10x\,\mathrm{d}x$$

using a Chebyshev series to approximate $x\cos x$ up to and including the term $T_2(x)$.

(6.23) Derive the recurrence relation used in Patterson's method, namely

$$R_{k+2,j} = \left(\frac{j^2-k^2}{j^2-(k+3)^2}\right) R_{k,j}$$

with

$$R_{k,j} = \int_{-1}^1 P_k(x)T_j(x)\mathrm{d}x.$$

(6.24) Evaluate the irregularly oscillatory integral

$$\int_0^1 e^{-x}\sin 10(x+x^2)\,\mathrm{d}x$$

by the transformation $y = x + x^2 = g(x)$, obtaining the relevant inverse function $x = g^{-1}(y)$ so that the integral can be evaluated by a standard oscillatory method.

(6.25) A set of quite unpleasant integrals which exhibit more than one problem area, including oscillations and singularities are:

$$\int_0^1 \frac{\cos ax - \cos \dfrac{a}{x}}{1 - x^2} \,\mathrm{d}x = \frac{\pi}{2} \sin a \qquad a > 0$$

$$\int_0^{\pi/2} \cos(a \tan x) \frac{x \,\mathrm{d}x}{\sin 2x} = -\frac{\pi}{4} \mathrm{Ei}(-a)$$

$$\int_0^{\infty} \ln x \ln \left(1 + \frac{b^2}{x^2}\right) \mathrm{d}x = \pi b (\ln b - 1)$$

Attempt to evaluate these integrals using combinations of methods as required. In the first instance it may be wise to try fairly crude methods to get some feel for the problems to be encountered.

(6.26) Use the Shanks' acceleration process to evaluate the integral

$$\int_0^{\infty} e^{-x} \sin(10x) \,\mathrm{d}x$$

using in the first instance the intervals $[0, \pi/10]$, $[\pi/10, \pi/5]$ and $[\pi/5, 3\pi/10]$ for the sub-integrals with Aitken's extrapolation process. Generate a full Shanks' table and use analytic formulae for the integration.

(6.27) Use the v transform of Levin to find the value of the integral

$$\int_0^{\infty} x \sin 10x \,\mathrm{d}x$$

by using analytic values over the separate cycles. Interpret the result in the light of the divergence of the integral.

(6.28) Use the v transform of Levin to evaluate the integral

$$\int_0^{\infty} e^{-\sqrt{x}} J_0(10x) \,\mathrm{d}x$$

evaluating the integrals over each cycle using a low-order Gauss–Legendre formula.

(6.29) A formula for two-dimension quadrature with a logarithmic singularity at the origin has the form

$$\int_0^1 \int_0^1 \ln(xy) f(x, y) \,\mathrm{d}x \mathrm{d}y = \sum_{i=1}^{N} W_i f(x_i, y_i) + E$$

Given $N = 4$ and the abscissae $(x_1, y_1) = (0.1, 0.1)$, $(x_2, y_2) = (0.1, 0.9)$, $(x_3, y_3) = (0.9, 0.1)$ and $(x_4, y_4) = (0.9, 0.9)$, find the values of W_i which force the formula to be exact for the monomials $f(x, y) = 1,\ x,\ y$ and xy.

Hence obtain a value for

$$\int_0^1 \int_0^1 \frac{\ln(xy)}{(1 + x^2 + y^2)^{1/2}}\, \mathrm{d}x\mathrm{d}y$$

(6.30) Obtain the weights W_i, $i = 1, 4$ for the double quadrature rule

$$\int\int_C \frac{f(x, y)}{(x^2 + y^2)^{1/4}}\, \mathrm{d}x\mathrm{d}y = \sum_{i=1}^{4} W_i f_i + E$$

to be exact for $f(x, y) = 1, \theta, x$ and y, where the region C is the circle of radius unity, θ is defined by the polar coordinates $x = r\cos\theta$ and $y = r\sin\theta$, and $f_i = f(x_i, y_i)$ where (x_i, y_i) are the polar defined points $r_i = \frac{1}{2}$, $i = 1, 4$ and $\theta_i = (2i - 1)\pi/4$, $i = 1, 4$. Find an estimate for the error E. Explain why the above form cannot be made exact for $f = r^n$ for more than one value of n. Evaluate the integral

$$\int\int_C \frac{(\arctan(y/x))^2}{(x^2 + y^2)^{1/4}}\, \mathrm{d}x\mathrm{d}y$$

and compare the result with the exact value.

(6.31) Apply the transformation $x = \tanh t$ and the corresponding y transformation to the integral

$$I = \int_{-1}^1 \int_{-1}^1 \frac{\mathrm{d}x\mathrm{d}y}{\sqrt{1 - x^2 y^2}} = 2\pi \ln 2$$

and hence evaluate the integral numerically by the trapezoidal rule. Initially use a step of 2.0 in each direction. Apply one step of a progressive doubling of the number of quadrature points.

(6.32) One way of generating a sequence of pseudo-random numbers is to use the recurrence relation

$$x_{n+1} = 4x_n(1 - x_n)$$

with the first value x_1 chosen so that $0 < x_1 < 1$ and so that $x_1 \neq k/4$ for $k = 0(1)4$. Employ these pseudo-random numbers to evaluate

$$\int_0^1 \frac{\mathrm{d}x}{1 + x^2} \sim \frac{1}{N} \sum_{i=1}^{N} \frac{1}{1 + x_i^2}$$

to obtain estimates for the integral for differing values of N. Compare the errors obtained with the theoretical estimates.

(6.33) Repeat the previous exercise with random numbers taken from a table such as that in Abramowitch and Stegun, Table 26.11, and compare the results with the self-generated numbers above.

(6.34) Use a Monte Carlo approach to evaluate the integral

$$\int_0^{\pi/2}\int_0^{\pi/2} \frac{\sin y\sqrt{1-k^2\sin^2 x\sin^2 y}}{1-k^2\sin^2 y}\,\mathrm{d}x\mathrm{d}y = \frac{\pi}{2\sqrt{1-k^2}}$$

and compare the results for different numbers of sampling points with the exact value. How does this method compare with the use of a product rule?

7

Ordinary Differential Equations

7.1 Introduction

In terms of usefulness and research effort, the numerical solution of ordinary differential equations must rank amongst the most important of numerical topics. In the previous chapters, the tools which come together for the solution of real problems have been set up. Now, aspects of linear algebra, quadrature and approximation theory can be applied to the three major types of equation arising in applied work, namely, ordinary differential equations, integral equations and partial differential equations. Before considering details of various numerical methods for the ordinary differential equation problem, it is worth thinking about the solutions of these problems as presented at an elementary level. In reality very few ordinary differential equations can be solved analytically. For example the equation

$$\frac{d^2y}{dx^2} + k^2y = 0 \tag{7.1.1}$$

has the well-known general solution

$$y = A\sin kx + B\cos kx. \tag{7.1.2}$$

But what does this solution mean? Surely the so-called solution is just a way of moving the problem into a different one. The solution to (7.1.1) is not immediately written in terms of powers of x and y, so we give the solution a fancy name such as sin or cos. In effect we are defining the solution to be the definition of the now newly invented transcendental functions sin and cos. From this definition the properties of the two functions follow very easily, for example a series solution of the ordinary differential equation yields the accepted series for the functions and the process of analysis can begin to establish the properties of the new function. It may seem a bit far-fetched to approach this problem in such a manner, but if attention is given to a second problem such as

$$\frac{d^2y}{dx^2} = xy \tag{7.1.3}$$

then the solution is not well-known at an elementary level and indeed the properties of the solution follow from the differential equation and yield the Airy functions. In fact the so-called analytic solutions of ordinary differential equations are predominantly transcendental functions which are listed in large numbers in reference books such as Abramowitch and Stegun(1965). Hence the problem has not really been solved in numerical terms. A name has been given to the solution and some of its properties found by applying analytic techniques. It is still necessary to tabulate, or in modern terms, find an algorithm for the newly named functions.

In this introduction some standard methods will be derived to set the scene for later analysis. Two main branches will be pursued in more detail as a result of this preliminary study, and analysis of the stability of the methods made. After much verbiage, this study will lead to the vexed question of stiffness, and special methods to handle such problems will be given. Research is still progressing in this area and inevitably some compromise is necessary in the choice of recent research which should be reported here. What looks like an encouraging method at the current time, could soon be found less useful than it originally appeared.

With a few exceptions, the methods and analysis to be presented here will allow the solution of sets of first-order ordinary differential equations of the form

$$\mathbf{y}' = \mathbf{f}(x, \mathbf{y}) \tag{7.1.4}$$

with

$$\mathbf{y} = (y_1, \dots, y_n)$$

$$\mathbf{f} = (f_1, \dots, f_n)$$

The vector function $\mathbf{f}$ can be any function of the independent variable x and the dependent vector $\mathbf{y}$ within the context that the ordinary differential equation requires a unique existing solution. The normal condition for this is that $\mathbf{f}$ is Lipschitzian, and a suitable set of boundary conditions are defined. If all the boundary conditions are defined at one point say x_0, then the problem is called an initial value problem. If the boundary conditions are specified at more than one point the problem becomes a boundary value problem, or possibly an eigenvalue problem. These last two possibilities are pursued in a paragraph of their own, as the main development will be to solve initial value problems.

Commonly equations are given which are not first order, but involve higher derivatives. Again with a few exceptions the usual approach is to reduce such equations to sets of first-order equations of the form given in equation (7.1.4) by the simple device of introducing new variables equal to the higher derivatives. This is most easily seen by an example. Hence for the second-order equation

$$y'' + f(x)y' + yg(x) = h(x) \tag{7.1.5}$$

the additional variable

$$y' = u \tag{7.1.6}$$

is defined to give the set of first-order equations:

$$y' = u$$

$$u' = h(x) - yg(x) - uf(x) \tag{7.1.7}$$

Clearly this process is easily generalized to more complicated situations, and allows the numerical study to concentrate on equations of the form of (7.1.4).

The simplest equation of this type is

$$\frac{dy}{dx} = f(x, y) \tag{7.1.8}$$

and if methods can be found for this equation it is usually very easy to generalize them to the system in (7.1.4). Hence this approach will be used, and as methods are found an eye will be kept open for any difficulties associated with the generalization. Once the solution of (7.1.8) is fully understood, it is then feasible to reconsider the analysis for sets of equations and to see if further issues arise. Indeed this is how the problem of stiffness first shows up.

The simplest method is no more than an attempt at integrating both sides of (7.1.8) over a short interval (x_n, x_{n+1}). Defining a step size h as $h = x_{n+1} - x_n$ then integrating (7.1.8) gives

$$\int_{x_n}^{x_{n+1}} \frac{dy}{dx}\, dx = \int_{x_n}^{x_{n+1}} f(x, y) dx \tag{7.1.9}$$

to give

$$y(x_{n+1}) - y(x_n) = \int_{x_n}^{x_{n+1}} f(x, y) dx \tag{7.1.10}$$

Now define $y_n = y(x_n)$ and $f_n = f(x_n, y_n)$. The integral on the right-hand side of (7.1.10) can be very crudely approximated by the rectangle of height $f(x_n, y_n)$ and width h. Certainly as h tends to zero this rectangle will tend to the correct integral. This is in fact one of the primitive quadrature rules of the previous chapter denoted by the left-hand Riemann sum. The result is Euler's formula

$$y_{n+1} = y_n + hf(x_n, y_n) \tag{7.1.11}$$

which allows a step-by-step solution to be found with the boundary condition $x = x_0,\ y = y_0$ being the first point. The solution is a table of values of y against x evaluated at the tabular points which are a distance h apart. By taking h to have a smaller step size then more accurate solutions are obtained. Just what the errors are in such a method will be left to the paragraph which treats this and related methods in more detail.

The application of this method to an example is very easy and is illustrated by the problem

$$\frac{dy}{dx} = \frac{y - x}{y + x} = f(x, y) \tag{7.1.12}$$

with the boundary conditions $y = 1$ when $x = 0$. The object is to compute y for $x = 0.1$. The solution is laid out in Table 7.1.

Table 7.1

x	y	$y+x$	$y-x$	f	hf
0.0	1.0000	1.0000	1.0000	1.0000	0.0200
0.02	1.0200	1.0400	1.0000	0.962	0.0192
0.04	1.0392	1.0792	0.9992	0.926	0.0185
0.06	1.0577	1.1177	0.9977	0.893	0.0179
0.08	1.0756	1.1556	0.9956	0.862	0.0172
0.10	1.0928	1.1928	0.9928	0.832	0.0166

As this algorithm is so simple, it would be very easy on a modern computer to gain high accuracy by just using a very small step size h. This yields a dichotomy for the user. Should a simple method be used with a very small step size, or a more complicated method which will take much more coding and de-bugging, but which will run with a less small step? The final workload may not be far different. It is also interesting to solve this equation analytically to see if the required value of $y(0.1)$ is forthcoming. The analytic solution is

$$\tfrac{1}{2}\ln(x^2+y^2)+\arctan(y/x)=\pi/2 \tag{7.1.13}$$

and so to get $y(0.1)$ requires the solution of a non-linear equation as in Chapter 1. Direct iteration of

$$y_{n+1}=\tfrac{1}{10}\sqrt{\exp[\ln 100+\pi-2\arctan 10y_n]-1} \tag{7.1.14}$$

gives the result 1.091 1268 after nine iterations. This confirms the considerable simplicity of Euler's method.

The method also generalizes very easily to a set of equations so that to solve

$$\begin{aligned} y' &= f(x, y, z) \\ z' &= g(x, y, z) \end{aligned} \tag{7.1.15}$$

the algorithm

$$\begin{aligned} y_{n+1} &= y_n + hf(x_n, y_n, z_n) \\ z_{n+1} &= z_n + hg(x_n, y_n, z_n) \\ x_{n+1} &= x_n + h \end{aligned} \tag{7.1.16}$$

is applicable with boundary conditions again giving the initial values. Euler's method can be summarized as being simple, self-starting (later methods can require special procedures to get them started), but inaccurate, unless a very small step size h is used.

An obvious improvement to Euler's method would be to increase the accuracy of the integral in (7.1.10) by using a more refined rule. Such a rule is the trapezoidal rule which would give instead of equation (7.1.11), the update formula

$$y_{n+1}=y_n+\tfrac{1}{2}h[f(x_n, y_n)+f(x_{n+1}, y_{n+1})] \tag{7.1.17}$$

which is called the modified formula of Euler. There is now a new problem as this formula is implicit rather than explicit. Hence the rule represents a non-linear equation for the required value y_{n+1}. Because the step-size h is small, the method of direct iteration from chapter one may be applied with good convergence rates in the form

$$y_{n+1}^{[k]} = y_n + \tfrac{1}{2}h[f(x_n, y_n) + f(x_{n+1}, y_{n+1}^{[k-1]})] \tag{7.1.18}$$

There is a practical issue here which is often overlooked at first. There is nothing gained iterating this rule to full machine accuracy. It has to be remembered that equation (7.1.18) is itself only yielding an approximation to y_{n+1} whose accuracy depends on the truncation error in using the trapezoidal rule. Hence the useful accuracy obtained from the direct iteration is often obtained in no more than one iteration. To achieve this a reasonable value for $y_{n+1}^{[0]}$ is required, and this is furnished by the original Euler formula. It is therefore normal to employ these methods together and such a combination is called a predictor–corrector pair. In general any explicit formula can be used as a predictor, but because the formula is explicit it will have a greater truncation error than the corresponding implicit rule. This extra accuracy can then be taken up with a single iteration of the corrector formula.

The convergence of the direct iteration is established by considering the difference

$$|y_{n+1}^{[k]} - y_{n+1}| = \tfrac{1}{2}h|f(x_{n+1}, y_{n+1}^{[k-1]}) - f(x_{n+1}, y_{n+1})| \tag{7.1.19}$$

of the iterated rule and its converged limit. For a solution of the original differential equation to exist, the function f must be Lipschitzian and hence

$$|f(x_{n+1}, y_{n+1}^{[k-1]}) - f(x_{n+1}, y_{n+1})| < K_n|y_{n+1}^{[k-1]} - y_{n+1}| \tag{7.1.20}$$

to give

$$|y_{n+1}^{[k]} - y_{n+1}| < \tfrac{1}{2}hK_n|y_{n+1}^{[k-1]} - y_{n+1}| \tag{7.1.21}$$

or

$$|y_{n+1}^{[k]} - y_{n+1}| < (\tfrac{1}{2}hK_n)^{k-1}|y_{n+1}^{[1]} - y_{n+1}| \tag{7.1.22}$$

and hence convergence if $(1/2)hK_n < 1$, that is for sufficiently small h. It will be seen later than when systems are stiff, one of the many difficulties which arise is that K_n becomes large and direct iteration is no longer feasible to iterate correctors. A full-scale Newton iteration is then required.

As an example, consider

$$\frac{dy}{dx} = \log(x + y) \tag{7.1.23}$$

with $y = 1$ when $x = 0$, finding y at $x = 0.2$ and 0.5. Applying Euler's predictor formula yields $y_1^{[1]} = 1 + hf(x_0, y_0) = 1$ and the modified formula iterates to give:

$$\begin{aligned} y_1^{[2]} &= 1 + \tfrac{1}{2}h[f(x_0, y_0) + f(x_1, y_1^{[1]})] \\ &= 1 + \tfrac{1}{2}[\log 1 + \log(0.2 + 1)] \\ &= 1.0079 \end{aligned} \tag{7.1.24}$$

A further iteration then gives $y_1^{[3]} = 1.0082$ which is the converged value to five figures. It can be seen that the gain in accuracy after the first iteration is probably fortuitous and within the underlying truncation error of the modified process itself.

For the second step with h now set at 0.3, Euler's predictor gives $y_2^{[1]} = 1.0082 + 0.3\log(0.2 + 1.0082) = 1.0328$, and modified Euler gives $y_2^{[2]} = 1.0082 + 0.3/2[\log 1.2082 + \log(0.5 + 1.0328)] = 1.104\,83$ at the first iterate. The second iterate $y_2^{[3]}$ yields 1.0490 as does the third.

The whole subject of methods of this type which are classified as linear multi-step methods, and of which Euler and modified Euler constitute just two simple examples, will be pursued in its own right later in the chapter. This detailed view will include extensive analysis of stability and truncation errors. In the meantime other approaches will be introduced.

In this initial survey of elementary methods, there are some techniques which give approximate analytic solutions rather than tables of results in the manner of Euler's method. Though not necessarily in the main-stream of ordinary differential equation methods, these approximate analytic methods have their relevance in starting other more refined processes, and in other branches of numerical analysis such as in integral equations. Such a method is Picard's method. This method is based on the proof of the uniqueness and existence for the solution of ordinary differential equations. The proof involves converting the ordinary differential equation into an integral equation and solving this equation iteratively. The convergence of the iterates then establishes the theorem. This process can be used to obtain approximate analytic solutions, and indeed if the iterated functions are stored as tables, can be the basis of a computerized iterative solution which will be pursued more deeply in Chapter 8.

The principle in Picard's method is to form a Volterra integral equation by simply integrating both sides of (7.1.8) to give

$$\int_{x_0}^{x} y(x)\,\mathrm{d}x = \int_{x_0}^{x} f(x, y)\,\mathrm{d}x \tag{7.1.25}$$

and the integral equation

$$y(x) = y_0 + \int_{x_0}^{x} f(x, y(x))\,\mathrm{d}x \tag{7.1.26}$$

This non-linear integral equation is solved by direct iteration to yield Picard's method in the form

$$y^{[k+1]}(x) = y_0 + \int_{x_0}^{x} f(x, y^{[k]}(x))\,\mathrm{d}x \tag{7.1.27}$$

where $y^{[k]}(x)$ is the kth iterated approximation to the true solution $y(x)$. For example for the equation

$$\frac{\mathrm{d}y}{\mathrm{d}x} = x^2 - y \tag{7.1.28}$$

with $y = 1$ when $x = 0$ the exact solution is $y = 2 - 2x + x^2 - e^{-x}$ and Picard gives $y^{[0]}(x) = 1$, using the tradition of starting the process with y_0. Then

$$\begin{aligned} y^{[1]}(x) &= 1 + \int_0^x f(x, y^{[0]}(x))\,\mathrm{d}x \\ &= 1 + \int_0^x (x^2 - 1)\,\mathrm{d}x = 1 - x + \frac{x^3}{3} \end{aligned} \tag{7.1.29}$$

and

$$\begin{aligned} y^{[2]}(x) &= 1 + \int_0^x f(x, y^{[1]}(x))\,\mathrm{d}x \\ &= 1 + \int_0^x \left(x^2 - 1 + x - \frac{x^3}{3}\right)\mathrm{d}x \\ &= 1 - x + \frac{x^2}{2} + \frac{x^3}{3} - \frac{x^4}{12} \end{aligned} \tag{7.1.30}$$

The method will clearly come to grief when the analytic integrals cannot be resolved, but the importance of the method lies in both its theoretical value and in the solution of non-linear Volterra equations.

On a different tack, but still looking for approximate analytic methods, is the method based on Taylor series. Equation (7.1.8) gives a formula for the first derivative. It is feasible to differentiate this formula to obtain higher derivatives, and then using

$$\begin{aligned} y(x) = y(x_0) + (x - x_0)y'(x_0) + \frac{(x - x_0)^2}{2!}y''(x_0) + \cdots \\ + \frac{(x - x_0)^n}{n!}y^{(n)}(x_0) + \cdots \end{aligned} \tag{7.1.31}$$

gives the approximate analytic form of $y(x)$. The initial condition yields $y(x_0)$. Illustrating this method on the above example gives the derivatives:

$$\begin{aligned} y' &= x^2 - y \\ y'' &= 2x - y' \\ y''' &= 2 - y'' \\ y^{iv} &= -y''' \\ y^{v} &= -y^{iv} \end{aligned} \tag{7.1.32}$$

from which the series solution follows as

$$y(x) = 1 - x + \frac{x^2}{2!} + \frac{x^3}{3!} - \frac{x^4}{4!} + \frac{x^5}{5!} - \tag{7.1.33}$$

There is a circle of convergence about the initial point x_0 in which the resulting series is valid. If a point near to this circle is used as a new initial point then a second expansion can be made about this new point to progress the solution beyond the initial region, continuing as far as is necessary. One problem with this method is difficulty in finding higher derivatives (suppose f is only known as a table), but its importance is in the basis it serves for the development of other methods. It is also commonly used as a starting procedure for methods which require such a device (see linear multistep paragraph). This is also one of the methods for which the reduction of high-order equations to sets of first order, as in (7.1.7), is not necessary or advantageous. Hence for the equation

$$xy'' + 2y' + xy^2 = 0 \tag{7.1.34}$$

with $y = 1$ and $y' = 0$ when $x = 0$ the derivatives are:

$$\begin{aligned} xy'' + 2y' + xy^2 &= 0 \\ xy''' + 3y'' + y^2 + 2xyy' &= 0 \\ xy^{iv} + 4y''' + 4yy' + 2xy'^2 + 2xyy'' &= 0 \\ xy^{v} + 5y^{iv} + 6y'^2 + 6yy'' + 2xyy''' + 6xy'y'' &= 0 \\ xy^{vi} + 6y^{v} + 24y'y'' + 8yy''' + 8xy'y''' + 2xyy^{iv} + 6xy''^2 &= 0 \end{aligned} \tag{7.1.35}$$

with the resulting series

$$y(x) = 1 - \frac{x^2}{6} + \frac{x^4}{60} - \frac{11x^6}{7560} + \cdots \tag{7.1.36}$$

after putting $x = 0$ in the expressions for the derivatives.

The final method in this introductory overview of methods is one of a family of methods of the Runge–Kutta type. These methods were first developed by Runge(1895), Kutta(1901) and Heun (1900). Effectively this method attempts to overcome the necessity for analytic knowledge of the higher derivatives for the implementation of the Taylor's series method. Instead the function f is evaluated at a sequence of points neighbouring the current point, say (x_n, y_n). These evaluations are used to estimate the required derivatives and allow progress to (x_{n+1}, y_{n+1}). The general form of a Runge–Kutta method is

$$y_{n+1} = y_n + b_1k_1 + b_2k_2 + \cdots + b_Rk_R \tag{7.1.37}$$

where

$$\begin{aligned} k_1 &= hf(x_n, y_n) \\ k_2 &= hf(x_n + hc_2, y_n + a_{21}k_1) \\ k_3 &= hf(x_n + hc_3, y_n + a_{31}k_1 + a_{32}k_2) \\ &\vdots \\ k_R &= hf\left(x_n + hc_R, y_n + \sum_{s=1}^{R-1} a_{Rs}k_s\right) \end{aligned} \tag{7.1.38}$$

Hence once the constants of the method are derived, that is values for c_r, a_{rs} and b_r are known, it is very easy to compute y_{n+1} by computing firstly k_1, then k_2 and so on to k_R and finally the required y_{n+1} from (7.1.37). It is this simplicity of approach which gives this method appeal to engineering and scientific users. The process of deriving specific formulae is somewhat tedious and more will be said on this in the main paragraph on this method. At this stage a simple low-order formula will be derived to illustrate the procedure.

Hence consider

$$y_{n+1} = y_n + b_1k_1 + b_2k_2 \tag{7.1.39}$$

with

$$k_1 = hf(x_n, y_n) \tag{7.1.40}$$

and

$$k_2 = hf(x_n + \mu h, y_n + \lambda k_1) \tag{7.1.41}$$

For compactness, use $f = f(x_n, y_n)$ and $f_x = \partial f/\partial x$ and so on to give, by Taylor's series

$$\begin{aligned} k_2 &= hf + h^2(\mu f_x + \lambda f f_y) \\ &\quad + \tfrac{1}{2}h^3(\mu^2 f_{xx} + 2\mu\lambda f f_{xy} + \lambda^2 f^2 f_{yy}) + O(h^4) \end{aligned} \tag{7.1.42}$$

which with equation (7.1.39) yields

$$\begin{aligned} y_{n+1} &= y_n + h(b_1 + b_2)f + h^2 b_2(\mu f_x + \lambda f f_y) + \\ &\quad + \tfrac{1}{2}h^3 b_2(\mu^2 f_{xx} + 2\mu\lambda f f_{xy} + \lambda^2 f^2 f_{yy}) + O(h^4) \end{aligned} \tag{7.1.43}$$

By contrast, a Taylor expansion of the solution gives

$$\begin{aligned} y_{n+1} &= y_n + hf + \tfrac{1}{2}h^2(f_x + f f_y) \\ &\quad + \tfrac{1}{6}h^3[f_{xx} + 2f f_{xy} + f^2 f_{yy} + f_y(f_x + f f_y)] + O(h^4) \end{aligned} \tag{7.1.44}$$

where

$$\frac{\mathrm{d}y}{\mathrm{d}x} = f(x, y) \tag{7.1.45}$$

is differentiated to give

$$\frac{\mathrm{d}^2y}{\mathrm{d}x^2} = f_x + f f_y \tag{7.1.46}$$

and

$$\frac{\mathrm{d}^3y}{\mathrm{d}x^3} = f_{xx} + 2f f_{xy} + f^2 f_{yy} + f_y(f_x + f f_y) \tag{7.1.47}$$

It is easy to see at this stage that the algebra will get quite heavy for higher order terms than those considered here.

Comparing coefficients gives the following set of equations for the various free constants:

$$b_1 + b_2 = 1 \qquad \mu b_2 = \tfrac{1}{2} \qquad \lambda b_2 = \tfrac{1}{2} \tag{7.1.48}$$

and hence

$$b_1 = 1 - c \quad b_2 = c \quad \mu = 1/2c \quad \lambda = 1/2c \tag{7.1.49}$$

for any $c \neq 0$. These values give

$$y_{n+1} = y_n + hf + \frac{h^2}{2}(f_x + ff_y) + \frac{h^3}{8c}(f_{xx} + 2ff_{xy} + f^2 f_{yy}) \tag{7.1.50}$$

which will agree with (7.1.44) if the truncation error T_n given by

$$T_n = -\left(\frac{h^3}{8c} - \frac{h^3}{6c}\right)[(f_{xx} + 2ff_{xy} + f^2 f_{yy})$$

$$+ f_y(f_x + ff_y)] + \frac{h^3}{8c} f_y(f_x + ff_y) + O(h^4) \tag{7.1.51}$$

is added on. This truncation error then reduces to

$$T_n = -\frac{h^3}{24c}[(3 - 4c)y_n''' - 3f_y y_n''] + O(h^4) \tag{7.1.52}$$

The next common property of Runge–Kutta methods is now apparent. Derivations generate families of methods with the same order of truncation error. The final choice within these families, in this case the choice of c, is made on different grounds than forcing agreement with Taylor's series. These can include minimizing the trunction error T_n, or attempting to improve stability conditions, or quite simply the choice might be to make the resulting coefficients as simple as possible. In this example a common choice is to put $c = 1/2$ to give the method

$$\begin{aligned} y_{n+1} &= y_n + (k_1 + k_2)/2 + T_n \\ k_1 &= hf(x_n, y_n) \\ k_2 &= hf(x_n + h, y_n + k_1) \\ T_n &= -\frac{h^3}{12}[y_n''' - 3f_y(x_n, y_n)y_n''] + O(h^4) \end{aligned} \tag{7.1.53}$$

Implementation of this rule is very straightforward, this being one of its appeals. The constants k_1 and k_2 are first computed and followed by the first formula of (7.1.53) to give y_{n+1}. Hence the method is self-starting, involves R function evaluations per step, and has an error form T_n which as it stands is not unduly useful.

This introductory section will be concluded by quoting some further commonly used Runge–Kutta methods and by completing a worked example using one of them.

A typical third-order formula is

$$\begin{aligned} y_{n+1} &= y_n + \tfrac{1}{6}(k_1 + 4k_2 + k_3) + O(h^4) \\ k_1 &= hf(x_n, y_n) \\ k_2 &= hf(x_n + h/2, y_n + k_1/2) \\ k_3 &= hf(x_n + h, y_n + 2k_2 - k_1) \end{aligned} \tag{7.1.54}$$

and a typical fourth-order rule, often called the standard Runge–Kutta method has the equations

$$\begin{aligned} y_{n+1} &= y_n + (k_1 + 2k_2 + 2k_3 + k_4)/6 \\ k_1 &= hf(x_n, y_n) \\ k_2 &= hf(x_n + h/2, y_n + k_1/2) \\ k_3 &= hf(x_n + h/2, y_n + k_2/2) \\ k_4 &= hf(x_n + h, y_n + k_3) \end{aligned} \tag{7.1.55}$$

As an example of this last standard Runge–Kutta method consider its application to

$$\frac{dy}{dx} = x^2 + y^2 \tag{7.1.56}$$

with boundary conditions $y = 0$ when $x = 0$. These results are illustrated in Table 7.2.

There is a class of problems which still does not allow of solution by the methods considered so far. These are problems in which the boundary conditions are given at more than one point, and are called boundary value problems. Problems with all the boundary conditions given at a single point are called initial value problems.

Table 7.2

x	y	k_1	k_2	k_3	k_4
0.0	0.000 000	0.000 000	0.000 250	0.000 250	0.001 000
0.1	0.000 333	0.001 000	0.002 250	0.002 250	0.004 000
0.2	0.002 667	0.004 001	0.006 252	0.006 253	0.009 008
0.3	0.009 004	0.009 008	0.012 268	0.012 273	0.016 045
0.4	0.021 359	0.016 046	0.020 336	0.020 349	0.025 174
0.5	0.041 791	0.025 175	0.030 545	0.030 576	0.036 524
0.6	0.072 448	0.036 525	0.043 073	0.043 133	0.050 336
0.7	0.115 660	0.050 338	0.058 233	0.058 346	0.067 278
0.8	0.174 081	0.067 030	0.076 559	0.076 760	0.087 292
0.9	0.250 908	0.872 955	0.098 926	0.099 272	0.112 263
1.0	0.350 233				

7.2 Boundary value problems

The simplest boundary value problems are second order. Consider an equation such as

$$y'' + P(x)y' + Q(x)y = F(x) \tag{7.2.1}$$

with boundary conditions $y(0) = a$ and $y'(0) = b$. Then the reduction to a pair of first-order equations in (7.1.7) leaves the boundary conditions $y(0) = a$ and $u(0) = b$, so that a method such as Euler's method, or a Runge–Kutta method could effect a solution starting from the initial point at zero and moving in a step-by-step fashion. This is an initial value problem. Suppose that instead of the condition $y'(0) = b$, the condition was $y(1) = b$. Now there would not be a suitable initial value for y' and the methods suggested so far need further thought before they are applicable. Two cases arise depending on whether the differential equation is linear or non-linear. Take first the linear case. Let the boundary conditions be $y(a) = A$ and $y(b) = B$, then two associated problems can be set up both of which are initial value problems. Let these have solutions $u(x)$ and $v(x)$ and let $u(x)$ satisfy

$$u'' + Pu' + Qu = F \tag{7.2.2}$$

with boundary conditions $u(a) = A$ and $u'(a)$ is taken arbitrarily say $u'(a) = 0$. Further let $v(x)$ satisfy

$$v'' + Pv' + Qv = 0 \tag{7.2.3}$$

with boundary conditions $v(a) = 0$ and $v'(a)$ again arbitrary but not zero. (This would lead to an identically zero solution.) The combination

$$y(x) = u(x) + cv(x) \tag{7.2.4}$$

satisfies the original differential equation for any c and the boundary condition $y(a) = A$. But c is free and can be used to force the second boundary condition to hold. Hence

$$u(b) + cv(b) = B \tag{7.2.5}$$

to give

$$c = \frac{B - u(b)}{v(b)} \tag{7.2.6}$$

The method of solution is to solve numerically by any convenient method, the two initial value problems for u and v from $x = a$ to $x = b$. The free constant c can then be found from (7.2.6) and the table of values for the required solution $y(x)$ follows from (7.2.4) using the tables for $u(x)$ and $v(x)$.

In the non-linear case, a sequence of initial value problems needs to be solved. Suppose the problem had the form

$$y'' = f(y', y, x) \tag{7.2.7}$$

with boundary conditions $y(a) = A$ and $y(b) = B$. Then an initial value problem will be obtained for any initial derivative condition such as $y'(a) = \alpha$. However the end outcome will not in general be the required value of $y(b)$ but $y(b) = B_1$. A sequence of such problems can be solved to build up knowledge of how B_1 varies with α. One strategy would be to take two α values, α_1 and α_2 and obtain B_1 and B_2. Linear interpolation will then yield a better estimate of the initial α which will hit B using the formula

$$\alpha_3 = \alpha_1 + \frac{\alpha_2 - \alpha_1}{B_2 - B_1}(B - B_1) \tag{7.2.8}$$

The process can then be repeated using α_2 and α_3 successively until B_i is within the required tolerance of B. This method is the shooting method named after the classic gunnery problem of finding the gun elevation for a given range. This is a boundary value problem. A shot is tried with an estimated elevation and the fall of shot observed. A converging sequence can then be set up. Numerically there are variations of this theme. Instead of two shots followed by a linear interpolation, m shots may be made followed by m-point interpolation. If the problem is linear then the two-shot approach with linear interpolation yields the solution immediately. In fact the u and v problem are no more than individual shots.

Boundary value problems can become much more intricate than these second-order problems suggest. Consider for example the equation

$$2f''' + ff'' = 0 \tag{7.2.9}$$

which arises in boundary layer theory. The usual boundary conditions are now $f(0) = 0 = f'(0)$ and $f'(\infty) = 1$. Now the initial value problems need to be integrated until $f'(x)$ settles down to its large-value limit. With a fourth-order problem there may well be two guessed boundary conditions which will require two-dimensional interpolation at each major step.

A different consideration arises with the boundary value problem

$$\frac{1}{P}F'' - F' - R(F) = 0 \tag{7.2.10}$$

with boundary conditions $F - (1/P)F' = 1$ at $x = 0$ and $F' = 0$ at $x = 1$. The function $R(F)$ is typically $2F^2$, and the obvious method is the shooting method starting at $x = 0$, so that the most complicated boundary condition is built-in. The result is most unsatisfactory. The reason is that stability considerations which will be covered in the next sections dictate that the step-size h has to be negative. Hence the shooting method has to be used in the reverse direction for good results. This problem will be used as an example of how boundary value problems can be converted to integral equations in Chapter 8, with the boundary conditions now built naturally into the equation.

As an example of the solution of a boundary value problem, consider

$$y'' + xy' + y = e^{-x} \tag{7.2.11}$$

with boundary conditions $y(0) = 0$ and $y(1) = 0$. The two associated problems are

$$u'' + xu' + u = e^{-x} \tag{7.2.12}$$

with boundary conditions $u(0) = 0$ and $u'(0) = 0$, and

$$v'' + xv' + v = 0 \tag{7.2.13}$$

with $v(0) = 0$ and $v'(0) = 1$. These equations can be solved by any of the standard methods. For simplicity the Taylor series has been used here to give

$$u(x) = \frac{x^2}{2} - \frac{x^3}{6} - \frac{x^4}{12} + \cdots \tag{7.2.14}$$

and

$$v(x) = x - \frac{x^3}{3} + \frac{x^5}{15} + \cdots \tag{7.2.15}$$

with

$$y(1) = 0 = \frac{x^2}{2} - \frac{x^3}{6} - \frac{x^4}{12} + \cdots + c\left(x - \frac{x^3}{3} + \frac{x^5}{15} + \cdots\right) \tag{7.2.16}$$

Hence $c = \frac{15}{44}$ to give the series

$$y(x) = \frac{15x}{44} + \frac{x^2}{2} - \frac{37x^3}{44} - \frac{x^4}{12} + \frac{x^5}{44} + \cdots \tag{7.2.17}$$

to the number of terms taken. To demonstrate how the shooting method gives the correct solution after just two shots in the linear case, consider the problem

$$y'' + ty' + y = 3t^2 + 2 \tag{7.2.18}$$

with boundary conditions $y(0) = 0$ and $y(1) = 1$. Using the standard fourth-order Runge–Kutta method to tabulate the two shots for the initial conditions $y'(0) = 0.4$ and $y'(0) = 0.1$ gives the values in Table 7.3.

Linear interpolation gives

$$y'(0) = 0.1 - \frac{0.4 - 0.1}{1.2900 - 1.0725}(0.0725) = 0.00 \tag{7.2.19}$$

and a shot with $y'(0) = 0$ gives tabulated values commensurate with the true solution $y = t^2$.

There is a third approach to boundary value problems which is used extensively in the solution of partial differential equations. The idea is to replace the derivatives in the differential equation by finite differences to convert the problem to one of solving linear algebraic equations. Simple differences as given in Chapter 2 are

Table 7.3

t	y with $y'(0) = 0.4$	y with $y'(0) = 0.1$
0.0	0.0	0.0
0.1	0.0498	0.0199
0.2	0.1188	0.0597
0.3	0.2063	0.1191
0.4	0.3115	0.1979
0.5	0.4339	0.2960
0.6	0.5729	0.4132
0.7	0.7283	0.5496
0.8	0.8996	0.7049
0.9	1.0869	0.8792
1.0	1.2900	1.0725

used, namely

$$y' = \frac{y_{i+1} - y_{i-1}}{2h} \tag{7.2.20}$$

$$y'' = \frac{y_{i+1} - 2y_i + y_{i-1}}{h^2} \tag{7.2.21}$$

where the range of y has been divided into steps h. For the previous example with boundary conditions $y(0) = y_1 = 0$ and $y(1) = y_n = 1$ for some n where $h = 1/n$, putting $n = 5$ gives the linear equations

$$(2 - ht_i)y_{i-1} + (2h^2 - 4)y_i + (2 + ht_i)y_{i+1} = 6h^2t_i^2 + 4h^2 \tag{7.2.22}$$

with $t_i = (i - 1)h$ and $h = 1/4$. The matrix for the unknowns y_2, y_3 and y_4 is

$$\begin{bmatrix} -3.875 & 2.062 & 0 \\ 1.875 & -3.875 & 2.125 \\ 0 & 1.812 & -3.875 \end{bmatrix} \begin{bmatrix} y_2 \\ y_3 \\ y_4 \end{bmatrix} = \begin{bmatrix} 0.273 \\ 0.344 \\ -1.727 \end{bmatrix} \tag{7.2.23}$$

with solution $y_2 = 0.062$, $y_3 = 0.250$ and $y_4 = 0.562$. As the matrix size is increased so the results become more accurate. Note that the resulting matrix will be tri-diagonal and lend itself to the iterative methods of Chapter 3.

This method can also be of value with eigenvalue problems which arise in the differential equation setting. Like the algebraic problem of Chapter 4, the eigenvalue problem is recognizable by having the trivial solution $y(x) \equiv 0$, and a possible parameter such as k in the example:

$$y'' - k^2y = 0 \tag{7.2.24}$$

with boundary conditions $y(0) = 0$ and $y(1) = 0$. For certain eigenvalues k there will exist non-trivial solutions. These problems are very common in applied mathematics, especially in problems involving vibrations where the eigenvalues are the allowed frequencies. There is an infinite set of eigenvalues in general. In quantum mechanics, the eigensolutions are the wave functions for the energy levels of the atomic system being considered.

By applying simple finite difference forms to the derivatives of the defining differential equation, the problem reduces immediately to an algebraic eigenvalue problem of the type considered in Chapter 4. The finer the step size, the larger the matrix becomes, and the more eigenvalues are then obtained. Some examples of these problems are given in the exercises.

These problems can also be solved by using variational methods of the type discussed in Section 8.5 in the integral equation chapter. The switch from integral operators to differential operators, the relation between equivalent differential and integral problems through Green's functions and the solutions via variational methods, are covered in this section.

The scene is now set for a more detailed analysis of the methods which have been introduced in this section. The main object will be to place each class of methods into an analytic framework, beginning with the linear multi-step methods of which Euler's method is the simplest. The Runge–Kutta class will then be considered, and in the final paragraphs, methods for treating the problem of stiff systems will be covered.

7.3 Linear multistep methods — order, zero stability

All linear multistep methods have the same general form which is defined by the formula

$$\sum_{j=0}^{k} \alpha_j y_{n+j} = h \sum_{j=0}^{k} \beta_j f_{n+j} \tag{7.3.1}$$

where, as in the introduction, y_j is the calculated value of $y(x)$ at the point x_j, and f_{n+j} is the calculated value of $f(x_{n+j})$. The corresponding theoretical or exact solution will be referred to using the notation $y(x_j)$. The constant k is called the step number, and such a method is termed k-step. Once the constants α_j and β_j have been fixed, a method is defined. Indeed, not any combination of α's and β's will give an authentic method and hence the first question to be asked is, what are the conditions on the α's and β's for a valid procedure? In effect the answer to this question will illuminate the truncation error problem and will leave discussion to be continued on rounding and stability difficulties later.

From an implementation point of view, the method in (7.3.1) is used in the form

$$\alpha_k y_{n+k} + \sum_{j=0}^{k-1} \alpha_j y_{n+j} = h\beta_k f(x_{n+k}, y_{n+k}) + h \sum_{j=0}^{k-1} \beta_j f_{n+j} \tag{7.3.2}$$

where the computation has already progressed through to x_{n+k-1} in that all the set $\{y_{n+j}\}$ for $j = 0, k-1$ are known from previous steps. Hence the formula (7.3.2) yields y_{n+k} in terms of these known quantities. If $\beta_k = 0$ then the formula is *explicit* and is used directly as in Euler's method of Section 7.1. For Euler, $k = 1$, and such a formula is often called a *predictor* formula. If $\beta_k \neq 0$ then y_{n+k} appears on both sides of (7.3.2) which is in reality a non-linear equation for y_{n+k}. Such a method is implicit, and is called a *corrector*. Direct iteration in the style of (7.1.18) is used to extract a solution, which in the more general case has the form

$$\alpha_k y_{n+k}^{[m+1]} + \sum_{j=0}^{k-1} \alpha_j y_{n+j} = h\beta_k f(x_{n+k}, y_{n+k}^{[m]}) + h\sum_{j=0}^{k-1} \beta_j f_{n+j} \tag{7.3.3}$$

and again few iterations are normally employed as the over-computed accurate root will still contain the underlying truncation error of the method. All that is needed is to raise the accuracy from that of the predictor to that of the slightly more accurate corrector. A more detailed analysis of the number of sensible iterations required appears later when predictor–corrector pairs are considered more fully.

To avoid ambiguity it is common to set $\alpha_k = 1$, so fixing the multiplicative constant, and if in addition, $\alpha_k = -\alpha_{k-1} = 1$ with other α's zero then the method is of the Adams type, whereas if $\alpha_k = -\alpha_{k-2} = 1$ with the other α's again zero then the method is of the Nyström type. Various names are associated with the higher order methods, but commonly an explicit Adams formula is an Adams–Bashforth method whereas an implicit Adams formula is an Adams–Moulton formula. Clearly for $k > 1$ the boundary conditions which fix (x_0, y_0) are not sufficient to get the method started. Such methods require an additional starting procedure to compute $y_1, \ldots, y_{k-1}$. One of the Runge–Kutta methods, a Taylor series method or Euler's method with a specially reduced step size h, are possibilities in this role. The important thing is to match the accuracy of the starting procedure with that of the main method.

One way to generate higher order methods is to use finite difference formulae from Chapter 2. Hence if the integration rules of (2.1.48) and (2.1.49) are used in (7.1.10) the two rules

$$y_{n+1} = y_n + h\left(1 + \tfrac{1}{2}\nabla + \tfrac{5}{12}\nabla^2 + \cdots\right) y_n' \tag{7.3.4}$$

$$y_{n+1} = y_n + h\left(1 - \tfrac{1}{2}\nabla - \tfrac{1}{12}\nabla^2 - \cdots\right) y_{n+1}' \tag{7.3.5}$$

are generated, the first being a predictor formula and the second a corrector. This pair is sometimes referred to as the Adams–Bashforth method. It is not difficult to truncate the series of backward differences at some suitable point and convert the difference terms into f_{n+j}'s to recover formulae of the type in (7.3.1) as $y_j' = f_j$. The more general approach being used here will consider the form of (7.3.1) rather than use differences.

The first stage in this process is to find conditions on α_j and β_j so that (7.3.1) is a method for solving $y' = f(x, y)$. If the whole equation in (7.3.1) is written on the left-hand side, and if the function $y(x)$ is substituted into the result with

$y' = f(x, y)$, then this expression will be a measure of the truncation error in using (7.2.1) to represent a solution to $y' = f(x, y)$. Hence define the linear difference operator

$$\mathcal{L}[y(x), h] = \sum_{j=0}^{k} [\alpha_j y(x + jh) - h\beta_j y'(x + jh)] \tag{7.3.6}$$

The expansion of this operator as a Taylor series will allow the order of the error to be determined. Hence

$$\mathcal{L}[y(x), h] = C_0 y(x) + C_1 h y'(x) + C_2 h^2 y''(x) + \cdots + C_q h^q y^{(q)}(x) + \cdots \tag{7.3.7}$$

will be zero to order p if $C_0 = \cdots = C_p = 0$ and $C_{p+1} \neq 0$. These conditions constitute linear algebraic equations for the constants α_j and β_j to force (7.3.1) to be of order p. Comparing coefficients gives the defining equations

$$C_0 = \alpha_0 + \alpha_1 + \cdots + \alpha_k \tag{7.3.8}$$

$$C_1 = \alpha_1 + 2\alpha_2 + \cdots + k\alpha_k - (\beta_0 + \beta_1 + \cdots + \beta_k) \tag{7.3.9}$$

$$\vdots$$

$$C_q = \frac{1}{q!}(\alpha_1 + 2^q\alpha_2 + \cdots + k^q\alpha_k) - \frac{1}{(q-1)!}(\beta_1 + 2^{q-1}\beta_2 + \cdots + k^{q-1}\beta_k) \tag{7.3.10}$$

with the definition that a method is of order p if $C_0 = C_1 = \cdots = C_p = 0$ and $C_{p+1} \neq 0$.

Hence a pth order formula has as its first non-zero coefficient C_{p+1} which gives a measure of the local truncation error and is called the error constant. The word local is used to indicate that this is the error due to this step only of the overall process. If errors already exist in earlier calculated values then the accumulated error is called the global error. This local trunction error can be defined as $\mathcal{L}[y(x_n), h]$ where $y(x)$ is the theoretical or exact solution, and this error is denoted by T_{n+k}.

These formulae can be used in one of two ways. Given a method, the C's can be evaluated to determine the order of the method, or conversely, methods of a given order can be generated by solving the resulting linear equations for the α's and β's which result in setting $C_j = 0$ for the suitable range of j values. It is easily seen that an implicit method has exactly one extra degree of freedom over a corresponding explicit method and in theory this allows an implicit method to be just one order higher than its explicit counterpart. Hence the largest order attainable by a k-step method is $2k$ for an implicit method and $2k - 1$ for an explicit method. Such a method is called *maximal*.

Two theoretical pursuits can be considered to elaborate the concept of order. The definition of local truncation error is not the most obvious and a link will be made between the quoted definition and a more conventional form. The second issue is to answer the question as to what the minimum requirements are for a method to be a valid one to solve $y' = f(x, y)$. In other words how many of the C_j coefficients need to be made zero?

The conventional form for the local truncation error is

$$T_{n+k} = y(x_{n+k}) - y_{n+k} \tag{7.3.11}$$

namely the difference between the exact solution and the calculated solution at the point (x_{n+k}, y_{n+k}). In addition for a *local* error the condition

$$y_{n+j} = y(x_{n+j}) \qquad j = 0, 1, \ldots, k-1 \tag{7.3.12}$$

needs to hold, in that all the previously calculated y values are assumed accurate. Then (7.3.6) can be written as

$$\sum_{j=0}^{k} \alpha_i y(x_n + jh) = h \sum_{j=0}^{k} \beta_j f(x_n + jh, y(x_n + jh)) + \mathcal{L}[y(x_n), h] \tag{7.3.13}$$

and y_{n+k} satisfies

$$\sum_{j=0}^{k} \alpha_j y_{n+j} = h \sum_{j=0}^{k} \beta_j f(x_{n+j}, y_{n+j}) \tag{7.3.14}$$

Subtracting (7.3.13) and (7.3.14) and using (7.3.12) gives

$$y(x_{n+k}) - y_{n+k} = h\beta_k[f(x_{n+k}, y(x_{n+k})) - f(x_{n+k}, y_{n+k})] + \mathcal{L}[y(x_n), h] \tag{7.3.15}$$

The difference in the two neighbouring f values can be resolved using the differential mean value theorem in the form:

$$\begin{aligned} f(x_{n+k}, y(x_{n+k})) &- f(x_{n+k}, y_{n+k}) \\ &= [y(x_{n+k}) - y_{n+k}]\frac{\partial f}{\partial y}(x_{n+k}, \eta_{n+k}) \end{aligned} \tag{7.3.16}$$

where $\eta_{n+k} \in [y_{n+k}, y(x_{n+k})]$. Hence the definition of $\mathcal{L}$ and the intuitive meaning of local truncation error are related by

$$\left[1 - h\beta_k \frac{\partial f(x_{n+k}, \eta_{n+k})}{\partial y}\right][y(x_{n+k}) - y_{n+k}] = \mathcal{L}[y(x_n), h] = T_{n+k} \tag{7.3.17}$$

For an explicit method with $\beta_k = 0$, both definitions are identical, but for an implicit method the two approaches tend to the same result when $h \to 0$. For a method of order p,

$$y(x_{n+k}) - y_{n+k} = T_{n+k} = C_{p+1}h^{p+1}y^{(p+1)}(x_n) + O(h^{p+2}) \tag{7.3.18}$$

and the expression on the right-hand side of (7.3.18) is called the principal local truncation error.

Inevitably errors accumulate from step to step and the resulting global error is defined as

$$e_{n+k} = y(x_{n+k}) - y_{n+k} \tag{7.3.19}$$

and now no assumption is made that $y_i = 0; \quad i = n, \ldots, n+k-1$.

The second theoretical result to be pursued is to discover what the least conditions are for a method to be valid. These are the conditions of *consistency* and the definition is that a method is *consistent* if its order p is greater than or equal to unity, or $C_0 = C_1 = 0$. In terms of the constants α and β these conditions are:

$$\sum_{j=0}^{k} \alpha_j = 0 \tag{7.3.20}$$

$$\sum_{j=0}^{k} j\alpha_j = \sum_{j=0}^{k} \beta_j \tag{7.3.21}$$

which can be very easily interpreted by considering the following two derivations.

Suppose y_n tends to some function $y(x_n)$ as $h \to 0$, then

$$y(x_n) = y_{n+j} + \phi_{n,j}(h) \tag{7.3.22}$$

where $\phi_{n,j}(h) \to 0$ as $h \to 0$. Hence

$$\sum_{j=0}^{k} \alpha_j y(x_n) = \sum_{j=0}^{k} \alpha_j y_{n+j} + \sum_{j=0}^{k} \alpha_j \phi_{n,j}(h) \tag{7.3.23}$$

or using (7.3.1):

$$y(x_n) \sum_{j=0}^{k} \alpha_j = h \sum_{j=0}^{k} \beta_j f_{n+j} + \sum_{j=0}^{k} \alpha_j \phi_{n,j}(h) \tag{7.3.24}$$

of which the right-hand side tends to zero as $h \to 0$. But in general $y(x_n) \neq 0$ and hence the first consistency condition is reproduced, namely equation (7.3.20). This condition holds if y_n tends to some function, not necessarily the solution of $y' = f(x, y)$. The second consistency condition covers this shortcoming. Consider y' being written in the form

$$y'(x_n) = \operatorname*{Lt}_{h \to 0} \frac{y_{n+j} - y_n}{jh} \tag{7.3.25}$$

or equivalently

$$y_{n+j} - y_n = jhy'(x_n) + jh\phi_{n,j}(h) \tag{7.3.26}$$

Then

$$\sum_{j=0}^{k} \alpha_j y_{n+j} - \sum_{j=0}^{k} \alpha_j y_n = h \sum_{j=0}^{k} j\alpha_j y'(x_n) + h \sum_{j=0}^{k} j\alpha_j \phi_{n,j}(h) \tag{7.3.27}$$

and using the original method in equation (7.3.1) gives

$$h \sum_{j=0}^{k} \beta_j f_{n+j} - y_n \sum_{j=0}^{k} \alpha_j = hy'(x_n) \sum_{j=0}^{k} j\alpha_j + h \sum_{j=0}^{k} j\alpha_j \phi_{n,j}(h) \tag{7.3.28}$$

But the second term is zero from the first consistency condition to leave

$$\sum_{j=0}^{k} \beta_j f_{n+j} = y'(x_n) \sum_{j=0}^{k} j\alpha_j + \sum_{j=0}^{k} j\alpha_j \phi_{n,j}(h) \tag{7.3.29}$$

and now letting $h \to 0$ gives just

$$f(x_n, y(x_n)) \sum_{j=0}^{k} \beta_j = y'(x_n) \sum_{j=0}^{k} j\alpha_j \tag{7.3.30}$$

and hence $y(x)$ satisfies $y' = f(x, y)$ if the second consistency condition holds.

So far all attention has revolved around the evaluation of truncation error. But the other major area of interest must be what effect rounding errors will have on a linear multistep method. A first-order ordinary differential equation is being solved numerically by using a non-linear difference equation of order k defined in (7.3.1). It is not possible to write down the analytic solution for this non-linear difference equation, but as $h \to 0$ this difference equation tends to one which is linear. This can be solved analytically to give some measure of the behaviour of such difference equations in these circumstances. Such analysis is called *zero stability analysis*. The linear difference equation is

$$\sum_{j=0}^{k} \alpha_j y_{n+j} = 0 \tag{7.3.31}$$

as $h \to 0$, and such an equation has the solution

$$y_n = A_1 \zeta_1^n + A_2 \zeta_2^n + \cdots + A_k \zeta_k^n \tag{7.3.32}$$

where $\zeta_1, \ldots, \zeta_k$ are roots of the auxiliary polynomial

$$\rho(\zeta) = \sum_{j=0}^{k} \alpha_j \zeta^j \tag{7.3.33}$$

For readers not familiar with difference equations, this solution can be easily seen by substituting the trial solution $y_n = A\zeta^n$ into (7.3.31). The process is exactly analogous to the trial solution $y = Ae^{mx}$ used to solve ordinary differential equations with constant coefficients. The polynomial $\rho(\zeta)$ is called the first characteristic polynomial and for completeness the second characteristic polynomial is defined as

$$\sigma(\zeta) = \sum_{j=0}^{k} \beta_j \zeta^j \tag{7.3.34}$$

Hence the conditions of consistency can be written as $\rho(1) = 0$ and $\sigma(1) = \rho'(1)$.

To return to the zero stability analysis and the solution (7.3.32), it is immediately obvious that there is a problem. The first-order differential equation has only one constant of integration. The kth order difference equation has k. Hence only one of the k solutions will correspond to the solution of the differential equation, and the others can be described as spurious. In practice the coefficients A_j of the spurious solutions will be zero to *machine accuracy*, and their values are effectively fixed from the extra boundary conditions furnished by the starting procedure. Hence these constants are not precisely zero if rounding errors are present. Therefore if any of the roots ζ_j has modulus greater than unity, when raised to a power of n after many steps, there could be a term which will ultimately dominate the true solution. This is instability.

With this problem in mind, a linear multistep method is said to be *zero stable* if no root of the first characteristic polynomial $\rho(\zeta)$ has modulus greater than unity and if every root of modulus unity is simple. The latter condition deals with the case of multiple roots when the solution to (7.3.31) is now of the form $(A_1 + nA_2)\zeta^n$ for a pair of equal roots. (Compare this with the differential equation case when the trial solution $(A_1 + xA_2)e^{mx}$ is used). The root which does correspond to the solution of the ordinary differential equation is called the *principal root*, and from the first consistency condition $\rho(1) = 0$, this root must be unity.

Dahlquist has a theorem which states that a necessary and sufficient condition for a linear multistep method to be convergent is that it be both consistent and zero stable. Earlier in the discussion on the order a method could attain, the concept of a maximal method was defined. In the light of the zero stability condition above it would seem that for a method to be useful it also needs to be, at worst, zero stable. It can be proved that no zero stable linear multistep method of stepnumber k can have order exceeding $k + 1$ for k odd and $k + 2$ for k even. Such a method is called *optimal*. An example of such a rule is Simpson's rule defined by

$$y_{n+2} - y_n = \tfrac{1}{3}h(f_{n+2} + 4f_{n+1} + f_n) \tag{7.3.35}$$

which, uniquely, happens to be also maximal.

Zero stability does not depend on the β's, and it would be unexpected if stability in general was to be also independent of these constants. A more refined analysis is required, but one which in some way still yields a linear difference equation. Such analysis is the first topic of the next paragraph, and this leads into considering both combinations of methods being used in tandem and the effect of applying these methods to sets of first-order differential equations.

7.4 Linear multistep methods — weak stability

The principle involved in weak stability analysis is to linearize the difference equation which represents the linear multistep method, not by forcing h to zero but by applying the method to a linear equation. Hence consider a linear multistep method defined by

$$\sum_{j=0}^{k} \alpha_j y_{n+j} = h \sum_{j=0}^{k} \beta_j f_{n+j} \tag{7.4.1}$$

and the corresponding exact solution which will satisfy

$$\sum_{j=0}^{k} \alpha_i y(x_{n+j}) = h \sum_{j=0}^{k} \beta_j f(x_{n+j}, y(x_{n+j})) + T_{n+k} \tag{7.4.2}$$

with $T_{n+k} = \mathcal{L}[y(x_n), h]$. If round-off is included in (7.4.1), then the solution $\tilde{y}_n$ satisfies

$$\sum_{j=0}^{k} \alpha_j \tilde{y}_{n+j} = h \sum_{j=0}^{k} \beta_j f(x_{n+j}, \tilde{y}_{n+j}) + R_{n+k} \tag{7.4.3}$$

and the global error is given by $\tilde{e}_n = y(x_n) - \tilde{y}_n$. Hence subtracting (7.4.2) and (7.4.3) gives

$$\sum_{j=0}^{k} \alpha_j \tilde{e}_{n+j} = h \sum_{j=0}^{k} \beta_j [f(x_{n+j}, y(x_{n+j})) - f(x_{n+j}, \tilde{y}_{n+j})] + \phi_{n+k} \tag{7.4.4}$$

with $\phi_{n+k} = T_{n+k} - R_{n+k}$. But the differential mean value theorem of equation (7.3.16) gives

$$\sum_{j=0}^{k} \alpha_j \tilde{e}_{n+j} = h \sum_{j=0}^{k} \beta_j \frac{\partial f(x_{n+j}, \eta_{n+j})}{\partial y} + \phi_{n+k} \tag{7.4.5}$$

It is at this stage that the weak stability assumption is made that the differential equation is linear, or equivalently that

$$\frac{\partial f}{\partial y} = \lambda = \text{constant} \tag{7.4.6}$$

and ϕ_{n+k} is assumed constant and equal to ϕ. Hence the linearized difference equation is

$$\sum_{j=0}^{k} (\alpha_j - h\lambda\beta_j)\tilde{e}_{n+j} = \phi \tag{7.4.7}$$

with the analytic solution

$$\tilde{e}_n = \sum_{s=1}^{k} a_s r_s^n - \frac{\phi}{h\lambda} \sum_{j=0}^{k} \beta_j \tag{7.4.8}$$

in which the a_s are the arbitrary constants and the r_s are the roots of

$$\sum_{j=0}^{k} (\alpha_j - h\lambda\beta_j) r^j = 0 \tag{7.4.9}$$

Now using the definitions of the first and second characteristic polynomials as defined in (7.3.33) and (7.3.34) the characteristic or stability polynomial can be defined as

$$\Pi(r, \bar{h}) = \rho(r) - \bar{h}\sigma(r) \tag{7.4.10}$$

with $\bar{h} = h\lambda$ and it is the roots of this polynomial which define the r_s.

A method is said to be absolutely stable if $|r_s| < 1$ for all s for a given $\bar{h}$. The range of $\bar{h}$ for which absolute stability occurs is called the interval of absolute stability. In general, if the step size is such that $\bar{h}$ falls into this interval then there can be no growth of the spurious solutions for large n. A very simple example is provided by the two-step Adams–Bashforth method

$$y_{n+2} - y_{n+1} = \tfrac{1}{2}h(3f_{n+1} - f_n) \tag{7.4.11}$$

for which the stability polynomial is

$$r^2 - \left(\tfrac{3}{2}\bar{h} + 1\right) r + \tfrac{1}{2}\bar{h} = 0 \tag{7.4.12}$$

with roots

$$r = \left(\frac{3\bar{h}}{4} + \frac{1}{2}\right) \pm \frac{1}{2}\sqrt{\frac{9\bar{h}^2}{4} + \bar{h} + 1} \tag{7.4.13}$$

Firstly note that the principal root is the one with the positive sign as this root tends to unity as $\bar{h} \to 0$. The ends of the interval of absolute stability occur when these roots are either $+1$ or -1 which yields the equations:

$$\frac{9\bar{h}^2}{4} + \bar{h} + 1 = \begin{cases} \left(-\tfrac{3}{2}\bar{h} + 1\right)^2 \\ \left(-\tfrac{3}{2}\bar{h} - 3\right)^2 \end{cases} \tag{7.4.14}$$

with roots at $\bar{h} = 0$ and $\bar{h} = -1$. A quick spot check then shows that the interval of absolute stability is indeed in the interior of the interval $(-1, 0)$.

Hence for positive λ and positive h it appears that the method is not useful, as it is certainly absolutely unstable. However, the solution of $y' = \lambda y$ for positive λ grows exponentially, that is $y = Ae^{\lambda x}$, and so the principle root will be $r_1 = e^{\bar{h}}$.

Hence a weaker stability definition is in order in such a case. It is surely acceptable if the errors grow less rapidly than the solution. Hence the definition that a linear multistep method is relatively stable if, for a given $\bar{h}$, $|r_s| < |r_1|$ for $s = 2, k$, and again the range of $\bar{h}$ for which relative stability occurs is called the interval of relative stability. Interest now lies in determining intervals of relative and absolute stability, as it is important that as a solution takes place the local value for λ, which will vary for a non-linear problem, should be such that the corresponding $\bar{h}$ is kept within the relevant interval of stability. The problem is now that of being given a polynomial with coefficients which are functions of $\bar{h}$, and requiring conditions on $\bar{h}$ to ensure that the roots of the polynomial satisfy the stability conditions. Four methods will be given here to resolve this problem.

The first of these is the root locus method which is simply to make repeated solutions for the roots of the polynomial for varying $\bar{h}$. Sub-tabulation or interpolation can then be used to determine more precisely the borders of the regions of the required intervals, in which at least one r_s just reaches unity. This is predominantly a computer solution and will require an effective polynomial solver, possibly based on Bairstow's method and deflation (see Chapter 1). To obtain analytic inequalities for the conditions on $\bar{h}$, attention is turned to the second method based on Schur polynomials (Schur, 1916).

A polynomial $\phi(r)$ is said to be a Schur polynomial if its roots satisfy $|r_s| < 1$ for $s = 1, \ldots, k$. This is precisely the requirement for absolute stability. Moreover the Schur criterion allows the degree of the polynomial to be reduced, and simultaneously yields an inequality on $\bar{h}$. Suppose

$$\phi(r) = c_k r^k + \cdots + c_1 r + c_0 \tag{7.4.15}$$

then define $\hat{\phi}(r)$ as

$$\hat{\phi}(r) = c_0^* r^k + \cdots + c_{k-1}^* r + c_k^* \tag{7.4.16}$$

where c_j^* is the complex conjugate of c_j. Then $\phi(r)$ is a Schur polynomial iff

$$|\hat{\phi}(0)| > |\phi(0)| \tag{7.4.17}$$

and $\phi_1(r)$ is Schur where

$$\phi_1(r) = (1/r)[\hat{\phi}(0)\phi(r) - \phi(0)\hat{\phi}(r)] \tag{7.4.18}$$

This criterion can be used repeatedly until the degree of the final polynomial is just one and hence the condition that its root has modulus less than unity is immediate.

This process is well illustrated using the simple example from (7.4.11) with stability polynomial (7.4.12). Then $\hat{\Pi}(r)$ is given by

$$\hat{\Pi}(r) = \tfrac{1}{2}\bar{h}r^2 - \left(\tfrac{3}{2}\bar{h} + 1\right) r + 1 \tag{7.4.19}$$

and $\Pi_1(r)$ reduces to

$$\Pi_1(r) = \left(\tfrac{1}{2}\bar{h} + 1\right) r - \left(\tfrac{3}{2}\bar{h} + 1\right) \tag{7.4.20}$$

leaving the inequality

$$1 > |\bar{h}/2| \tag{7.4.21}$$

Hence the root of $\Pi_1(r)$ has modulus less than unity if $\bar{h} \in (-1, 0)$ and the inequality in (7.4.21) is satisfied automatically.

This method is also possible for relative stability determination as the requirement is now

$$|r_s| < e^{\bar{h}} \tag{7.4.22}$$

Hence let $r = Re^{\bar{h}}$ and use the polynomial

$$\rho(Re^{\bar{h}}) - \bar{h}\sigma(Re^{\bar{h}}) = 0 \tag{7.4.23}$$

to seek the condition $|R_s| < 1$ for relative stability.

The third method for determining intervals of stability is the Routh–Hurwitz criterion, based on Fraboul (1962), in which the interior of the unit circle inside which the roots of the characteristic polynomial must lie for absolute stability is mapped into the negative half-plane by

$$r = \frac{(1+z)}{(1-z)} \tag{7.4.24}$$

to give

$$\rho\left(\frac{1+z}{1-z}\right) - \bar{h}\sigma\left(\frac{1+z}{1-z}\right) = 0 \tag{7.4.25}$$

By multiplying throughout by $(1-z)^k$ the resulting polynomial is

$$a_0 z^k + \cdots + a_k = 0 \tag{7.4.26}$$

In terms of these coefficients the Routh–Hurwitz criterion is that for the roots of (7.4.23) to lie in $Re\, z < 0$, all the principle minors of the matrix

$$\begin{pmatrix} a_1 & a_3 & a_5 & \cdots & a_{2k-1} \\ a_0 & a_2 & a_4 & \cdots & a_{2k-2} \\ 0 & a_1 & a_3 & \cdots & a_{2k-3} \\ 0 & a_0 & a_2 & \cdots & a_{2k-4} \\ \vdots & \vdots & \vdots & & \vdots \\ 0 & 0 & 0 & \cdots & a_k \end{pmatrix} \tag{7.4.27}$$

are positive. Care is needed in interpreting this matrix as the polynomial has only coefficients from a_0 to a_k and any elements with indices greater than k are taken as zero. Also it is conventional to treat $a_0 > 0$ to fix the overall sign of the polynomial. In particular for $k = 2$ these conditions reduce to $a_0 > 0$, $a_1 > 0$ and $a_2 > 0$; for $k = 3$, $a_0 > 0$, $a_1 > 0$, $a_2 > 0$, $a_3 > 0$ and $a_1a_2 - a_3a_0 > 0$ and for $k = 4$, $a_1 > 0$, $a_2 > 0$, $a_3 > 0$, $a_4 > 0$ and $a_1a_2a_3 - a_0a_3^2 - a_4a_1^2 > 0$. Again relative stability can be handled by the same substitution as for Schur.

The fourth method is called the boundary locus method in which a region $\mathcal{R}$ is defined such that if $\bar{h}$ is inside $\mathcal{R}$ then the roots of $\Pi(r, \bar{h}) = 0$ lie within the required unit circle. The boundary of $\mathcal{R}$ is denoted by $\partial\mathcal{R}$, and as the roots of $\Pi(r, \bar{h}) = 0$ are continuous functions of $\bar{h}$, then $\bar{h}$ lies on $\partial\mathcal{R}$ when the roots lie on the unit circle. Hence

$$\Pi(e^{i\theta}, \bar{h}) = \rho(e^{i\theta}) - \bar{h}\sigma(e^{i\theta}) = 0 \tag{7.4.28}$$

and the locus of $\partial\mathcal{R}$ is

$$\bar{h} = \frac{\rho(e^{i\theta})}{\sigma(e^{i\theta})} \tag{7.4.29}$$

Hence the end-points of the interval of stability occur where $\partial\mathcal{R}$ cuts the real axis which can be determined as the solution of a non-linear equation. Just which side of these boundary points has roots with modulus less than unity can be determined by a spot check.

This section is completed by studying the above analysis for the more realistic situation when methods are used in combination, commonly in predictor–corrector pairs, of which the simple combination of Euler's method and modified Euler is an example. The first stage of this discussion is to have a notation to define the exact combination being used. This is achieved by letting P represent an application of a predictor formula, and C represent an application of a corrector formula. The number of evaluations of the function f can also vary, and is a measure of the workload. Hence an evaluation is represented by E. For a predictor-corrector pair the two obvious modes of operation are $P(EC)^m$ and $P(EC)^mE$. If the predictor is defined by the polynomials

$$\rho^*(\zeta) = \sum_{j=0}^{k} \alpha_j^* \zeta^j \tag{7.4.30}$$

$$\sigma^*(\zeta) = \sum_{j=0}^{k} \beta_j^* \zeta^j \tag{7.4.31}$$

and the corrector by

$$\rho(\zeta) = \sum_{j=0}^{k} \alpha_j \zeta^j \tag{7.4.32}$$

$$\sigma(\zeta) = \sum_{j=0}^{k} \beta_j \zeta^j \tag{7.4.33}$$

then the full formula for $P(EC)^mE$ mode is:

$$y_{n+k}^{[0]} + \sum_{j=0}^{k-1} \alpha_j^* y_{n+j}^{[m]} = h \sum_{j=0}^{k-1} \beta_j^* f_{n+j}^{[m]} \tag{7.4.34}$$

$$y_{n+k}^{[s+1]} + \sum_{j=0}^{k-1} \alpha_j y_{n+j}^{[m]} = h\beta_k f_{n+k}^{[s]} + h\sum_{j=0}^{k-1} \beta_j f_{n+j}^{[m]} \tag{7.4.35}$$

with $s = 0, \ldots, m-1$ and

$$f_{n+k}^{[s]} = f(x_{n+k}, y_{n+k}^{[s]}) \tag{7.4.36}$$

The alternative mode of operation $P(EC)^m$ has the equations:

$$y_{n+k}^{[0]} + \sum_{j=0}^{k-1} \alpha_j^* y_{n+j}^{[m]} = h\sum_{j=0}^{k-1} \beta_j^* f_{n+j}^{[m-1]} \tag{7.4.37}$$

$$y_{n+k}^{[s+1]} + \sum_{j=0}^{k-1} \alpha_j y_{n+j}^{[m]} = h\beta_k f_{n+k}^{[s]} + h\sum_{j=0}^{k-1} \beta_j f_{n+j}^{[m-1]} \tag{7.4.38}$$

The most commonly used mode is $PECE$, and for these combinations it is of interest to determine the local truncation error involved, especially how this relates to the individual errors of the predictor and corrector alone. Clearly there is little gained in using a very accurate predictor, and then throwing this gain away with an inaccurate corrector.

To investigate this effect let $\mathcal{L}^*$ and $\mathcal{L}$ be the linear difference operators for the predictor and corrector respectively, and let the respective orders be p^* and p with error constants $C_{p^*+1}^*$ and C_{p+1}. Then

$$\mathcal{L}^*[y(x), h] = C_{p^*+1}^* h^{p^*+1} y^{(p^*+1)}(x) + O(h^{p^*+2}) \tag{7.4.39}$$

and

$$\mathcal{L}[y(x), h] = C_{p+1} h^{p+1} y^{(p+1)}(x) + O(h^{p+2}) \tag{7.4.40}$$

and hence for the predictor using (7.3.18)

$$y(x_{n+k}) - y_{n+k}^{[0]} = C_{p^*+1}^* h^{p^*+1} y^{(p^*+1)}(x_n) + O(h^{p^*+2}) \tag{7.4.41}$$

Similarly the corrector gives

$$\sum_{j=0}^{k} \alpha_i y(x_{n+j}) = h\sum_{j=0}^{k} \beta_j f(x_{n+j}, y(x_{n+j})) + \mathcal{L}[y(x_n), h] \tag{7.4.42}$$

and

$$y_{n+k}^{[s+1]} + \sum_{j=0}^{k-1} \alpha_j y_{n+j}^{[m]} = h\beta_k f_{n+k}^{[s]} + h\sum_{j=0}^{k-1} \beta_j f_{n+j}^{[m-t]} \tag{7.4.43}$$

where $t = 0$ in $P(EC)^m E$ mode and $t = 1$ in $P(EC)^m$ mode. Subtracting (7.4.42) and (7.4.43) and assuming no previous errors to make the local error assumption gives

$$y(x_{n+k}) - y_{n+k}^{[s+1]} = h\beta_k[f(x_{n+k}, y(x_{n+k})) - f(x_{n+k}, y_{n+k}^{[s]})] + \mathcal{L}[y(x_n), h] \tag{7.4.44}$$

Applying the differential mean value theorem then gives

$$y(x_{n+k}) - y_{n+k}^{[s+1]} = h\beta_k \frac{\partial f}{\partial y}(x_{n+k}, \eta_{n+k,s})[y(x_{n+k}) - y_{n+k}^{[s]}] + \mathcal{L}[y(x_n), h] \tag{7.4.45}$$

A number of possibilities now arises. If $p^* \geq p$, then putting $s = 0$ gives

$$y(x_{n+k}) - y_{n+k}^{[1]} = C_{p+1}h^{p+1}y^{(p+1)}(x_n) + O(h^{p+2}) \tag{7.4.46}$$

and with (7.3.40) successive substitution gives

$$y(x_{n+k}) - y_{n+k}^{[m]} = C_{p+1}h^{p+1}y^{(p+1)}(x_n) + O(h^{p+2}) \tag{7.4.47}$$

Hence if $p^* \geq p$ and $m \geq 1$ then the local error in both $P(EC)^m$ and $P(EC)^m E$ modes is that of the corrector alone.

If $p^* = p - 1$, substitution as above now yields:

$$y(x_{n+k}) - y_{n+k}^{[1]} = \left[\beta_k \frac{\partial f}{\partial y} C_p^* y^{(p)}(x_n) + C_{p+1}y^{(p+1)}(x_n)\right] h^{p+1} + O(h^{p+2}) \tag{7.4.48}$$

and successive substitution gives for $m \geq 2$

$$y(x_{n+k}) - y_{n+k}^{[m]} = C_{p+1}h^{p+1}y^{(p+1)}(x_n) + O(h^{p+2}) \tag{7.4.49}$$

Hence again the error is that of the corrector alone as long as the extra corrector step is executed to force $m \geq 2$. In fact this rule continues to hold with, for example, $p^* = p - 2$ then the requirement is that $m \geq 3$.

When $p^* = p$ a special case arises with

$$C_{p+1}h^{p+1}y^{(p+1)}(x_n) = y(x_{n+k}) - y_{n+k}^{[m]} + O(h^{p+2}) \tag{7.4.50}$$

and

$$C_{p+1}^* h^{p+1}y^{(p+1)}(x_n) = y(x_{n+k}) - y_{n+k}^{[0]} + O(h^{p+2}) \tag{7.4.51}$$

which when subtracted gives

$$(C_{p+1}^* - C_{p+1})h^{p+1}y^{(p+1)}(x_n) = y_{n+k}^{[m]} - y_{n+k}^{[0]} + O(h^{p+2}) \tag{7.4.52}$$

and hence the principle local error is now

$$C_{p+1}h^{p+1}y^{(p+1)}(x_n) = \frac{C_{p+1}}{(C_{p+1}^* - C_{p+1})}(y_{n+k}^{[m]} - y_{n+k}^{[0]}) \tag{7.4.53}$$

This local error can be employed in two ways. Either the calculated values can be improved, or the step can be adjusted. Hence if the new estimates are indicated by $\hat{y}_{n+k}^{[m]}$ then for the predictor the new adjusted value is

$$\hat{y}_{n+k}^{[0]} = y_{n+k}^{[0]} + \frac{C_{p+1}^*}{(C_{p+1}^* - C_{p+1})}(y_{n+k-1}^{[m]} - y_{n+k-1}^{[0]}) \tag{7.4.54}$$

where y_{n+k-1} is used as y_{n+k} is not known at the predictor stage. The equivalent formula for the corrector is

$$\hat{y}_{n+k}^{[m]} = y_{n+k}^{[m]} + \frac{C_{p+1}}{(C_{p+1}^* - C_{p+1})}(y_{n+k}^{[m]} - y_{n+k}^{[0]}) \tag{7.4.55}$$

The final topic of this paragraph is to consider what happens to weak stability when methods are used in combination in predictor–corrector pairs. Clearly if the corrector is iterated to convergence then the weak stability characteristics are those of the corrector alone. However in the light of the previous discussion, iteration to convergence is a waste of effort, and in general, modes such as $PECE$ are commonly employed. The technique for obtaining the stability polynomial is illustrated by considering the specific case of $PECE$ mode. The predictor and corrector are from (7.4.34) and (7.4.35)

$$y_{n+k}^{[0]} + \sum_{j=0}^{k-1} \alpha_j^* y_{n+j}^{[1]} = h \sum_{j=0}^{k-1} \beta_j^* f_{n+j}^{[0]} \tag{7.4.56}$$

$$y_{n+k}^{[1]} + \sum_{j=0}^{k-1} \alpha_j y_{n+j}^{[1]} = h\beta_k f_{n+k}^{[0]} + h \sum_{j=0}^{k-1} \beta_j f_{n+j}^{[1]} \tag{7.4.57}$$

The theoretical solution $y(x)$ satisfies

$$\sum_{j=0}^{k} \alpha_j^* y(x_{n+j}) = h \sum_{j=0}^{k-1} \beta_j^* f(x_{n+j}, y(x_{n+j})) \tag{7.4.58}$$

$$\sum_{j=0}^{k} \alpha_j y(x_{n+j}) = h \sum_{j=0}^{k} \beta_j f(x_{n+j}, y(x_{n+j})) \tag{7.4.59}$$

and again defining $e_n^{[0]} = y(x_n) - y_n^{[0]}$ and $e_n^{[1]} = y(x_n) - y_n^{[1]}$, the previous argument for a single method can be used to give the linearized equations

$$e_{n+k}^{[0]} + \sum_{j=0}^{k-1} \alpha_j^* e_{n+j}^{[1]} = \bar{h} \sum_{j=0}^{k-1} \beta_j^* e_{n+j}^{[1]} \tag{7.4.60}$$

$$\sum_{j=0}^{k} \alpha_j e_{n+j}^{[1]} = \bar{h}\beta_k e_{n+k}^{[0]} + \bar{h} \sum_{j=0}^{k-1} \beta_j^* e_{n+j}^{[1]} \tag{7.4.61}$$

where again $\bar{h} = h\lambda$ and the method has been applied to $y' = \lambda y$. With any number of combinations of methods the predicted errors and indeed any intermediate corrector errors can be eliminated to give a single difference equation in terms of $e_{n+j}^{[m]}$. In this case $e_{n+k}^{[0]}$ is eliminated to leave the required linear difference equation

$$\sum_{j=0}^{k} \alpha_j e_{n+j}^{[1]} - \bar{h}\sum_{j=0}^{k-1} \beta_j e_{n+j}^{[1]} = -\bar{h}\beta_k \left[\sum_{j=0}^{k-1} \alpha_j^* e_{n+j}^{[1]} - \bar{h}\sum_{j=0}^{k-1} \beta_j^* e_{n+j}^{[1]}\right] \tag{7.4.62}$$

which by adding $-\bar{h}\beta_k e_{n+k}^{[1]}$ to both sides reduces to a stability polynomial

$$\Pi_{PECE}(r, \bar{h}) = \rho(r) - \bar{h}\sigma(r) + \bar{h}\beta_k[\rho^*(r) - \bar{h}\sigma^*(r)] = 0 \tag{7.4.63}$$

and any of the standard methods can now be applied to discover the intervals of absolute or relative stability.

Some useful examples of pairs of methods which can be operated in *PECE* mode include:

(i) Adams–Bashforth–Moulton fourth-order method

$$\text{P} \qquad y_{n+4} - y_{n+3} = (h/24)(55f_{n+3} - 59f_{n+2} + 37f_{n+1} - 9f_n) \tag{7.4.64}$$

$$\text{C} \qquad y_{n+4} - y_{n+3} = (h/24)(9f_{n+4} + 19f_{n+3} - 5f_{n+2} + f_{n+1}) \tag{7.4.65}$$

with

$$C_5 h^5 y^{(5)}(x_n) \sim -\tfrac{19}{720}(y_{n+4}^{[1]} - y_{n+4}^{[0]}) \tag{7.4.66}$$

(ii) Crane and Klopfenstein (1965)

$$\begin{aligned} \text{P} \qquad & y_{n+4} - 1.547\,652 y_{n+3} + 1.867\,503 y_{n+2} - 2.017\,204 y_{n+1} + 0.697\,353 y_n \\ & = h(2.002\,247 f_{n+3} - 2.031\,690 f_{n+2} + 1.818\,609 f_{n+1} - 0.714\,320 f_n) \end{aligned} \tag{7.4.67}$$

C equation (7.4.65)

with

$$C_5 h^5 y^{(5)}(x_n) \sim -(y_{n+4}^{[1]} - y_{n+4}^{[0]})/16.219\,66 \tag{7.4.68}$$

(iii) Klopfenstein and Millman (1968)

$$\begin{aligned} \text{P} \qquad & y_{n+4} + 0.29 y_{n+3} + 15.39 y_{n+2} - 12.13 y_{n+1} - 4.5 y_n \\ & = h(2.27 f_{n+3} + 6.65 f_{n+2} + 13.91 f_{n+1} + 0.69 f_n) \end{aligned} \tag{7.4.69}$$

C equation (7.4.65)

with

$$C_5 h^5 y^{(5)}(x_n) \sim -(y_{n+4}^{[1]} - y_{n+4}^{[0]})/18.0274 \tag{7.4.70}$$

Linear multistep methods will be discussed further in the light of new problems which arise when systems of equations are solved. In the mean time, the other large class of methods, Runge–Kutta methods will be treated with a view to stability considerations.

7.5 Runge–Kutta methods

Some simple Runge–Kutta methods have been introduced in the first section, together with the method of derivation which naturally yields the order of the method. Hence to find the order of a given method, requires the expansion of the method in the form of equation (7.1.44) to discover the number of terms which agree with the Taylor expansion. This is equivalent to the expansions in equations (7.3.6) and (7.3.7) for the linear multistep methods.

By their nature there is no equivalent of zero stability in the Runge–Kutta case, because as $h \to 0$ there is only one root to the auxiliary polynomial which is the principle root at $r = 1$. Weak stability does require some attention in this case. As with the linear multi-step work the difference equations are linearized using $y' = \lambda y$. For the three-stage method

$$\begin{aligned} k_1 &= \lambda y = f(x, y) \\ k_2 &= \lambda y(1 + a_{21}h\lambda) = f(x_n + hc_2, y_n + a_{21}k_1) \\ k_3 &= \lambda y(1 + a_{31}h\lambda + a_{32}h^2\lambda^2) \end{aligned} \tag{7.5.1}$$

with $\phi(x, y, h) = b_1k_1 + b_2k_2 + b_3k_3$ giving

$$\begin{aligned} y_{n+1} - y_n &= h\phi(x_n, y_n, h) \\ &= h\lambda[(b_1 + b_2 + b_3) + (b_2a_{21} + b_3a_{31})h\lambda + b_3a_{32}h^2\lambda^2]y_n \end{aligned} \tag{7.5.2}$$

the difference equation has only one auxiliary root at

$$r_1 = 1 + (b_1 + b_2 + b_3)\bar{h} + (b_2a_{21} + b_3a_{31})\bar{h}^2 + b_3a_{32}\bar{h}^3 \tag{7.5.3}$$

The method is absolutely stable if $|r_1| < 1$, and it is easy to find the condition on $\bar{h}$ for stability.

In the case of a p-stage method of order p, a special situation arises as

$$\begin{aligned} y_{n+1} - y_n &= h\phi(x_n, y_n, h) + O(h^{p+1}) \\ &= hf(x_n, y_n) + \frac{h^2}{2!}f^{(1)}(x_n, y_n) + \cdots \\ &\quad + \frac{h^p}{p!}f^{(p-1)}(x_n, y_n) + O(h^{p+1}) \end{aligned} \tag{7.5.4}$$

But $y' = \lambda y$ and hence

$$y_{n+1} - y_n = \left(h\lambda + \frac{h^2\lambda^2}{2!} + \cdots + \frac{h^p\lambda^p}{p!} \right) y_n + O(h^{p+1}) \tag{7.5.5}$$

to give the root of the auxiliary equation as

$$r_1 = 1 + \bar{h} + \frac{\bar{h}^2}{2!} + \cdots + \frac{\bar{h}^p}{p!} \tag{7.5.6}$$

Hence any p-stage p-order method has the same interval of absolute stability, and these can be listed for the first few stage numbers R:

R_1	$(-2.0, 0)$
R_2	$(-2.0, 0)$
R_3	$(-2.51, 0)$
R_4	$(-2.78, 0)$

Much of the difficulty with Runge–Kutta methods lies in the heavy algebra required to generate a new method especially for the higher orders. Runge–Kutta methods use R function evaluations per step against the one or two for a linear multistep method, whatever its order. In general therefore, Runge–Kutta methods are more expensive than the corresponding linear multistep method. The intervals of stability are also quite restricted as can be seen from the above list. This will be a particular problem when stiffness is encountered in the next section. There are some advantages however. The methods are self-starting and very easy to programme. There is also no problem in changing the step size as the solution progresses, unlike the corresponding linear multistep work where the earlier tabulated values will no longer be at the correct points if h is altered.

The Runge–Kutta methods considered so far are exemplified by equations (7,1.37) and (7.1.38), and are explicit formulae in that the set $\{k_j\}$ can be computed in the order $k = 1, \ldots, R$. A more general set of formula is the implicit formulae with the general form:

$$y_{n+1} = y_n + h\phi(x_n, y_n, h) \tag{7.5.7}$$

$$\phi(x, y, h) = \sum_{r=1}^{R} b_r k_r \tag{7.5.8}$$

$$k_r = f\left(x + hc_r,\ y + h\sum_{s=1}^{R} a_{rs}k_s \right) \qquad r = 1, \ldots, R \tag{7.5.9}$$

and often the row condition is also imposed in the form

$$c_r = \sum_{s=1}^{R} a_{rs} \tag{7.5.10}$$

Such a method is often represented compactly using the Butcher array

$$\begin{array}{c|cccc} c_1 & a_{11} & a_{12} & \cdots & a_{1R} \\ c_2 & a_{21} & a_{22} & \cdots & a_{2R} \\ \vdots & \vdots & & & \vdots \\ c_s & a_{s1} & a_{s2} & \cdots & a_{sR} \\ \hline & b_1 & b_2 & \cdots & b_s \end{array}$$

Now the k_r's need to be found implicitly as a set of non-linear equations but the extra degrees of freedom do indeed allow methods with extended ranges of absolute stability, albeit at the cost of a difficult and expensive algorithm to implement. The algebra needed to establish agreement with Taylor's expansion is also considerably complicated. By comparison with (7.1.39) and the following algebra, consider a two-stage method. Then ϕ expands as in (7.1.44) to give

$$\begin{aligned} \phi(x, y, h) = f &+ \tfrac{1}{2}hF + \tfrac{1}{6}h^2(Ff_y + G) \\ &+ \tfrac{1}{24}h^3[(3f_{xy} + 3ff_{yy} + f_y^2)F + Gf_y + H] + O(h^4) \end{aligned} \tag{7.5.11}$$

with

$$F = f_x + ff_y \tag{7.5.12}$$

$$G = f_{xx} + 2ff_{xy} + f^2 f_{yy} \tag{7.5.13}$$

$$H = f_{xxx} + 3ff_{xxy} + 3f^2 f_{xyy} + f^3 f_{yyy} \tag{7.5.14}$$

The equivalent to (7.1.42) is the expansion of k_r to give

$$\begin{aligned} k_r = f &+ h[c_r f_x + (a_{r1}k_1 + a_{r2}k_2)f_y] \\ &+ \tfrac{1}{2}h^2[c_r^2 f_{xx} + 2c_r(a_{r1}k_1 + a_{r2}k_2)f_{xy} + (a_{r1}k_1 + a_{r2}k_2)^2 f_{yy}] \\ &+ \tfrac{1}{6}h^3[c_r^3 f_{xxx} + 3c_r^2(a_{r1}k_1 + a_{r2}k_2)f_{xxy} + 3c_r(a_{r1}k_1 + a_{r2}k_2)^2 f_{xyy} \\ &+ (a_{r1}k_1 + a_{r2}k_2)^3 f_{yyy}] + O(h^4) \end{aligned} \tag{7.5.15}$$

but now successive substitution to get the expansion of each k_r is no longer possible as in moving from (7.1.41) to (7.1.42). Instead each k_r has to be expressed as an expansion in the form

$$k_r = A_r + hB_r + h^2C_r + h^3D_r + O(h^4) \tag{7.5.16}$$

which when substituted into (7.5.15), and after equating coefficients of like terms of powers of h, yields

$$A_r = f$$

$$B_r = c_r F$$

$$C_r = (a_{r1}c_1 + a_{r2}c_2)Ff_y + \tfrac{1}{2}c_r^2 G$$

$$\begin{aligned} D_r = {} & [a_{r1}(a_{11}c_1 + a_{12}c_2) + a_{r2}(a_{21}c_1 + a_{22}c_2)]Ff_y^2 \\ & + a_r(a_{r1}c_1 + a_{r2}c_2)F(f_{xy} + ff_{yy}) \\ & + \tfrac{1}{2}(a_{r1}c_1^2 + a_{r2}c_2^2)Gf_y + \tfrac{1}{6}c_r^3 H \quad r = 1, 2 \end{aligned} \tag{7.5.17}$$

and

$$\begin{aligned} \phi(x, y, h) = {} & b_1A_1 + b_2A_2 + h(b_1B_1 + b_2B_2) \\ & + h^2(b_1C_1 + b_2C_2) + h^3(b_1D_1 + b_2D_2) + O(h^4) \end{aligned} \tag{7.5.18}$$

These expansions then give the equations

$$b_1 + b_2 = 1 \tag{7.5.19}$$

for order one,

$$b_1c_1 + b_2c_2 = \tfrac{1}{2} \tag{7.5.20}$$

for order two,

$$b_1(a_{11}c_1 + a_{12}c_2) + b_2(a_{21}c_1 + a_{22}c_2) = \tfrac{1}{6} \tag{7.5.21}$$

$$b_1c_1^2 + b_2c_2^2 = \tfrac{1}{3} \tag{7.5.22}$$

for order three, and

$$(b_1a_{11} + b_2b_{21})(a_{11}c_1 + a_{12}c_2) + (b_1a_{12} + b_2a_{22})(a_{21}c_1 + a_{22}c_2) = \tfrac{1}{24} \tag{7.5.23}$$

$$b_1c_1(a_{11}c_1 + a_{12}c_2) + b_2c_2(a_{21}c_1 + a_{22}c_2) = \tfrac{1}{8} \tag{7.5.24}$$

$$b_1(a_{11}c_1^2 + a_{12}c_2^2) + b_2(a_{21}c_1^2 + a_{22}c_2^2) = \tfrac{1}{12} \tag{7.5.25}$$

$$b_1c_1^3 + b_2c_2^3 = \tfrac{1}{4} \tag{7.5.26}$$

for order four. There are eight equations for six unknowns, but the choice

$$b_1 = b_2 = \frac{1}{2}, \qquad c_1 = \frac{1}{2} \pm \frac{\sqrt{3}}{6}, \qquad c_2 = \frac{1}{2} \mp \frac{\sqrt{3}}{6} \tag{7.5.27}$$

will satisfy the complete set to give the implicit method

$$\begin{aligned} y_{n+1} &= y_n + \tfrac{1}{2}h(k_1 + k_2) \\ k_1 &= f\left(x_n + \left(\frac{1}{2} + \frac{\sqrt{3}}{6}\right)h,\ y_n + \frac{1}{4}hk_1 + \left(\frac{1}{4} + \frac{\sqrt{3}}{6}\right)hk_2\right) \\ k_2 &= f\left(x_n + \left(\frac{1}{2} - \frac{\sqrt{3}}{6}\right)h,\ y_n + \left(\frac{1}{4} - \frac{\sqrt{3}}{6}\right)hk_1 + \frac{1}{4}hk_2\right) \end{aligned} \tag{7.5.28}$$

Stability analysis applied to this method by applying the method to the equation $y' = \lambda y$ gives

$$\frac{y_{n+1}}{y_n} = \frac{1 + \bar{h}/2 + \bar{h}^2/12}{1 - \bar{h}/2 + \bar{h}^2/12} \tag{7.5.29}$$

which gives an interval of stability $(-\infty, 0)$.

It is clear that algebra of this complexity for even low-order formulae on scalar equations will become completely untenable for higher orders and for systems of equations. To this end Butcher, in a series of papers summarized in Butcher (1987), introduced a structured approach to the problem which will be discussed briefly here. The underlying idea is the analogy between the concept of rooted trees from graph theory and the higher derivatives which are required in Runge–Kutta expansions. Consider the case when the function f is such that

$$f : \mathcal{R}^m \to \mathcal{R}^m \qquad m > 1 \tag{7.5.30}$$

then the object is to write down the total derivatives as in (7.1.46) and (7.1.47). The partial derivatives can be represented in the notation typified by

$${}^i f_{jk} = \frac{\partial^2({}^i f)}{\partial({}^j y)\partial({}^k y)} \tag{7.5.31}$$

where ${}^i f$ is the ith component of f. Hence for the simple case of $m = 2$,

$$y^{(2)} = \begin{pmatrix} {}^1 f_1 & {}^1 f_2 \\ {}^2 f_1 & {}^2 f_2 \end{pmatrix} \tag{7.5.32}$$

which can be written as

$$y^{(2)} = \frac{\partial f}{\partial y} f \tag{7.5.33}$$

where $\partial f/\partial y$ is the Jacobian matrix. This is analogous with $y^{(2)} = f_y f$ in the scalar case. However $y^{(3)}$ does not have a matrix representation, and it is necessary to pursue Butcher's approach. In order to get a generalization of the scalar case, it is necessary to define the Nth Fréchet derivative of f which is an operator on the space $\mathcal{R}^m \times \cdots \times \mathcal{R}^m$, ($N$ times) which in the above notation for partial derivatives takes the value

$$f^{(N)}(y)(K_1, \ldots, K_N) = \sum_{i=1}^{m} \sum_{j_1=1}^{m}, \cdots \sum_{j_N=1}^{m} {}^i f_{j_1 j_2, \ldots j_m} \, {}^{j_1}K_1, \ldots, \; {}^{j_N}K_N \mathbf{e}_i \tag{7.5.34}$$

where

$$K_j = [{}^1K_j, \ldots, {}^mK_j]^T \in \mathcal{R}^m \qquad t = 1, \ldots, N \tag{7.5.35}$$

the K's representing the operands. The vector $\mathbf{e}_i$ is the usual vector with zero components except for a unit component in the ith element. With this notation, and using f as the operand instead of K, then for $m = 2$

$$y^{(2)} = f^{(1)}(y)(f) = \begin{pmatrix} {}^1f_1({}^1f) + {}^1f_2({}^2f) \\ {}^2f_1({}^1f) + {}^2f_2({}^2f) \end{pmatrix} \tag{7.5.36}$$

or in simple terms

$$y^{(2)} = f^{(1)}(f) \tag{7.5.37}$$

which is equivalent to (7.5.32). Moving up an order gives the result

$$y^{(3)} = f^{(2)}(f, f) + f^{(1)}(f^{(1)}(f)) \tag{7.5.38}$$

which compares with the scalar case

$$y^{(3)} = f_{yy}f^2 + f_y^2 f \tag{7.5.39}$$

In this way Butcher demonstrates an equivalence between the scalar case and the Fréchet derivatives for general m so that $y^{(p)}$ is a linear combination of Fréchet derivatives of orders up to $p - 1$, and the components of these sums are called elementary differentials which are to be compared with F and G in (7.5.11) and (7.5.12) in the scalar case. The concept of an elementary differential F_s which maps $\mathcal{R}^m$ to $\mathcal{R}^m$ of f is defined by first noting that the only differential of order one is f itself. Then the Fréchet derivative

$$f^{(M)}(F_1, F_2, \ldots, F_M) \tag{7.5.40}$$

is an elementary differential with an order

$$1 + \sum_{s=1}^{M} r_s \tag{7.5.41}$$

where F_s has order r_s. The notation is made even more compact by writing

$$f^{(1)}(f) = \{f\} \tag{7.5.42}$$

and

$$f^{(2)}(f, f) = \{f f\} \tag{7.5.43}$$

Repeated operators can also be represented compactly by

$$f^{(1)}(f^{(1)}(f)) = \{\{f\}\} = \{_2 f\}_2 \tag{7.5.44}$$

Note that the order of the operand is always the number of operands within the brackets. The position has been reached that the total derivatives can be written down very succinctly. Moreover the elementary differentials correspond with rooted trees in graph theory, and this link allows order conditions to be written down for general Runge–Kutta methods.

A small diversion into rooted trees is now needed to yield the required conditions. A rooted tree of order n is a set of n nodes linked by lines in such a way that there is only one lower node called the root and the branches never grow together again. Hence typical trees are

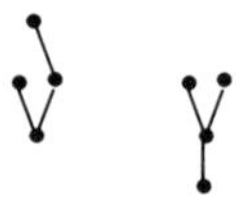 (7.5.45)

It turns out that the number of elementary differentials of a given order is precisely the number of possible rooted trees of the same order. There is in fact a one-to-one correspondence between the two. Both differentials and trees can be grafted together to generate new differentials or trees by the same rule. Two trees are grafted together by joining each of their roots down to a new root. By analogy with the notation using { and } for differentials, the notation [and] can be used for trees. Hence any trees appearing within the symbols [and] are grafted together according to the above rule. The first four orders with the complete set of possibilities are shown in Table 7.4. The tree consisting of a single node is denoted by τ.

The function $r(t)$ gives the order of the tree or number of nodes, $\sigma(t)$ is called the symmetry of the tree, and $\gamma(t)$ is the density of the tree and these are defined by

$$r([t_1^{n_1}t_2^{n_2}\cdots]) = 1 + n_1 r(t_1) + n_2 r(t_2) + \cdots \tag{7.5.46}$$

$$\sigma([t_1^{n_1}t_2^{n_2}\cdots]) = n_1!n_2!\cdots(\sigma(t_1))^{n_1}(\sigma(t_2))^{n_2}\cdots \tag{7.5.47}$$

$$\gamma([t_1^{n_1}t_2^{n_2}\cdots]) = r([t_1^{n_1}t_2^{n_2}\cdots])(\gamma(t_1))^{n_1}(\gamma(t_2))^{n_2}\cdots \tag{7.5.48}$$

Table 7.4

Tree	Name	$r(t)$	$\sigma(t)$	$\gamma(t)$	$\alpha(t)$
•	τ	1	1	1	1
	$[\tau]$	2	1	2	1
	$[\tau^2]$	3	2	3	1
	$[[\tau]]$	3	1	6	1
	$[\tau^3]$	4	6	4	1
	$[\tau[\tau]]$	4	1	8	3
	$[[\tau^2]]$	4	12	12	1
	$[[[\tau]]]$	4	1	24	1

and $r(\tau) = \sigma(\tau) = \gamma(\tau) = 1$. The function $\alpha(t)$ is the number of ways of labelling a tree excluding permutations at the same level from the root to give

$$\alpha(t) = \frac{r(t)!}{\sigma(t)\gamma(t)} \tag{7.5.49}$$

Butcher proved that the total derivative $y^{(q)}$ is a linear combination of elementary differentials of order q with the form

$$y^{(q)} = \sum_{r(t)=q} \alpha(t)F(t) \tag{7.5.50}$$

with

$$F([t_1 t_2, \ldots, t_M]) = \{F(t_1)F(t_2), \ldots, F(t_M)\} \tag{7.5.51}$$

and $F(\tau) = f$. In this form (7.5.38) becomes

$$y^{(3)} = \{f^2\} + \{_2 f\}_2 \tag{7.5.52}$$

and

$$y^{(4)} = \{f^3\} + 3\{f\{f\}_2 + \{_2 f^2\}_2 + \{_3 f\}_3 \tag{7.5.53}$$

Returning to the derivation of a Runge–Kutta method, the object is to force agreement between the expansion

$$y(x_{n+1}) = y(x_n) + hy'(x_n) + \tfrac{1}{2}h^2 y''(x_n) + \cdots \tag{7.5.54}$$

and

$$y_{n+1} = y_n(0) + h\frac{\mathrm{d}}{\mathrm{d}h}y_n(h)\bigg|_{h=0} + \frac{1}{2}h^2\frac{\mathrm{d}^2}{\mathrm{d}h^2}y_n(h)\bigg|_{h=0} + \cdots \tag{7.5.55}$$

In order to formalize the link between the rooted trees and the notation using [and], with the elementary differentials and the notation { and } define the functions ψ_i in terms of the elements of the Butcher array by

$$\psi_i(\tau) = \sum_{j=1}^{s} a_{ij} \tag{7.5.56}$$

and

$$\psi_i([t_1 \cdots t_N]) = \sum_{j=1}^{s} a_{ij}\psi_j(t_1)\cdots\psi_j(t_N) \tag{7.5.57}$$

for $i = 1, 2, \ldots, s+1$ and define

$$\psi(t) = \psi_{s+1}(t) \tag{7.5.58}$$

Table 7.5

Tree	Name	$\psi_i(t)$ $i = 1, \ldots, s$	$\psi(t)$
	τ	$\sum_j a_{ij} (= c_i)$	$\sum_i b_i$
	$[\tau]$	$\sum_j a_{ij} c_j$	$\sum_i b_i c_i$
	$[\tau^2]$	$\sum_j a_{ij} c_j^2$	$\sum_i b_i c_i^2$
	$[[\tau]]$	$\sum_j a_{ij} \sum_k a_{jk} c_k$	$\sum_{ij} b_i a_{ij} c_j$
	$[\tau^3]$	$\sum_j a_{ij} c_j^2$	$\sum_i b_i c_i^3$
	$[\tau[\tau]]$	$\sum_j a_{ij} c_j \sum_k a_{jk} c_k$	$\sum_{ij} b_i c_i a_{ij} c_j$
	$[[\tau^2]]$	$\sum_j a_{ij} \sum_k a_{jk} c_k^2$	$\sum_{ij} b_i c_{ij} c_j^2$
	$[[[\tau]]]$	$\sum_j a_{ij} \sum_k a_{jk} \sum_n a_{kn} c_n$	$\sum_{ijk} b_i a_{ij} a_{jk} c_k$

Here b_i has been set as $a_{s+1,i}$. Table 7.5 now lists the possibilities for the functions $\psi_i(t)$ and $\psi(t)$ for trees to order four.

The elementary differentials are now given by

$$\frac{d^q y(h)}{dh^q} = \sum_{r(t)=q} \alpha(t)\gamma(t)\psi(t)F(t) \tag{7.5.59}$$

with

$$F(\tau) = f \tag{7.5.60}$$

and

$$F([t_1, \ldots, t_N]) = \{F(t_1), \ldots, F(t_N)\} \tag{7.5.61}$$

the function F providing the formal link between the rooted trees and the elementary differentials.

With this background in place a Runge–Kutta method can be said to have order p if $\psi(t) = 1/\gamma(t)$ holds for all trees of order $r(t) \leq p$ and not for some tree of order $p + 1$. Using Table 7.5 it becomes easy to establish the order of a method given its Butcher array, or conversely to use the conditions as defining equations to find the Butcher array elements and derive a method.

The stability characteristics of a general Runge–Kutta method also need to be considered. To obtain linear difference equations the method is applied to the test equations

$$\mathbf{y}' = \mathbf{C}\mathbf{y} \tag{7.5.62}$$

to give

$$y_{n+1} = y_n + h \sum_{i=1}^{s} b_i k_i \tag{7.5.63}$$

with

$$k_i = \mathbf{C}\left[y_n + h \sum_{j=1}^{s} a_{ij} k_j\right] \tag{7.5.64}$$

for $i = 1, 2, \ldots, s$. If $\mathbf{D}$ has distinct eigenvalues λ_i, the diagonalization theorem implies the existence of a matrix $\mathbf{H}$ such that

$$\mathbf{H}^{-1}\mathbf{C}\mathbf{H} = \mathbf{D} = \text{diag}(\lambda_1, \lambda_2, \ldots, \lambda_m) \tag{7.5.65}$$

Transforming y_n and k_i gives

$$y_n = \mathbf{C}z_n \quad \text{and} \quad k_i = \mathbf{C}l_i \tag{7.5.66}$$

with the resulting equations

$$z_{n+1} = z_n + h \sum_{i=1}^{s} b_i l_i \tag{7.5.67}$$

and

$$l_i = \mathbf{D}\left[z_n + h \sum_{j=1}^{s} a_{ij} l_j\right] \tag{7.5.68}$$

for $i = 1, 2, \ldots, s$. Hence the transformation uncouples the difference equation and justifies the use of the scalar test problem $y' = \lambda y$. The form of the result is then immediate as being

$$y_{n+1} = R(\hat{h}) y_n \tag{7.5.69}$$

the function $R(\hat{h})$ being called the stability function, and a method is absolutely stable if $R(\hat{h}) < 1$.

The general Runge–Kutta equation can be reformulated from (7.5.7), (7.5.8) and (7.5.9) as

$$Y_i = y_n + \hat{h} \sum_{j=1}^{s} a_{ij} Y_j \qquad i = 1, 2, \ldots, s$$

$$y_{n+1} = y_n + \hat{h} \sum_{i=1}^{s} b_i Y_i \tag{7.5.70}$$

which can be concisely written as

$$Y = y_n e + \hat{h}\mathbf{A}Y \tag{7.5.71}$$

$$y_{n+1} = y_n + \hat{h} b^T Y \tag{7.5.72}$$

by letting $Y = (Y_1, Y_2, \ldots, Y_s)^T$ and $e = (1, 1, \ldots, 1)^T$. Hence

$$R(\hat{h}) = 1 + \hat{h} b^T (I - \hat{h}\mathbf{A})^{-1} e \tag{7.5.73}$$

yields the stability function. There is also a determinantal form of the stability function by Dekker and Verwer (1984) which is obtained by using Cramer's rule and yields

$$R(\hat{h}) = \frac{\det[I - \hat{h}\mathbf{A} + \hat{h} e b^T]}{\det[I - \hat{h}\mathbf{A}]} \tag{7.5.74}$$

For explicit methods the numerator in (7.5.74) is unity and hence the stability function is a polynomial in $\hat{h}$. The implicit methods yield a rational function, and it is this form that allows $R(\hat{h})$ to remain less than unity as $\hat{h} \to \infty$ as in (7.5.29).

This development of Runge–Kutta methods opens the way to consider stiff systems and how they may be solved. Firstly, the earlier work on linear multistep methods will also be extended to sets of equations.

7.6 Systems of equations and stiffness

In the discussion so far on differential equations, the linear multistep analysis has been restricted to the solution of a single equation of the form $y' = f(x, y)$, though in general, solutions are required for systems of equations of the type given in (7.1.4). The question then arises as to how much of the work already covered will carry over to the many equation situation.

The basic problem is now

$$\mathbf{y}' = \mathbf{f}(x, \mathbf{y}) \tag{7.6.1}$$

with boundary condition

$$\mathbf{y}(x_0) = \mathbf{y}_0 \tag{7.6.2}$$

where $\mathbf{y}$ is the vector $(y_1, \ldots, y_m)$ and $\mathbf{f}$ is the vector of functions $(f_1, \ldots, f_m)$. The linear difference operator can be defined in a manner analogous to (7.3.6) to give

$$\mathcal{L}[\mathbf{y}(x), h] = \sum_{j=0}^{k} [\alpha_j \mathbf{y}(x + jh) - h\beta_j \mathbf{y}'(x + jh)] \tag{7.6.3}$$

and so zero stability and consistency will still imply convergence.

Weak stability analysis requires some re-thought. For the method

$$\sum_{j=0}^{k} \alpha_j \mathbf{y}_{n+j} = h \sum_{j=0}^{k} \beta_j \mathbf{f}_{n+j} \tag{7.6.4}$$

the error $\tilde{e}_n$ satisfies a vector equation of the form of (7.4.4), namely

$$\sum_{j=0}^{k} \alpha_j \tilde{\mathbf{e}}_{n+j} = h \sum_{j=0}^{k} \beta_j [\mathbf{f}(x_{n+j}, \mathbf{y}(x_{n+j})) - \mathbf{f}(x_{n+j}, \tilde{\mathbf{y}}_{n+j})] + \boldsymbol{\phi}_{n+k} \tag{7.6.5}$$

and now the mean value theorem for several variables is used to give

$$\sum_{j=0}^{k} \alpha_j \tilde{\mathbf{e}}_{n+j} = h \sum_{j=0}^{k} \beta_j \frac{\partial \mathbf{f}}{\partial \mathbf{y}} \tilde{\mathbf{e}}_{n+j} + \boldsymbol{\phi}_{n+k} \tag{7.6.6}$$

where the notation $\partial \mathbf{f}/\partial \mathbf{y}$ is the $m \times m$ Jacobian matrix

$$\mathbf{J} = \left\{ \frac{\partial f_i}{\partial x_j} \right\} \tag{7.6.7}$$

The weak stability assumption is to make the matrix $\mathbf{J}$ a constant, and to make $\boldsymbol{\phi}_{n+k}$ equal to the constant vector $\boldsymbol{\phi}$. Hence the system of linear difference equations is

$$\sum_{j=0}^{k} (\alpha_j \mathbf{I} - h\beta_j \mathbf{J}) \tilde{\mathbf{e}}_{n+j} = \boldsymbol{\phi} \tag{7.6.8}$$

with $\mathbf{I}$ the unit matrix. These difference equations can be uncoupled by using the diagonalization theorem. Hence for distinct eigenvalues λ_i of $\mathbf{J}$ there exists a non-singular matrix $\mathbf{H}$ such that

$$\mathbf{H}^{-1}\mathbf{J}\mathbf{H} = \boldsymbol{\Lambda} = \begin{bmatrix} \lambda_1 & & \\ & \ddots & \\ & & \lambda_m \end{bmatrix} \tag{7.6.9}$$

Hence pre-multiplying (7.6.8) by $\mathbf{H}^{-1}$ and letting $\tilde{\mathbf{e}}_n = \mathbf{H}\mathbf{d}_n$ gives

$$\sum_{j=0}^{k} (\alpha_j \mathbf{I} - h\beta_j \boldsymbol{\Lambda}) \mathbf{d}_{n+j} = \psi \tag{7.6.10}$$

with $\psi = \mathbf{H}^{-1}\phi$ which uncouples into

$$\sum_{j=0}^{k} (\alpha_j - h\beta_j \lambda_i) d_{i,n+j} = \psi_i \qquad i = 1, \ldots, m \tag{7.6.11}$$

which is now a set of uncoupled difference equations of the form of (7.4.7) but now each eigenvalue of $\mathbf{J}$ replaces the earlier λ. Immediately a new problem arises, as the stability criterion on $\bar{h}$ must now apply simultaneously to all the set of eigenvalues. Hence if one eigenvalue is very large and another very small, there may be no h which allows both $h\lambda$'s to lie in the interval of stability together. This is the problem of stiffness. It looks as though the only solution to this dilemma is to have a stability interval which stretches across the whole of the left-hand half-plane, and few of the methods so far discussed have other than quite limited finite stability regions.

A classical example used by Lambert(1973) is

$$\mathbf{y}' = \mathbf{A}\mathbf{y} \tag{7.6.12}$$

with

$$\mathbf{y}(0) = (1, 0, -1)^T \tag{7.6.13}$$

and

$$\mathbf{A} = \begin{bmatrix} -21 & 19 & -20 \\ 19 & -21 & 20 \\ 40 & -40 & -40 \end{bmatrix} \tag{7.6.14}$$

which has the analytic solution

$$\mathbf{y}(x) = \begin{bmatrix} \frac{1}{2}e^{-2x} + \frac{1}{2}e^{-40x}(\cos 40x + \sin 40x) \\ \frac{1}{2}e^{-2x} - \frac{1}{2}e^{-40x}(\cos 40x + \sin 40x) \\ -e^{-40x}(\cos 40x - \sin 40x) \end{bmatrix} \tag{7.6.15}$$

A direct application of Euler's method to these equations with $h = 0.04$ simply gives wildly inaccurate results, and indeed the largest eigenvalue is 40 and the smallest is 2. Hence the combinations $h\lambda$ fall outside the interval of stability. This can be looked at in two ways, either, as has been done, from the point of view of keeping each $h\lambda$ inside the interval of stability simultaneously, or from a more physical standpoint that there is a transient solution which damps out very quickly with increasing x and also a slower decaying solution. The numerical process cannot cope with the fast decaying transient, and so fails to 'pick up' the slower decaying long-term trend. With an interval of stability which is say (a, b) with $a, b < 0$ and $a > b$ then the step sizes to keep $h\lambda$ inside b for the largest eigenvalue will push $\bar{h}$ outside a for the smallest eigenvalue and so there is no possible h for stability. Even if, as is often the case, $a = 0$, then so small an h is required to get into the stability range that the method becomes impractical.

A number of definitions are employed to try to encompass these concepts. One is to define the stiffness ratio. A system

$$\mathbf{y}' = \mathbf{A}\mathbf{y} + \phi(x) \tag{7.6.16}$$

is said to be stiff if

(i) $Re\,\lambda_t < 0$
(ii) $\max_t |Re\,\lambda_t| \gg \min_t |Re\,\lambda_t|$

where the λ_t are the eigenvalues of **A**. The ratio

$$\frac{\max_t |Re\,\lambda_t|}{\min_t |Re\,\lambda_t|} \tag{7.6.17}$$

is called the stiffness ratio. For the example above the stiffness ratio has the value 20, and such a value causes the stability problems encountered.

To overcome this difficulty the region of absolute stability needs to be the whole left-hand half plane. Such a method is called A-stable. However, Dahlquist(1963) showed that

(i) an explicit method cannot be A-stable
(ii) the order of an implicit linear multistep method is ≤ 2
(iii) the second-order method with the smallest error constant is the trapezoidal rule.

These results make the earlier work on high-order linear multistep methods seem very limited in the stiff situation. Either a wider class of methods needs to be considered which will exhibit the required characteristics or the definition of A-stability needs weakening to include more existing methods without sacrificing their ability to solve stiff systems. One such definition, Widlund (1967), is that a method is A(α)-stable for $\alpha \in (0, \pi/2)$ if its region of stability contains the infinite wedge

$$W_\alpha = \{h\lambda \mid -\alpha < \pi - \arg h\lambda < \alpha\} \tag{7.6.18}$$

A method is $A(0)$-stable if it is $A(\alpha)$-stable as $\alpha \to 0$. Even with this restricted definition it can be shown that

(i) no explicit method can be $A(0)$-stable.
(ii) there is only one $A(0)$-stable method with k steps and order exceeding $k+1$, namely the trapezoidal rule
(iii) $\exists\ A(\alpha)$-stable methods $\forall\ \alpha \to [0, \pi/2)$ for order $p = k = 3$ and $k = p = 4$.

Unfortunately, an A-stable method is not necessarily going to solve all the problems associated with a stiff system. A further difficulty is illustrated by considering a simple example. Consider

$$y' = \lambda y \qquad Re\,\lambda < 0 \tag{7.6.19}$$

for which the trapezoidal rule has the form

$$y_{n+1} = y_n \left(\frac{1 + \frac{h\lambda}{2}}{1 - \frac{h\lambda}{2}} \right) \tag{7.6.20}$$

and hence

$$\left| \frac{y_{n+1}}{y_n} \right| = \left| \frac{1 + \frac{1}{2}h\lambda}{1 - \frac{1}{2}h\lambda} \right| = \left[\frac{1 + h\,Re\,\lambda + \frac{1}{4}h^2|\lambda|^2}{1 - h\,Re\,\lambda + \frac{1}{4}h^2|\lambda|^2} \right]^{1/2} \tag{7.6.21}$$

Suppose now that $|Re\,\lambda| \gg 0$, then if h is not excessively small then

$$\left|\frac{y_{n+1}}{y_n}\right| \to +1 \tag{7.6.22}$$

and so the method converges *very slowly*. However the method is A-stable. This example encourages yet a further definition.

A method is L-stable (Ehle (1969)) if it is A-stable and if it is applied to $y' = \lambda y$ with $Re\,\lambda < 0$ then

$$y_{n+1} = R(h\lambda)y_n \tag{7.6.23}$$

where $|R(h\lambda)| \to 0$ as $Re\,(h\lambda) \to -\infty$. Clearly L-stability $\Rightarrow$ A-stability $\Rightarrow$ $A(0)$-stability.

It appears from the above investigations that the linear multistep methods considered in this chapter are going to prove unsatisfactory for stiff systems. Implicit methods seem to be a necessity, but the usual direct iterative solutions used for these methods converge because the relevant Lipschitz constant M has a modulus less than unity. The smallness of h then gives the convergence. However for a stiff system at least one of the relevant M's will be large. Hence it becomes necessary to employ a full Newton's method for the solution of the implicit equation as a non-linear equation. Again the computational load is increased.

A broader range of methods would seem to be required.

In Section 7.5 on Runge–Kutta methods, the implicit methods such as (7.5.28) have been shown to have intervals of absolute stability of $(-\infty, 0)$ which will be suitable for stiff systems. Like all methods for stiff systems the computational load is high in that the values of k_1 and k_2 have to be extracted from the defining formulae as solutions of non-linear equations. Nevertheless implicit Runge–Kutta formulae are commonly used methods for dealing with stiff systems. There is a half-way house in methods classified as semi-explicit such as:

$$\begin{aligned} y_{n+1} &= y_n + \tfrac{1}{6}h(k_1 + 4k_2 + k_3) \\ k_1 &= f(x_n, y_n) \\ k_2 &= f\left(x_n + \tfrac{1}{2}h, y_n + \tfrac{1}{4}hk_1 + \tfrac{1}{4}hk_2\right) \\ k_3 &= f(x_n + h, y_n + hk_2) \end{aligned} \tag{7.6.24}$$

In this method, the implicit part of the definition of the k's is restricted to just one equation which can be handled with a one-variable non-linear equation solver rather than (7.5.28) where a full set of non-linear equations requires solution.

Another alternative is to combine linear multistep methods with Runge–Kutta methods to get hybrid methods. The most commonly successful of these has the form

$$\sum_{j=0}^{k} \alpha_j y_{n+j} = h\sum_{j=0}^{k} \beta_j f_{n+j} + h\beta_\nu f_{n+\nu} \tag{7.6.25}$$

and $\nu \notin \{0, 1, \ldots, k\}$ which adds further degrees of freedom to the conventional linear multistep form. The definitions of order and zero stability still follow through from the work of Section 7.3.

In particular the linear difference operator is now

$$\mathcal{L}[y(x), h] = \sum_{j=0}^{k} [\alpha_j y(x + jh) - h\beta_j y'(x + jh)] - h\beta_\nu y'(x + \nu h)$$

$$= C_0 y(x) + C_1 y(1)(x) + \cdots + C_q h^q y^{(q)}(x) + \cdots \tag{7.6.26}$$

and a method has order p if $C_0 = C_1 = \cdots = C_p = 0$, and $C_{p+1} \neq 0$. Similar expressions for the constants C_i hold to those in (7.3.8) to (7.3.10) namely:

$$C_0 = \alpha_0 + \alpha_1 + \cdots + \alpha_k \tag{7.6.27}$$

$$C_1 = \alpha_1 + 2\alpha_2 + \cdots + k\alpha_k - (\beta_0 + \beta_1 + \cdots + \beta_k + \beta_\nu) \tag{7.6.28}$$

$$\vdots$$

$$C_q = \frac{1}{q!}(\alpha_1 + 2^q\alpha_2 + \cdots + k^q\alpha_k)$$

$$- \frac{1}{(q-1)!}(\beta_1 + 2^{q-1}\beta_2 + \cdots + k^{q-1}\beta_k + \nu^{q-1}\beta_\nu) \tag{7.6.29}$$

Gragg and Stetter (1964) show that the introduction of the parameters ν and β_ν can be used to increase the order two above the corresponding linear multistep method. A typical two-step method which is zero stable and has order five has

$$\begin{aligned}
\alpha_0 &= \alpha \qquad \alpha_1 = -(1 + \alpha) \\
\beta_0 &= -\tfrac{1}{12}(5\alpha + 1) + \tfrac{1}{8}(1 + \alpha)/(s + 1) \\
\beta_1 &= \tfrac{2}{3}(1 - \alpha) - \tfrac{1}{4}(1 + \alpha)/s \\
\beta_2 &= \tfrac{1}{12}(5 + \alpha) + \tfrac{1}{8}(1 + \alpha)/(s - 1) \\
\beta_\nu &= -\tfrac{1}{4}(1 + \alpha)/s(s^2 - 1) \\
s &= \frac{8}{15}\frac{(1 - \alpha)}{(1 + \alpha)}
\end{aligned} \tag{7.6.30}$$

for the range of α such that $-1 \leq \alpha < 1$. The corresponding error constant has the value

$$C_6 = \frac{1}{2.6!}(1 + \alpha)(3s^2 - 1) \tag{7.6.31}$$

so that as long as α is chosen to keep the zero stability condition satisfied, it is possible to squeeze out a further order. The choice $s = 1/\sqrt{3}$ gives such a sixth order, two-stage zero stable method.

From a practical point of view there is nothing gained by analyzing the weak stability characteristics of these methods in isolation as their implementation requires a number of associated formulae. Consider one of the Butcher–Gear formulae (Butcher, 1965; Gear, 1965),

$$y_{n+2} - \frac{1}{31}(32y_{n+1} - y_n) = \frac{h}{93}(15f_{n+2} + 12f_{n+1} - f_n + 64f_{n+3/2}) \quad (7.6.32)$$

which is of order five and zero stable. Both y_{n+2} and $y_{n+\nu}$ need to be predicted. A conventional predictor copes with y_{n+2} but a special hybrid predictor is required for $y_{n+\nu}$. A suitable formula would be

$$y_{n+3/2} - y_n = \tfrac{1}{8}h(9f_{n+1} + 3f_n) \quad (7.6.33)$$

The notation of P for a predictor and C for a corrector can be extended to deal with hybrid combinations by using P_ν for a hybrid predictor, P_H for an explicit hybrid rule, C_H for a hybrid corrector and as before E for a function evaluation. Hence modes such as $P_\nu EP_HE$ and $P_\nu EP_kEC_HE$ become the combinations that are employed. Weak stability can still be analysed by applying the combination to $y' = \lambda y$ and substituting out the two predicted y values to get a difference equation whose auxiliary polynomial gives the stability criteria as in $PECE$ mode. In general these methods exhibit the required stability range to cope with stiff systems and like all other methods for these problems they involve a much higher computational load for their implementation.

Another group of methods finding current favour for stiff systems are those based on linear-multistep methods. Two obvious groups of methods appear available. The first group have the form

$$\mathbf{y}_{n+1} - \mathbf{y}_n = h[(1-\theta)\mathbf{f}_{n+1} + \theta\mathbf{f}_n] \quad (7.6.34)$$

and these methods are A-stable for $\theta \leq 1/2$ and are termed the theta method. Low order is clearly the main problem with this approach, but on the other hand simplicity over say an implicit Runge–Kutta method will allow large numbers of steps to be performed relatively economically.

A second group of methods are those with the form

$$\sum_{j=0}^{k} \alpha_j \mathbf{y}_{n+j} = h\beta_k \mathbf{f}_{n+k} \quad (7.6.35)$$

These methods are called backward differentiation methods. These methods are $A(\alpha)$ stable, and may be combined with a predictor in an Adams–Moulton formula, but with the corrector replaced by (7.6.35). Some typical formulae are

$$y_{n+2} - \tfrac{4}{3}y_{n+1} + \tfrac{1}{3}y_n = \tfrac{2}{3}hf_{n+2} \quad (7.6.36)$$

$$y_{n+3} - \frac{18}{11}y_{n+2} + \frac{9}{11}y_{n+1} - \frac{2}{11}y_n = \frac{6h}{11}f_{n+3} \quad (7.6.37)$$

$$y_{n+4} - \frac{48}{25}y_{n+3} + \frac{36}{25}y_{n+2} - \frac{16}{25}y_{n+1} + \frac{3}{25}y_n = \frac{12h}{25}f_{n+4} \quad (7.6.38)$$

each of which has a region of stability which includes all the left-hand half plane, except for (7.6.38) which has just a small region of this plane omitted.

The numerical solution of ordinary differential equations presents the user with an extensive spectrum of methods. A glance at a computer library such as NAG will show that there is a flow-chart to enable a choice of method to be made. Almost certainly a user needs to establish whether the equations being solved are stiff. For a typical engineering problem where the right-hand side functions can be very obtuse, it is not usually easy to analyse the stiffness criteria. The defining functions may themselves be the result of considerable computation and possibly even arise from other numerical methods.

If a non-stiff approach appears to give numerical instability in practice then converting to a stiff method would seem appropriate. In either case there is a choice between a linear multistep method and a Runge–Kutta method. The pros and cons are considerable. A Runge–Kutta method in its explicit form is easy to code, can have high order, allows easy step adjustment, but requires many function evaluations per step. A linear multistep method requires storing and updating previous y values, and allows a less easy step adjustment strategy, but requires few function evaluations per step (only two in PECE mode per independent variable). In the stiff case, both methods are more difficult to implement as the Runge–Kutta method will require the solution of non-linear equations with the added function evaluations, and the corrector formulae will also need to be so treated in the linear multistep methods. The Hybrid methods face similar difficulties as well as requiring special predictors.

MAPLE contains routines for the solution of ordinary differential equations with options to allow series solutions, analytic solutions and numerical solutions. A Runge–Kutta method with step-size adjustment is used in the numerical case.

Further reading

Lambert J. D. (1991) *Numerical Methods for Ordinary Differential Systems*, John Wiley and Sons, Chichester.

Butcher J. C. (1987) *The Numerical Analysis of Ordinary Differential Equations: Runge–Kutta and General Linear Methods*, John Wiley and Sons, Chichester.

Gear C. W. (1971) *Numerical Initial Value Problems in Ordinary Differential Equations*, Prentice-Hall, Englewood Cliffs, New Jersey.

Exercises

(7.1) Use Euler's method to tabulate y against x for values of x ranging from 0 to 1 in steps of 0.1 for the differential equation

$$\frac{dy}{dx} = x^2 + y^2$$

with boundary conditions $y = 0$ when $x = 0$. Compare the resulting table of values with that obtained for a step-size of 0.05.

(7.2) Using the initial solution $y^{[0]}(x) = 0$, use Picard's method to obtain an approximate analytic solution to the differential equation in exercise (7.1). Find approximations up to and including x^{11}, and compare the solution with the table from Euler's method.

(7.3) Solve the differential equation

$$\frac{dy}{dx} = 0.1(x^3 + y^2)$$

with $y(0) = 1$ using a Taylor series method up to powers of x^6.

(7.4) Use the standard fourth-order Runge–Kutta method for the solution of exercise (7.1), and compare the workload and accuracy with that obtained by Euler's method.

(7.5) Compare the use of Modified Euler's predictor–corrector method with a Taylor series and with the standard Runge–Kutta method on the problem

$$\frac{dy}{dx} = y \sin x$$

with $y(0) = 1/e$, and use the analytic result as a standard. Integrate out to $x = \pi$.

(7.6) Apply the three methods of exercise (7.5) to the problem

$$\frac{dy}{dt} = x$$

$$\frac{dx}{dt} = 2 - 4y - x \sin t$$

with $y = x = 0$ at $t = 0$.

(7.7) Solve the examples in exercises (7.1), (7.3), (7.5), and (7.6) using a predictor–corrector method such as Adams–Bashforth, and again compare the workload with the previous methods. In particular compare the fourth-order Runge–Kutta approach with a corresponding fourth-order predictor–corrector. A list of formulae is given in equations (7.4.64) and following.

(7.8) Using the standard Runge–Kutta method solve the equation for the non-linear motion of a simple pendulum given by

$$\frac{d^2\theta}{dt^2} = -\frac{g}{l} \sin \theta$$

with $d\theta/dt = 0$ and $\theta = \pi/4$ at $t = 0$. Take the physical constant $g/l = 30.0$. Find the period of oscillation and compare this with the small-amplitude result which can be found analytically.

(7.9) The equations of a projectile with air resistance proportional to the square of the velocity are

$$\frac{d^2x}{dt^2} = -k\left(\left(\frac{dx}{dt}\right)^2 + \left(\frac{dy}{dt}\right)^2\right)$$

$$\frac{d^2y}{dt^2} = -g - k\left(\left(\frac{dx}{dt}\right)^2 + \left(\frac{dy}{dt}\right)^2\right)$$

with $g = 9.81$ m/s/s, $k = 0.01$, x is the horizontal coordinate and y the vertical coordinate. Given an initial velocity of 20 m/s and a projection angle of 50°, find the range. For a given g and k, find the angle for a maximum range.

(7.10) Use a Taylor series method and Euler's method to solve the linear boundary value problems

$$\text{(i)}\quad \frac{d^2y}{dt^2} + y = t^3 \qquad y(0) = 0 \quad y(\pi/2) = 1$$

$$\text{(ii)}\quad \frac{d^2y}{dt^2} + t^2y = t \qquad y(0) = 0 \quad y(1) = 1$$

(7.11) Use the standard Runge–Kutta method to solve the boundary value problem

$$\frac{d^2y}{dt^2} + y\sin(t) = \tan(t)$$

with $y(0) = 0$ and $y(1) = 1.106$ as a pair of associated initial value problems.

(7.12) Solve the following non-linear boundary value problems:

$$\text{(i)}\quad \frac{d^2y}{dt^2} + y^2\frac{dy}{dt} + y = \sin t \qquad y(0) = 0,\ \ y(1) = 0.5584$$

$$\text{(ii)}\quad \frac{d^2y}{dt^2} + \sin y\frac{dy}{dt} + y^2 = t \qquad y(0) = 0,\ \ y(1) = 0.389\,67$$

$$\text{(iii)}\quad \frac{d^2y}{dt^2} + \left(\frac{dy}{dt}\right)^2 + y\sin t = t^3 \qquad y(0) = 0,\ \ y(1) = 0.562\,83$$

$$\text{(iv)}\quad \frac{d^3y}{dt^3} + y^2\frac{d^2y}{dt^2} = y\sin t \qquad y(0) = 0,\ \ y'(0) = 2,\ \ y(1) = 2.2736$$

(7.13) Convert the eigenvalue problem

$$y'' + 2y' + ky = 0$$

with $y(0) = y(1) = 0$ to an algebraic eigenvalue problem using finite differences, and find the first three eigenvalues. Take a step size of $h = 0.1$ initially and then compare the result with that obtained with $h = 0.05$.

(7.14) Repeat the procedure of exercise (7.13) for the eigenvalue problems:

$$\text{(i)}\quad y'' - k^2 y = 0$$

$$\text{(ii)}\quad x^2 y'' + xy' + (k^2x^2 - 1)y = 0$$

each with the boundary conditions $y(0) = y(1) = 0$.

(7.15) Show that the order of the multistep method

$$y_{n+2} + (b-1)y_{n+1} - by_n = \tfrac{1}{4}h[(b+3)f_{n+2} + (3b+1)f_n]$$

is 2 if $b \neq -1$ and 3 if $b = -1$.

(7.16) Use the linear difference operator to derive a set of implicit two-step methods with $\alpha_0 = a$, a being a free parameter. That is, take $k = 2$, $\alpha_2 = 1$, and set $C_0 = C_1 = C_2 = C_3 = 0$ to obtain equations for α_1, β_0, β_1 and β_2 with the standard notation.

Consider the cases $a \neq 1$, $a = -1$, $a = 0$ and $a = -5$ from the point of view of recognizing standard methods.

(7.17) Find the order and error constant for Quadé's method defined by

$$y_{n+4} - \frac{8}{19}(y_{n+3} - y_{n+1}) - y_n = \frac{6h}{19}(f_{n+4} + 4f_{n+3} + 4f_{n+1} + f_n)$$

(7.18) Investigate the zero stability characteristics of Quadé's method.

(7.19) Consider the various methods of exercise (7.16) from the point of view of zero stability.

(7.20) Show that Simpson's rule:

$$y_{n+2} - y_n = \tfrac{1}{3}h(f_{n+2} + 4f_{n+1} + f_n)$$

is both optimal and maximal with respect to order.

(7.21) Find the intervals of absolute and relative stability for Simpson's rule defined in exercise (7.20) directly from the characteristic equation.

(7.22) Use the methods of Schur and Routh–Hurwitz to find the regions of absolute stability for the methods:

$$\text{(i)}\qquad y_{n+2} - y_{n+1} = \tfrac{1}{2}h(3f_{n+1} - f_n)$$

$$\text{(ii)}\qquad y_{n+3} - y_{n+2} = \tfrac{1}{24}h(9f_{n+3} + 19f_{n+2} - 5f_{n+1} + f_n)$$

Investigate the regions of relative stability. These methods are the two-step Adams–Bashforth and three-step Adams–Moulton methods respectively.

(7.23) Investigate the region of relative stability for the predictor–corrector method

$$y_{n+2} + 4y_{n+1} - 5y_n = h(4f_{n+1} - 2f_n)$$
$$y_{n+2} - y_n = \tfrac{1}{3}h(f_{n+2} + 4f_{n+1} + f_n)$$

in PECE mode.

(7.24) Investigate the region absolute stability for the predictor–corrector method in PECE mode which is defined by the characteristic polynomials

$$\rho(\zeta) = \zeta^4 - 1 \qquad \sigma(\zeta) = \tfrac{4}{3}(2\zeta^3 - \zeta^2 + 2\zeta)$$

for the predictor and

$$\rho(\zeta) = \zeta^3 - \tfrac{9}{8}\zeta^2 + \tfrac{1}{8} \qquad \sigma(\zeta) = \tfrac{3}{8}(\zeta^3 + 2\zeta^2 - \zeta)$$

for the corrector.

(7.25) Investigate the region of absolute stability for the Kutta–Nyström fifth-order six-stage method defined by

$$\begin{aligned}
y_{n+1} - y_n &= \frac{h}{192}(23k_1 + 125k_3 - 81k_5 + 125k_6)\\
k_1 &= f(x_n, y_n)\\
k_2 &= f\left(x_n + \tfrac{1}{3}h,\, y_n + \tfrac{1}{3}hk_1\right)\\
k_3 &= f\left(x_n + \tfrac{2}{5}h,\, y_n + \tfrac{1}{25}h(4k_1 + 6k_2)\right)\\
k_4 &= f\left(x_n + h,\, y_n + \tfrac{1}{4}h(k_1 - 12k_2 + 15k_3)\right)\\
k_5 &= f\left(x_n + \tfrac{2}{3}h,\, y_n + \tfrac{1}{81}h(6k_1 + 90k_2 - 50k_3 + 8k_4)\right)\\
k_6 &= f\left(x_n + \tfrac{4}{5}h,\, y_n + \tfrac{1}{75}h(6k_1 + 36k_2 + 10k_3 + 8k_4)\right)
\end{aligned}$$

(7.26) Consider the A-stability and L-stability and the method defined by

$$y_{n+1} - y_n = h[(1-\theta)f_{n+1} + \theta f_n]$$

for varying θ.

(7.27) Establish that the hybrid method

$$y_{n+2} - \frac{1}{31}(32y_{n+1} - y_n) = \frac{h}{93}(15f_{n+2} + 12f_{n+1} - f_n + 64f_{n+3/2})$$

is zero stable and has order five.

(7.28) Find the values of β_0, β_ν and ν so that the hybrid method

$$y_{n+1} - y_n = h(\beta_0 f_n + \beta_\nu f_{n+\nu})$$

has order three.

(7.29) Analyse the weak stability properties of the hybrid predictor–corrector defined by

$$P_\nu \qquad y_{n+3/2} - y_n = \tfrac{1}{8}h(9f_{n+1} + 3f_n)$$

$$P_H \qquad y_{n+2} - \tfrac{1}{31}(32y_{n+1} - y_n) = \tfrac{1}{93}h(15f_{n+2} + 12f_{n+1} - f_n + 64f_{n+3/2})$$

in $P_\nu EP_H E$ mode.

(7.30) Find the stability polynomial for the hybrid method

$$P_\nu \qquad y_{n+3/2} - y_n = \tfrac{3}{8}h(3f_{n+1} + f_n)$$

$$P_k \qquad y_{n+2} + 4y_{n+1} - 5y_n = h(4f_{n+1} + 2f_n)$$

$$C_H \qquad y_{n+2} - y_{n+1} = \tfrac{1}{6}h(f_{n+2} + f_{n+1} + 4f_{n+1/2})$$

in $P_\nu EP_k EC_H E$ mode.

(7.31) Determine the order of the implicit Runge–Kutta method whose Butcher array is

$$\begin{array}{c|ccc} 0 & 0 & 0 & 0 \\ \frac{1}{2} & \frac{5}{24} & \frac{1}{3} & -\frac{1}{24} \\ 1 & \frac{1}{6} & \frac{2}{3} & \frac{1}{6} \\ \hline & \frac{1}{6} & \frac{2}{3} & \frac{1}{6} \end{array}$$

8

Integral Equations

8.1 Introduction

Despite the fundamental importance of integral equations in mathematics, the subject is one which is commonly omitted from both analytic and numerical courses. Most students study catalogues of methods to solve standard differential equations, but few seem to have seen any equivalent methods for integral equations. On the other hand there is a huge literature of classical analysis for these equations based round the traditional split between Fredholm equations and Volterra equations. As is often the case, non-linear problems prove far less tractable. Because the analytic work may be unfamiliar, some of the relevant topics will be included here. In the first instance, consider the linear case only.

The two classical forms are

$$f(x) = \phi(x) + \int_a^b K(x, y)f(y)\,dy \tag{8.1.1}$$

which is a Fredholm equation as the limits of integration are fixed, and

$$f(x) = \phi(x) + \int_a^x K(x, y)f(y)\,dy \tag{8.1.2}$$

where variable limits occur. This type is classed as a Volterra equation. The functions $K(x, y)$ and $\phi(x)$ are given, and it is $f(x)$ which is required. The function $K(x, y)$ is called the kernel and the function $\phi(x)$ is called the driving function. Equations for which $\phi(x) = 0$ are termed first kind equations, whereas those with a non-zero ϕ are second kind equations.

There is a close relationship between differential and integral equations. Any differential equation may be converted into any number of integral equations, and indeed one method for achieving this will be covered later in this chapter. The reverse process is not always possible. Hence in a rough sense, integral equations exhibit a broader solution space than differential equations which makes them of so much interest, theoretically. The split of ordinary differential equations into initial value problems and boundary value problems, and the existence of eigenvalue problems has its equivalent with integral equations. Equation (8.1.2) can be solved trivially at $x = a$ where $f(a) = \phi(a)$. Hence Volterra equations are analogous to initial value problems in ordinary differential equations, that is all the boundary conditions are given at one point and the numerical solution progresses from this point in a step-by-step fashion. The same is true of Volterra equations as the initial function value at a is the initial point and again the solution progresses in a step-by step fashion as will be apparent in the following sections. On the other hand, Fredholm integral equations are equivalent to boundary value problems as is clear from the form of equation (8.1.1) that it is necessary to 'know all of f' to be able to evaluate the integral, just like the situation in a boundary value problem with boundary conditions at more than one point, and therefore there is missing initial information to prevent a step-by-step approach. Hence like the ordinary differential equation analogue, solutions often reduce to a set of linear algebraic equations for the whole interval. It is not possible to compute an f value at one point without also calculating it at all the relevant points in the interval of definition.

In the case when $\phi(x) = 0$, the solution $f(x) \equiv 0$ is clearly one possibility, and for equations of the form

$$f(x) = \lambda \int_a^b K(x, y) f(y)\, dy \tag{8.1.3}$$

the tell-tale signs of an eigenvalue problem are present. Indeed there are values of λ for which non-trivial solutions exist, and the corresponding λ's are called eigenvalues and the solutions, eigensolutions. Unlike the differential case where the number of eigenvalues is often infinite, in the integral equation case the number is usually finite. Hence the three types of problem arise with integral equations as they did with differential equations.

8.2 Numerical methods for linear integral equations

Following the discussion in the previous paragraph, consider a second-kind Fredholm integral equation such as that in (8.1.1). The first method for a numerical solution of this equation is the collocation method, where the integral is replaced by a quadrature rule to give

$$f(x) = \phi(x) + \sum_{i=1}^{N} K(x, y_i) f(y_i) W_i + E(x) \tag{8.2.1}$$

where the quadrature rule is of the general form discussed at length in Chapter 6 namely

$$\int_a^b F(y) dy = \sum_{i=1}^{N} W_i F(y_i) \tag{8.2.2}$$

with weights W_i and abscissae y_i. Any suitable rule may be used with he expected increase in accuracy from the more refined rules for which the error term $E(x)$ is small. By ignoring $E(x)$, an approximate solution for $f(x)$ can be obtained by the simple device of letting x take on the values y_i for $i = 1, N$ to generate a set of linear algebraic equations for the N unknowns $f(y_i)$. Hence a table of values for $f(x)$ is found at each of the points of collocation of the quadrature rule.

Consider a simple example of this procedure with the equation

$$f(x) = 1 + 2\int_0^1 e^{x-t} f(t)\mathrm{d}t \tag{8.2.3}$$

This equation allows of a very simple analytic solution as it belongs to the family of problems with degenerate kernels. Such kernels have the form

$$K(x, y) = \sum_{i=1}^{M} F_i(x)G_i(y) \tag{8.2.4}$$

Hence the general equation can be written as

$$\begin{aligned} f(x) &= \phi(x) + \int_a^b \sum_{i=1}^{M} F_i(x)G_i(y)f(y)\,\mathrm{d}y \\ &= \phi(x) + \sum_{i=1}^{M} F_i(x)\int_a^b G_i(y)f(y)\,\mathrm{d}y \\ &= \phi(x) + \sum_{i=1}^{M} F_i(x)C_i \end{aligned} \tag{8.2.5}$$

where the definite integrals from a to b are just constants denoted by C_i. Immediately the general form for the solution $f(x)$ is known. By substituting this form for $f(x)$ into the original equation a set of linear algebraic equations is obtained for the constants C_i to complete the solution. For eigenvalue problems of this type it is clear that there will be M eigenvalues as a matrix of order M will result.

To return to the specific example, the analytic solution is easily derived, M being one in this case. Hence by analogy with the above general degenerate kernel case

$$\begin{aligned} f(x) &= 1 + 2\int_0^1 e^{x-t} f(t)\,\mathrm{d}t \\ &= 1 + 2e^x \int_0^1 e^{-t} f(t)\,\mathrm{d}t \\ &= 1 + 2e^x C \end{aligned} \tag{8.2.6}$$

where

$$C = \int_0^1 e^{-t} f(t)\,\mathrm{d}t$$

Substituting this form for $f(x)$ into the original yields

$$C = \int_0^1 e^{-t}\,dt + 2C\int_0^1 dt \tag{8.2.7}$$

and hence

$$C = -\left(1 - \frac{1}{e}\right) \tag{8.2.8}$$

to give the analytic solution

$$f(x) = 1 - 2e^{-x}\left(1 - \frac{1}{e}\right) \tag{8.2.9}$$

To implement (8.2.1) for this example, a five-point Gauss–Legendre rule is utilized so that

$$\begin{aligned}\int_0^1 F(t)\,dt &= \frac{1}{2}\int_{-1}^1 F\left(\frac{1}{2}s + \frac{1}{2}\right) ds\\ &= \frac{1}{2}\sum_{i=1}^{5} W_i F\left(\frac{s_i + 1}{2}\right)\end{aligned} \tag{8.2.10}$$

which accounts for the change of interval to the standard range $(-1, 1)$ for Gauss–Legendre. The weights and abscissae for the five-point rule are reproduced for convenience in Table 8.1.

The set of resulting linear algebraic equations is then generated as in (8.2.1) to give

$$f(x) = 1 + \sum_{i=1}^{5} W_i e^x e^{-t_i} f(t_i) \tag{8.2.11}$$

with

$$t_i = \tfrac{1}{2}(s_1 + 1) \qquad x = t_j = x_j, \quad j = 1, \ldots, 5$$

By letting $f_i = f(t_i)$, the equations

$$-\sum_{i=1}^{5} W_i e^{x_j - t_i} f(t_i) + f(x_j) = 1 \tag{8.2.12}$$

Table 8.1

i	W_i	s_i
1	0.236 93	0.906 18
2	0.478 63	0.538 47
3	0.568 89	0.00 00
4	0.478 63	−0.538 47
5	0.236 93	−0.906 18

are formed and these expand to give the equations

$$
\begin{aligned}
(1 - W_1)f_1 - W_2 e^{t_1-t_2} f_2 - W_3 e^{t_1-t_3} f_3 - W_4 e^{t_1-t_4} f_4 - W_5 e^{t_1-t_5} f_5 &= 1 \\
-W_1 e^{t_2-t_1} f_1 + (1 - W_2)f_2 - W_3 e^{t_2-t_3} f_3 - W_4 e^{t_2-t_4} f_4 - W_5 e^{t_2-t_5} f_5 &= 1 \\
&\vdots \\
-W_1 e^{t_5-t_1} f_1 - W_2 e^{t_5-t_2} f_2 - W_3 e^{t_5-t_3} f_3 - W_4 e^{t_5-t_4} f_4 + (1 - W_5)f_5 &= 1
\end{aligned}
\tag{8.2.13}
$$

In matrix form the numerical values are

$$
\begin{bmatrix}
0.763\,07 & -0.575\,24 & -0.894\,96 & -0.985\,60 & -0.586\,37 \\
-0.197\,14 & 0.521\,37 & -0.744\,66 & -0.820\,08 & -0.487\,89 \\
-0.150\,61 & -0.365\,66 & 0.431\,11 & -0.626\,51 & -0.372\,73 \\
-0.115\,06 & -0.279\,35 & -0.434\,62 & 0.521\,37 & -0.284\,75 \\
-0.095\,735 & -0.232\,43 & -0.361\,62 & -0.398\,25 & 0.763\,07
\end{bmatrix}
\begin{bmatrix} f_1 \\ f_2 \\ f_3 \\ f_4 \\ f_5 \end{bmatrix}
=
\begin{bmatrix} 1 \\ 1 \\ 1 \\ 1 \\ 1 \end{bmatrix}
\tag{8.2.14}
$$

which solves using any of the standard methods of Chapter 3 to give the values shown in Table 8.2 where a comparison with the analytic values is made.

The accuracy of this method depends entirely on the accuracy of the quadrature rule employed. The resulting function values will be found at the collocation points of the quadrature rule and it may be necessary to employ interpolation if other values are required.

A somewhat different approach to the problem of solving a linear second-kind Fredholm equation is to express the unknown function $f(x)$ as a linear combination of a given set of basis functions, $\{p_i(x)\}$. These functions are chosen to be suitable in some way for a given problem. A commonly used set is the set $p_i(x) = x^i$, but physical insight may suggest functions which will enhance the emergence of a solution. Hence $f(x)$ is expressed as

$$
f(x) = \sum_{i=0}^{N} a_i p_i(x) \tag{8.2.15}
$$

for some N. This form is substituted into equation (8.1.1) to give

$$
\sum_{i=0}^{N} a_i p_i(x) = \phi(x) + \int_a^b K(x, y) \sum_{i=0}^{N} a_i p_i(y)\,\mathrm{d}y \tag{8.2.16}
$$

Table 8.2

x	$f(x)$	True
0.953 09	−2.279 06	−2.279 08
0.769 23	−1.728 36	−1.728 38
0.500 00	−1.084 37	−1.084 3812
0.230 76	−0.592 38	−0.592 392
0.046 91	−0.324 95	−0.324 960

or

$$\sum_{i=0}^{N} a_i \left[p_i(x) - \int_a^b K(x, y) p_i(y)\, dy \right] = \phi(x) \tag{8.2.17}$$

If x is set to x_j for $j = 0, \ldots, N$, a set of linear algebraic equations results for the unknowns a_i from which equation (8.2.15) can be used to compute values of $f(x)$ at any required point. The resulting accuracy depends on how well the choice of basis set is able to fit the solution function, and on how large a value of N is employed. A poor choice of basis set will never yield a good solution however large N is taken.

As an example of this method, consider again the example defined in (8.2.3). Consider the choice of $N = 4$ and the basis set $p_i(x) = x^i$. Let the set $x_j = \{0.0, 0.25, 0.5, 0.75, 1.0\}$ to give the five equations derived from (8.2.17) as

$$\sum_{i=0}^{4} a_i \left[x_j^i - 2 \int_0^1 e^{x_j - t} t^i\, dt \right] = 1 \tag{8.2.18}$$

or

$$\sum_{i=0}^{4} a_i \left[x_j^i - 2e^{x_j} \int_0^1 t^i e^{-t}\, dt \right] = 1 \tag{8.2.19}$$

The required integrals would normally be found numerically using a suitable method from Chapter 6, but here analytic values are available. Define

$$I_i = \int_0^1 t^i e^{-t}\, dt \tag{8.2.20}$$

then integration by parts yields the recurrence relation

$$I_i = -[t^i e^{-t}]_0^1 + i \int_0^1 t^{i-1} e^{-t}\, dt = -\frac{1}{e} + iI_{i-1} \tag{8.2.21}$$

which gives

$$\begin{aligned} I_0 &= 1 - e^{-1} \\ I_1 &= 1 - 2e^{-1} \\ I_2 &= 2 - 5e^{-1} \\ I_3 &= 6 - 16e^{-1} \\ I_4 &= 24 - 65e^{-1} \end{aligned} \tag{8.2.22}$$

Hence (8.2.19) can be written as

$$\sum_{i=0}^{4} a_i [x_j^i - 2e^{x_j} I_i] = 1 \tag{8.2.23}$$

with $x_j = j/4$. The linear equations for the unknowns a_i are

$$\begin{bmatrix} 1-2I_0 & -sI_1 & -2I_2 & -2I_3 & -2I_4 \\ 1-2e^{1/4}I_0 & \frac{1}{4}-2e^{1/4}I_1 & \frac{1}{4}^2-2e^{1/4}I_2 & \frac{1}{4}^3-2e^{1/4}I_3 & \frac{1}{4}^4-2e^{1/4}I_4 \\ 1-2e^{1/2}I_0 & \frac{1}{2}-2e^{1/2}I_1 & \frac{1}{2}^2-2e^{1/2}I_2 & \frac{1}{2}^3-2e^{1/2}I_3 & \frac{1}{2}^4-2e^{1/2}I_4 \\ 1-2e^{3/4}I_0 & \frac{3}{4}-2e^{3/4}I_1 & \frac{3}{4}^2-2e^{3/4}I_2 & \frac{3}{4}^3-2e^{3/4}I_3 & \frac{3}{4}^4-2e^{3/4}I_4 \\ 1-2eI_0 & 1-2eI_1 & 1-2eI_2 & 1-2eI_3 & 1-2eI_4 \end{bmatrix} \begin{bmatrix} a_0 \\ a_1 \\ a_2 \\ a_3 \\ a_4 \end{bmatrix} = \begin{bmatrix} 1 \\ 1 \\ 1 \\ 1 \\ 1 \end{bmatrix} \tag{8.2.24}$$

with the solution

$$\begin{gathered} a_0 = -0.264\,25 \qquad a_1 = -1.262\,73 \\ a_2 = -0.644\,50 \qquad a_3 = -0.177\,34 \\ a_4 = -0.087\,76 \end{gathered} \tag{8.2.25}$$

These values yield the values shown in Table 8.3, again with the true values for comparison.

For the value at $x = 0.5$ there is a direct comparison with the results of the collocation method in Table 8.2. In many ways this is a more hit-and-miss method as its success depends on making a good choice for the basis set $p_i(x)$. Typical approximating functions include powers of x, series of Chebyshev polynomials, trigonometric functions and splines. The set of functions has to be linearly independent.

There are variations to the selection process for the basis functions which have a more theoretical background. Consider the residual

$$g(x) = f(x) - \int_a^b K(x, y) f(y)\,\mathrm{d}y - \phi(x) \tag{8.2.26}$$

then one objective is to select a solution of the form (8.2.15) which minimizes $g(x)$ in some sense. A number of named methods fall into this category. The usual criterion is to force to zero the inner products of $g(x)$ with each of the fitting functions $p_i(x)$ yielding equations for the fitting coefficients a_i. Hence the

Table 8.3

x	$f(x)$	True
0.0	−0.264 25	−0.264 241
0.25	−0.623 33	−0.623 318
0.5	−1.08 439	−1.084 381
0.75	−1.676 413	−1.676 398
1.0	−2.436 58	−2.436 564

condition is

$$(g, p_i) = \int_a^b w(x)g(x)p_i(x)\,\mathrm{d}x = 0 \qquad i = 0, 1, \ldots, n \tag{8.2.27}$$

where $w(x)$ is a convenient weight function. In this form the method is called the method of moments. If $w(x) = 1$, the classical Galerkin method is recovered. Advantage can be taken of the orthogonality of the fitting functions, which will give some simplifications later. Indeed, if $(p_i, p_j) = \delta_{ij}$, then (8.2.27) implies that the generalized Fourier expansion of $g(x)$

$$\hat{g}(x) = \sum_{i=0}^{N} (g, p_i)p_i(x) \tag{8.2.28}$$

is zero. Hence a 'minimum requirement' on the residual $g(x)$ is imposed.

Substituting $g(x)$ from (8.2.26) into (8.2.27) gives the set of linear algebraic equations for the required coefficients a_i in the form

$$\sum_{j=0}^{N} C_{ij}a_j - \sum_{j=0}^{N} D_{ij}a_j = b_i \qquad 1 = 0, 1, \ldots, N \tag{8.2.29}$$

where

$$C_{ij} = \int_a^b w(x)p_j(x)p_i(x)\,\mathrm{d}x \tag{8.2.30}$$

$$b_i = \int_a^b w(x)\phi(x)p_i(x)\,\mathrm{d}x \tag{8.2.31}$$

$$D_{ij} = \int_a^b \int_a^b w(x)K(x, y)p_j(y)p_i(x)\,\mathrm{d}x\,\mathrm{d}y \tag{8.2.32}$$

The matrix $\mathbf{C}$ is the Gram matrix and is symmetric and positive definite. If the p_i's are orthogonal then this matrix is the unit matrix and one set of integrals is avoided. The implementation of this method is however quite laborious in that $(N + 1)^2$ double integrals are required.

Unlike the Fredholm methods above, methods for Volterra equations commonly employ a step-by-step procedure which makes these equations easier to solve. If in addition an *equal-step* quadrature rule is used to replace the integral then in effect the matrix generated in the collocation method is lower triangular so allowing the step-by-step solution. Hence for the general equation

$$f(x) = \phi(x) + \int_a^x K(x, y)f(y)\,\mathrm{d}y \tag{8.2.33}$$

the trapezoidal rule gives

$$f(x) = \phi(x) + \frac{h}{2}\left[K(x,a)f(a) + 2\sum_{i=1}^{j-1} K(x,y_i)f(y_i) + K(x,y_j)f(y_j)\right] \tag{8.2.34}$$

where $y_i = a + ih$ for a suitable step size h. By now setting $x = x_j = y_j$ a formula is obtained for the computation of $f(y_j)$, that is

$$f(x_j)\left(1 - \frac{h}{2}K(x_j, x_j)\right) = \phi(x_j) + \frac{h}{2}\left[K(x_j,a)f(a) + 2\sum_{i=1}^{j-1} K(x_j,y_i)f(y_i)\right] \tag{8.2.35}$$

Successive use of this formula with increasing j yields the required f values for as large a j as is required. This is very much analogous to initial value ordinary differential equation solving.

If a quadrature rule with higher accuracy, but which does not have equally spaced abscissae is used, then a complication does indeed arise. At each step the integral

$$\int_a^{x_j} K(x,y)f(y)\,\mathrm{d}y \tag{8.2.36}$$

is required. To use the Gauss–Legendre rule for example would mean changing the interval to the standard $(-1, 1)$ and using the Gauss–Legendre abscissae. These would not coincide with the abscissae used at any of the previous steps (which is the advantage of equal spacing). Hence recourse would need to be made to interpolation rules to compute the f values at the new abscissae. This throws up another anomaly. The reason for using Gauss–Legendre is to take advantage of the high accuracy the formula gives, namely polynomials to order x^{2n-1} will be handled exactly with an n-point rule. This is the equivalent, from Chapter 6, of using Hermite interpolation. But the derivatives of $f(x)$ will not be known for the equivalent interpolator to be implemented (it is the choice of abscissae which avoids these derivatives in the integration rule). The next best interpolator is Lagrange's formula which is exact to only x^n. Therefore the high-order integration accuracy is just thrown out at the interpolation stage. Hence a Newton–Cotes rule, based on a Lagrangian fit, may as well be employed from the start. But, as explained in Chapter 6, the high-order ones are unstable, and so the use of the trapezoidal and Simpson's rules is very common in the solution of Volterra equations.

Applying the method of (8.2.35) to the example

$$f(x) = \frac{1}{(x+2)^2} - 2\int_0^x \frac{f(y)}{(x-y+2)^2}\,\mathrm{d}y \tag{8.2.37}$$

with $h = 1/256$ gives the results shown in Table 8.4.

Table 8.4

x	$f(x)$
0.0	0.2500
0.125	0.208 05
0.25	0.174 42
0.375	0.147 26
0.50	0.125 15
0.625	0.107 04
0.75	0.092 118
0.875	0.069 411

Only every 32nd step is shown in Table 8.4. It is common to use a very small step size such as in this example as the formula for each step is very simple and easily computed for small h.

A further method which is applicable to both Volterra and Fredholm equations has already been aired in Chapter 7 under the guise of Picard's method (7.1.27). The ordinary differential equation was converted into a Volterra equation and this was solved by direct iteration to give for (8.2.1)

$$f^{[k+1]}(x) = \phi(x) + \int_a^x K(x, y) f^{[k]}(y)\,\mathrm{d}y \tag{8.2.38}$$

with $f^{[0]}(x) = \phi(x)$. A numerical process can be formed from this formula by representing the iterates $f^{[k]}(x)$ as tables of values at a set of fixed x values. These values are arranged to coincide with the abscissae of the quadrature rule to be employed, so allowing the new iterates to be computed. In theory this iteration converges under quite general conditions but the rate of convergence may be slow.

Applying this method to a Fredholm equation gives

$$f^{[k+1]}(x) = \phi(x) + \lambda \int_a^b K(x, y) f^{[k]}(y)\,\mathrm{d}y \tag{8.2.39}$$

and it can be shown that now the iteration only converges if

$$\lambda < \frac{(b-a)}{\displaystyle\int_a^b \int_a^b K^2(x, y)\,\mathrm{d}x\,\mathrm{d}y} \tag{8.2.40}$$

and so for many practical examples this method is not feasible.

Consider the example

$$y(x) = \frac{2x}{\pi} + \int_0^{\pi/2} K(s, x) y(s)\,\mathrm{d}s \tag{8.2.41}$$

with the kernel

$$K(s, x) = \begin{cases} s - \frac{2xs}{\pi} & s < x \\ x - \frac{2xs}{\pi} & s > x \end{cases} \tag{8.2.42}$$

By substitution it is relatively easy to confirm that the solution is $y = \sin x$. Taking $y^{[0]} = 2x/\pi$, and applying one step of an iterative solution gives

$$
\begin{aligned}
y^{[1]}(x) &= \frac{2x}{\pi} + \int_0^x \left(s - \frac{2xs}{\pi}\right)\frac{2s}{\pi}\,\mathrm{d}s + \int_x^{\pi/2}\left(x - \frac{2xs}{\pi}\right)\frac{2s}{\pi}\,\mathrm{d}s \\
&= \frac{2x}{\pi} + \left(1 - \frac{2x}{\pi}\right)\frac{2x^3}{3\pi} + \frac{2x}{\pi}\left(\frac{\pi^2}{8} - \frac{x^2}{2}\right) - \frac{4x}{\pi^2}\left(\frac{\pi^3}{24} - \frac{x^3}{3}\right) \\
&= x\left(\frac{2}{\pi} + \frac{\pi}{12}\right) - \frac{x^3}{3\pi}
\end{aligned}
\tag{8.2.43}
$$

giving a non-standard series for $\sin x$ namely $0.8984x - x^3/9.42$. For this example, convergence does occur and increased accuracy can be expected for further iterations.

8.3 Non-linear integral equations

A range of methods is available for linear integral equations, but the non-linear case provides further difficulties, as the earlier methods do not always generalize well. Before pursuing these numerical details, a method to forge a link between ordinary differential equations and integral equations will be given. This is an analytic technique, but converts a given ordinary differential equation into any number of equivalent integral equations. In practice, it will often be the case that the equivalent integral equation is easier to solve than the original ordinary differential equation. This is particularly the case for boundary value problems. In Chapter 7 some methods for such problems were given and it is clear that considerably more work is needed to solve a boundary value problem than a similar initial value problem. Because the boundary conditions become part of the resulting integral equation, a standard integral equation method can be used. Commonly, however, the resulting integral equations are non-linear, and hence the inclusion of this work here.

The conversion procedure is often called Fubini's method, and is based on the method of variation of parameters used in ordinary differential equation work to find a particular integral. The procedure will be illustrated by an example which demonstrates the sort of problem which can arise using the ordinary differential equation approach, and then shows the simplifications which occur on conversion to integral equation form. The original ordinary differential equation arises in the study of the reaction rates in a chemical reactor, Pakes and Storey (1967) and is defined by

$$\frac{1}{P}F'' - F' - R(F) = 0 \tag{8.3.1}$$

with boundary conditions

$$F - \frac{1}{P}F' = 1 \quad \text{at} \quad x = 0$$

$$F' = 0 \quad \text{at} \quad x = 1 \tag{8.3.2}$$

and typically $R(F) = 2F^2$. The natural approach using ordinary differential methods is to apply the shooting method starting with the most complicated boundary condition, that is the one at $x = 0$. The 'shots' are carried out for varying F or F' at $x = 0$ with the aim of reaching $x = 1$ with $F' = 0$. In general F' will be other than zero and interpolation can be used as in Chapter 7 to move towards an initial F or F' which will give the required F' at $x = 1$. This process proves to be less easy than might be expected. If, say, a standard Runge–Kutta method is used for the shots, it turns out that the interval of stability requires negative h. Hence shooting in the forward direction gives unstable results. Therefore the process has to be worked starting at $x = 1$ and computing back to $x = 0$. The larger the value of P, the worse the problem. There is a case here for conversion to integral equation form.

The method of Fubini (Tricomi, 1961) is based on isolating part of the given equation on the left-hand side and solving this part ignoring the right-hand side. Different left-hand side possibilities yield different integral equations, and there is scope to absorb difficulties by this choice of left-hand side. Hence in this example set

$$\frac{1}{P}F'' - F' = R(F) \tag{8.3.3}$$

with the left-hand side solving to give

$$F = C_1 + e^{Px}C_2 \tag{8.3.4}$$

The method of Fubini, and indeed the method of variation of parameters lets the two constants of integration C_1 and C_2 be functions of x. Hence consider

$$F = C_1(x) + e^{Px}C_2(x) \tag{8.3.5}$$

Substituting back into the original equation will give one equation for the new unknown functions, and there will be a degree of freedom left over which can be used in some manner to enhance the required solution. Equation (8.3.5) differentiates to give

$$F' = C_2(x)Pe^{Px} + C_1'(x) + C_2'e^{Px} \tag{8.3.6}$$

and the extra degree of freedom is used to make all the derivative terms in C_1 and C_2 equal to zero. Therefore

$$F' = C_2(x)Pe^{Px} \tag{8.3.7}$$

with the extra condition

$$C_1'(x) + C_2'e^{Px} = 0 \tag{8.3.8}$$

This leaves

$$F'' = C_2'(x)Pe^{Px} + C_2(x)P^2e^{Px} \tag{8.3.9}$$

and now substituting F, F' and F'' into the original, causes cancellation of the terms involving just $C_1(x)$ and $C_2(x)$ (because of the original choice of $F(x)$ as a solution of the left-hand side), and the remaining terms are

$$C_2'(x) = R(F)e^{-Px} \tag{8.3.10}$$

$$C_1'(x) = -R(F) \tag{8.3.11}$$

which integrate to give

$$C_2(x) = \gamma_1 + \int_0^x R(F(t))e^{-Pt}\,\mathrm{d}t \tag{8.3.12}$$

$$C_1(x) = \gamma_2 - \int_0^x R(F(t))\,\mathrm{d}t \tag{8.3.13}$$

where the two constants γ_1 and γ_2 are the constants of integration and will be fixed in the next part of the process which builds in the boundary conditions. The choice of lower limit for the integrals is not critical as a different choice will simply alter the constants of integration accordingly. It is usual to choose the lower limit to correspond with the more difficult boundary condition as the integrals will become zero when this limit is selected in fixing the boundary conditions. The dummy variable t is also introduced so that confusion does not arise in using x for two purposes.

The integral equation is

$$F(x) = \gamma_2 + \gamma_1 e^{Px} - \int_0^x R(F)\,\mathrm{d}t + e^{Px}\int_0^x R(F)e^{-Pt}\,\mathrm{d}t \tag{8.3.14}$$

and the boundary conditions give

$$F'(1) = \gamma_1 P e^P + P e^P \int_0^1 e^{-Pt} R(F)\,\mathrm{d}t = 0 \tag{8.3.15}$$

and

$$F - \frac{1}{P}F' = \gamma_2 + \gamma_1 - \gamma_1 = \gamma_2 = 1 \tag{8.3.16}$$

at $x = 0$, so fixing γ_1 and γ_2. The final equation can be written in conventional form to give

$$F(x) = 1 + \int_0^1 K(x,t)R(f(t))\,\mathrm{d}t \tag{8.3.17}$$

where

$$K(x,t) = \begin{cases} -1 & t < x \\ -e^{P(x-t)} & t > x \end{cases} \tag{8.3.18}$$

which is a non-linear Fredholm equation in conventional form.

By starting with different left-hand sides, alternative integral equations can be found. Hence starting with

$$F'' = P(F' + 2R(F)) \tag{8.3.19}$$

gives the trial form

$$F(x) = A(x) + xB(x) \tag{8.3.20}$$

with $F' = B(x)$ and $A'(x) + xB'(x) = 0$, and $F'' = B'(x)$. By the same process as above, $A(x)$ and $B(x)$ satisfy

$$B(x) = \gamma_1 + P\int_0^x (F' + R(F))\,\mathrm{d}t \tag{8.3.21}$$

and

$$A(x) = \gamma_2 - P\int_0^x t(F' + R(F))\,\mathrm{d}t \tag{8.3.22}$$

to give the integral equation

$$F(x) = \gamma_2 - P\int_0^x t(F' + R(F))\,\mathrm{d}t + x\gamma_1 + Px\int_0^x (F' + R(F))\,\mathrm{d}t \tag{8.3.23}$$

Strictly this is an integro-differential equation, but in this case it is not difficult to integrate by parts to remove the terms in F'. As in the previous example, the boundary conditions give

$$F(0) - \frac{F'(0)}{P} = \gamma_2 - \frac{1}{P}\gamma_1 = 1 \tag{8.3.24}$$

and

$$F'(1) = \gamma_1 + P\int_0^1 (F' + R(F))\,\mathrm{d}t = 0 \tag{8.3.25}$$

which fix γ_1 and γ_2 to give the integro-differential equation

$$F(x) = 1 - \int_0^1 (F' + R(F))\,\mathrm{d}t - P\int_0^1 t(F' + R(F))\,\mathrm{d}t - Px\int_0^1 (F' + R(F))\,\mathrm{d}t \tag{8.3.26}$$

To convert to an integral equation, integration by parts gives

$$\begin{aligned} F(x) = 1 - (F(1) - F(0)) - \int_0^1 R(F)\,\mathrm{d}t - P\int_0^x t\,\mathrm{d}F - P\int_0^x tR(F)\,\mathrm{d}t \\ - PxF(1) + PxF(x) - Px\int_x^1 R(F)\,\mathrm{d}t \end{aligned} \tag{8.3.27}$$

$$\begin{aligned} = 1 - (F(1) - F(0)) - \int_0^1 R(F)\,\mathrm{d}t - P[xF(x)] + P\int_0^x F\,\mathrm{d}t \\ - P\int_0^x tR(F)\,\mathrm{d}t - PxF(1) + PxF(x) - Px\int_x^1 R(F)\,\mathrm{d}t \end{aligned} \tag{8.3.28}$$

But

$$F(1) = 1 - \int_0^1 (F' + R(F))\,\mathrm{d}t - P\int_0^1 t(F' + R(F))\,\mathrm{d}t \qquad (8.3.29)$$

$$F(0) = 1 - \int_0^1 (F' + R(F))\,\mathrm{d}t \qquad (8.3.30)$$

and again integrating these by parts to remove F' terms gives

$$(2 + P)F(1) - F(0) = 1 + P\int_0^1 F\,\mathrm{d}t - \int_0^1 R(F)\,\mathrm{d}t - P\int_0^1 tR(F)\,\mathrm{d}t \quad (8.3.31)$$

$$F(1) = 1 - \int_0^1 R(F)\,\mathrm{d}t \qquad (8.3.32)$$

Hence $F(1)$ and $F(0)$ can be substituted out of equation (8.3.28), and collection of terms gives the alternative integral equation form

$$F(x) = P(\alpha - x) - P\int_x^1 F(t)\,\mathrm{d}t + \int_0^1 K(x, t)R(F)\,\mathrm{d}t \qquad (8.3.33)$$

where

$$K(x, t) = \begin{cases} P(x - \alpha) & t < x \\ P(t - \alpha) & t > x \end{cases} \qquad (8.3.34)$$

and

$$\alpha = 1 + \frac{1}{P} \qquad (8.3.35)$$

With this preamble complete, attention now returns to the numerical solution of non-linear integral equations. As in the linear case, there is a natural split into Fredholm and Volterra equations. Consider first the solution of a non-linear Volterra equation with the general form

$$f(x) = \phi(x) - \int_a^x G(x, t, f(t))\,\mathrm{d}t \qquad (8.3.36)$$

The method of using an equally spaced quadrature rule to replace the integral to generate a step-by-step solution is still valid. In the linear case in equation (8.2.34), by putting $x = x_j = y_j$ the leading $f(x)$ and the final $f(y_j)$ become the same and combine to give the term $f(x_j)(1 - h/2K(x_j, x_j))$ in equation (8.2.35). In the non-linear case the second $f(y_j)$ will be inside the function G. Hence the equivalent of equation (8.2.35) will be a non-linear algebraic equation whose solution will give $f(x_j)$. This equation has the form:

$$f(x_j) = \phi(x_j) - \frac{h}{2}\left[G(x_j, a, f(a)) + 2\sum_{i=1}^{j-1} G(x_j, y_i, f(y_i)) + G(x_j, y_j, f(y_j))\right] \qquad (8.3.37)$$

with $y_i = a + ih$ once again. Hence setting $x_j = y_j$ gives the non-linear equation for the next function point at $f(y_j)$. This equation can be solved by any of the standard methods of Chapter 1, of which Newton's method is an obvious candidate. The required derivatives for Newton's method can be approximated using the small size of h to advantage as in the following example.

Consider the example

$$f(x) = \frac{1}{(x+2)^2} - \int_0^x \frac{e^{-f(t)}}{(x-t+2)^2}\,\mathrm{d}t \tag{8.3.38}$$

then the trapezoidal approximation akin to (8.3.37) is

$$f_j = \frac{1}{(x_j+2)^2} - \frac{h}{2}\left[\frac{e^{-f(0)}}{(x_j+2)^2} + 2\sum_{i=1}^{j-1}\frac{e^{-f(x_i)}}{(x_j-x_i+2)^2} + \frac{e^{-f(x_j)}}{4}\right] \tag{8.3.39}$$

which can be reorganized into the convenient form

$$f_j + \frac{h}{8}e^{-f_j} = \frac{1}{(x_j+2)^2}\left[1 - \frac{h}{2}e^{-f(0)}\right] - h\sum_{i=1}^{j-1}\frac{e^{-f(x_j)}}{(x_j-x_i+2)^2} = S, \text{ say} \tag{8.3.40}$$

The non-linear equation is then simply written down as

$$F(f_j) = f_j + \tfrac{1}{8}he^{-f_j} - S = 0 \tag{8.3.41}$$

with the first derivative of F being

$$F'(f_j) = 1 - \tfrac{1}{8}he^{-f_j} \sim 1 \text{ for small } h \tag{8.3.42}$$

for use in Newton's formula. The simplified iteration is now

$$f_j^{(n+1)} = f_j^{(n)} - F(f_j^{(n)}) \tag{8.3.43}$$

which is iterated at each step. The result for this particular example is shown in Table 8.5.

In Table 8.5 the step size used was $h = 1/250$ and only every 25th f value has been displayed.

In the Volterra case, the alternative of direct iteration as defined in equation (8.2.38) is effective, as the presence of a non-linearity does not now make the method for linear problems any harder. The rule would therefore be

$$f^{(k+1)}(x) = \phi(x) + \int_a^x G(x, y, f^{(k)}(x))\,\mathrm{d}y \tag{8.3.44}$$

and it is in general no more difficult to integrate $G(x, y, f^{(k)}(x))$ than $K(x, y)f^{(k)}(y)$ in equation (8.2.38). The technique is to iterate for each $f(x_j)$ value in turn. If,

Table 8.5

x	$f(x)$
0.0	0.2500
0.1	0.207 80
0.2	0.169 65
0.3	0.134 91
0.4	0.103 07
0.5	0.073 711
0.6	0.046 501
0.7	0.021 161
0.8	−0.025 404
0.9	−0.024 798
1.0	−0.045 776

therefore, $f(x_0), \ldots, f(x_j)$ are known from previous iterates then writing $f(x_j) = f_j$ gives the iteration

$$f_{j+1}^{(k+1)} = \phi(x_{j+1}) + \int_a^{x_{j+1}} G(x_{j+1}, y, f(y))\,\mathrm{d}y \qquad (8.3.45)$$

and to evaluate $G(x_{j+1}, y, f(y))$ in the quadrature uses the known values for $f(x_0), \ldots, f(x_j)$ and the last iterated value for $f_{j+1}(y)$ at $y = x_{j+1}$. It is common to set the initial value to

$$f_{j+1}^{(0)} = \phi(x_{j+1}) \qquad (8.3.46)$$

The method was implemented on the above example and a similar table of values was obtained to that in Table 8.5 by employing Clenshaw–Curtis quadrature using a 16-point set of cosine weighted abscissae and just three iterations per point.

The Fredholm problem is less easy to resolve. Direct iteration in the manner of equation (8.2.39) has the same convergence uncertainty that dogs the linear case, and now the condition for convergence may not be known. In some circumstances the method can be very effective but care is needed. Applying the collocation method will result, not in a set of linear algebraic equations as in (8.2.1), but an equivalent set of non-linear algebraic equations. If one was prepared to solve a set of such equations using one of the techniques from Chapter 1 then a method is found. It is common to use physical knowledge to allow direct iteration to be used rather than some form of n-dimensional Newton. The presence of a small parameter may help to make it obvious which terms to isolate on the left-hand side to generate a suitable iteration. Either way, the process here is quite tedious for any but very small sets of unknowns.

One effective alternative has been called the δf method (Evans, Hyslop and Morgan 1983) and uses perturbation methods to arrive at a solution. Consider the non-linear Fredholm equation rewritten in the form

$$g(f; x) = \phi(x) + \int_a^b G(x, y, f(y))\,\mathrm{d}y - f(x) \qquad (8.3.47)$$

The object of the exercise is to find an f which will force g to zero. The function g is a residual. Suppose $f^{(k)}$ is known from a previous step, then let

$$f^{(k+1)}(y) = f^{(k)} + \delta f^{(k)}(y) \tag{8.3.48}$$

to give

$$g(f^{(k+1)}(y); x) = \phi(x) + \int_a^b G(x, y, f^{(k)} + \delta f^{(k)})\,\mathrm{d}y - f^{(k)}(x) - \delta f^{(k)}(x) \tag{8.3.49}$$

Now force $g(f^{(k+1)}(y); x)$ to be zero by the choice of $\delta f^{(k)}$. First expand for small δf to give

$$0 = \phi(x) + \int_a^b \left\{ G(x, y, f^{(k)}) + \delta f^{(k)} \frac{\partial G}{\partial f}(x, y, f^{(k)}) + \cdots \right\} \mathrm{d}y - f^{(k)}(x) - \delta f^{(k)}(x) \tag{8.3.50}$$

where $\partial G/\partial f$ is the Fréchet derivative of G with respect to f. At this stage the expression can be linearized by ignoring terms of order δf^2 and over. The result is a linear Fredholm equation for δf

$$0 = g(f^{(k)}; x) + \int_a^b \delta f^{(k)}(y) \frac{\partial G}{\partial f}(x, y, f^{(k)})\,\mathrm{d}y - \delta f^{(k)}(x) \tag{8.3.51}$$

This equation can now be processed using a conventional method for linear problems. In fact because δf is a small correction to $f^{(k)}$ only a low-order method is required. It has been found that the basis set method with

$$\delta f(x) = \sum_{i=0}^{M} a_i x^i \tag{8.3.52}$$

with M as small as 4, 6 or 8 is very effective. As the accuracy of the iterates for f improves then the value of M can be increased accordingly. An alternative basis set for approximating $\delta f(x)$ which is based on approximation theory is

$$\delta f(x) = \sum_{i=0}^{M} a_i T_i(x) \tag{8.3.53}$$

where a Chebyshev expansion is employed. Such an expansion tends to have coefficients a_i which fall off for increasing i with a resulting improvement in stability.

The process involved in this perturbation is no more than a generalized Newton's method to solve the non-linear equation $g(f : x) = 0$ in function space. The Newton iteration is formally

$$f^{n+1} = f^n - D^{-1} f^n (f^n) \tag{8.3.54}$$

where $D^{-1} f^n$ is the inverse Fréchet derivative which operates on f^n. The practical process of finding this inverse derivative operator is equivalent to solving the δf integral equation.

As an example of this procedure consider the non-linear integral equation

$$f(x) = 2 - \frac{1}{\pi}\int_{-1}^{1} \frac{f^2(y)}{1+(x-y)^2}\,\mathrm{d}y \tag{8.3.55}$$

which can be rewritten as

$$g(f : x) = 2 - \frac{1}{\pi}\int_{-1}^{1} \frac{f^2(y)}{1+(x-y)^2}\,\mathrm{d}y - f(x) \tag{8.3.56}$$

and the object is to find $f(x)$ which forces $g(f : x)$ to zero. Hence if $f^{(n)}(x)$ is a known iterate then

$$g(f^n + \delta f^n; x) = 2 - \frac{1}{\pi}\int_{-1}^{1} \frac{(f^n + \delta f^n)^2}{1+(x-y)^2}\,\mathrm{d}y - f(x) \tag{8.3.57}$$

which to first order gives

$$g(f^n + \delta f^n; x) = 0 = g(f^n; x) - \frac{1}{\pi}\int_{-1}^{1} \frac{2f^n \delta f^n}{1+(x-y)^2}\,\mathrm{d}y - \delta f^n \tag{8.3.58}$$

This equation is a linear Fredholm equation for each iterate, which for the purposes of illustration is solved here using a simple collocation method based on Simpson's rule.

Taking $f^{(0)}(x) = 2$, then

$$g(2 : x) = 2 - \frac{4}{\pi}\int_{-1}^{1} \frac{\mathrm{d}y}{1+(x-y)^2} \tag{8.3.59}$$

and using Simpson's rule for the quadrature in (8.3.58) gives

$$\delta f(x) = g(2; x) - \frac{4}{\pi}\frac{1}{3}\left\{\frac{\delta f(-1)}{1+(x+1)^2} + \frac{4\delta f(0)}{1+x^2} + \frac{\delta f(1)}{1+(x-1)^2}\right\} \tag{8.3.60}$$

The resulting 3×3 matrix obtained by setting $x = -1$, $x = 0$ and $x = 1$ successively is

$$\begin{pmatrix} 1+\frac{4}{3\pi} & \frac{8}{3\pi} & \frac{4}{15\pi} \\ \frac{2}{3\pi} & 1+\frac{16}{3\pi} & \frac{2}{3\pi} \\ \frac{4}{15\pi} & \frac{8}{3\pi} & 1+\frac{4}{3\pi} \end{pmatrix} \begin{pmatrix} \delta f(-1) \\ \delta f(0) \\ \delta f(1) \end{pmatrix} = \begin{pmatrix} -\frac{4}{\pi}\arctan 2 \\ -\frac{8}{\pi}\arctan 1 \\ -\frac{4}{\pi}\arctan 2 \end{pmatrix} \tag{8.3.61}$$

with a solution

$$\begin{aligned} \delta f(1) &= -0.5672 \\ \delta f(0) &= -0.6522 \\ \delta f(-1) &= -0.5672 \end{aligned} \tag{8.3.62}$$

Improved accuracy would be obtained by employing a basis-set approximation of the type suggested in (8.3.53), and continuing the iterations to convergence.

8.4 Eigenvalue problems

It would be wrong to omit some mention of the eigenvalue problem as it specifically involves the integral equation formulation. This topic forms the complete content of some books and has application in areas such as quantum mechanics where techniques have been developed to compute energy values for atomic systems from integral equation formulations of Schrödinger's equation, such as the Hartree–Fock equations. This work is closely related to the content of the final chapter on optimization, and involves the use of variational principles which are only lightly touched upon in this work, primarily in the discussion on finite elements in Chapter 9.

Consider the integral eigenvalue problem written in operator form

$$y = \lambda \mathbf{K} y \tag{8.4.1}$$

with

$$\mathbf{K}y(x) = \int_a^b K(x, t)y(y)\,\mathrm{d}t \tag{8.4.2}$$

If the inner product of (8.4.1) is taken with y, the relation

$$\lambda = \frac{(y, y)}{(y, \mathbf{K}y)} \tag{8.4.3}$$

is obtained, provided $(y, \mathbf{K}y) \neq 0$, which is true if $\mathbf{K}$ is a positive definite operator satisfying

$$(\psi, \mathbf{K}\psi) \geq 0 \quad \forall \quad \psi \neq 0 \tag{8.4.4}$$

As it stands, equation (8.4.3) has little value as the eigensolution $y(x)$ is not known and hence cannot be used to compute λ. However (8.4.3) forms the basis for considering

$$\Lambda(\phi) = \frac{(\phi, \phi)}{(\phi, \mathbf{K}\phi)} \tag{8.4.5}$$

as a variational principle for λ. By writing $\phi(x) = y(x) + \delta y(x)$ and the corresponding Λ as $\lambda + \delta\lambda$ and expanding to first order gives

$$\begin{aligned}(y + \delta y, y + \delta y) &= (y, y) + (\delta y, y) + (y, \delta y) + (\delta y, \delta y)\\ &= (y, y) + 2(\delta y, y) + (\delta y, \delta y)\end{aligned} \tag{8.4.6}$$

for the numerator, and

$$(y + \delta y, \mathbf{K}(y + \delta y)) = (y, \mathbf{K}y) + (\delta y, \mathbf{K}y) + (y, \mathbf{K}\delta y) + (\delta y, \mathbf{K}\delta y)$$

$$= (y, \mathbf{K}y) + 2(\delta y, \mathbf{K}y) + (\delta y, \mathbf{K}\delta y)$$

$$= \frac{1}{\lambda}(y, y) + \frac{2}{\lambda}(\delta y, y) + (\delta y, \mathbf{K}\delta y) \tag{8.4.7}$$

for the denominator using $y = \lambda \mathbf{K} y$ and hence $\mathbf{K}y = y/\lambda$. Hence the first variation of Λ satisfies

$$\delta\lambda = \frac{(\delta y, \delta y) - \lambda(\delta y, \frac{\mathbf{K}}{\delta} y)}{\frac{1}{\lambda}(y, y) + \frac{2}{\lambda}(\delta y, y) + (\delta y, \mathbf{K}\delta y)} \tag{8.4.8}$$

which is second order in δy. Hence the functional (8.4.5) is stationary for the required eigenvalues λ.

It can be further shown that Λ is an upper bound for the set of eigenvalues λ_i. Suppose $\mathbf{K}$ has a complete set of eigenfunctions $\{y_i\}$ with eigenvalues λ_i so that

$$y_i = \lambda_i \mathbf{K} y_i \tag{8.4.9}$$

with the properties that $\lambda_i \geq 0$ and the eigensolutions $\{y_i\}$ are orthonormal or $(y_i, y_j) = \delta_{ij}$ as

$$(y_j, y_i) = \lambda_i (y_j, \mathbf{K} y_i) \tag{8.4.10}$$

$$(y_i, y_j) = \lambda_j (y_i, \mathbf{K} y_j) \tag{8.4.11}$$

Subtracting gives

$$0 = (\lambda_i - \lambda_j)(y_j, \mathbf{K} y_i) \tag{8.4.12}$$

but $\lambda_i \neq \lambda_j$ and therefore the orthogonal result follows as

$$(y_i, \mathbf{K} y_j) = \left(y_i, \frac{1}{\lambda} y_j \right) = \frac{1}{\lambda}(y_i, y_j) \tag{8.4.13}$$

To establish the upper bound property of Λ, expand $\phi(x)$ as a series of basis functions to give

$$\phi(x) = \sum_i a_i y_i(x) \tag{8.4.14}$$

then

$$\Lambda = \frac{(\phi, \phi)}{(\phi, \mathbf{K}\phi)} = \frac{\left(\sum_i a_i y_i, \sum_i a_i y_i \right)}{\left(\sum_i a_i y_i, \mathbf{K} \sum_i a_i y_i \right)}$$

$$= \frac{\left(\sum_i a_i y_i, \sum_i a_i y_i \right)}{\left(\sum_i a_i y_i, \sum_i \frac{a_i}{\lambda_i} y_i \right)} = \frac{\sum_i a_i^2}{\sum_i \frac{a_i^2}{\lambda_i}} \tag{8.4.15}$$

using the orthonormal property. Hence

$$\Lambda - \lambda_1 = \frac{\sum_i a_i^2}{\sum_i \frac{a_i^2}{\lambda}} = \lambda_1$$

$$= \frac{\sum_i a_i^2\left(1 - \frac{\lambda_1}{\lambda_i}\right)}{\sum_i \frac{a_i^2}{\lambda_i}} \tag{8.4.16}$$

But $\lambda_i \geq \lambda_1$ for $i = 1, 2, \ldots$ and hence $\Lambda \geq \lambda_1$.

The well-known Rayleigh–Ritz procedure is based on this variational principle. The function $\phi(x)$ is chosen to include a set of free parameters $c_1, c_2, \ldots c_N$ and the functional (8.4.5) is minimized with respect to these constants. The minimization is usually completed numerically using one of the many optimization routines described in Chapter 10. However, for the purpose of illustration an analytic example is completed here. Take the eigenvalue problem

$$y(x) = \lambda \int_0^1 e^{x+t} y(t)\,\mathrm{d}t \tag{8.4.17}$$

which has a positive definite kernel (prove it). Then taking $\phi(x) = 1$ with no free parameters gives

$$\Lambda = \frac{(\phi, \phi)}{(\phi, \mathbf{K}\phi)}$$

$$= \frac{\int_0^1 \mathrm{d}x}{\int_0^1 \int_0^1 e^{x+t}\,\mathrm{d}x\,\mathrm{d}t}$$

$$= \frac{1}{(e-1)^2} \tag{8.4.18}$$

which has a numerical value of 0.338 697, compared with the analytic result

$$\lambda = \frac{2}{(e^2 - 1)} = 0.313\,035 \tag{8.4.19}$$

Even with this very crude trial function and with no optimization, quite a reasonable upper bound is generated.

The process illustrated here is often put into a more general context by replacing (8.4.5) by the stationary values of $(\phi, \mathbf{K}\phi)$ subject to the constraint $(\phi, \phi) = 1$. Using Lagrange multipliers gives the equivalent minimization problem

$$\delta\{(\phi, \phi) - \lambda(\phi, \mathbf{K}\phi)\} = 0 \tag{8.4.20}$$

which for the trial function

$$\phi = \sum_{i=1}^{N} c_i \phi_i \tag{8.4.21}$$

results in the equations

$$\frac{\partial}{\partial c_i}\{(\phi, \phi) - \lambda(\phi, \mathbf{K}\phi)\} = 0 \tag{8.4.22}$$

which reduces to the algebraic eigenvalue problem

$$\sum_{j=1}^{N} c_j(\phi_{ij} - \lambda K_{ij}) = 0 \tag{8.4.23}$$

where

$$\phi_{ij} = (\phi_i, \phi_j) \tag{8.4.24}$$

and

$$K_{ij} = (\phi_i, \mathbf{K}\phi_j) \tag{8.4.25}$$

There is no need with this form to find the free constants, c_i, explicitly. This process is illustrated by considering the integral equation

$$y(x) = 2\lambda \int_0^x (1 - \sqrt{x})t\sqrt{t}\,y(t)\,\mathrm{d}t + 2\lambda \int_x^1 (1 - \sqrt{t})\sqrt{x}t\,y(t)\,\mathrm{d}t \tag{8.4.26}$$

This integral equation arises as a Green's function problem (see Roach(1970)), and the first stage in applying the above method is to symmetrize the kernel. Multiply both sides of (8.4.26) by $\sqrt{x}$ to give

$$\sqrt{x}y(x) = 2\lambda \int_0^x (1 - \sqrt{x})t\sqrt{t}\,y(t)\sqrt{x}\,\mathrm{d}t + 2\lambda \int_x^1 (1 - \sqrt{t})\sqrt{x}t\,y(t)\sqrt{x}\,\mathrm{d}t \tag{8.4.27}$$

and put $Y(x) = \sqrt{x}y(x)$ to give

$$Y(x) = 2\lambda \int_0^x (1 - \sqrt{x})t\sqrt{x}\,Y(t)\,\mathrm{d}t + 2\lambda \int_x^1 (1 - \sqrt{t})x\sqrt{t}\,Y(t)\,\mathrm{d}t \tag{8.4.28}$$

Now try a two-parameter trial function $\phi_1 = 1$ and $\phi_2 = x$. The relevant operators and inner products are

$$\begin{aligned}\mathbf{K}\phi_1 &= 2\int_0^x (1 - \sqrt{x})t\sqrt{x}\,\mathrm{d}t + 2\int_x^1 (1 - \sqrt{t})x\sqrt{t}\,\mathrm{d}t \\ &= \frac{x}{3} - \frac{x^{5/2}}{3}\end{aligned} \tag{8.4.29}$$

$$\mathbf{K}\phi_2 = 2\int_0^x (1-\sqrt{x})t^2\sqrt{x}\,\mathrm{d}t + 2\int_x^1 (1-\sqrt{t})xt^{3/2}\,\mathrm{d}t$$

$$= \frac{2x}{15} - \frac{2}{15}x^{7/2} \tag{8.4.30}$$

$$(\phi_1, \phi_1) = \int_0^1 1\,\mathrm{d}x = 1 \tag{8.4.31}$$

$$(\phi_1, \phi_2) = \int_0^1 x\,\mathrm{d}x = \frac{1}{2} = (\phi_2, \phi_1) \tag{8.4.32}$$

$$(\phi_2, \phi_2) = \int_0^1 x^2\,\mathrm{d}x = \frac{1}{3} \tag{8.4.33}$$

$$(\phi_1, \mathbf{K}\phi_1) = \int_0^1 \frac{x}{3} - \frac{x^{5/2}}{3}\,\mathrm{d}x = \frac{1}{14} \tag{8.4.34}$$

$$(\phi_1, \mathbf{K}\phi_2) = \int_0^1 \frac{2x}{15} - 2\frac{x^{7/2}}{15}\,\mathrm{d}x = \frac{1}{27} \tag{8.4.35}$$

$$(\phi_2, \mathbf{K}\phi_1) = \int_0^1 \frac{x^2}{3} - \frac{x^{7/2}}{3}\,\mathrm{d}x = \frac{1}{27} \tag{8.4.36}$$

$$(\phi_2, \mathbf{K}\phi_2) = \int_0^1 \frac{2}{15}(x^2 - x^{9/2})\,\mathrm{d}x = \frac{2}{99} \tag{8.4.37}$$

Hence the matrix in (8.4.23) is

$$\begin{bmatrix} 1 - \frac{\lambda}{14} & \frac{1}{2} - \frac{\lambda}{27} \\ \frac{1}{2} - \frac{\lambda}{27} & \frac{1}{3} - \frac{2\lambda}{99} \end{bmatrix} \tag{8.4.38}$$

with characteristic equation

$$\frac{4\lambda^2}{56133} - \frac{29\lambda}{4158} + \frac{1}{12} = 0 \tag{8.4.39}$$

and eigenvalue estimates of 83.943 and 13.931.

There is a powerful extension of the variational approach in which an iteration is added. The method is called the variation–iteration method and is defined by

$$\phi^{(n+1)} = \mathbf{K}\phi^{(n)} \tag{8.4.40}$$

with $\phi^{(0)}$ being some suitable trial function. A sequence of approximate eigenvalues is then generated from

$$\lambda^{(n)} = \frac{(\phi^{(n)}, \phi^{(n)})}{(\phi^{(n)}, \mathbf{K}\phi^{(n)})}$$

$$= \frac{(\phi^{(n)}, \phi^{(n)})}{(\phi^{(n)}, \phi^{(n+1)})}$$

$$= \frac{(n, n)}{(n, n+1)} \tag{8.4.41}$$

with the abbreviated notation

$$(n, m) = (\phi^{(n)}, \phi^{(m)}) = \int_a^b \phi^{(n)}\phi^{(m)}\,\mathrm{d}x \tag{8.4.42}$$

In addition the intermediate iterate

$$\lambda^{(n+1/2)} = \frac{(n, n+1)}{(n+1, n+1)} \tag{8.4.43}$$

can be defined with the property

$$\lambda^{(n)} \geq \lambda^{(n+1/2)} \geq \lambda^{(n+1)} \tag{8.4.44}$$

To establish these inequalities, take the Cauchy–Schwartz inequality in the form

$$|(\phi, \psi)| \leq (\phi, \phi)(\psi, \psi) \tag{8.4.45}$$

with

$$(\phi, \psi) = \int_a^b \phi(x)\psi(x)\,\mathrm{d}x \tag{8.4.46}$$

Then taking $\phi = \phi^{(n)}$ and $\psi = \phi^{(n+1)}$ gives

$$|(n, n+1)^2| \leq (n, n)(n+1, n+1) \tag{8.4.47}$$

and hence

$$\lambda^{(n)} = \frac{(n, n)}{(n, n+1)} \geq \frac{(n, n+1)}{(n+1, n+1)} = \lambda^{(n+1/2)} \tag{8.4.48}$$

Now taking the inner product $(\phi, \mathbf{K}\psi)$, Cauchy–Schwartz gives

$$|(\phi, \mathbf{K}\psi)|^2 \leq (\phi, \mathbf{K}\phi)(\psi, \mathbf{K}\psi) \tag{8.4.49}$$

and now using $\phi = \phi^{(n)}$ and $\psi = \phi^{(n+1)}$ with $\phi^{(n+1)} = \mathbf{K}\phi^{(n)}$ gives

$$|(n, n+2)|^2 \leq (n, n+1)(n+1, n+2) \tag{8.4.50}$$

Hence by the symmetry of $\mathbf{K}$

$$|(n, n+2)|^2 = (n+1, n+1)^2 \leq (n, n+1)(n+1, n+2) \tag{8.4.51}$$

to give

$$\lambda^{(n+1)} = \frac{(n+1, n+1)}{(n+1, n+2)} \leq \frac{(n, n+1)}{(n+1, n+1)} \tag{8.4.52}$$

Hence (8.4.52) and (8.4.48) combine to yield (8.4.44).

This process is illustrated in the example

$$y(x) = \lambda \int_0^x \left(\frac{t}{x} - xt\right) y(t)t \, \mathrm{d}t + \lambda \int_x^1 \left(\frac{x}{t} - xt\right) y(t)t \, \mathrm{d}t \tag{8.4.53}$$

using the variation–iteration process with $\phi^{(0)}(x) = 1$ to give

$$\begin{aligned}\phi_1 = \mathbf{K}\phi^{(0)} &= \int_0^x \left(\frac{t}{x} - xt\right) t \, \mathrm{d}t + \int_x^1 \left(\frac{x}{t} - xt\right) t \, \mathrm{d}t \\ &= \tfrac{2}{3}x(1-x)\end{aligned} \tag{8.4.54}$$

Hence

$$\begin{aligned}\lambda^{(0)} &= \frac{(\phi^{(0)}, \phi^{(0)})}{(\phi^{(0)}, \phi^{(1)})} \\ &= \frac{\int_0^1 \mathrm{d}x}{\int_0^1 \frac{2}{3}x(1-x)\,\mathrm{d}x} = 9\end{aligned} \tag{8.4.55}$$

Continuing the process gives

$$\begin{aligned}\lambda^{(1/2)} &= \frac{(\phi^{(0)}, \phi^{(1)})}{(\phi^{(1)}, \phi^{(2)})} \\ &= \frac{\int_0^1 \frac{2}{3}x(1-x)\,\mathrm{d}x}{\frac{4}{9}\int_0^1 x^2(1-2x+x^2)\,\mathrm{d}x} \\ &= \frac{\frac{2}{3}\left[\frac{1}{2} - \frac{1}{3}\right]}{\frac{4}{9}\left[\frac{1}{3} - \frac{2}{4} + \frac{1}{5}\right]} \\ &= \frac{15}{2} = 7.5\end{aligned} \tag{8.4.56}$$

and

$$\begin{aligned}\phi^{(2)} &= \int_0^x \left(\frac{1}{x} - x\right) \frac{2}{3}t^3(1-t)\,\mathrm{d}t + \int_x^1 \left(\frac{x}{t} - xt\right) \frac{2}{3}t^2(1-t)\,\mathrm{d}t \\ &= \frac{2}{3}\left(\frac{2x^4}{15} - \frac{x^3}{4} + \frac{7x}{60}\right)\end{aligned} \tag{8.4.57}$$

which has itself a variational principle defined by

$$\Lambda_I = \frac{-(\phi, W\phi)}{(W\phi, \mathcal{G}\, W\phi)} \tag{8.5.13}$$

Which of these principles gives the best results? Expanding the two forms for Λ_D and Λ_I as in (8.4.15) gives

$$\Lambda_D = \frac{\sum_i \lambda_i a_i^2}{\sum_i a_i^2} \tag{8.5.14}$$

and with (8.4.15) gives

$$\Lambda_D - \Lambda_I = \frac{\left(\sum_i \lambda_i a_i^2\right)\left(\sum_i \frac{a_i^2}{\lambda_i}\right) - \left(\sum_i a_i^2\right)^2}{\left(\sum_i a_i^2\right)^2\left(\sum_i \frac{a_i^2}{\lambda_i}\right)} \tag{8.5.15}$$

which has a positive numerator from the Cauchy–Schwartz inequality to render

$$\Lambda_D \geq \Lambda_I \tag{8.5.16}$$

Hence the integral formulation gives the tighter bounds. An example of this inequality in practice is Bessel's equation in the form

$$\frac{\mathrm{d}}{\mathrm{d}x}\left(x\frac{\mathrm{d}y}{\mathrm{d}x}\right) + \lambda x y = 0 \tag{8.5.17}$$

with $y(0)$ finite and $y(1) = 0$. For Λ_D consider the trial function $\phi(x) = 1 - x$ which gives

$$(\phi, W\phi) = \int_0^1 (1-x)^2 x\, \mathrm{d}x = 1/12 \tag{8.5.18}$$

and

$$-(\phi, L\phi) = -\int_0^1 (1-x)(-1)\, \mathrm{d}x = 1/2 \tag{8.5.19}$$

leaving $\Lambda = 6$. The Green's function from conditions (i)–(v) above is

$$\mathcal{G}_1(x, t) = \ln(t) \qquad x < t \tag{8.5.20}$$

$$\mathcal{G}_2(x, t) = \ln(x) \qquad x > t \tag{8.5.21}$$

as the fundamental solution $y = a\ln(x) + b$ satisfies

$$\frac{\mathrm{d}}{\mathrm{d}x}(xy') = 0 \tag{8.5.22}$$

Then for Λ_I the numerator is the same as for Λ_D and the denominator is

$$(W\phi, \mathcal{G}W\phi) = \int_0^1 x(1-x)\left\{\int_0^x t(1-t)\ln x\,dt + \int_x^1 t(1-t)\ln t\,dt\right\} dx$$

$$= \int_0^1 (x-x^2)\left\{\ln x\left(\frac{x^3}{3}-\frac{x^2}{2}\right) - \frac{1}{9} + \frac{1}{4} - \frac{x^3}{3}\ln x\right.$$

$$\left. + \frac{x^3}{9} + \frac{x^2}{2}\ln x - \frac{x^2}{4}\right\} dx$$

$$= \frac{31}{36.60} \tag{8.5.23}$$

Hence $\Lambda_I = 36.60/12/31 = 5.806$ and the exact eigenvalue is $\lambda_1 = 5.783$ demonstrating the superiority of the integral formulation. Clearly by using more sophisticated trial functions, and then optimizing the resulting function of these variables, accurate results can be obtained. Chapter 10 deals with optimization, and the work in this section is a good example of how such work is applied.

Further reading

Delves L. M. and Walsh J. (1974) *Numerical Solution of Integral Equations*, Clarendon Press, Oxford.

Morse P. M. and Feshbach H. (1953) *Methods of Theoretical Physics, Part II*, McGraw-Hill, New York.

Roach G. F. (1970) *Green's Functions*, Van Nostrand Reinhold Company, London.

Exercises

(8.1) Consider the application of direct iteration, basis set (using the set x^i) and collocation using Simpson's rule as a quadrature rule on the integral equation

$$f(x) = 1 + \lambda \int_0^{\pi/2} \cos(x-y) f(y)\,dy$$

for $\lambda = \frac{1}{2}$ and $\lambda = 1$. Compare the results with the analytic solution. Are there values of λ for which no solution exists?

(8.2) Use a five-point generalized Simpson's rule to convert the integral equation

$$f(x) = 2 + \lambda \int_0^1 e^{xy} f(y)\,dy$$

into a set of simultaneous algebraic equations using $\lambda = \frac{1}{32}$ and taking the collocation points to be at $x = 0,\ \frac{1}{4},\ \frac{1}{2},\ \frac{3}{4}$ and 1.

(8.3) Solve numerically the integral equation

$$x(t) = 1 + \frac{1}{t}\int_0^t s(s-t)x^{\nu}(s)\,\mathrm{d}s$$

for the linear case $\nu = 1$. What modifications are needed for the case when $\nu = 2$? Consider a range of t from 0 to 1.

(8.4) Solve numerically the integral equation

$$\phi(x) = (x+2)^{-2} - 2\int_0^x \frac{\phi}{(x-t+2)^2}\,\mathrm{d}t$$

for the range $x \in [0, 1]$.

(8.5) Solve numerically the integral equation

$$y(x) + \frac{2}{\pi}\int_0^{2\pi} \frac{y(s)}{5 - 3\cos(x+s)}\,\mathrm{d}s = \begin{cases} \frac{\sin x}{\pi} & 0 \le x \le \pi \\ 0 & -\pi \le x \le 0 \end{cases}$$

(8.6) Confirm that the integral equation

$$F(x) = 1 + 2\int_0^1 K(x,\xi)F(\xi)\,\mathrm{d}\xi$$

with the kernel

$$K(x,\xi) = \begin{cases} -1 & \xi < x \\ -e^{P(x-\xi)} & \xi \ge x \end{cases}$$

satisfies the boundary value problem

$$\frac{1}{P}F'' - F' - 2F = 0$$

with boundary conditions

$$F - \frac{1}{P}F' = 1$$

at $x = 0$ and $F' = 0$ at $x = 1$. Hence solve the boundary value problem using a numerical solution to the equivalent integral equation.

(8.7) Solve the problem in exercise (8.6) using the equivalent but alternative integral equation

$$\frac{1}{P}F(x) = (\alpha - x) - \int_x^1 F(s)\,\mathrm{d}s + 2\int_0^1 F^2(s)K(x,s)\,\mathrm{d}s$$

with the kernel

$$K(x,s) = \begin{cases} x - \alpha & s < x \\ s - \alpha & s \ge x \end{cases}$$

and $\alpha = 1 + 1/P$.

(8.8) A non-linear spring gives rise to the differential equation

$$x'' = -ax - bx^3$$

with $x = x_0,\ x' = 0$ at $t = 0$. Establish that an equivalent Volterra integral equation is

$$x(t) = x_0 \cosh t + \int_0^t K(t, s)x^3(s)\,\mathrm{d}s$$

where $K(t, s) = \sinh(s - t)$ for $a = -1$ and $b = 1$. Find a numerical solution to this problem.

(8.9) Use Fubini's method to convert the boundary value problem

$$\frac{\mathrm{d}^2y}{\mathrm{d}x^2} + \frac{\mathrm{d}y}{\mathrm{d}x} + y^3 = x$$

with boundary conditions $y(0) = 0$ and $y(1) = 1$ to a non-linear Fredholm integral equation of the form

$$y(x) = g(x) + \int_0^1 K(x, t)y^3(t)\,\mathrm{d}t$$

using

$$\frac{\mathrm{d}^2y}{\mathrm{d}x^2} + \frac{\mathrm{d}y}{\mathrm{d}x} = 0$$

to establish the starting function for the method. Hence apply the δf method to obtain a numerical solution.

(8.10) Apply direct iteration to the Volterra integral equation

$$f(x) = 1 + \int_0^x (1 + y^2)^{1/2} f^2(y)\,\mathrm{d}y$$

using $f^{[0]} = 1$. Compare the solution with that obtained by using a collocation method, based on say Simpson's rule.

(8.11) Use Fubini's method to convert the boundary value problem

$$yy'' + y'^2 + y^3 = x$$

with boundary conditions $y(0) = 0$ and $y(1) = 1$ into the non-linear integral equation

$$y^2(x) = x + \int_0^1 (y^3(t) - t)G(x, t)\,\mathrm{d}t$$

where

$$G(x, t) = \begin{cases} 2t(1 - x) & t \le x \\ 2x(1 - t) & t \ge x \end{cases}$$

Use $yy'' + y'^2 = 0$ for the initial Fubini form. Obtain a conventional linear Fredholm integral equation for δf, and hence start off a numerical solution.

(8.12) By using the separability of the kernel show that the integral equation

$$y(x) = \lambda \int_0^1 e^{x+t} \phi(t)\,\mathrm{d}t$$

has an eigenvalue $\lambda = 2/(e^2 - 1)$.

(8.13) Show that the integral equation

$$y(x) = \lambda \int_0^{\pi/2} \cos(x - t) y(t)\,\mathrm{d}t$$

has eigenvalues $4/(\pi + 2)$ and $4/(\pi - 2)$ analytically. Using $\phi(x) = 1$ show that $\Lambda = \pi/4 = 0.785\,398$, noting that this is indeed an upper bound on λ_1.

(8.14) Apply the variation–iteration method to the integral equation of the previous exercise. Show that $(0, 0) = \pi/2$, $(0, 1) = 2$ and $(1, 1) = \pi/2 + 1$, and hence demonstrate that $\lambda^{(1/2)}$ gives the exact result.

(8.15) Apply the variation–iteration method to the integral equation

$$y(x) = \lambda \int_0^1 (t + x) y(t)\,\mathrm{d}t$$

which has a non-positive definite operator. Take $\phi^{(0)}(x) = 1$ and obtain $\lambda^{(0)}$, $\lambda^{(1/2)}$, $\lambda^{(1)}$, and $\lambda^{(3/2)}$. Confirm now that $\lambda^{(1/2)}$ is not an upper bound as the positive definite condition in the theory is violated.

(8.16) Apply the variation–iteration method to the integral equation

$$y(x) = \lambda \int_0^x t y(t)\,\mathrm{d}t + \lambda \int_x^1 x y(t)\,\mathrm{d}t$$

using $\phi^{(0)}(x) = 1$ and proceeding to $\lambda^{(3/2)}$.

(8.17) Show that the Green's function for the differential equation

$$y'' + \lambda y = 0$$

with boundary conditions $y(0) = y(1) = 0$ is

$$\mathcal{G}_1(x, t) = -x(1 - t) \qquad x < t$$

$$\mathcal{G}_2(x, t) = -t(1 - x) \qquad x > t$$

and use the trial function $x(1 - x)$ to find Λ_D, Λ_I and compare these values with the true result of π^2.

(8.18) Find a value for λ_1 for the eigenvalue problem

$$u'' + \lambda u = 0$$

with $u(0) = u(1) = 0$ using the trial function

$$\phi(x) = x(1 - x)(1 + cx)$$

optimizing with respect to c. You will get two minima, one giving an approximation to $\lambda_1 = \pi^2$ and the other to $\lambda_2 = 4\pi^2$. Higher accuracy can be obtained by taking quadratic and cubic terms instead of $(1 + cx)$.

(8.19) Solve exercise (8.18) using the trial function

$$\phi = c_1\phi_1(x) + c_2\phi_2(x) + c_3\phi_3(x)$$

with

$$\phi_1(x) = x(1 - x)$$
$$\phi_2(x) = x^2(1 - x)$$
$$\phi_3(x) = x^3(1 - x)$$

(8.20) Use the trial function $\psi = e^{-cx}$ to estimate the fundamental eigenvalue λ_1 for the differential equation

$$x^2\frac{\mathrm{d}^2\psi}{\mathrm{d}x^2} + 2x\frac{\mathrm{d}\psi}{\mathrm{d}x} + \lambda x e^{-x}\psi = 0$$

with $\psi \to 0$ as $x \to \infty$ on $[0, \infty]$. Obtain a better estimate using the combination

$$\phi(x) = Ae^{-x} + Bxe^{-x}.$$

9

Partial Differential Equations

9.1 Introduction

The various branches of applied mathematics can be simply considered as investigations of the underlying partial differential equations. Hence quantum mechanics depends on Schrödinger's equations, fluid mechanics on various forms of Navier–Stokes' equations and electromagnetic theory on Maxwell's equations. Partial differential equations form a very large area of study in mathematics, and are therefore an important aspect of numerical analysis. Partial differential equations result from mathematical models of most real-life situations. They rarely permit analytic evaluation except under very idealized boundary conditions. The *order* of a partial differential equation is the order of the highest derivative. First-order equations can often be reduced to the solution of ordinary differential equations, which will be seen later in the considerations of characteristics. Second-order equations tend to demonstrate the numerical methods applicable to partial differential equations in general, and these are implementable for other orders. For the most part, consideration here is limited to linear problems — the non-linear ones constituting current research problems. To begin the discussion, three specific physical applications which prove typical of the problems to be solved are introduced. A classification of second-order equations is then covered, and each of the three physical problems falls into a different class in this categorization.

The first of these physical problems is the heat or diffusion equation which can be derived by considering an arbitrary volume V. The heat crossing the boundary will equate to the change of heat within the solid, which results in the equation

$$\int_V \rho c \frac{\partial \theta}{\partial t} \mathrm{d}V = \int_S k \,\mathrm{grad}\,\theta \;\mathrm{d}\mathbf{S} + \int_V H(\mathbf{r}, \theta, t)\mathrm{d}V \qquad (9.1.1)$$

where $\mathrm{d}\mathbf{S}$ is an element normal to the surface of V, θ is the temperature, k is the thermal conductivity, ρ the density and c the specific heat. H represents any heat generated in the volume by such action as radioactive decay, electrical heating or chemical action. The integral over the surface S can be converted to a volume integral by the divergence theorem (1.2.33) to give

$$\int_V \rho c \frac{\partial \theta}{\partial t} \mathrm{d}V = \int_V \mathrm{div}(k \,\mathrm{grad}\,\theta)\mathrm{d}V + \int_V H(\mathbf{r}, \theta, t)\;\mathrm{d}V \qquad (9.1.2)$$

But this balance is valid for an arbitrary volume and therefore the integrands themselves must satisfy

$$\rho c \frac{\partial \theta}{\partial t} = \operatorname{div}(k \operatorname{grad} \theta) + H(\mathbf{r}, \theta, t) \tag{9.1.3}$$

In the special but common case in which $k =$ constant, the diffusion equation reduces to

$$\frac{\partial \theta}{\partial t} = K\nabla^2\theta + Q(\mathbf{r}, \theta, t) \tag{9.1.4}$$

with

$$K = \frac{k}{\rho c}$$

$$Q = \frac{H}{\rho c}$$

Typical boundary conditions for this equation would include:

(i) $\theta(\mathbf{r}, t)$ is a prescribed function of t for every point $\mathbf{r}$ on the surface,

(ii) the flux over boundary $\partial\theta/\partial n$ is prescribed where $\mathbf{n}$ is a normal vector to the surface element $d\mathbf{S}$, that is $d\mathbf{S} = \mathbf{n} dS$, and

(iii) the surface radiation is defined over S for example by

$$\frac{\partial \theta}{\partial n} = -a(\theta - \theta_0) \tag{9.1.5}$$

which is Newton's law of radiation.

The heat or diffusion equation applies to a very large number of other physical situations. The diffusion of matter such as smoke in the atmosphere, or a dye or pollutant in a liquid is covered by Fink's law

$$\mathbf{J} = -D \operatorname{grad} c \tag{9.1.6}$$

where D is the coefficient of diffusion and c the concentration. The vector $\mathbf{J}$ is the diffusion current vector, and therefore c satisfies

$$\frac{\partial c}{\partial t} = \operatorname{div}(D \operatorname{grad} c) \tag{9.1.7}$$

or

$$\frac{\partial c}{\partial t} = D\nabla^2 c \tag{9.1.8}$$

Other physical situations, in which the diffusion equation underlies the situation include neutron slowing, vorticity diffusion and propagation of long electromagnetic waves in a good conductor such as an aerial.

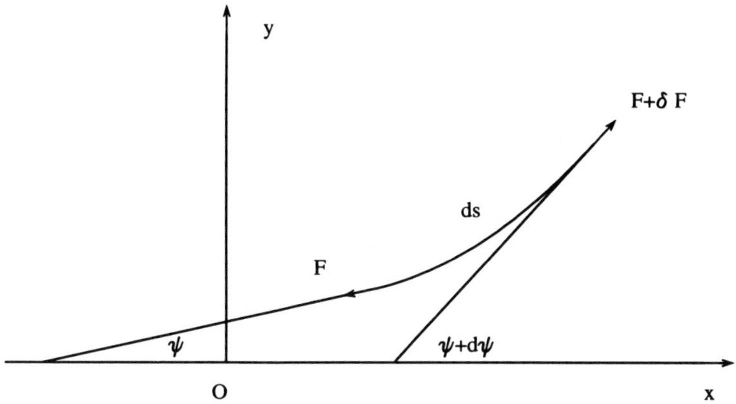

Figure 9.1

The second of the fundamental physical equations is the wave equation. Consider a small length of a stretched string as shown in Figure 9.1.

Then Newton's second law applied to the small length gives

$$F\sin(\psi + \mathrm{d}\psi) - F\sin\psi = \rho\ \mathrm{d}s\frac{\partial^2 y}{\partial t^2} = F\cos\psi\mathrm{d}\psi \tag{9.1.9}$$

where F is the tension in the string, ρ is the string density and y the displacement from the neutral position. From differential geometry

$$\tan\psi = \frac{\partial y}{\partial x} \tag{9.1.10}$$

and hence

$$\sec^2\psi\mathrm{d}\psi = \frac{\partial^2 y}{\partial x^2}\mathrm{d}x \tag{9.1.11}$$

which yields

$$\begin{aligned}\rho\frac{\partial^2 y}{\partial t^2} &= F\cos^3\psi\frac{\partial^2 y}{\partial x^2}\frac{\partial x}{\partial s}\\ &= F\cos^4\psi\frac{\partial^2 y}{\partial x^2}\end{aligned} \tag{9.1.12}$$

However for oscillations of small amplitude

$$\cos^2\psi = \left\{1 + \left(\frac{\partial y}{\partial x}\right)^2\right\}^{-1} \sim 1 \tag{9.1.13}$$

to yield the wave equation in the form

$$\frac{\partial^2 y}{\partial x^2} = \frac{1}{c^2}\frac{\partial^2 y}{\partial t^2} \tag{9.1.14}$$

with

$$c^2 = \frac{F}{\rho} \tag{9.1.15}$$

The third important example is Laplace's equation which is based on Gauss's law in electromagnetism (see Atkin, 1962) namely

$$\int_S \mathbf{E}.\mathbf{dS} = \rho \tag{9.1.16}$$

where $\mathbf{E}$ is the electric field, and ρ the charge density. If $\rho = 0$, then an application of the divergence theorem (1.2.33), together with S being the surface of an arbitary volume, gives the differential equation

$$\operatorname{div}\mathbf{E} = 0 \tag{9.1.17}$$

But for fields for which

$$\operatorname{curl}\mathbf{E} \neq 0 \tag{9.1.18}$$

there exists a ϕ such that

$$\mathbf{E} = \operatorname{grad}\phi \tag{9.1.19}$$

and hence

$$\operatorname{div}\operatorname{grad}\phi = 0 \tag{9.1.20}$$

or

$$\nabla^2\phi = 0 \tag{9.1.21}$$

which is Laplace's equation. The same equation holds for the flow of an ideal fluid. Such a fluid has no viscosity or compressibility factors and the equation of continuity is $\operatorname{div}\mathbf{q} = 0$, where $\mathbf{q}$ is the flow velocity vector (see Acheson, 1990). For irrotational flows the equivalent of (9.1.18) holds to allow the use of the potential function $\mathbf{q} = \operatorname{grad}\phi$ and again the defining equation is (9.1.21).

These three major physical problems are typical of the main types of second-order linear partial differential equations, and in the next paragraph mathematical arguments will be used to establish a classification of such problems.

9.2 Characteristics and classification

Consider a general second-order quasi-linear equation defined by the equation

$$Rr + Ss + Tt = W \tag{9.2.1}$$

where

$$p = \frac{\partial z}{\partial x}, \quad q = \frac{\partial z}{\partial y}, \quad r = \frac{\partial^2 z}{\partial x^2}, \quad s = \frac{\partial^2 z}{\partial x \partial y} \quad \text{and } t = \frac{\partial^2 z}{\partial y^2} \tag{9.2.2}$$

with

$$R = R(x, y), \quad S = S(x, y), \quad T = T(x, y), \qquad W = W(x, y, z, p, q). \tag{9.2.3}$$

Then the characteristic curves for this equation are defined as curves along which highest partial derivatives are not uniquely defined. In this case these derivatives are the second-order derivatives r, s and t. The set of linear algebraic equations which these derivatives satisfy can be written down in terms of differentials, and the condition for this set of linear equations to have a non-unique solution will yield the equations of the characteristics, whose significance will then become more apparent. Hence the linear equations follow as $\mathrm{d}z = p\mathrm{d}x + q\mathrm{d}y$ and hence

$$\begin{aligned} \mathrm{d}p &= r\mathrm{d}x + s\mathrm{d}y \\ \text{and } \mathrm{d}q &= s\mathrm{d}x + t\mathrm{d}y \end{aligned} \tag{9.2.4}$$

to give the linear equations

$$\left.\begin{aligned} Rr + Ss + Tt &= W \\ r\mathrm{d}x + s\mathrm{d}y &= \mathrm{d}p \\ s\mathrm{d}x + t\mathrm{d}y &= \mathrm{d}q \end{aligned}\right\} \tag{9.2.5}$$

and there will be no unique solution when

$$\begin{vmatrix} R & S & T \\ \mathrm{d}x & \mathrm{d}y & 0 \\ 0 & \mathrm{d}x & \mathrm{d}y \end{vmatrix} = 0 \tag{9.2.6}$$

which expands to give the differential equation

$$R\left(\frac{\mathrm{d}y}{\mathrm{d}x}\right)^2 - S\left(\frac{\mathrm{d}y}{\mathrm{d}x}\right) + T = 0 \tag{9.2.7}$$

But when the determinant in (9.2.6) is zero, the other determinants in Cramer's rule for the solution of (9.2.5) will also be zero as the condition for no solution is not required. Hence the alternative determinantal condition

$$\begin{vmatrix} R & T & W \\ \mathrm{d}x & 0 & \mathrm{d}p \\ 0 & \mathrm{d}y & \mathrm{d}q \end{vmatrix} = 0 \tag{9.2.8}$$

also holds, and gives an equation which holds along a characteristic, namely

$$-R\mathrm{d}y\,\mathrm{d}p - T\,\mathrm{d}x\,\mathrm{d}q + W\,\mathrm{d}x\mathrm{d}y = 0 \tag{9.2.9}$$

or

$$R\frac{\mathrm{d}p}{\mathrm{d}x}\frac{\mathrm{d}y}{\mathrm{d}x}+T\frac{\mathrm{d}q}{\mathrm{d}x}-W\frac{\mathrm{d}y}{\mathrm{d}x}=0 \tag{9.2.10}$$

Returning now to (9.2.6), this equation is a quadratic in $\mathrm{d}y/\mathrm{d}x$ and there are three possible cases which arise. If the roots are real the characteristics form two families of real curves. A partial differential equation resulting in real characteristics is said to be hyperbolic. The condition is that

$$S^2 - 4RT > 0 \tag{9.2.11}$$

The second case is when the roots are equal to give the parabolic case and the condition

$$S^2 - 4RT = 0 \tag{9.2.12}$$

and when the roots are complex the underlying equation is said to be elliptic with the condition

$$S^2 - 4RT < 0 \tag{9.2.13}$$

The importance of characteristics only becomes apparent at this stage. The first feature is the use of characteristics to classify equations. The methods that will be used subsequently are quite different from type to type. In the case of hyperbolic equations, the characteristics are real and are used directly in the solution. Characteristics also play a role in reducing equations to a standard or canonical form. Consider the operator

$$R\frac{\partial^2}{\partial x^2}+S\frac{\partial^2}{\partial x\partial y}+T\frac{\partial^2}{\partial y^2} \tag{9.2.14}$$

and put $\xi=\xi(x,y)$, $\eta=\eta(x,y)$ and $z=\zeta$ to see what a general change of variable yields. The result is the operator

$$A(\xi_x,\xi_y)\frac{\partial^2\zeta}{\partial\xi^2}+2B(\xi_x,\xi_y,\eta_x\eta_y)\frac{\partial^2\zeta}{\partial\xi\partial\eta}$$
$$+A(\eta_x,\eta_y)\frac{\partial^2\zeta}{\partial\eta^2}=F(\xi,\eta,\zeta,\zeta_\xi,\zeta_n) \tag{9.2.15}$$

where

$$A(u,v)=Ru^2+Suv+Tv^2 \tag{9.2.16}$$

and

$$B(u_1,v_1,u_2,v_2)=Ru_1u_2+\tfrac{1}{2}S(u_1v_2+u_2v_1)+Tv_2v_2 \tag{9.2.17}$$

The question is now asked, for what ξ and η do we get the simplest form? Certainly if ξ and η can be found to make the coefficients A equal to zero, then a simplified form will result. However the condition that A should be zero is a partial differential equation of first order which can be solved analytically (Sneddon 1957). Different cases arise in the three classifications. In the hyberbolic case when $S^2-4RT>0$, let $R\alpha^2 + S\alpha + T = 0$ have roots λ_1 and λ_2 then $\xi = f_1(x, y)$ and $\eta = f_2(x, y)$ where $f_1(x, y)$ and $f_2(x, y)$ are the solutions of the two factors in the related ordinary differential equations

$$\left[\frac{\mathrm{d}y}{\mathrm{d}x} + \lambda_1(x, y)\right]\left[\frac{\mathrm{d}y}{\mathrm{d}x} + \lambda_2(x, y)\right] = 0 \qquad (9.2.18)$$

Therefore the required transformations are precisely the defining functions of the characteristic curves. Hence with this change of variable the partial differential equation becomes

$$\frac{\partial^2 \zeta}{\partial \eta \partial \xi} = \phi(\xi, \eta, \zeta, \zeta_\xi, \zeta_n) \qquad (9.2.19)$$

which is the canonical form for the hyperbolic case.

In the parabolic case there is now only one root, and any independent function is used for the other variable in the transformation. Hence when $S^2 = 4RT$, the transformation $\xi = f_1(x, y)$ and $\eta =$ any independent function yields

$$\frac{\partial^2 \zeta}{\partial \eta^2} = \phi_1(\xi, \eta, \zeta, \zeta_\xi, \zeta_n) \qquad (9.2.20)$$

which is the canonical form for a parabolic equation.

In the elliptic case there are again two sets of characteristics but they are now complex. Writing $\xi = \alpha + \mathrm{i}\beta$ and $\eta = \alpha - \mathrm{i}\beta$ gives the real form

$$\frac{\partial^2 \zeta}{\partial \xi \partial \nu} = \frac{1}{4}\left(\frac{\partial^2 \zeta}{\partial \alpha^2} + \frac{\partial^2 \zeta}{\partial \beta^2}\right) \qquad (9.2.21)$$

and hence the elliptic canonical form

$$\frac{\partial^2 \zeta}{\partial \alpha^2} + \frac{\partial^2 \zeta}{\partial \beta^2} = \psi(\alpha, \beta, \zeta, \zeta_\alpha, \zeta_\beta) \qquad (9.2.22)$$

Note that Laplace's equation is in canonical form as is the heat equation, but the wave equation is not. As an example of reduction to canonical form, consider the quasi-linear second-order partial differential equation

$$\frac{\partial^2 u}{\partial x^2} + 2\frac{\partial^2 u}{\partial x \partial y} + \frac{\partial^2 u}{\partial y^2} + c^2\frac{\partial u}{\partial y} = 0 \qquad (9.2.23)$$

Then the equation of the characteristic curves is

$$\left(\frac{\mathrm{d}y}{\mathrm{d}x}\right)^2 - 2\frac{\mathrm{d}y}{\mathrm{d}x} + 1 = 0 \qquad (9.2.24)$$

or factorizing

$$\left(\frac{\mathrm{d}y}{\mathrm{d}x} - 1\right)^2 = 0 \qquad (9.2.25)$$

Therefore the transformation for the canonical form is:

$$\left.\begin{aligned} p &= x - y \\ q &= x \end{aligned}\right\} \tag{9.2.26}$$

and the required partial derivatives are:

$$\frac{\partial^2 u}{\partial x^2} = \frac{\partial^2 u}{\partial p^2}\left(\frac{\partial p}{\partial x}\right)^2 + \frac{\partial^2 u}{\partial q^2}\left(\frac{\partial q}{\partial x}\right)^2 \tag{9.2.27}$$

and

$$\frac{\partial^2 u}{\partial x \partial y} = \frac{\partial^2 u}{\partial p^2}\left(\frac{\partial p}{\partial y}\frac{\partial p}{\partial x}\right) \tag{9.2.28}$$

which yields the reduced form

$$\begin{aligned} \frac{\partial^2 u}{\partial x^2} + 2\frac{\partial^2 u}{\partial x \partial y} + \frac{\partial^2 u}{\partial y^2} &= \frac{\partial^2 u}{\partial p^2} + \frac{\partial^2 u}{\partial q^2} - 2\frac{\partial^2 u}{\partial p^2} + \frac{\partial^2 u}{\partial p^2} \\ &= \frac{\partial^2 u}{\partial q^2} \end{aligned} \tag{9.2.29}$$

with the transformed equation being

$$\frac{1}{c^2}\frac{\partial^2 u}{\partial q^2} = \frac{\partial u}{\partial p} \tag{9.2.30}$$

From a numerical point of view, the canonical forms reduce the number of different types of equation for which solutions need to be found. Effectively, effort can be concentrated on the canonical forms alone, and in this spirit the parabolic type will now be considered in detail.

9.3 Parabolic equations

In the light of the canonical forms derived above, the numerical solution of the heat equation

$$\frac{\partial u}{\partial t} = \frac{1}{c^2}\frac{\partial^2 u}{\partial x^2} \qquad \text{with } c \text{ a constant} \tag{9.3.1}$$

is considered. The (x, t) plane is divided into a grid as shown in Figure 9.2, and elementary finite difference forms are used to represent the partial derivatives.

Hence the first-order derivative with respect to t is represented by

$$\frac{\partial u}{\partial t} = \frac{u_{i,j+1} - u_{i,j}}{k} \tag{9.3.2}$$

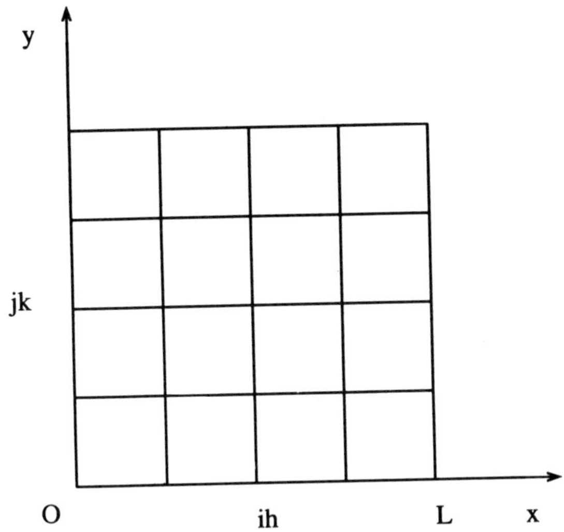

Figure 9.2

and the second derivative in x by

$$\frac{\partial^2 u}{\partial x^2} = \frac{u_{i+1,j} - 2u_{i,j} + u_{i-1,j}}{h^2} \tag{9.3.3}$$

where $u_{i,j}$ is the calculated value of $u(x, t)$ at the grid point (ih, jk). It would be possible here to use higher differences as derived in Chapter 2, but it seems more effective in practice to use a finer grid with the cruder representation. Hence the first method for the numerical solution of the heat equation is called the explicit method and has the finite difference form:

$$\frac{u_{i,j+1} - u_{i,j}}{k} = \frac{1}{c^2}\frac{u_{i+1,j} - 2u_{i,j} + u_{i-1,j}}{h^2} \tag{9.3.4}$$

or in the form used computationally

$$u_{i,j+1} = u_{i,j} + \frac{r}{c^2}\left(u_{i-1,j} - 2u_{i,j} + u_{i+1,j}\right) \tag{9.3.5}$$

where $r = k/h^2$. The boundary conditions for the heat equation will consist of a given $u(x, t)$ for $t = 0$, and values for $u(x, t)$ for all $t > 0$ at $x = a$ and $x = b$ for some constants a and b. Hence (9.3.5) can be used to progress the solution into the region $t > 0$ in a step-by-step manner which is often illustrated by the computational molecule in Figure 9.3.

This process is illustrated by the following example.

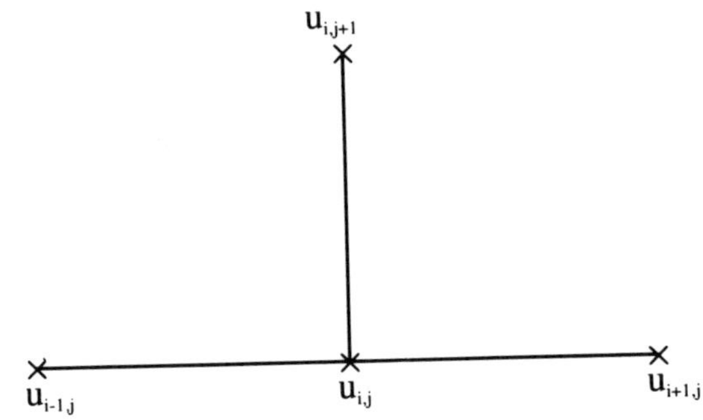

Figure 9.3

Example

A hollow tube 20 cm long is initially filled with air containing 2% ethyl alcohol vapour. At the bottom of the tube is a pool of alcohol which evaporates into the stagnant gas above. Heat transfers to the alcohol from its surroundings to maintain a constant 30°C temperature at which the vapour pressure of the alcohol is 0.1 atmospheres. At the upper end of the tube, vapours dissipate to the air making the concentration zero.

Find the concentration c as a function of x and t where c satisfies

$$\frac{\partial c}{\partial t} = D\frac{\partial^2 c}{\partial x^2} \qquad \text{and } D = 0.119 \text{ cm}^2/\text{s at } 30^\circ\text{C} \tag{9.3.6}$$

Let $r = \frac{1}{4}$ and $h = 4$ cm to give $k = 33.6$; then the rule becomes:

$$c_{i,j+1} = \tfrac{1}{4}\left(c_{i-1,j} + c_{i+1,j}\right) + \tfrac{1}{2}c_{i,j} \tag{9.3.7}$$

and initially on $0 < x < 20$ for $t = 0$, $c = 2$ and for $t \geq 0$ and $x = 0$, $c = 0$. Also from the vapour pressure of alcohol being 0.1 at 30°C on $x = 20$, the concentration is given by $c = 10$. The results are shown in Table 9.1.

This gives good accuracy compared with the analytic solution which has the form

$$c(x,t) = 5x + \frac{200}{\pi}\sum_{n=1}^{\infty} e^{-0.01175n^2t}\sin\frac{n\pi x}{10}$$

$$-\frac{120}{\pi}\sum_{n=1}^{\infty} e^{-0.00294(2n-1)^2t}\sin\frac{(2n-1)\pi x}{20} \tag{9.3.8}$$

where we see the solution tends to the expected linear function for large t.

The explicit method exhibits a stability problem when the step in the time direction, k, is made too large. The effect manifests itself with large inaccurate values appearing in the solution. This is due to instability which will be analysed

Table 9.1

t	x 0	4	8	12	16	20
0.0	0.0	2.0	2.0	2.0	2.0	10.0
33.6	0.0	1.5	2.0	2.0	4.0	10.0
67.2	0.0	1.25	1.875	2.5	5.0	10.0
100.8	0.0	1.094	1.875	2.969	5.625	10.0

later, and it will be shown that for stability, the ratio r has to satisfy $r < \frac{1}{2}$, and hence the step k has to be of order h^2 which gives a very slow progression in the t direction. It would be much better not to be restricted in this way. One such unrestricted method is the implicit method of Crank and Nicolson (1947).

In this method, $\partial^2 u/\partial x^2$ is approximated by its average finite difference form on the $j+1$ and jth rows. Hence

$$\frac{u_{i,j+1} - u_{i,j}}{k} = \frac{1}{2}\left\{\frac{u_{i+1,j+1} - 2u_{i,j+1} + u_{i-1,j+1}}{h^2} + \frac{u_{i+1,j} - 2u_{i,j} + u_{i-1,j}}{h^2}\right\} \tag{9.3.9}$$

which in terms of $r = k/h^2$ has the form

$$-ru_{i-1,j+1} + (2+2r)u_{i,j+1} - ru_{i+1,j+1} = ru_{i-1,j} + (2-2r)u_{i,j} + ru_{i+1,j} \tag{9.3.10}$$

in which the left-hand side contains all the unknown grid points and the right-hand side all the known points. If there are N internal mesh points along each time row, then there are also N equations to be solved simultaneously for that row of values. Also the resulting linear algebraic equations have diagonal dominance as each diagonal coefficient will have the value $(2+2r)$, and the equations are sparse having a non-zero row length of just three. Hence the iterative methods of Chapter 3 are directly applicable. Moreover this method is completely stable for all r.

Using the same example as for the explicit method to illustrate this implicit approach gives the equations:

$$-u_{i-1,j+1} + 4u_{i,j+1} - u_{i+1,j+1} = u_{i-1,j} + u_{i+1,j} \tag{9.3.11}$$

with $r = 1$ and results in a set of four linear equations for the first row of values with the form

$$\left.\begin{aligned} -0.0 + 4u_1 - u_2 &= 0.0 + 2.0 \\ -u_1 + 4u_2 - u_3 &= 2.0 + 2.0 \\ -u_2 + 4u_3 - u_4 &= 2.0 + 2.0 \\ -u_3 + 4u_4 - 10.0 &= 2.0 + 10.0 \end{aligned}\right\} \tag{9.3.12}$$

Two approaches are now possible. Either a method such as Jacobi can be used to give the iteration

$$4u_{i,j+1}{}^{(n+1)} = u_{i-1,j} + u_{i+1,j} + u_{i-1,j+1}{}^{(n)} + u_{i+1,j+1}{}^{(n)} \quad i = 1, 4 \qquad (9.3.13)$$

Or as the left-hand side is always the same for each successive row, the alternative is to invert the matrix involved, and then use the inverse to compute successive rows as required. If the matrix is denoted by **A**, then

$$A = \begin{bmatrix} 4 & -1 & 0 & 0 \\ -1 & 4 & -1 & 0 \\ 0 & -1 & 4 & -1 \\ 0 & 0 & -1 & 4 \end{bmatrix} \qquad (9.3.14)$$

with

$$A^{-1} = \begin{bmatrix} 0.2554 & 0.0718 & 0.019\,14 & 0.004\,78 \\ 0.0718 & 0.2871 & 0.076\,55 & 0.019\,14 \\ 0.019\,14 & 0.076\,55 & 0.2871 & 0.0718 \\ 0.004\,78 & 0.019\,14 & 0.0718 & 0.2554 \end{bmatrix} \qquad (9.3.15)$$

and now $x = A^{-1}b$ which for the first row gives $b^T = (2, 4, 4, 22)$ and for the second row $b^T = (2.019, 4.052, 8.011, 23.072)$. The results are shown in Table 9.2.

Compared with the explicit method the freedom to choose r allows reasonable progress in the direction of positive t to be made.

A more general form of finite difference representation is

$$\begin{aligned} u_{i,j+1} - u_{i,j} = r\,\{&\lambda(u_{i-1,j+1} - 2u_{i,j+1} + u_{i+1,j+1}) \\ &+(1-\lambda)(u_{i-1,j} - 2u_{i,j} + u_{i+1,j})\} \quad 0 \le \lambda \le 1 \end{aligned} \qquad (9.3.16)$$

where specific choices of λ give the following methods

$\lambda = \frac{1}{2}$ gives Crank–Nicholson

$\lambda = 0$ gives explicit method

$\lambda = 1$ gives O'Brien's method (O'Brien, Hyman and Kaplan 1951)

For $\lambda > 0$ these methods involve the solution of simultaneous linear equations for all the x-values along a line of t values and more will be said about the stability

Table 9.2

			x			
t	0	4	8	12	16	20
0	0	2.0	2.0	2.0	2.0	10.0
134.4	0	0.980	2.019	3.072	5.992	10.0
268.8	0	1.070	2.363	4.305	6.555	10.0
403.2	0	1.276	2.861	4.762	6.962	10.0

of these methods later. For heat equations in non-canonical form, it is usual to apply the characteristic transformation to get the standard form and then apply the above methods to this form. Attention will now turn to the second of the classes of equation.

9.4 Hyperbolic equations

Hyperbolic equations have real characteristics. We can ignore this fact and use difference approximations as for parabolic equations, but the theoretical nature of the solution of hyperbolic equations makes this process very risky. The boundary conditions are known along a curve, and the solution propagates outwards from this curve. If the boundary conditions are known for a particular length of curve, the characteristics through the extreme points will form an outer boundary. Inside this region the two families of characteristics will intersect, and will be used shortly to obtain the required solution. Outside this region, the solution is not known, and would depend on boundary values beyond those given. The region involved is called the region of influence, and the approach using finite differences would not expose the extent of the valid solution region. There is a further potential trap in that hyperbolic equations can have solutions with discontinuities which represent physical effects such as shock waves in compressible flow problems. These discontinuities show up as intersections of characteristics of the same family, and hence the numerical process based on characteristics can predict such regions.

The canonical form for the wave equation

$$\frac{\partial^2 y}{\partial t^2} = c^2 \frac{\partial^2 y}{\partial x^2} \tag{9.4.1}$$

is

$$\frac{\partial^2 y}{\partial \xi \partial \eta} = 0 \tag{9.4.2}$$

where

$$\begin{aligned} \xi &= x + ct \\ \eta &= x - ct \end{aligned} \tag{9.4.3}$$

and a simple analytic general solution can be written down as

$$y = f(x + ct) + g(x - ct) \tag{9.4.4}$$

This solution is no more useful than the original form as far as the problem has just been moved from solving a partial differential equation to solving a functional equation in order to put in the boundary conditions. This will also generally prove somewhat difficult.

Hence for hyperbolic equations the most successful approach is to use the characteristic curves. Along these curves there are the ordinary differential equations

(9.2.10). Hence the substance of the method is to use finite differences to find the approximate path of the characteristics and then use the ordinary differential equations along these curves, also in finite difference form, to yield the required solution. Consider a general second-order equation

$$a\frac{\partial^2 u}{\partial x^2} + b\frac{\partial^2 u}{\partial x \partial y} + c\frac{\partial^2 u}{\partial y^2} + c = 0 \tag{9.4.5}$$

for which the characteristics satisfy the differential equation

$$a\left(\frac{\mathrm{d}y}{\mathrm{d}x}\right)^2 - b\left(\frac{\mathrm{d}y}{\mathrm{d}x}\right) + c = 0 \tag{9.4.6}$$

as derived in (9.2.7). Along these curves the relation

$$a\frac{\mathrm{d}y}{\mathrm{d}x}\frac{\mathrm{d}p}{\mathrm{d}x} + c\frac{\mathrm{d}q}{\mathrm{d}x} + e\frac{\mathrm{d}y}{\mathrm{d}x} = 0 \tag{9.4.7}$$

or

$$a\frac{\mathrm{d}y}{\mathrm{d}x}\mathrm{d}p + c\mathrm{d}q + e\mathrm{d}y = 0 \tag{9.4.8}$$

holds. As the equation is hyperbolic, the roots of (9.4.6) are real, and hence may be factorized to give two first-order equations for the two required families of curves, namely

$$\frac{\mathrm{d}y}{\mathrm{d}x} = f \qquad \frac{\mathrm{d}y}{\mathrm{d}x} = g \tag{9.4.9}$$

where f and g are known functions, being the solutions of the quadratic (9.4.6).

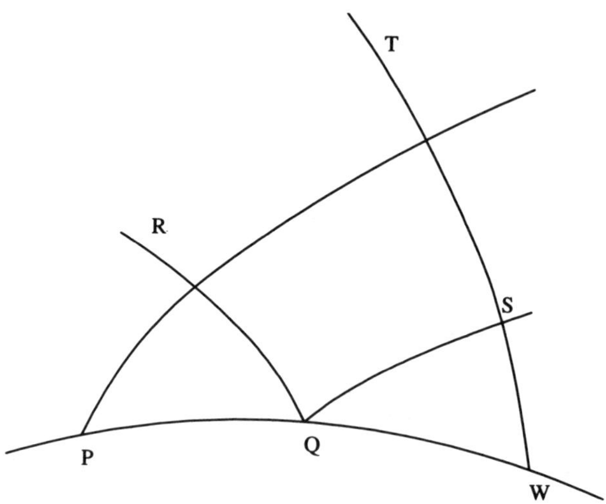

Figure 9.4

Let Γ be a non-characteristic curve on which initial values are known for u, p and q, as shown in Figure 9.4.

Let P and Q be typical points on γ and let the f characteristic through P intersect the g characteristic through Q at $R(x_R, y_R)$. As a first approximation, take PR and QR as straight lines of gradient f_P and g_Q. Then equation (9.4.9) gives:

$$y_R - y_P = f_P(x_R - x_P) \tag{9.4.10}$$

$$y_R - y_Q = g_Q(x_R - x_Q) \tag{9.4.11}$$

for the equations of the two lines PR and QR, and equation (9.4.8) gives:

$$af\mathrm{d}p + c\mathrm{d}q + e\mathrm{d}y = 0 \tag{9.4.12}$$

$$ag\mathrm{d}p + c\mathrm{d}q + e\mathrm{d}y = 0 \tag{9.4.13}$$

for the difference relation holding along these lines. In terms of the simple finite differences, these equations reduce to

$$a_P f_P(p_R - p_P) + c_P(q_R - q_P) + e_P(y_R - y_P) = 0 \tag{9.4.14}$$

and

$$a_Q q_Q(p_R - p_Q) + c_Q(q_R - q_Q) + e_Q(y_R - y_Q) = 0 \tag{9.4.15}$$

Hence x_R, y_R follow from (9.4.10) and (9.4.11) by the simple solution of a pair of simultaneous equations, and similarly equations (9.4.14) and (9.4.15) give p_R and q_R. It remains to find u_R from the relation

$$\mathrm{d}u = \frac{\partial u}{\partial x}\mathrm{d}x + \frac{\partial u}{\partial y}\mathrm{d}y = p\mathrm{d}x + q\mathrm{d}y \tag{9.4.16}$$

which in finite difference form gives

$$u_R - u_P = \tfrac{1}{2}(p_P + p_R)(x_R - x_P) + \tfrac{1}{2}(q_P + q_R)(y_R - y_P) \tag{9.4.17}$$

where averaging of the now known p and q values at the points P and Q, and R, is used to improve accuracy. Hence the solution progresses along intersecting characteristics from the initial curve of boundary values. The region of influence becomes naturally defined depending on the range of points like P and Q used along Γ. The values so obtained can be improved by simple iterative refinements with very little extra work. These first values can be used to make better estimates of the gradients which appear in the underlying formulae, using the averaging process of the type used in (9.4.17). Hence the gradient of PR can be estimated using the average of f_P and f_R, the latter being available from the previous calculation. Hence improved values for y_R and x_R are found from the solution of the simultaneous equations

$$\begin{aligned} y_R - y_P &= \tfrac{1}{2}(f_P + f_R)(x_R - x_P) \\ y_R - y_Q &= \tfrac{1}{2}(g_Q + g_R)(x_R - x_Q) \end{aligned} \tag{9.4.18}$$

In a similar vein, equations (9.4.14) and (9.4.15) have the improved forms

$$\begin{aligned}&\tfrac{1}{2}(a_P + a_R)\tfrac{1}{2}(f_P + f_R)(p_R - p_P) + \tfrac{1}{2}(c_P + c_R)(q_R - q_P)\\&\quad + \tfrac{1}{2}(e_P + e_R)(y_R - y_P) = 0\end{aligned} \tag{9.4.19}$$

and

$$\begin{aligned}&\tfrac{1}{2}(a_Q + a_R)\tfrac{1}{2}(g_Q + g_R)(p_R - p_Q) + \tfrac{1}{2}(c_Q + c_R)(q_R - q_Q)\\&\quad + \tfrac{1}{2}(e_Q + e_R)(y_R - y_Q) = 0\end{aligned} \tag{9.4.20}$$

These yield better improved values for p_R and q_R and then an improved u_R follows as before from equation (9.4.17).

To see an example of this approach, consider the solution of the partial differential equation

$$\frac{\partial^2 u}{\partial x^2} - u^2 \frac{\partial^2 u}{\partial y^2} = 0 \tag{9.4.21}$$

with boundary conditions

$$u = 0.2 + 5x^2$$

$$\frac{\partial u}{\partial y} = 3x$$

along $y = 0$ for $0 \le x \le 1$. This equation demonstrates the power of the characteristic method, as the equation is non-linear. It might be added that care is needed with problems of this type to ensure that the equations remains hyperbolic throughout the solution range. Clearly the classification of this equation depends on the solution, and such equations in general exhibit all three classifications in various regions of their solution. Hence the limits of the hyperbolic region can be found from the characteristic method and the solution on these extremes will become the boundary conditions for the parabolic and elliptic sections. However this illustration is limited to the hyperbolic region, and indeed to two typical points P and Q.

Consider the characteristics through $x = 0.2$ and $x = 0.3$ on the initial curve $y = 0$. From $u = 0.2 + 5x^2$, $\partial u/\partial x = p = 10x$ on Γ, the initial line. The slopes of the characteristics, m, satisfy

$$m^2 - u^2 = 0 \tag{9.4.22}$$

yielding

$$f = u = -g \tag{9.4.23}$$

The initial values of the variables are given by:

$$u = 0.2 + 5x^2 = f = -g$$

$$p = 10x, \quad q = 3x$$

$$a - 1, \quad b = e = 0, \quad c = -u^2$$

with the specific values at the points P and Q being

$$f_P = 0.4, \quad g_Q = -0.65, \quad p_P = 2.0, \quad p_Q = 3.0,$$
$$u_P = 0.4, \quad u_Q = 0.65, \quad q_P = 0.6, \quad q_Q = 0.9,$$
$$c_P = -0.16 \text{ and } c_Q = -0.4225.$$

Hence the equations for y_R and x_R are

$$y_R = 0.4(x_R - 0.2) \text{ and } y_R = -0.65(x_R - 0.3)$$

to give

$$x_R = 0.261\,90 \text{ and } y_R = 0.024\,762.$$

Along these characteristics the equations

$$0.4(p_R - 2.0) - 0.16(q_R - 0.6) = 0$$

and

$$-0.65(p_R - 3.0) - 0.425(q_R - 0.9) = 0$$

hold to give $p_R = 2.455\,24$ and $q_R = 1.738\,10$. Then u_R follows from

$$\begin{aligned} u_R &= 0.4 + \tfrac{1}{2}(2.0 + 2.455\,24)(0.0619) + \tfrac{1}{2}(1.738\,10 + 0.6)(0.024\,762) \\ &= 0.566\,84 \end{aligned}$$

to complete the initial stage of the process. The improvement stage then uses the updated equation for the characteristics

$$y_R = \tfrac{1}{2}(0.4 + 0.566\,84)(x_R - 0.2)$$

and

$$y_R = -\tfrac{1}{2}(0.65 + 0.566\,84)(x_R - 0.3)$$

to yield the improved values $x_R = 0.255\,72$ and $y_R = 0.026\,938$. The improved equations along the characteristics are

$$\tfrac{1}{2}(0.4 + 0.566\,84)(p_R - 2.0) - \tfrac{1}{2}(0.16 + 0.321\,31)(q_R - 0.6) = 0$$

and

$$-\tfrac{1}{2}(0.65 + 0.566\,84)(p_R - 3.0) - \tfrac{1}{2}(0.4225 + 0.321\,31)(q_R - 0.9) = 0$$

which yield

$$p_R = 2.531\,17 \quad \text{and} \quad q_R = 1.667\,00$$

with the new value for u_R being given by

$$u_R = 0.4 + \tfrac{1}{2}\{(2 + 2.531\,17)(0.055\,72) + (0.6 + 1.667\,00)(0.026\,983)\}$$
$$= 0.556\,77$$

A further iteration gives $x_R = 0.255\,78$, $y_R = 0.026\,68$, $p_R = 2.528\,76$, $q_R = 1.678\,37$ and $u_R = 0.556\,67$ which shows no further gain in accuracy. Hence it is common to apply only one refinement step. If further accuracy is required then the distance apart of the typical points P and Q on the initial curve is reduced to give a finer characteristic grid. One problem which then arises is that the points with the new grid will not coincide with the original set of solution points as the new approximate characteristic curves will be different from the original curves. In order to facilitate a comparison so that some idea of the attained accuracy can be made, it is necessary to employ a two-dimensional interpolation procedure. Such a procedure can be used to advantage if graphical output of the results is required, where some interface will be required with a standard contour drawing routine.

This completes a first look at two of the standard classes of partial differential equations, and leaves the elliptic class to be completed.

9.5 Elliptic equations

In a similar approach to the parabolic and hyperbolic equations, the canonical form derived in Section 9.2 allows the study of elliptic equations to be typified by the real canonical form

$$\frac{\partial^2 u}{\partial x^2} + \frac{\partial^2 u}{\partial y^2} = 0 \tag{9.5.1}$$

with boundary conditions given on a *closed* boundary. The solution is then computed for interior points. The method of solution is again to set up a finite difference representation, but now, with the boundary curve being necessarily closed no explicit method can be found. In fact we will end up with a set of algebraic equations with one unknown for each interior grid point. Hence generally, the problem reduces to the solution of a vast number of linear algebraic equations. Further the boundary conditions will be generally given on a curved boundary which is unlikely to coincide with the grid points. This does not usually occur for parabolic equations and the method of characteristics is such that it is no problem for hyperbolic equations. Hence, approximating partial derivatives with the usual sort of finite difference approximation, yields the difference approximations:

$$\begin{aligned}\frac{\partial^2 u}{\partial y^2} &= \frac{u_{i,j+1} - 2u_{i,j} + u_{i,j-1}}{h^2}\\ \frac{\partial^2 u}{\partial x^2} &= \frac{u_{i+1,j} - 2u_{i,j} + u_{i-1,j}}{h^2}\end{aligned} \tag{9.5.2}$$

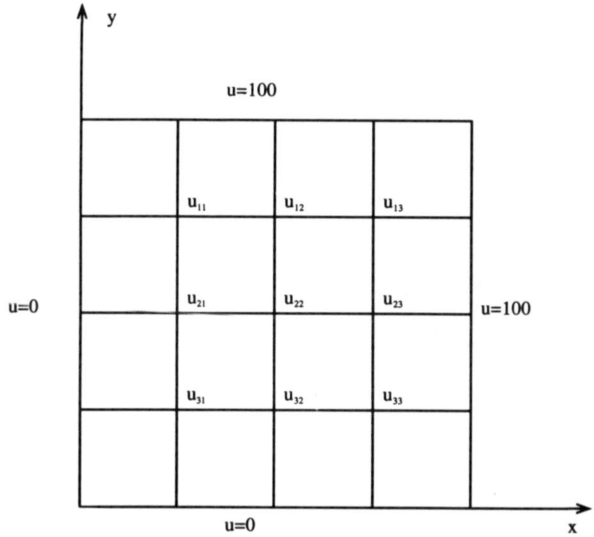

Figure 9.5

As an example consider the solution of Laplace's equation on the problem illustrated in Figure 9.5, with boundary conditions specified over a rectangular region to give one equation for each interior grid point. Hence the resulting nine equations are:

$$\frac{\partial^2 u}{\partial x^2} + \frac{\partial^2 u}{\partial y^2} = \nabla^2 u = \frac{1}{h^2}\left[u_{i+1,j} + u_{i-1,j} + u_{i,j+1} + u_{i,j-1} - 4u_{i,j}\right] = 0 \quad (9.5.3)$$

with i and j taking the values 1, 2 and 3. The resulting equations are sparse in the sense of Chapter 3 with, in general, three non-zero elements per row. Moreover the system is diagonally dominant with a diagonal value of four on each row. In this simple example the solution can be obtained by any of the methods discussed in Chapter 3 to yield

$$\begin{array}{ccccccccccc}
0 & + & 100 & + & u_{12} & + & u_{21} & - & 4u_{11} & = & 0 \\
u_{11} & + & 100 & + & u_{13} & + & u_{22} & - & 4u_{12} & = & 0 \\
u_{12} & + & 100 & + & 100 & + & u_{23} & + & 4u_{13} & = & 0 \\
0 & + & u_{11} & + & u_{22} & + & u_{31} & - & 4u_{21} & = & 0 \\
u_{21} & + & u_{12} & + & u_{23} & + & u_{32} & - & 4u_{22} & = & 0 \\
u_{22} & + & u_{13} & + & 100 & + & u_{33} & - & 4u_{23} & = & 0 \\
0 & + & u_{21} & + & u_{32} & + & 0 & - & 4u_{31} & = & 0 \\
u_{31} & + & u_{22} & + & u_{33} & + & 0 & - & 4u_{32} & = & 0 \\
u_{32} & + & u_{23} & + & 100 & + & 0 & - & 4u_{33} & = & 0
\end{array} \quad (9.5.4)$$

and hence

$$\begin{aligned}
u_{11} &= u_{33} = 50.00\\
u_{12} &= u_{23} = 71.43\\
u_{13} &= 85.72\\
u_{21} &= u_{32} = 28.57\\
u_{22} &= 50.00\\
u_{31} &= 14.29
\end{aligned}$$

A number of computational issues arise in this type of problem. Firstly the unknowns are doubly subscripted as u_{ij}, and these unknowns would normally be represented as a single vector $\mathbf{x}$ in the linear equations algorithms of Chapter 3. Hence there needs to be a correspondence between the x_i's and the u_{ij}'s. This ordering is often completed in a row-by-row approach, but an improved strategy is to alternate a row with a column. Such an arrangement helps to improve the convergence rate of the iterative method used to solve the resulting linear equations, and is called an alternating direction implicit method, (Peaceman and Rachford, 1955).

Most problems do not have nice rectangular boundaries, but rather involve curvilinear forms. The result is that near the boundary the simple computational molecule cannot be applied and special equations will be derived in a later paragraph to deal with these problems.

So far, simple finite difference approximations have been illustrated for the solution of the three major classes of quasi-linear second-order partial differential equations. This is the stage to consider some further theoretical considerations which will answer some of the questions which arose earlier. In particular, why is there a limit in the stability range for the explicit solution of the parabolic heat equation? As part of this discussion a number of other issues concerning convergence of methods will be pursued.

9.6 Convergence, stability and compatability

An introduction to partial differential equations and simple finite difference methods has been made. Classifications into parabolic, elliptic and hyperbolic equations using characteristics resulted in a method for hyperbolic equations. The heat equation was tackled using a finite difference representation of the relevant partial derivatives, and the explicit method and the implicit Crank–Nicolson method were used. A similar approach for elliptic equations yields a large set of linear algebraic equations. Three theoretical concepts — convergence, stability and compatability arise with partial differential equations, and the study of these concepts throws light on the methods of solution which can be utilized.

Suppose $u(x, t)$ is the solution of a finite difference approximation to a partial differential equation. Let an error be introduced (possibly due to round-off) at (x_0, t_0) so that $u(x_0, t_0) + E$ replaces $u(x_0, t_0)$. Let $u^*(x, t)$ be the resulting value at the point (x, t). The difference $u^*(x, t) - u(x, t)$ is called the *departure* of the solution due to E. If errors are introduced at more than one stage *cumulative departures* arise which are not in general additive.

If $|E(x, t)| < \delta$, that is δ is the maximum absolute error at any stage, then the method is *point wise stable* if the cumulative departure $\to 0$ as $\delta \to 0$ and if it does not increase faster than some power of h^{-1} as $h \to 0$.

If the corresponding continuous solution is bounded and if a process in a semi-infinite strip is considered, $0 < x < 1$, $t > 0$ say, then if a fixed grid is taken and if $\exists$ disturbances at $t = 0$ for which the finite difference solutions $u_{i,j}$ became unbounded as $j \to \infty$ then the method is *step-wise unstable*.

A method is *unstable according to John* under the following conditions: given $0 < x < 1$ and $0 < t < T$ from some T, then if as $h \to 0$, the finite difference solutions at $t = T$ become unbounded, the process is *unstable*.

Convergence is simply a matter of whether the *exact* finite difference solution tends to the true continuous solution as the grid size $h \to 0$.

For parabolic equations the explicit scheme can be shown to converge by applying *maximum principle analysis*. Consider the normalized equations

$$\frac{\partial u}{\partial t} = \frac{\partial^2 u}{\partial x^2} \quad 0 < x < 1 \tag{9.6.1}$$

where the physical constant c of (9.3.1) has been removed by scaling the variables, then the simple explicit representation is

$$u_{i,j+1} = ru_{i-1,j} + (1 + 2r)u_{i,j} + ru_{i+1,j} \quad i = 1, \ldots, M - 1 \tag{9.6.2}$$

with

$$r = \frac{k}{h^2} = \frac{\Delta t}{(\Delta x)^2} \tag{9.6.3}$$

In this approximation, $\partial u/\partial t$ is represented by:

$$\frac{u_{i,j+1} - u_{i,j}}{k} + 0(k) \tag{9.6.4}$$

and $\partial^2 u/\partial x^2$ by

$$\frac{u_{i-1,j} - 2u_{i,j} + u_{i+1,j}}{h^2} + 0(h^2) \tag{9.6.5}$$

Hence (9.6.2) with local truncation error included is:

$$U_{i,j+1} = rU_{i-1,j} + (1 - 2r)U_{i,j} + rU_{i+1,j} + 0[k^2 + kh^2] \tag{9.6.6}$$

where $U_{i,j}$ is the true solution at the point (i, j). Define $z_{i,j} = U_{i,j} - u_{i,j}$ so representing the error at each point, then subtracting (9.6.2) and (9.6.6) yields

$$z_{i,j+1} = rz_{i-1,j} + (1 - 2r)z_{i,j} + rz_{i+1,j} + 0[k^2 + kh^2] \quad i = 1, \ldots, M - 1 \tag{9.6.7}$$

But u and U agree on the boundaries assuming the boundary conditions are exact and hence

$$z_{i,0} = 0, \quad i = 0, \dots, M \quad \text{and} \quad z_{o,j} = z_{M_{i,j}} = 0, \quad j = 0, \dots, N \tag{9.6.8}$$

Equation (9.6.7) has non-negative terms in its right-hand side if

$$0 < r \leq \tfrac{1}{2} \tag{9.6.9}$$

to give

$$|z_{i,j+1}| \leq r|z_{i-1,j}| + (1-2r)|z_{i,j}| + r|z_{i+1,j}| + A[k^2 + kh^2] \tag{9.6.10}$$

where the definition of order is

$$f(s) = 0[\phi(s)] \tag{9.6.11}$$

if $\exists\ A > 0$ s.t $|f(s)| \leq A|\phi(s)|\ \forall\ s$. Further

$$|z_{i,j+1}| \leq \|z_j\| + A[k^2 + kh^2] \quad i = 1, \dots, M-1 \tag{9.6.12}$$

where $\|z_j\| = \max_{i=0,M} |z_{i,j}|$. Hence

$$\|z_{j+1}\| \leq \|z_j\| + A[k^2 + kh^2] \tag{9.6.13}$$

but $\|z_0\| = 0$, and hence

$$\begin{aligned} \|z_j\| &\leq A_j[k^2 + kh^2] \\ &\leq AT[k + h^2] \end{aligned} \tag{9.6.14}$$

where $jk \leq T$ and T is the upper limit of t. Hence the error $\to 0$ as k and $h \to 0$ to establish convergence. Further this result is precisely that of stability in the sense of John with the stability criteria $0 < r < \frac{1}{2}$.

There are a number of generally applicable ways of analyzing the second of the theoretical concepts, namely stability, for the numerical solution of partial differential equations. Of these, Fourier stability analysis and matrix stability analysis are treated here. Firstly, consider *Fourier stability analysis* applied to the parabolic equation method defined in (9.6.2).

The approach is to introduce errors along $t = 0$ which can be represented by a Fourier Series of the form

$$\sum_i A_i\, e^{-\hat{\imath}\lambda i x} \tag{9.6.15}$$

As i has been used as an index point in the grid system it is convenient to represent $\sqrt{(-1)}$ by $\hat{\imath}$ in the next few lines to avoid notational confusion. Because this analysis applies to linear problems only, one term such as $e^{-\hat{\imath}\lambda x}$ may be considered and then superposition used to find the effect of a sum of terms such as in (9.6.15). Let $e^{-\hat{\imath}\lambda x} = e^{-\hat{\imath}\lambda ih} = u_{i,0}$ on $t = 0$, then

$$u_{i,j} = e^{\alpha jk} e^{-\hat{\imath}\lambda ih} \tag{9.6.16}$$

for other t values by direct solution of the difference equations (9.6.2) by separation of the variables. Substituting this solution into (9.6.2) gives

$$\begin{aligned} e^{\alpha k} &= r\left\{e^{\hat{\imath}\lambda h} + e^{-\hat{\imath}\lambda h}\right\} + (1-2r) \\ &= 1 - 4r\sin^2 \tfrac{1}{2}\lambda h \end{aligned} \tag{9.6.17}$$

where r is non-negative. Hence $e^{\alpha k} < 1$ and if $e^{\alpha k} \geq 0$ then the solution will decay as $j \to \infty$. If $-1 < e^{\alpha k} < 0$ then the solution will be a decaying oscillation. Finally for $e^{\alpha k} < -1$ the solution increases and is unstable.

Hence stability occurs if $-1 \leq 1 - 4r\sin^2 \frac{1}{2}\lambda h$ or

$$r \leq \frac{1}{2\sin^2 \frac{1}{2}h} = \frac{1}{1 - \cos\lambda h} \tag{9.6.18}$$

Hence for non-oscillatory decay

$$1 - 4r\sin^2 \tfrac{1}{2}\lambda h \geq 0 \tag{9.6.19}$$

or

$$\text{and } r \leq \frac{1}{2(1-\cos\lambda h)} \tag{9.6.20}$$

Inequality (9.6.18) can be eased because $1/(1 - \cos\lambda h) < \frac{1}{2}$ giving

$$r \leq \tfrac{1}{2} \tag{9.6.21}$$

for stability which is the condition used in Section 9.3.

A second method of analysis of stability is the *matrix stability analysis* approach which will be illustrated for both the explicit and Crank–Nicolson methods.

For the explicit method the finite difference form is

$$u_{i,j+1} = ru_{i-1,j} + (1-2r)u_{i,j} + ru_{i+1,j} \tag{9.6.22}$$

which gives

$$\begin{pmatrix} u_{1,j+1} \\ u_{2,j+1} \\ \vdots \\ u_{N-1,j+1} \end{pmatrix} = \begin{pmatrix} (1-2r) & r & & & \\ r & (1-2r) & r & & \\ & r & (1-2r) & & \\ & & \ddots & & \\ & & & \ddots & \\ & & & r & (1-2r) \end{pmatrix} \begin{pmatrix} u_{1,j} \\ u_{2,j} \\ \vdots \\ u_{N-1,j} \end{pmatrix} \tag{9.6.23}$$

or in matrix form $\mathbf{u}_{j+1} = \mathbf{A}\mathbf{u}_j$. Hence $\mathbf{u}_j = \mathbf{A}\mathbf{u}_{j-1} = \mathbf{A}(\mathbf{A}\mathbf{u}_{j-2}) = \cdots = \mathbf{A}^j\mathbf{u}_0$ where $\mathbf{u}_0$ is the initial value. Suppose small errors are introduced so that the starting vector is $\mathbf{u}_0{}^*$. Then the equation being solved is

$$\mathbf{u_j}^* = \mathbf{A}^j\mathbf{u_0}^* \tag{9.6.24}$$

Let $\mathbf{e} = \mathbf{u} - \mathbf{u}^*$, then

$$\mathbf{e_j} = \mathbf{u_j} - \mathbf{u_j}^* = \mathbf{A}^j\left(\mathbf{u_0} - \mathbf{u_0}^*\right) = \mathbf{A}^j\mathbf{e_0} \tag{9.6.25}$$

and hence errors propagate in exactly the same way as the solution $\mathbf{u}$.

If $\mathbf{e_j}$ is bounded as j increases, then the solution is stable. In order to investigate the behaviour of $\mathbf{e}_j$, let the eigenvalues of $\mathbf{A}$ be λ_s and the eigenvectors be $\mathbf{v_s}$ so that $\mathbf{A}\mathbf{v_s} = \lambda_s\mathbf{v_s}$. Then as $\mathbf{v}_s$ forms a basis set for the underlying vector space, $\mathbf{e}_0$ can be expressed as a linear combination of these vectors to give

$$\mathbf{e_0} = \sum_{s=1}^{N-1} c_s\mathbf{v_s} \tag{9.6.26}$$

and hence

$$\begin{aligned}\mathbf{e_1} &= \mathbf{A}\mathbf{e_0} = \mathbf{A}\sum c_s\mathbf{v_s}\\ &= \sum c_s\mathbf{A}\mathbf{v_s} = \sum c_s\lambda_s\mathbf{v_s}\end{aligned} \tag{9.6.27}$$

Similarly

$$\mathbf{e_j} = \sum c_s\lambda_s{}^j\mathbf{v_s} \tag{9.6.28}$$

Therefore stability is achieved if the eigenvalue with largest modulus has modulus ≤ 1. The remaining problem is to find the eigenvalues of $\mathbf{A}$. Let $\mathbf{A} = \mathbf{I} + r\mathbf{T}$ where $\mathbf{I}$ is the unit matrix and

$$\mathbf{T} = \begin{bmatrix} -2 & 1 & & & & & \\ 1 & -2 & 1 & & & & \\ & & \cdots & \cdots & & & \\ & & & & 1 & -2 & 1 \\ & & & & & 1 & -2 \end{bmatrix} \tag{9.6.29}$$

which is an $(N-1, N-1)$ matrix. It is left as an exercise to show that the eigenvalues of $\mathbf{T}$ are λ_s where

$$\lambda_s = -4\sin^2\left(\frac{s\pi}{2N}\right) \qquad s = 1, \ldots N-1 \tag{9.6.30}$$

and its eigenvectors are

$$\nu_\mathrm{s} = \left(\sin\frac{s\pi}{N}, \ldots, \sin\frac{(N-1)s\pi}{N}\right) \tag{9.6.31}$$

Therefore the eigenvalues of $\mathbf{A}$ are

$$1 + r\left\{-4\sin^2\frac{s\pi}{2N}\right\} \tag{9.6.32}$$

Hence the explicit scheme is stable if

$$\left|1 - 4r\sin^2\left(\frac{s\pi}{2N}\right)\right| \le 1 \tag{9.6.33}$$

or

$$-1 \le 1 - 4r\sin^2\left(\frac{s\pi}{2N}\right) \tag{9.6.34}$$

leading to the condition $r \le \frac{1}{2}$ for stability.

The equivalent analysis for the Crank–Nicolson method simply requires the relevant matrix $\mathbf{A}$. Crank–Nicolson's method has the form

$$(2\mathbf{I} - r\mathbf{T})\mathbf{u}_{j+1} = (2\mathbf{I} + r\mathbf{T})\mathbf{u}_j \tag{9.6.35}$$

which in terms of $\mathbf{T}$ gives

$$\mathbf{u}_{j+1} = (2\mathbf{I} - r\mathbf{T})^{-1}(2\mathbf{I} + r\mathbf{T})\mathbf{u}_j \tag{9.6.36}$$

Hence stability will occur when the eigenvalues of $\mathbf{A} = (2\mathbf{I} - r\mathbf{T})^{-1}(2\mathbf{I} + r\mathbf{T})$ have modulus less than unity. These eigenvalues are

$$\frac{2 - 4r\sin^2\left(\frac{s\pi}{2N}\right)}{2 + 4r\sin^2\left(\frac{s\pi}{2N}\right)} \qquad s = 1, \ldots, N-1 \tag{9.6.37}$$

which are < 1 $\forall$ positive r, giving the unrestricted stability reported in Section 9.3.

The third and rather different topic to be considered in this section is that of *compatibility*. This is the phenomenon of a finite difference scheme converging to a different partial differential equation for varying steps h. To illustrate this phenomenon consider the following methods for the parabolic heat equation. The methods discussed have the form

$$L\ u(x_r, y_s) = 0 \qquad \text{where } L \text{ is a linear operator} \tag{9.6.38}$$

whereas the true solution is

$$L\ u = c \tag{9.6.39}$$

where c is the local trunction error. Using a Taylor expansion gives for the explicit method:

$$\begin{aligned} c = &\left(\frac{\partial u}{\partial y} + \frac{1}{2}k\frac{\partial^2 u}{\partial y^2} + \frac{1}{8}k^2\frac{\partial^3 u}{\partial y^3} + \cdots\right) \\ &- \left(\frac{\partial^2 u}{\partial x^2} + \frac{1}{12}h^2\frac{\partial^4 u}{\partial x^4} + \frac{1}{360}h^4\frac{\partial^6 u}{\partial x^6} + \cdots\right) \end{aligned} \tag{9.6.40}$$

and for Crank–Nicolson

$$c = \left(\frac{\partial u}{\partial y} + \frac{1}{2}k\frac{\partial^2 u}{\partial y^2} + \frac{1}{6}k^2\frac{\partial^3 u}{\partial y^3} + \cdots\right) - \left(1 + \frac{1}{2}k\frac{\partial}{\partial y} + \frac{1}{4}k^2\frac{\partial^2}{\partial y^2} + \cdots\right)\left(\frac{\partial^2 u}{\partial x^2} + \frac{1}{12}h^2\frac{\partial^4 u}{\partial x^4} + \cdots\right) \tag{9.6.41}$$

Hence for the explicit method

$$Lu - \left(\frac{\partial u}{\partial y} - \frac{\partial^2 u}{\partial x^2}\right) = \left(\frac{1}{2}\frac{\partial^2 u}{\partial y^2} - \frac{1}{12}h^2\frac{\partial^4 u}{\partial x^4}\right) \tag{9.6.42}$$

and for Crank–Nicolson

$$Lu - \left(\frac{\partial u}{\partial y} - \frac{\partial^2 u}{\partial x^2}\right) = -\frac{1}{12}h^2\frac{\partial^4 u}{\partial x^4} + \frac{1}{6}k^2\left(\frac{\partial^3 u}{\partial y^3} - \frac{3}{2}\frac{\partial^4 u}{\partial x^2 \partial y^2}\right) \tag{9.6.43}$$

Hence these well-known methods are compatible as the right-hand sides $\to 0$ as $k, h \to 0$. To see a method which is not compatible consider the method of Du Fort and Frankel defined by

$$\frac{u_{i,j+1} - u_{i,j-1}}{2k} = \frac{u_{i+1,j} - u_{i,j+1} - u_{i,j-1} + u_{i-1,j}}{h^2} \tag{9.6.44}$$

where the normal second-derivative approximation of (9.3.3) has the term $2u_{i,j}$ replaced by $u_{i,j+1} + u_{i,j-1}$ to give a computational molecule shown in Figure 9.6

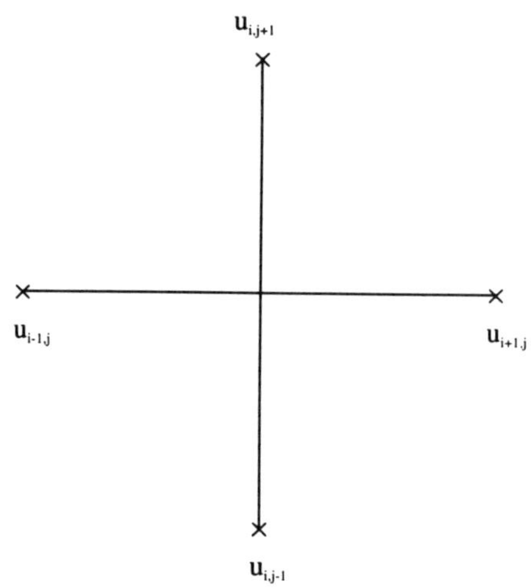

Figure 9.6

It is easy to show that this method is stable $\forall r$, and so, at first sight, looks like a good candidate. However, the compatibility expansion gives:

$$Lu - \left(\frac{\partial u}{\partial y} - \frac{\partial^2 u}{\partial x^2}\right) = \frac{1}{6}k^2\frac{\partial^3 u}{\partial y^3} - \frac{1}{12}h^2\frac{\partial^4 u}{\partial x^4} + \frac{k^2}{h^2}\frac{\partial^2 u}{\partial y^2} \tag{9.6.45}$$

and the limit now depends on the relative rates of convergence of h and k to zero as to whether the right-hand side $\to 0$ as $h, k \to 0$ or not. Indeed if $k/h = c$ convergence is to

$$\frac{\partial u}{\partial y} - \frac{\partial^2 u}{\partial x^2} + c^2\frac{\partial^2 u}{\partial y^2} = 0 \tag{9.6.46}$$

which is not even parabolic but hyperbolic! This type of situation needs to be guarded against.

9.7 An explicit stable parabolic scheme

The above stability analysis has demonstrated that the explicit scheme for parabolic equations is only stable if $r < \frac{1}{2}$. This restriction is quite tying in that the step size in the time direction is only the square of that in the x direction, forcing progress in the time step to be very slow. The root cause of this problem can be seen by considering the characteristics for the heat equation. The relevant differential equation is

$$\left(\frac{\mathrm{d}t}{\mathrm{d}x}\right)^2 = 0 \tag{9.7.1}$$

with solution $t =$ const. It is seen in the hyperbolic case that a solution can be found at the intersection of two characteristics by using the differential equations which holds along each. In effect information is transmitted along these lines from the boundary curve and at the point of intersection both pieces of information combine to give the required solution. In the parabolic case, these two separate characteristics have flattened out to give the same curve. However boundary information flows along this curve from the boundary conditions at each end, that is at $x = 0$ and $x = L$ say, and again combine to yield a solution. It is therefore expected that to get a solution at some interior point for a particular t value would require the use of the boundary conditions at $x = 0$ and $x = L$ for that t value. The explicit method uses the t value of $t - k$ and so blatently does not satisfy this simple test. Clearly if k is forced to be very small (through the stability condition) then the ideal predicted by the characteristic argument becomes more closely approached. On the other hand the method of Crank–Nicolson does use the required boundary points and would be expected to perform well.

The question then arises as to whether the only unrestricted methods will be those with an implicit form, or whether there is some device which can restore an explicit rule. One such approach is that due to Evans and Abdullah (1983). The idea stems from two formulae for parabolic equations by Saul'yev (1964) which are asymmetric with the form

$$\frac{u_{i-1,j+1} - u_{i,j+1} - u_{i,j} + u_{i+1,j}}{h^2} \tag{9.7.2}$$

being used for $\partial^2 u/\partial x^2$. Combined with the usual form for $\partial u/\partial t$ we get Saul'yev's formula

$$(1 + r)u_{i,j+1} - ru_{i-1,j+1} = ru_{i+1,j} + (1 - r)u_{i,j} \tag{9.7.3}$$

with the computational molecule shown in Figure 9.7.

Equally the formula with the form

$$-ru_{i+1,j+1} + (1 + r)u_{i,j+1} = ru_{i-1,j} + (1 - r)u_{i,j} \tag{9.7.4}$$

with the molecule in Figure 9.8 could be obtained by reversing the asymmetry. The method of approach is now to combine these two methods for pairs of points to give the computational molecule shown in Figure 9.9.

We now get two simultaneous equations for the two u values on time-line $j + 1$. Hence the method is explicit (strictly group explicit) as

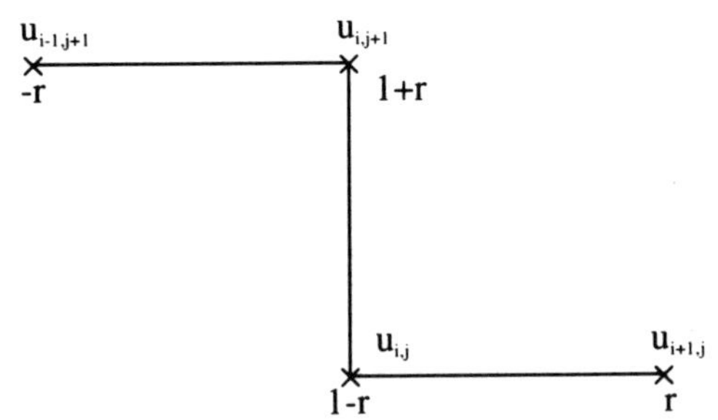

Figure 9.7

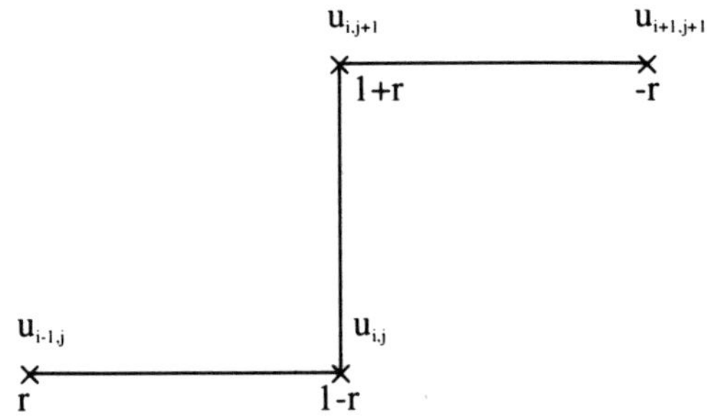

Figure 9.8

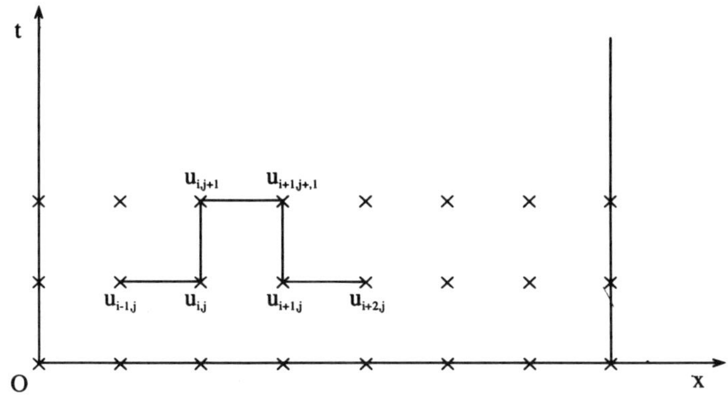

Figure 9.9

$$\begin{pmatrix} u_{i,j+1} \\ u_{i+1,j+1} \end{pmatrix} = \frac{1}{1+2r}\begin{pmatrix} 1+r & r \\ r & 1+r \end{pmatrix} \times \left\{ \begin{pmatrix} 1-r & 0 \\ 0 & 1-r \end{pmatrix} \begin{pmatrix} u_{i,j} \\ u_{i+1,j} \end{pmatrix} + \begin{pmatrix} ru_{i-1,j} \\ ru_{i+2,j} \end{pmatrix} \right\} \tag{9.7.5}$$

is the solution of the two equations. The term group explicit is used to indicate that a group of two points is found explicitly. The result is a method with unrestricted stability.

This method is stable for all r, as can be demonstrated using Fourier stability analysis by setting in the usual way:

$$\begin{aligned} u_{i,j} &= e^{\sqrt{-1}\beta x} e^{\alpha t} \\ &= e^{\sqrt{-1}\beta ih} e^{\alpha jk} \\ &= e^{\sqrt{-1}i\beta k} \xi^{j} \end{aligned} \tag{9.7.6}$$

Hence for stability $|\xi| \leq 1$.

Substituting for $u_{i,j}$ into

$$\begin{aligned} \frac{u_{i,j+1} - u_{i,j}}{k} = \frac{1}{h^2} &\left(\theta_1 \left(u_{i+1,j+1} - u_{i,j+1}\right) - \theta_2 \left(u_{i,j+1} - u_{i-1,j+1}\right)\right. \\ &\left. + \theta_1' \left(u_{i+1,j} - u_{i,j}\right) - \theta_2' \left(u_{i,j} - u_{i-1,j}\right)\right) \end{aligned} \tag{9.7.7}$$

from which Saul' yev's methods are obtained by setting

$$\left.\begin{matrix} \theta_1 = \theta_2{}' = 1 & \theta_2 = \theta_1{}' = 0 \\ \theta_2 = \theta_1{}' = 1 & \theta_1 = \theta_2{}' = 0 \end{matrix}\right\} \tag{9.7.8}$$

to give

$$\xi = \frac{-r\theta_1{}' e^{\sqrt{-1}\beta h} + [1 - r(\theta_1{}' + \theta_2{}')] + r\theta_2{}' e^{\sqrt{-1}\beta h}}{-r\theta_1 e^{\sqrt{-1}\beta h} + [1 + r(\theta_1 + \theta_2)] - r\theta_2 e^{-\sqrt{-1}\beta h}} \tag{9.7.9}$$

or

$$\therefore \ |\xi|^2 = \frac{1 + 16r^2\theta_1'\theta_2's^4 - 4r(\theta_1' + \theta_2')s^2 + 4r^2(\theta_1' - \theta_2')^2s^2}{1 + 16r^2\theta_1\theta_2s^4 + 4r(\theta_1 + \theta_2)s^2 + 4r^2(\theta_1 - \theta_2)^2s^2} \tag{9.7.10}$$

where $s = \frac{1}{2}\sin^2 \beta h$. Therefore for either form, the stability condition is

$$\left| \frac{1 - 4rs^2 + 4r^2s^2}{1 + 4rs^2 + 4r^2s^2} \right| < 1 \tag{9.7.11}$$

which is indeed satisfied for any r as required. This method has been extended to other co-ordinate systems such as cylindrical polars.

9.8 Curved boundaries

In the solution of Laplace's equation, a rectilinear grid was set up with boundary conditions conveniently at the given grid points. In general, the boundary conditions which are given in practical problems will be on curves which do not intersect the grid points exactly. There are a number of ways of handling curved boundaries. One approach is to avoid the problem by converting to a co-ordinate system which has the curved boundary as one of its constant surfaces. This is the approach used in the analytic solution of applied mathematical problems of this type.

Suppose we need to solve $\nabla^2 u = 0$ with u given on $y = 0$, $\theta = d$ and $r = a$ where r, θ are polar coordinates in the x, y plane as in Figure 9.10.

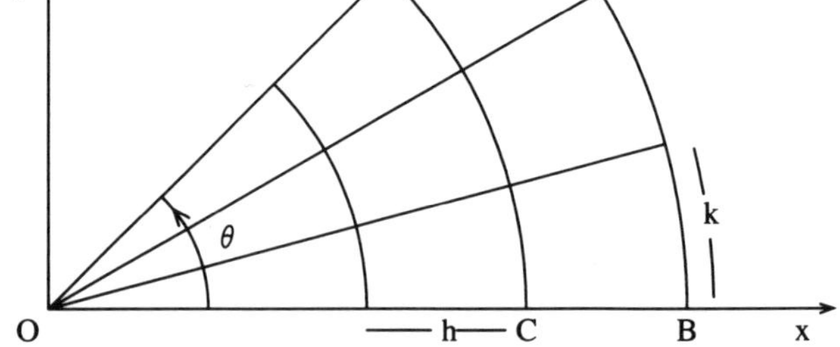

Figure 9.10

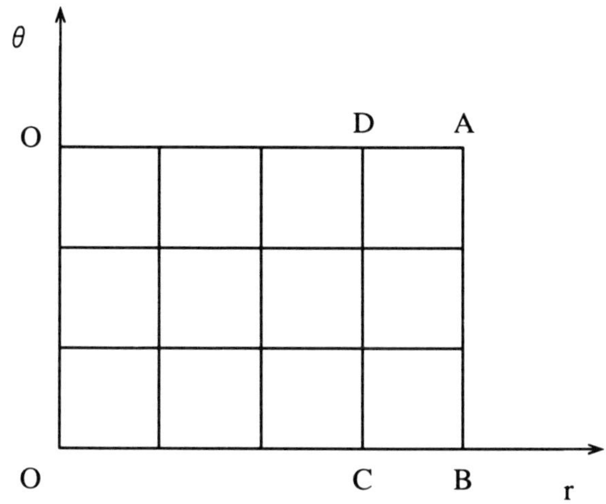

Figure 9.11

Instead of using a rectangular grid, a curvilinear grid is employed. Hence set $x = r\cos\theta$ and $y = r\sin\theta$ to get

$$\frac{\partial^2 u}{\partial r^2} + \frac{1}{r}\frac{\partial u}{\partial r} + \frac{1}{r^2}\frac{\partial^2 u}{\partial \theta^2} = 0 \tag{9.8.1}$$

or use the general curvilinear form of $\operatorname{div}(\operatorname{grad} u) = 0$ from (1.2.30) and (1.2.31). Then, in the (r, θ) plane a rectangular grid is recovered as shown in Figure 9.11 and using $r = ih$ and $\theta = jk$ in (9.8.1) gives the finite difference form

$$\frac{u_{i+1,j} - 2u_{i,j} + u_{i-1,j}}{h^2} + \frac{1}{ih}\frac{(u_{i+1,j} - u_{i-1,j})}{2h} + \frac{1}{(ih)^2}\frac{(u_{i,j+1} - 2u_{i,j} + u_{i,j-1})}{k^2} = 0 \tag{9.8.2}$$

and hence a set of linear algebraic equations in the usual way. This approach is limited to those curved boundaries which can be represented in a convenient co-ordinate system, and therefore lacks full generality. A more numerical approach is required.

Consider a typical curved boundary intersecting part of a rectangular grid system as shown in Figure 9.12

A Taylor expansion can be used to find u_A and u_3 in terms of the values of u at the grid-point u_0 to yield

$$u_A = u_0 + h\theta_1\frac{\partial u_0}{\partial x} + \frac{1}{2}h^2\theta_1^2\frac{\partial^2 u_0}{\partial x^2} + 0(h^3) \tag{9.8.3}$$

$$u_3 = u_0 - h\frac{\partial u_0}{\partial x} + \frac{1}{2}h^2\frac{\partial^2 u_0}{\partial x^2} + 0(h^3) \tag{9.8.4}$$

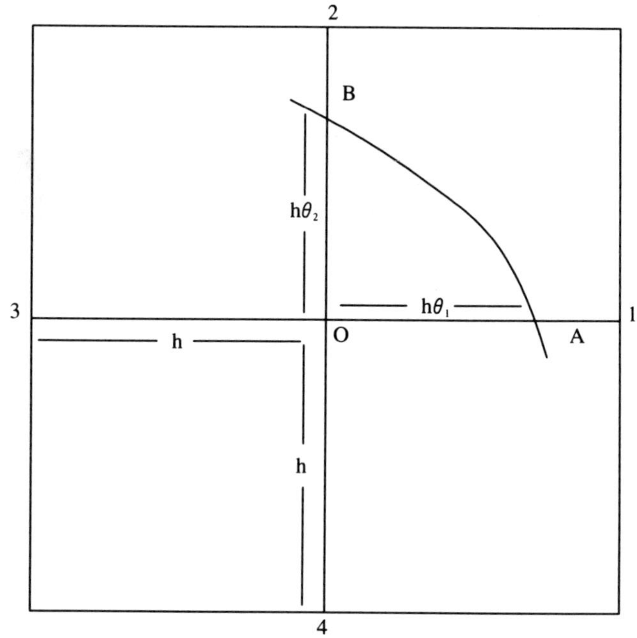

Figure 9.12

Eliminating $\partial^2 u_0/\partial x^2$ gives

$$\frac{\partial u_0}{\partial x} = \frac{1}{h}\left\{\frac{1}{\theta_1(1+\theta_1)}u_A - \frac{(1-\theta_1)}{\theta_1}u_0 - \frac{\theta_1}{(1+\theta_1)}u_3\right\} \tag{9.8.5}$$

and eliminating $\partial u_0/\partial x$ gives

$$\frac{\partial^2 u_0}{\partial x} = \frac{1}{h^2}\left\{\frac{2}{\theta_1(1+\theta_1)}u_A + \frac{2}{(1+\theta_1)}u_3 - \frac{2}{\theta_1}u_0\right\} \tag{9.8.6}$$

Hence approximate derivatives at points near to the boundary in terms of values within the boundary are available. The result is that amongst the defining equations these special ones for points near to the boundary are included. The boundary itself can be represented in the computer program by the values of θ_1 and θ_2 in the form of a data structure or Pascal RECORD. It is then possible to generate a quite elegant code which takes account of the curved boundary.

To illustrate the process on a simple example consider the solution of Laplace's equation with the grid and boundary conditions shown in Figure 9.13.

Conversion of the grid to polar coordinates gives again the transformed equation

$$\frac{\partial^2 \phi}{\partial r^2} + \frac{1}{r}\frac{\partial \phi}{\partial r} + \frac{1}{r^2}\frac{\partial^2 \phi}{\partial \theta^2} = 0 \tag{9.8.7}$$

with the new grid shown in Figure 9.14.

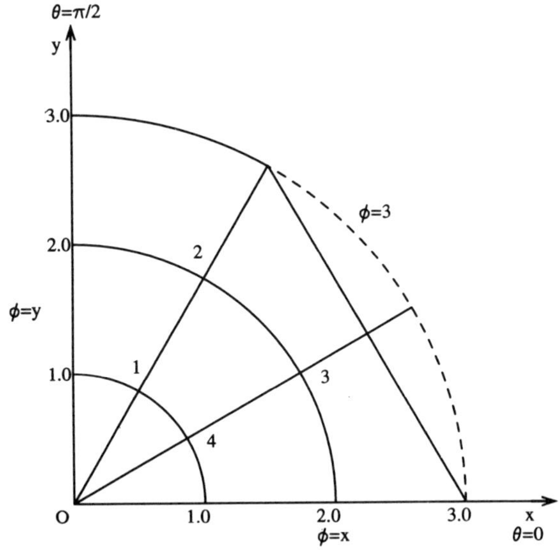

Figure 9.13

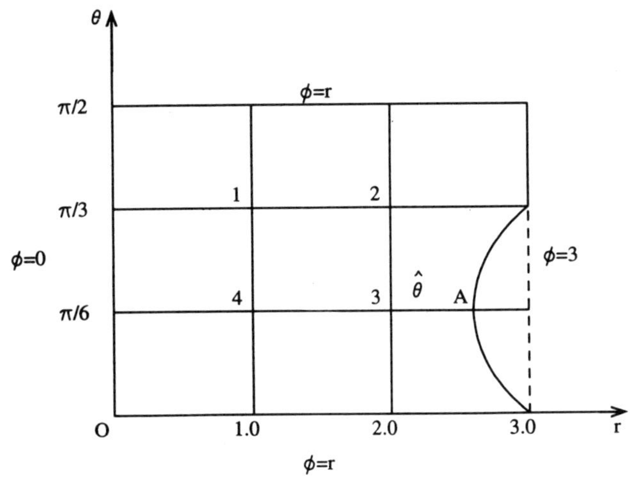

Figure 9.14

Now only grid-point 3 will require the curved approximations of (9.8.5), and four equations for four unknowns are easily found. The point A is defined by $\theta = \pi/6$ and $r = 3\sqrt{3}/2$, and hence the linear equations are

$$\phi_2 - 2\phi_1 + \frac{\phi_2}{2} + \frac{(1 - 2\phi_1 + \phi_4)}{\left(\frac{\pi}{6}\right)^2} = 0$$

$$\phi_3 - 2\phi_4 + \frac{\phi_3}{2} + \frac{(\phi_1 - 2\phi_4 + 1)}{\left(\frac{\pi}{6}\right)^2} = 0$$

$$3 - 2\phi_2 + \phi_1 + \frac{(3 - \phi_1)}{4} + \frac{(1 - 2\phi_1 + \phi_4)}{\left(4\frac{\pi}{6}\right)^2} = 0$$

$$\frac{2(3 - (1 + \hat{\theta})\phi_3 + \hat{\theta}\phi_4)}{\hat{\theta}(1 + \hat{\theta})} + \frac{(3 - (1 - \hat{\theta}^2)\phi_3 - \hat{\theta}^2\phi_4)}{\hat{\theta}(1 + \hat{\theta})} + \frac{(\phi_2 - 2\phi_3 + 2)}{4\left(\frac{\pi}{6}\right)^2} = 0$$

(9.8.8)

which reduce to

$$\begin{bmatrix} -9.295 & 1.5 & 0 & 3.647 \\ 3.647 & 0 & 1.5 & -9.295 \\ 0.75 & -3.824 & 0.9119 & 0 \\ 0 & 0.9119 & -1.978 & -0.1662 \end{bmatrix} \begin{bmatrix} \phi_1 \\ \phi_2 \\ \phi_3 \\ \phi_4 \end{bmatrix} = \begin{bmatrix} -3.647 \\ -3.647 \\ -5.5738 \\ -2.7865 \end{bmatrix} \qquad (9.8.9)$$

to yield by Gauss elimination, the results

$$\begin{aligned} \phi_4 &= 1.2625 \\ \phi_3 &= 2.3458 \\ \phi_2 &= 2.2627 \\ \phi_1 &= 1.2529 \end{aligned} \qquad (9.8.10)$$

with ϕ_2 and ϕ_3 being nearest to the $\phi = 3$ boundary giving the largest values.

9.9 Finite element method

There is a completely different approach to the solution of partial differential equations which has found ever increasing success in recent years. The method was first employed by Zienkiewicz (1977) in the solution of stress problems in civil engineering, but the technique is applicable to a wide range of problems. This is in its own right a large area of numerical analysis and the object here is to give an introduction and leave the advanced features to specialist text books which are listed in the bibliography.

The starting point is to find an equivalent variational principle to a given partial differential equation. The classic link between these two forms is the Euler–Lagrange formula. A variational rule for ordinary derivatives has the form

$$J = \int_V f(y_i, \dot{y}_i, x)\, \mathrm{d}x \qquad (9.9.1)$$

which can be shown to have a minimum when the Euler–Lagrange differential equation

$$\frac{\partial f}{\partial y_i} - \frac{\mathrm{d}}{\mathrm{d}x}\left(\frac{\partial f}{\partial \dot{y}_i}\right) = 0 \qquad (9.9.2)$$

is satisfied for $i = 1, \ldots, n$. Often the variational principle is a fundamental physical rule as in classical mechanics where Hamilton's principle leads directly to a variational rule from which the Euler–Lagrange equations give the usual Lagrange's equations. The principle is effectively a minimization of the total energy of the system being considered. The Euler–Lagrange equation extends to partial derivatives so that the variational principle

$$J = \int\int_V \left[\frac{1}{2}\left(\frac{\partial\phi}{\partial x}\right)^2 + \frac{1}{2}\left(\frac{\partial\phi}{\partial y}\right)^2 - c\phi\right] \mathrm{d}x\,\mathrm{d}y \tag{9.9.3}$$

yields the Euler–Lagrange formula

$$\frac{\partial^2\phi}{\partial x^2} + \frac{\partial^2\phi}{\partial y^2} + c = 0 \tag{9.9.4}$$

or Poisson's equation.

The problem normally presented to a numerical analysist is the reverse of the one solved here. Rather than being given a variational problem and being asked to convert it to an equivalent partial differential equation, it will be the partial differential equation which is given and the variational principle which is required. In simple cases it is possible to work backwards to find a principle which will reduce to the required partial differential equation, or it may be that a principle is immediate from physical considerations, but it would be very satisfying to be able to generate a principle without reference to the underlying physics. Certainly, if a general second-order operator L is defined by

$$L\phi = a\frac{\partial^2\phi}{\partial x^2} + b\frac{\partial^2\phi}{\partial x\partial y} + c\frac{\partial^2\phi}{\partial y^2} + F\left(x, y, \phi, \frac{\partial\phi}{\partial x}, \frac{\partial\phi}{\partial x}\right) \tag{9.9.5}$$

is given, then the functional

$$J(\phi) = \int\int_V \phi L\phi\,\mathrm{d}x\,\mathrm{d}y - 2\int\int_V \phi f\,\mathrm{d}x\mathrm{d}y \tag{9.9.6}$$

has a minimum which corresponds to the unique solution of $L\phi = f$ if the operator L is self-adjoint and positive definite.

If no functional of the above type can be found, then one approach is to use a weighted residual method. Defining a residual as

$$r(\phi) = L\phi - f \tag{9.9.7}$$

the idea is to find a value of ϕ which makes this residual zero. A trial function may be chosen in the form

$$\phi^{(n)} = \sum_{i=1}^{n} c_i\psi_i \tag{9.9.8}$$

which is then manipulated to force $r(\phi)$ to zero. The least squares method achieves this end by choosing the constants c_i to minimize

$$J(\phi) = \int\int_V r(\phi)^2 \, dx \, dy \tag{9.9.9}$$

This gives the set of linear equations

$$\frac{\partial J}{\partial c_i} = 0 \quad i = 1, \ldots, n \tag{9.9.10}$$

for the constants c_i. The Galerkin method uses the basis functions themselves as weight functions and so minimizes

$$J(\phi) = \int\int_V r(\phi)\psi_i \, dx \, dy \tag{9.9.11}$$

The obvious difficulty with these methods is the choice of trial functions which satisfy any essential boundary conditions. This choice becomes especially difficult with irregular boundaries. The minimization process is not a fundamental problem, and in Chapter 10 a survey is made of numerical optimization procedures which could be used.

The finite element method is based on splitting the region V into elements and taking the sum of the integrals over each element to make up the full integral over the whole of V. Because the elements are small, and can be made smaller to obtain increased accuracy, simple forms of ϕ can be used in the variational principle as applied locally to a given element. This simplicity will allow the analytical application of the variational principle and will yield a banded set of linear algebraic equations for the equivalent of the c_i's which represent ϕ in each element. These equations can be solved using the extensive methods of Chapter 3.

The simplest form of small element is the simplex, or in two dimensions the triangle as in Figure 9.15.

The value of ϕ inside the element can be expressed in the form

$$\phi = \mathbf{N}\phi^e \tag{9.9.12}$$

where $\mathbf{N}$ is called the shape function, and $\phi^e = (\phi_1, \phi_2, \phi_3)$ which are the ϕ values at the vertices. Ultimately these ϕ values will be the unknowns which will be found from the final banded set of linear algebraic equations. Because the element is small, it is reasonable to make a simple approximation to ϕ in the element, such as the linear form

$$\phi = \alpha_1 + \alpha_2 x + \alpha_3 y \tag{9.9.13}$$

Now forcing $\phi = \phi_i$ at each vertex (x_i, y_i) gives the equations

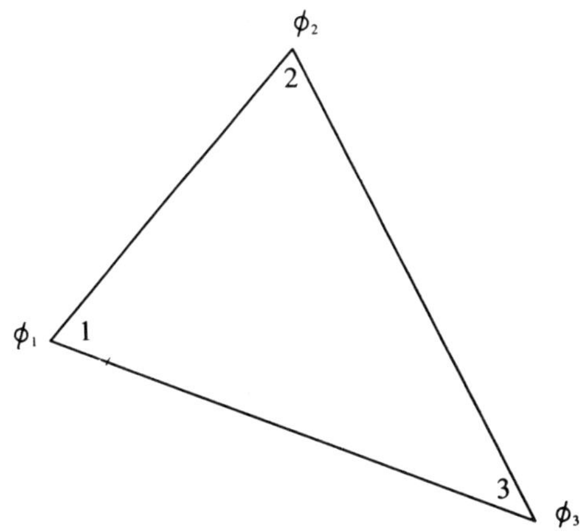

Figure 9.15

$$\begin{aligned}\phi_1 &= \alpha_1 + \alpha_2 x_1 + \alpha_3 y_1\\ \phi_2 &= \alpha_1 + \alpha_2 x_2 + \alpha_3 y_2\\ \phi_3 &= \alpha_1 + \alpha_2 x_3 + \alpha_3 y_3\end{aligned} \tag{9.9.14}$$

to give the solution

$$\phi = N_1\phi_1 + N_2\phi_2 + N_3\phi_3 \tag{9.9.15}$$

with

$$N_i = (a_i + b_i x + c_i y)/2\Delta \quad i = 1,\ 2,\ 3 \tag{9.9.16}$$

where

$$\begin{aligned}a_1 &= x_2 y_3 - x_3 y_2\\ b_1 &= y_2 - y_3\\ c_1 &= x_3 - x_2\end{aligned} \tag{9.9.17}$$

The constants a_2, a_3, b_2, b_3, c_2 and c_3 are defined by cyclic permutation and

$$2\Delta = \begin{vmatrix} 1 & x_1 & y_1 \\ 1 & x_2 & y_2 \\ 1 & x_3 & y_3 \end{vmatrix} \tag{9.9.18}$$

Hence the shape function is explicitly defined and it is clear that ϕ is continuously defined along each element boundary. If the method is extended to more complicated shapes then more free parameters are required to ensure the continuity of ϕ on the boundaries. It is the values ϕ^e which are the unknowns which will form the variational parameters. Any boundary conditions will yield known values for the corresponding ϕ^e and will form part of the right-hand side of the set of resulting linear equations.

The next stage is to substitute the form (9.9.15) into the variational principle. The steps involved are easily seen by following through an example. Consider the partial differential equation

$$\frac{\partial}{\partial x}\left(k_x\frac{\partial\phi}{\partial x}\right)+\frac{\partial}{\partial y}\left(k_y\frac{\partial\phi}{\partial y}\right)+\frac{\partial}{\partial z}\left(k_z\frac{\partial\phi}{\partial z}\right)+Q=0 \tag{9.9.19}$$

with variational form

$$J=\int\!\!\int\!\!\int_V\left[\frac{1}{2}\left\{k_x\left(\frac{\partial\phi}{\partial x}\right)^2+k_y\left(\frac{\partial\phi}{\partial y}\right)^2+k_z\left(\frac{\partial\phi}{\partial z}\right)^2\right\}-Q\phi\right]\mathrm{d}x\,\mathrm{d}y\,\mathrm{d}z \tag{9.9.20}$$

Then in an element e

$$\phi=[N_i,N_j,\ldots]\begin{bmatrix}\phi_i\\ \phi_j\\ \vdots\end{bmatrix}=\mathbf{N}\phi^e \tag{9.9.21}$$

and $J=J^e$. Then the derivatives with respect to ϕ_i of J^e are

$$\begin{aligned}\frac{\partial J^e}{\partial\phi_i}=\int_{V_e}&\left\{k_x\frac{\partial\phi}{\partial x}\frac{\partial}{\partial\phi_i}\left(\frac{\partial\phi}{\partial x}\right)+k_y\frac{\partial\phi}{\partial y}\frac{\partial}{\partial\phi_i}\left(\frac{\partial\phi}{\partial y}\right)\right.\\ &\left.+k_z\frac{\partial\phi}{\partial z}\frac{\partial}{\partial\phi_i}\left(\frac{\partial\phi}{\partial z}\right)-Q\frac{\partial\phi}{\partial\phi_i}\right\}\mathrm{d}x\,\mathrm{d}y\,\mathrm{d}z\end{aligned} \tag{9.9.22}$$

with the derivatives

$$\frac{\partial\phi}{\partial x}=\left[\frac{\partial N_i}{\partial x},\frac{\partial N_j}{\partial x},\ldots\right]\begin{bmatrix}\phi_i\\ \phi_j\\ \vdots\end{bmatrix} \tag{9.9.23}$$

$$\frac{\partial\phi}{\partial\phi_i}=N_i \tag{9.9.24}$$

$$\frac{\partial}{\partial\phi_i}\left(\frac{\partial\phi}{\partial x}\right)=\frac{\partial N_i}{\partial x} \tag{9.9.25}$$

Substituting into (9.9.22) gives

$$\frac{\partial J^e}{\partial\phi_i}=\mathbf{h}_i^e\begin{bmatrix}\phi_i\\ \phi_j\\ \vdots\end{bmatrix}+F^e \tag{9.9.26}$$

where

$$h_{ij}^e=\int_{V^e}\left\{k_x\frac{\partial N_i}{\partial x}\frac{\partial N_j}{\partial x}+k_y\frac{\partial N_i}{\partial y}\frac{\partial N_j}{\partial y}+k_z\frac{\partial N_i}{\partial z}\frac{\partial N_j}{\partial z}\right\}\mathrm{d}x\,\mathrm{d}y\,\mathrm{d}z \tag{9.9.27}$$

and

$$F^e = -\int_{V^e} QN_i \, dV \tag{9.9.28}$$

Hence the derivative of J^e with respect to ϕ_i is known explicitly as the N_i's are known explicitly. There is a contribution to J from each element e, and the full contribution to J can be found by summing the contributions from each element to give the form

$$J = \int_V f \, dV = \sum_e \int_{V_e} f^e \, dv_e \tag{9.9.29}$$

with the derivative

$$\frac{\partial J}{\partial \phi_i} = \sum_e \frac{\partial}{\partial \phi_i} \int_{V_e} f^e \, dv_e = \sum \frac{\partial J^e}{\partial \phi_i} \tag{9.9.30}$$

The required minimum will occur when

$$\frac{\partial J}{\partial \phi_i} = 0 \tag{9.9.31}$$

to give the equations

$$\frac{\partial J}{\partial \phi_i} = 0 = \mathbf{H}\phi_i + \mathbf{F} \tag{9.9.32}$$

where

$$\mathbf{H} = \sum_e h_{ij}^e \tag{9.9.33}$$

and

$$F_i = \sum_e F_i^e \tag{9.9.34}$$

These equations form a banded set of linear algebraic equations for the values of the unknowns ϕ_i at the nodes of the triangular elements. The process can be summed up by the following steps.

(i) Set up a suitable variational procedure.
(ii) Choose the shape of the finite element.
(iii) Calculate the shape function $\mathbf{N}$, making sure that continuity of the dependent variables is preserved along the element boundaries.
(iv) Substitute $\phi = \mathbf{N}\phi^e$ into the variational form to obtain h_{ij}^e.
(v) Sum h_{ij}^e over each element. This process is called assembly.
(vi) Solve the banded set of linear algebraic equations.

To illustrate the complete process an example will be worked in detail. Consider the solution of Laplace's equation

$$\frac{\partial^2 u}{\partial x^2} + \frac{\partial^2 u}{\partial y^2} = 0 \tag{9.9.35}$$

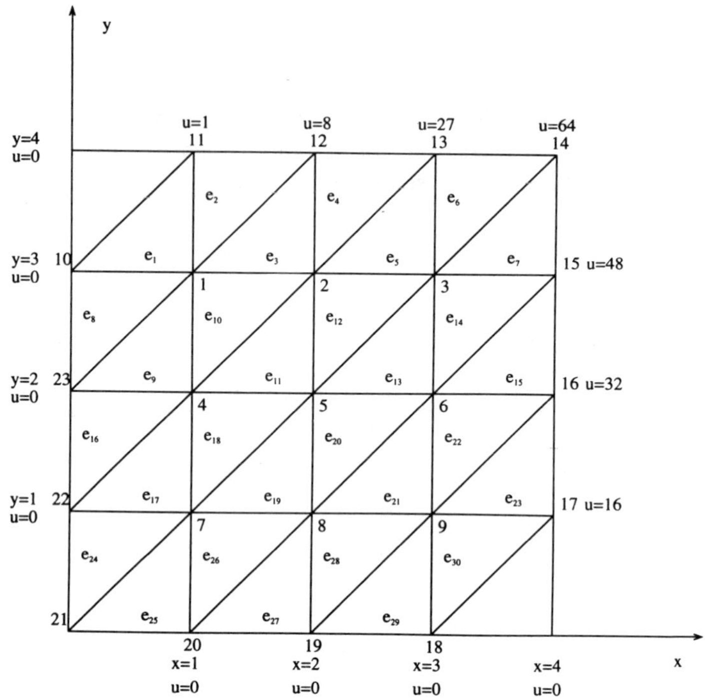

Figure 9.16

with boundary conditions

$$u = 0 \quad \text{when} \quad x = 0$$

$$u = 0 \quad \text{when} \quad y = 0$$

$$u = x^3 \quad \text{when} \quad y = 4$$

$$u = 16y \quad \text{when} \quad x = 4$$

The triangular grid of elements is shown in Figure 9.16 with the nodes numbered. Hence the unknowns are u_1 to u_9.

The variational principle is

$$J = \frac{1}{2} \int_0^4 \int_0^4 \left[\left(\frac{\partial u}{\partial x} \right)^2 + \left(\frac{\partial u}{\partial y} \right)^2 \right] \mathrm{d}x \, \mathrm{d}y \tag{9.9.36}$$

The equations to be applied are those from (9.9.13) through to (9.9.34) and in particular (9.9.16) gives

$$\frac{\partial N_i}{\partial x} = \frac{b_i}{2\Delta} \tag{9.9.37}$$

$$\frac{\partial N_i}{\partial y} = \frac{c_i}{2\Delta} \tag{9.9.38}$$

for each element. Therefore (9.9.27) gives

$$\begin{aligned} h_{ij}^e &= \int_{V_e} \{b_i b_j + c_i c_j\} \frac{1}{4\Delta^2}\,\mathrm{d}x\,\mathrm{d}y \\ &= \frac{b_i b_j + c_i c_j}{4\Delta} \end{aligned} \tag{9.9.39}$$

From the diagram it can be seen that there are 23 nodes of which nodes 1 to 9 are the non-boundary nodes at which u is required. There are 30 elements to be processed. Hence for example element e_{16} has nodes 22, 23 and 4 which will mean that h_{ij}^{16} will involve the elements

$$\begin{bmatrix} h_{22,22} & h_{22,23} & h_{22,4} \\ h_{23,22} & h_{23,23} & h_{23,4} \\ h_{4,22} & h_{4,23} & h_{4,4} \end{bmatrix} \tag{9.9.40}$$

which as they are processed are added into the matrix **H**. This process has been programmed, but the first steps consist of taking the element e^1 with nodes 10, 11 and 1. Then the values of a_i, b_i and c_i needed for (9.9.16) are shown in Table 9.3 and the contribution to h_{ij} is shown in Table 9.4, which will be added into the existing values of **H** at each step, **H** being set to **0** initially. For the second element the corresponding tables are Table 9.5 and the contribution to h_{ij} is shown in Table 9.6.

Hence after e_{30} is processed there will be a 23 × 23 matrix **H** in the form

$$\mathbf{H}\begin{bmatrix} u_1 \\ \vdots \\ u_{23} \end{bmatrix} = 0 \tag{9.9.41}$$

Table 9.3

	1	2	3
a	−1	−3	3
b	1	−1	0
c	0	1	−1

Table 9.4

		j		
		10	11	1
	10	−1/2	1/2	0
i	11	1/2	−1	1/2
	1	1/2	1	−1/2

Table 9.5

	1	2	3
a	−4	1	2
b	0	−1	1
c	1	0	−1

Table 9.6

		j		
		1	11	12
i	1	−1/2	0	1/2
	11	0	−1/2	1/2
	12	1/2	1/2	−1

but u_{10} to u_{23} are known to give the linear equations

$$\begin{bmatrix} h_{11} & \cdots & h_{19} \\ \vdots & & \vdots \\ h_{91} & \cdots & h_{99} \end{bmatrix} \begin{bmatrix} u_1 \\ \vdots \\ u_9 \end{bmatrix} = - \begin{bmatrix} h_{1,10} & \cdots & h_{1,23} \\ \vdots & & \vdots \\ h_{9,10} & \cdots & h_{9,23} \end{bmatrix} \begin{bmatrix} u_{10} \\ \vdots \\ u_{23} \end{bmatrix} = \mathbf{b} \tag{9.9.42}$$

The 9×9 matrix $\mathbf{H}$ has the elements

$$\begin{bmatrix} -4 & 1/2 & 0 & 1/2 & 0 & 0 & 0 & 0 & 0 \\ 0 & -4 & 1/2 & 0 & 1/2 & 0 & 0 & 0 & 0 \\ 0 & 0 & -4 & 0 & 0 & 1/2 & 0 & 0 & 0 \\ 0 & 1 & 0 & -4 & 1/2 & 0 & 1/2 & 0 & 0 \\ 0 & 0 & 1 & 0 & -4 & 1/2 & 0 & 1/2 & 0 \\ 0 & 0 & 0 & 0 & 0 & -4 & 0 & 0 & 1/2 \\ 0 & 0 & 0 & 0 & 1 & 0 & -4 & 1/2 & 0 \\ 0 & 0 & 0 & 0 & 0 & 1 & 0 & -4 & 1/2 \\ 0 & 0 & 0 & 0 & 0 & 0 & 0 & 0 & -4 \end{bmatrix} \tag{9.9.43}$$

and the solution vector has the $\mathbf{u}$ values

$$\{3.88, 10.88, 24.16, 4.20, 8.89, 17.25, 2.92, 5.56, 10.0\}$$

which are physically reasonable results. It can be seen how the matrix $\mathbf{H}$ is sparse with diagonal dominance which in a large problem would allow the easy use of iterative solutions with the economical storage methods which were discussed in Chapter 3.

Further reading

Ames W. F. (1969) *Numerical Methods for Partial Differential Equations*, Nelson, London.

Davies A. J. (1980) *The Finite Element Method: a First Approach*, Oxford University Press, Oxford.

Smith G. D. (1978) *Numerical Solution of Partial Differential Equations: Finite Difference Method*, Oxford University Press, Oxford.

Exercises

(9.1) Compute a solution of the equation

$$\frac{\partial^2 u}{\partial x^2} = \frac{\partial u}{\partial t}$$

for $0 \le x \le 1$, given that $u = \sin \pi x$ when $t = 0$, $u = 0$ for $x = 0$, $t > 0$ and $u = 1$ for $x = 1$, $t > 0$. Use an explicit method with the x step size at 0.1 and a suitable t step. Repeat the calculation using the Crank–Nicholson method.

(9.2) A bar of length one unit, radiates heat to an atmosphere of constant temperature. The ends of the bar are kept at a constant temperatures of zero. At $t = 0$ the temperature distribution in the bar is $u = \sin \pi x$. Given that the defining equation is

$$\frac{\partial^2 u}{\partial x^2} - k(u - u_0) = \frac{1}{c^2}\frac{\partial u}{\partial t}$$

set up and implement both an explicit and an implicit method for a numerical solution. Take $c = 2$, $k = 0.1$ and $u_0 = 0$.

(9.3) Carry out any necessary preliminary analysis to find a numerical method for the solution of the equations

$$4\frac{\partial^2 u}{\partial x^2} - 12\frac{\partial^2 u}{\partial x \partial t} + 9\frac{\partial^2 u}{\partial t^2} = \frac{\partial u}{\partial t}$$

with boundary conditions

$$\left.\begin{aligned} u &= 0 \quad \text{when} \quad x = 0 \\ u &= 1 \quad \text{when} \quad x = 1 \end{aligned}\right\} \quad \text{with} \quad t > 0$$

$$u = x^2 \quad \text{when} \quad 2t + 3x = 0 \quad \text{with} \quad 0 \le x \le 1$$

(9.4) Determine the regions of parabolicity, ellipticity and hyperbolicity for the equation

$$\frac{\partial^2 u}{\partial x^2} + xy\frac{\partial^2 u}{\partial x \partial y} + (4xy - y^3 - x^3)\frac{\partial^2 u}{\partial y^2} = \frac{\partial u}{\partial x} + \frac{\partial u}{\partial y}$$

(9.5) Use matrix stability analysis to show that the backward difference scheme

$$\frac{u_{i,j+1} - u_{i,j}}{k} = \frac{u_{i-1,j+1} - 2u_{i,j+1} + u_{i+1,j+1}}{h^2}$$

for the numerical discretization of

$$\frac{\partial u}{\partial t} = \frac{\partial^2 u}{\partial x^2}$$

is stable for all r where $r = k/h^2$. Implement the method with the boundary conditions $u = 0$ on $t = 0$ for $0 < x < 1$, $u = 1$ on $x = 0$ for $t > 0$ and $\partial u/\partial x = 0$ on $x = 1$ for $t > 0$. How does this method compare with that of Crank–Nicolson?

(9.6) A violin string is fixed at each end. Its length is one unit. When playing pizzicato, the string is plucked at a point a third of the way along, and the initial displacement is 0.1 units at this point. Solve numerically the equation for the subsequent displacement given by

$$\frac{\partial^2 y}{\partial x^2} = \frac{1}{c^2}\frac{\partial^2 y}{\partial t^2}$$

with $c = 1$.

(9.7) Use the characteristic method to find a solution of the equation

$$\frac{\partial^2 z}{\partial x^2} + \frac{\partial^2 z}{\partial x \partial y} - 2\frac{\partial^2 z}{\partial y^2} = 0$$

at the intersection of the characteristic curves through the points $(0, 0)$ and $(0.1, 0.0)$. The boundary conditions are that $z = x$ on $y = 0$ and

$$\frac{\partial z}{\partial x} = \frac{\partial z}{\partial y} = 1$$

on $y = 0$.

(9.8) Find the analytic form of the characteristic curves for the partial differential equation

$$\frac{\partial^2 u}{\partial x^2} + 2\left(x + \frac{1}{y}\right)\frac{\partial^2 u}{\partial x \partial y} + \frac{4x}{y}\frac{\partial^2 u}{\partial y^2} = xy$$

and hence set up a numerical scheme for the solution of the equation.

(9.9) Reduce the partial differential equation

$$9\frac{\partial^2 u}{\partial x^2} + 24\frac{\partial^2 u}{\partial x \partial y} + 16\frac{\partial^2 u}{\partial y^2} = 3\frac{\partial u}{\partial y} + 9u^2$$

to canonical form by first finding the characteristic curves. Hence implement a stable finite difference scheme with the boundary conditions

$$u = 0 \qquad \text{on} \qquad 3y = 4x$$

and

$$u = 10 \qquad \text{on} \qquad x = 0 \qquad \text{and} \qquad x = 1$$

How does the non-linearity affect the scheme?

(9.10) Solve numerically the partial differential equation

$$\frac{\partial^2 y}{\partial x^2} + 4\frac{\partial^2 y}{\partial x \partial t} + 3\frac{\partial^2 y}{\partial t^2} = 0$$

with boundary conditions

$$\left.\begin{array}{lll} y = 1 & \text{when} & t - 3x = 0 \\ y = 0 & \text{when} & t - 3x = 1 \end{array}\right\} \quad \text{with} \quad t > 0$$

$$y = (t - 3x)^2 \quad \text{when} \quad t - x = 0 \quad \text{with} \quad 0 \leq x + 3t \leq 1$$

(9.11) Solve numerically the equation

$$\frac{\partial^2 u}{\partial x^2} = x^2 u^2 \frac{\partial^2 u}{\partial y^2}$$

with boundary conditions $u = 2x$ and $\partial u/\partial y = x^2$ on $y = 0$ for $0 \leq x \leq 1$.

(9.12) Set up a numerical procedure to solve numerically the partial differential equation

$$\frac{\partial^2 \phi}{\partial x^2} + \frac{\partial^2 \phi}{\partial y^2} = 0$$

with boundary conditions

$$\phi = 0 \quad \text{on} \quad x = 0 \quad \forall\ y \in [0, 1]$$

$$\phi = 0 \quad \text{on} \quad y = 0 \quad \forall\ x \in [0, 1]$$

$$\phi = 100 \quad \text{on} \quad x = 1 \quad \forall\ y \in [0, 1]$$

$$\phi = 100 \quad \text{on} \quad y = 1 \quad \forall\ x \in [0, 1]$$

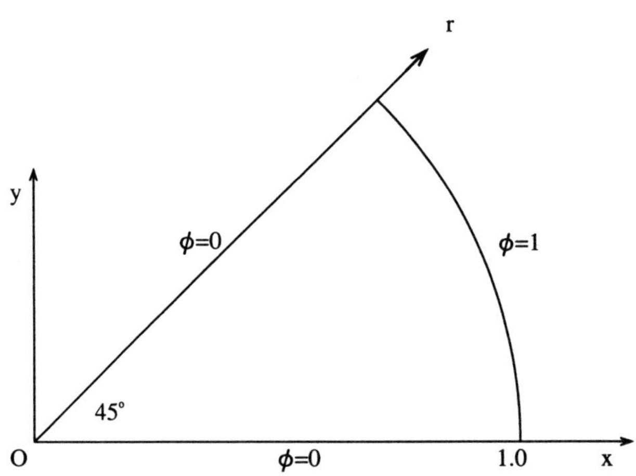

Figure 9.17

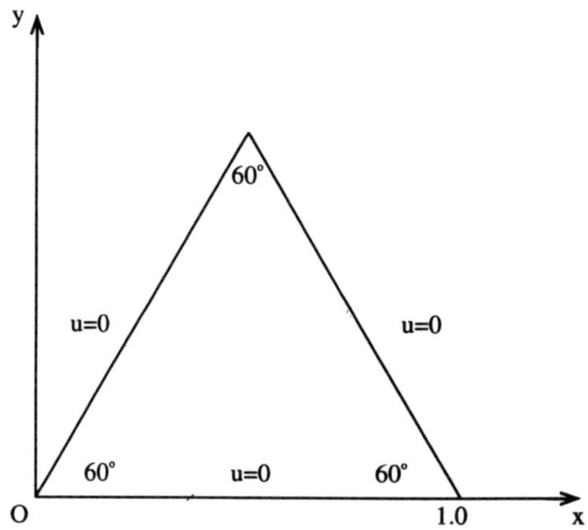

Figure 9.18

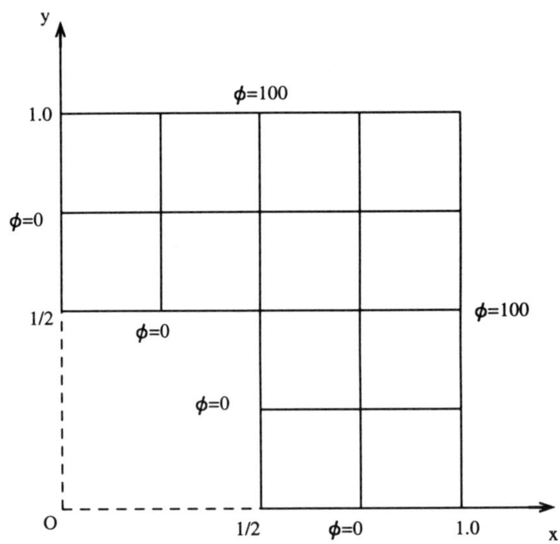

Figure 9.19

(9.13) Solve numerically Laplace's equation with the boundary conditions specified in Figure 9.17.

(9.14) Solve numerically Poisson's equation

$$\nabla^2 u = k$$

on the triangular region with boundary conditions specified in Figure 9.18, and with $k = 1$.

(9.15) Solve Laplace's equation given in exercise (9.7) with the boundary conditions specified in Figure 9.19.

(9.16) Repeat the solution of exercise (9.13), but replacing the boundary condition $\phi = 1$ on $r = 1$ by

$$\frac{\partial \phi}{\partial r} = 1.$$

(9.17) Use the curved boundary approximations of the relevant partial derivatives to set up a numerical solution to Laplace's equation with boundary conditions

$$\phi = y \quad \text{on} \quad x = 0 \quad \text{for} \quad 0 \le y \le 1$$

$$\phi = x \quad \text{on} \quad y = 0 \quad \text{for} \quad 0 \le x \le 2$$

and $\phi = 1$ on the positive quadrant of the ellipse

$$\tfrac{1}{4}x^2 + y^2 = 1$$

(9.18) Set up a finite element scheme to solve

$$\frac{\partial^2 u}{\partial x^2} + \frac{\partial^2 u}{\partial y^2} = 0$$

with the boundary conditions specified in Figure 9.20.

(9.19) Set up a finite element scheme for Laplace's equation on the region defined by Figure 9.21.

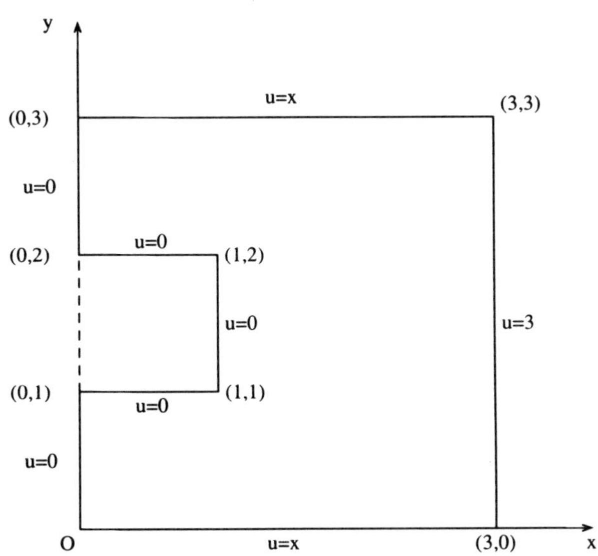

Figure 9.20

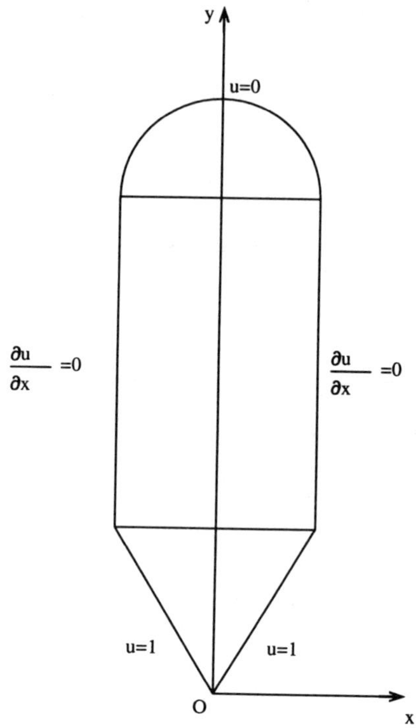

Figure 9.21

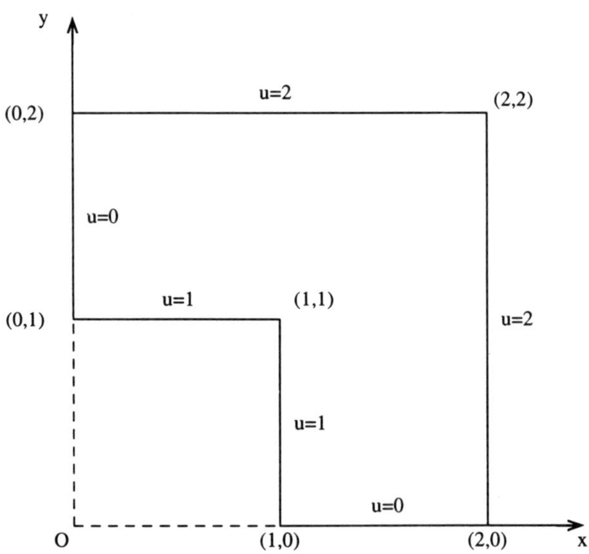

Figure 9.22

(9.20) Set up a finite element scheme for the equation

$$\frac{\partial}{\partial x}\left(k_x \frac{\partial u}{\partial x}\right) + \frac{\partial}{\partial y}\left(k_y \frac{\partial u}{\partial y}\right) = 0$$

with the boundary conditions specified in Figure 9.22 and $k_x = k_y = (x^2 + y^2)^{1/2}$.

10

Optimization Methods

10.1 Introduction

Optimization is an important practical area not always included in the general area of numerical analysis. Nevertheless the subject addresses a problem which is little different in philosophy from any of the other topics in this book. Mathematicians encounter integrals they cannot express in closed form, so they resort to numerical quadrature; similarly with ordinary differential equations or with partial differential equations. In optimization, a function $f(\mathbf{x})$ is given and the requirement is to find the vector $\mathbf{x}^*$ at which $f(\mathbf{x})$ is a minimum (or maximum). There are all manner of variations on this theme which include extra conditions on f. The simplest problem is when f is a function of only one variable. The classical analytic solution is to find the zeros of the first derivative of f and then classify the turning points according to the behaviour of the higher derivatives. It may be that the function is only available computationally, or that derivatives are not easily evaluated. In such a case, what is called a uni-variate optimization method is required. The study here will begin with some simple ways of effecting uni-variate optimization. These then form a basis for the later study of the many-variable problem.

These techniques address the problem of finding a local minimum given some starting criterion which restricts the region of interest. A more difficult problem is that of global optimization where an overall minimum is required from amongst potentially many local points. A further complication can be added with constraints on one or more variables. This is the topic of constrained optimization. The methods which follow vary in the level of availability of higher derivatives. Some methods will only need the evaluation of f itself, whereas others will require both first and second derivatives.

The practical applications of these techniques arise in all walks of applied mathematics. Obvious examples in which cost is minimized in a process by controlling the process parameters, whether it be a chemical process or an economic model, arise commonly. Variational solutions to mathematical problems can be solved numerically by introducing a set of parameters to define a solution function in a general sense and then optimizing the variational principle with respect to these parameters in the manner of the eigenvalue problems of Chapter 8 and the finite element work in Chapter 9. Solving a set of non-linear equations by minimizing their sum of squares is another obvious application. Operations research is littered with examples which can be solved numerically using optimization techniques of one sort or another.

10.2 Univariate optimization

A point x^* is a strong local minimum if

(i) $f(x)$ is defined on a neighbourhood, $N = (x^* - \delta, x^* + \delta)$.
(ii) $f(x^*) < f(y)\ \forall\ y \in N$ with $y \neq x*$.

Further, a minimum x^* is an isolated local minimizer if there is no other local minimizer in the neighbourhood of x^*. The importance of these conditions can be seen in the examples

$$f(x) = \begin{cases} 1 - x^2 & x < 0 \\ 0 & x = 0 \\ 1 - x^2 & x > 0 \end{cases} \tag{10.2.1}$$

in which $x = 0$ is a local minimizer, and a global maximizer, and

$$f(x) = x^4 + x^2 \sin(1/x) \qquad x \neq 0 \tag{10.2.2}$$

with $f(0) = 0$, where $x = 0$ is a local non-isolated minimum. The necessary conditions for a univariate minimum are that

(i) $f'(x^*) = 0$
and
(ii) $f''(x^*) \geq 0$
and the sufficient conditions for a strong local minimum are
(i) $f'(x^*) = 0$
and
(ii) $f''(x^*) > 0$
These conditions will only be valid if the relevant derivatives exist.

How can such minima be found numerically? One approach is an interval reduction method in which the required minimum is first located inside a given interval AB of length L, say. Let a point C divide this interval in the ratio $c : 1$ and a point D in the ratio $1 : c$. The objective function $f(x)$ can be evaluated at C and D and a comparison made with the values at A and B. Hence it can be seen whether the minimum now lies in the interval AD or CD and the superfluous interval can be rejected, so leaving a reduced interval for the next step. The reduction continues until the interval is squeezed down to a length less than the required accuracy. Increased efficiency would ensue if the ratio of $CD : DB$ was also $c : 1$ as then only one extra point is required in the new reduced interval in order to proceed to the next step as one of the original four points is re-used. Hence c must satisfy

$$\frac{c}{1+c} = \frac{1-c}{1} \tag{10.2.3}$$

or

$$c^2 + c - 1 = 0 \tag{10.2.4}$$

to give

$$c = \frac{\sqrt{5}-1}{2} = 0.618\,034 \tag{10.2.5}$$

Hence at every step the interval is reduced by precisely this ratio, called the golden section. Hence for a given accuracy say ε, and an initial interval length L, the number of steps is m where

$$\varepsilon = L(0.618\,034)^m \tag{10.2.6}$$

A number of illustrative examples of the application of this algorithm will now be considered. The six examples are:

$$f_1 = (1-x)^2 - 1, \quad x^* = 1, \quad f^* = -1 \tag{10.2.7}$$

$$f_2 = (1-x)^6 - 1, \quad x^* = 1, \quad f^* = -1 \tag{10.2.8}$$

$$f_3 = (1-x)^{10} - 1, \quad x^* = 1, \quad f^* = -1 \tag{10.2.9}$$

$$f_4 = e^{-2x} - e^{-x}, \quad x^* = \ln(2), \quad f^* = -0.25 \tag{10.2.10}$$

$$f_5 = \cos x - \cosh x - 1, \quad x^* = 0, \quad f^* = 1 \tag{10.2.11}$$

$$f_6 = \cos x, \qquad x^* = \pi, \quad f^* = -1 \tag{10.2.12}$$

in which x^* is the minimum point and f^* the value of f at this point. Starting with the interval [0, 2], the reduction with f_1 begins to close in on the minimum with the sequence:

$$[0, 2]\ [0.7639, 2.0]\ [0.7639, 1.5279]\ [0.7639, 1.2361]\ [0.9443, 1.236]$$

and after 31 steps the interval is

$$[0.999\,999\,462\,55, 1.000\,000\,5374] \text{ with } f_1 = -0.999\,999\,999\,999$$

The working precision is 14 digits and the asking precision was 12 figures. It is clear that the x values are correct only to half the accuracy of the f values. This is a normal optimization phenomenon and is explained by the following simple analysis.

All algorithms are attempting to find a value of $x = x^*$ to force $f'(x^*)$ to be zero. Hence computationally the function $f(x)$ is evaluated at or near to x^* and the resulting f values are compared. However

$$f(x) = f(x^*) + hf'(x^*) + \frac{h^2}{2!}f''(x^*) + \cdots \tag{10.2.13}$$

and $f'(x^*) = 0$ to leave

$$f(x) = f(x^*) + \frac{h^2}{2!}f''(x^*) + \cdots \tag{10.2.14}$$

where $h = x - x^*$. Note that the difference in neighbouring f values is of order h^2. Therefore if these values are computed to say m figures of accuracy, the corresponding x values can only be found to an accuracy of $m/2$ figures. Hence for any optimization process of this type, the expected accuracy at the minimum point is only half the working accuracy. The value of the objective function at this point will be known to full accuracy. If both f'' and f''' were zero, there would be an h^4 effect and only a quarter accuracy would be found. The next few examples illustrate this vividly.

Again starting with $[0, 2]$, example f_2 reduces in 11 steps to [0.996 89, 1.006 944] with the f values at each end both equal to $-1.000\,000\,000\,000$. This is the phenomenon with only a sixth accuracy to be expected in x as $f^{(vi)}(-1)$ is the first non-vanishing derivative, and two correct figures result. The function f_3 takes this still further to give [0.944 27, 1.0557] after six steps with f equal to -1 to 14 figures. This has only a tenth accuracy and gives the one expected figure.

Example f_4 has the normal second non-vanishing derivative starting from $[0, 2]$ and reaches [0.693 146 818 87, 0.693 147 893 76] in 31 iterations. The number of iterations is expected to be the same as for f_1, as the interval reduction is the same ratio in both cases. The low-accuracy examples take corresponding fewer steps as they fail to separate close f values when the x accuracy is much less. The accuracy attained by the low-accuracy examples is the same as that for f_1 and f_4 for *the same number of steps*.

Example f_5 is another low-accuracy example reaching [−7.331 37(−4), 1.639 34(−4)] after a start at $[0, 2]$ and with 14 steps. Here there are three correct figures and an expected accuracy of a quarter full accuracy. This is a good example to test stopping criteria, a topic which is dealt with in more detail shortly.

The final example has no special difficulties but yields a well-known result and gives [3.141 591 974, 3.141 593 381] after 31 iterations from $[3, 4]$. The f values are correct to full accuracy at -1.

The golden section search is not optimum in the sense that the interval reduction is not maximized for a given number of function evaluations. The strategy which achieves this is based on the Fibonacci numbers defined by

$$F_k = F_{k-1} + F_{k-2}, \qquad F_0 = F_1 = 1 \tag{10.2.15}$$

Consider an interval $[0, F_N]$ for some N. Let the objective function be evaluated at the Fibonacci points F_i then these N allowed function evaluations will reduce the original interval to $1/F_N$ of its original length. If the original interval was just $[0, F_2]$ then the interval reduction is the maximum possible, that is half. This reduction will reduce as higher sequences are generated with the limit

$$\lim_{n\to\infty} \frac{F_n}{F_{n+1}} = c = 0.618\,034 \tag{10.2.16}$$

which means that the golden section search uses the limiting interval reduction at every step. To use a Fibonacci search would require an initial scaling of the interval to the predetermined F_N, so it is more usual to apply the golden section search to convergence.

Interval reduction methods are not usually as effective as interpolation methods which are discussed next. With these methods the function $f(x)$ is represented locally in some other more convenient form whose local minimum is easily found analytically. An update of this local fitting of $f(x)$ then continues to give a sequence of points converging to the required minimum. Powell's quadratic method (1964) is the simplest of these methods. Suppose the function $f(x)$ is evaluated at $x = \alpha$, $x = \beta$ and $x = \gamma$. The quadratic $y = ax^2 + bx + c$ can be forced to pass through these points to fix the constants a, b and c. (This is just Lagrange interpolation of Chapter 2). The values of the constants are

$$a = [(\gamma - \beta)f_\alpha + (\alpha - \gamma)f_\beta + (\beta - \alpha)f_\gamma]/\Delta \tag{10.2.17}$$

$$b = [(\beta^2 - \gamma^2)f_\alpha + (\gamma^2 - \alpha^2)f_\beta + (\alpha^2 - \beta^2)f_\gamma]/\Delta \tag{10.2.18}$$

$$c = [\beta\gamma(\gamma - \beta)f_\alpha + \gamma\alpha(\alpha - \gamma)f_\beta + \alpha\beta(\beta - \alpha)f_\gamma]/\Delta \tag{10.2.19}$$

where

$$\Delta = (\alpha - \beta)(\beta - \gamma)(\gamma - \alpha) \tag{10.2.20}$$

and the local minimum of the quadratic is at the point x_m where

$$x_m = \frac{1}{2}\left[\frac{(\beta^2 - \gamma^2)f_\alpha + (\gamma^2 - \alpha^2)f_\beta + (\alpha^2 - \beta^2)f_\gamma}{(\beta - \gamma)f_\alpha + (\gamma - \alpha)f_\beta + (\alpha - \beta)f_\gamma}\right] \tag{10.2.21}$$

and x_m is a minimum if the following condition holds

$$\frac{(\beta - \gamma)f_\alpha + (\gamma - \alpha)f_\beta + (\alpha - \beta)f_\gamma}{(\alpha - \beta)(\beta - \gamma)(\gamma - \alpha)} < 0 \tag{10.2.22}$$

Hence the algorithm is as follows.

(i) Choose a step length h and initial point x_0.

(ii) Evaluate $f(x_0)$ and $f(x_0 + h)$.

(iii) If $f(x_0) < f(x_0 + h)$ evaluate $f(x_0 - h)$ otherwise evaluate $f(x_0 + 2h)$, to give the three points α, β and γ.

(iv) Find x_m from (10.2.21) and apply the test (10.2.22).

(v) If the point is a maximum or if it is a minimum, further than some prescribed limit H from the nearest of the original three points, then discard the point furthest from the turning point and replace it by the point obtained by taking the step H in the decreasing direction, making this point the new current point.

(vi) If the turning point is a minimum and is within the required accuracy of the nearest point then the required minimum is found.

(vii) If the turning point is a minimum and convergence has not occurred, nor has the limit H been violated, then discard the point with the highest function value and continue the process with the new set of three points.

This type of algorithm is very typical of optimization procedures with considerable scope for small variations in the decision making parts of the algorithm.

Encoding a suitable stopping criterion for these algorithms raises a number of important issues. In the examples above, the varying accuracy of the x values and the f values poses a problem. Hence care is needed to setting up a suitable stopping criterion which does not have expectations beyond that which is theoretically possible. If f'' and f''' also vanish then the x values are only determined to $m/4$ figures. The earlier examples which behave like this are often set for student projects and invariably puzzle the recipient and confuse the stopping criterion. This phenomenon does not occur if a sum of squares is minimized, for example, as can be seen by expanding this case out.

Suppose x_n and x_{n+1} are the last two points in an iteration then a suitable stopping criterion might be

$$|x_n - x_{n+1}| < \varepsilon(|x_{n+1}| + \delta) \qquad (10.2.23)$$

This will give a relative error for values of x_n not near zero, and will automatically give an absolute error based on the value of δ if x_n were small. To guard against the possibility of a case in which only $m/4$ figures could be found it is a safeguard also to test the f values, but with the error ε^2. Hence (10.2.23) is used with

$$|f(x_n) - f(x_{n+1})| < \varepsilon^2(|f(x_{n+1})| + \delta) \qquad (10.2.24)$$

If either condition holds then the attainable accuracy is achieved.

Powell's method was applied to the same six examples used to illustrate the golden section method. For f_1, immediate convergence from the starting point $x = 0$ to full accuracy occurs as this example is a quadratic and will be fitted exactly. This is a good starting example to validate computer code. For f_2 the inherent loss of accuracy is again noted, and a start at $x = 0$ gives f values of 0.999 999 999 992 with $x = 0.985\,956$ after 32 steps, the sixth of machine accuracy being well illustrated. The stopping criterion relied entirely on the f values in this case. Further effects of this type arose in f_3 where 36 steps reduced f to the same minimum as before but now x was 0.9233 to give the expected one-figure accuracy, with the accuracy factor here only one tenth. Only nine iterations are needed in the more conventional example f_4 to give $x = 0.693\,147\,18$ after a start at $x = 0$ with $f = 0.25$ to full accuracy. The x value is at the expected half accuracy with seven figures.

The example f_5 is a special case, giving full accuracy in x and f in one step. The reason is that to order four this function is a quadratic, as it behaves like $1.0 + x^4/12$. As far as machine accuracy is concerned, the fit is exact and no comparisons are required for the updating formulae. The final example f_6 is conventional and gives $3.141\,592\,66$ with $f = -1$ after 17 steps with the expected half accuracy of seven figures.

It is easy to extend the quadratic interpolation method to higher degree polynomials, and the method due to Davidon (1959) uses a cubic fit in which

$$y(x) = f_0 + G_0 x + cx^2 + dx^3 \tag{10.2.25}$$

and the fitting satisfies

$$y(0) = f_0, \quad y(\alpha) = f_\alpha, \quad y'(0) = G_0, \quad y'(\alpha) = G_\alpha \tag{10.2.26}$$

which is analogous to a spline fit. For a fit on a more general interval than $[0, \alpha]$, a simple linear shift may be employed. The equations for c and d are

$$\alpha^2 c + \alpha^3 d = f_\alpha - f_0 - G_0\alpha \tag{10.2.27}$$

$$2\alpha c + 3\alpha^2 d = G_\alpha - G_0 \tag{10.2.28}$$

which solve to give

$$c = -(G_0 + z)/\alpha \tag{10.2.29}$$

$$d = (G_0 + g_\alpha + 2z)/\alpha^2/3 \tag{10.2.30}$$

with

$$z = \frac{3}{\alpha}(f_0 - f_\alpha) + G_0 + G_\alpha \tag{10.2.31}$$

The minimum occurs for

$$x_m = \frac{\alpha(G_0 + z \pm w)}{G_0 + G_\alpha + 2z} \tag{10.2.32}$$

where

$$w = \sqrt{(z^2 - G_0 G_\alpha)} \tag{10.2.33}$$

For a minimum, the positive sign has to be taken as

$$y''(x_m) = \pm 2w/\alpha > 0 \tag{10.2.34}$$

To avoid excess cancellation (10.2.32) is used in the form

$$x_m = \alpha\left(1 - \frac{G_\alpha + w - z}{G_\alpha - G_0 + 2w}\right) \tag{10.2.35}$$

Hence an algorithm analogous to that of Powell can be set up.

(i) Evaluate f_0 and G_0. If $G_0 \geq 0$ then all the following steps proceed in the negative direction. The positive case is:

(ii) Fix an estimate f_e to the sought minimum, and find

$$\alpha = \min\left\{\kappa, -\frac{2(f_0 - f_e)}{G_0}\right\} \tag{10.2.36}$$

This value of α lies at the minimum were f a quadratic and were f_e exact. It is common to set $\kappa = 2$.

(iii) Evaluate f_α and G_α.

(iv) If $G_\alpha > 0$ or $f_\alpha > f_0$ then apply (10.2.32) for x_m. Continue the process with the smaller interval $[x_0, x_m]$ or $[x_m, \alpha]$ according to whether the positive or negative directions are being pursued.

(v) If (iv) does not apply, replace α by 2α and return to (iii).
Again a stopping criterion is used to determine when sufficient accuracy has been attained.

This algorithm has been applied to the six illustrative examples with the following results. The quadratic example f_1 requires a special clause in the code. The fitting cubic (10.2.25) yields $d = 0$, or $G_0 + g_\alpha + 2z = 0$. Hence the minimum x_m will no longer be the solution of the quadratic $y'(x) = 0$ which yielded (10.2.32), but will now be the equivalent linear result

$$x_m = \frac{G_0\alpha}{2(G_0 + z)} \tag{10.2.37}$$

With this addition immediate convergence as in the Powell case occurs to full accuracy. For f_2, eight steps gives 1.0008 with $f = -0.999\,999\,999\,998$. This appears to be four-figure accuracy rather than the expected three, but the final result taken is the average of the final interval end-points, and these are correct to the three figures and the extra figure is fluked on averaging. Similarly on f_3 the final x was 1.0024 from the end-point values of (0.9267, 1.0925) with f correct to 12 figures, again showing some extra accuracy. The function f_4 again behaves sufficiently like a quadratic to fall out with the modified code in one step to full accuracy. The fifth example gives 3.981 744(−5) after eight steps with $f = 1.0$ to 12 figures showing the inherent inaccuracy in the f values once again. The final example f_6 takes six steps to yield $x = 3.141\,592\,6535$ with $f = -1$ to machine accuracy.

Comparison between the Powell and Davidon methods is a little unfair as the Davidon approach uses knowledge of the derivatives to increase the local fitting accuracy and is expected to be faster. However, in a practical situation in which these derivatives are not readily found, the faster method is not applicable.

The most important use of these two last methods will be in the multivariate methods which follow in the next paragraph. These methods require local minima to be found along a given direction $\mathbf{d}$ from some initial point $\mathbf{x}_0$. Hence instead of using the variable x in the uni-variate case, the line search uses the variable s along the line $\mathbf{x}_0 + s\mathbf{d}$. The codes are easily modified for use as line-search methods.

10.3 Multivariate optimization

As with univariate optimization, a suitable starting point is to consider the theoretical conditions for a minimum. A point $\mathbf{x}^*$ is said to be a strong local minimum if $f(\mathbf{x}^*) < f(\mathbf{y})$ for all $\mathbf{y}$ in some neighbourhood of $\mathbf{x}^*$. This is just the vector equivalent of the univariate condition. In the smooth case when higher derivatives are available the univariate conditions can be generalized by first defining

$$\mathbf{g}(\mathbf{x}) = \nabla f(\mathbf{x}) \tag{10.3.1}$$

and

$$\mathbf{G}(\mathbf{x}) = \nabla^2 f(\mathbf{x}) \tag{10.3.2}$$

where $\nabla f(\mathbf{x})$ is the gradient vector $\partial f/\partial x_i$ and ∇^2 is the Hessian matrix of second derivatives $\partial^2 f/\partial x_i \partial x_j$. This matrix is clearly square and symmetric. For brevity, define f^* as $f(\mathbf{x}^*)$ and similarly $\mathbf{g}^*$ as $\mathbf{g}(\mathbf{x}^*)$. Again by analogy with the univariate case the sufficient condition for a local minimum is that $\mathbf{g}^* = \mathbf{0}$ and that $\mathbf{G}^*$ is positive definite.

Before considering specific algorithms, it is instructive to decide what useful properties such algorithms might exhibit. In rough terms an algorithm needs to generate a set of iterates $\mathbf{x}^{(k)}$ which converge to the minimum $\mathbf{x}^*$. If the error is given by $\boldsymbol{\varepsilon}^{(k)} = \mathbf{x}^{(k)} - \mathbf{x}^*$and if $\boldsymbol{\varepsilon}^{(k)} \to 0$ as $k \to \infty$, then the algorithm is said to converge. If there is no restriction on how close $\mathbf{x}^{(k)}$ needs to be to $\mathbf{x}^*$, then the convergence is global. If the convergence only occurs in some neighbourhood of $\mathbf{x}^*$ then the convergence is local. The rate of convergence can also be classified. If the limit

$$\frac{||\boldsymbol{\varepsilon}^{(k+1)}||}{|\boldsymbol{\varepsilon}^{(k)}||^p} \to \alpha > 0 \tag{10.3.3}$$

holds then p is the order of convergence and α the rate of convergence. In the linear case with $p = 1$, a satisfactory rate of convergence is required in that α needs to be very much less than unity for a practically acceptable outcome. Superlinear convergence occurs when this rate tends to zero. If $p = 2$ then quadratic convergence occurs.

A basic desirable property of any algorithm is that it should satisfy the descent property

$$f^{(k+1)} < f^{(k)} \tag{10.3.4}$$

By itself, this condition is very weak, as it allows very small reductions. The concept of a sufficient decrease is encapsulated in the Goldstein–Armijo principle

$$0 < -\mu_1\alpha^{(k)}\mathbf{g}^{(k)T}\mathbf{s}^{(k)} \leq f^{(k)} - f^{(k+1)} \leq -\mu_2\alpha^{(k)}\mathbf{g}^{(k)T}\mathbf{s}^{(k)} \tag{10.3.5}$$

where $\alpha^{(k)}$ is the step taken in a search direction $\mathbf{s}^{(k)}$. The constants μ_1 and μ_2 are suitable constants and restrict the steps taken from being either too large or too small.

Many of the algorithms to be discussed rely on the convenient properties of quadratic functions. Such functions are the simplest functions with well determined minima, and hence many algorithms are designed with this model in mind. An algorithm which will find the minimum of a quadratic function in a finite number of steps is said to have quadratic termination. Many practical algorithms have this property and may be applied iteratively to non-quadratic examples.

Most of the methods to be discussed now follow a general approach which is common to them all. Suppose the iterative process has reached the point $\mathbf{x}^{(k)}$ then the next steps in general terms are as follows.

(i) Determine a direction of search $\mathbf{s}^{(k)}$.
(ii) Find a step $\alpha^{(k)}$ which minimizes $f(\mathbf{x}^{(k)} + \alpha\mathbf{s}^{(k)})$ with respect to α.
(iii) Update $\mathbf{x}^{(k+1)}$ to $\mathbf{x}^{(k)} + \alpha^{(k)}\mathbf{s}^{(k)}$.

The various methods will differ in the manner in which the search direction is chosen, and the method for minimizing along the search direction, with the associated issue of how accurate these searches need to be. Clearly some way from the minimum it would seem unnecessary to insist on high-accuracy line searches. These line searches use the algorithms of the previous paragraph with $f(x)$ replaced by $f(\mathbf{x}^{(k)} + \alpha\mathbf{s}^{(k)})$, and α being the new one-dimensional variable.

The most simple minded of algorithms is to choose the search direction to be the line of steepest descent and so define

$$\mathbf{s}^{(k)} = -\mathbf{g}^{(k)} \tag{10.3.6}$$

It is easy to give counter-examples which show the shortcomings of this method. The region defined by

$$f(\mathbf{x}) = 100(x_2 - x_1^2)^2 + (1 - x_1)^2 \tag{10.3.7}$$

consists of a 'banana' shaped valley leading from the region around (1.0, 1.0) to the minimum at (−1.0, 1.0). The traditional starting point is (1.2, 1.0), a little up the valley wall. Steepest descent takes no account of second derivatives, and so moves down the line of steepest descent into the 'banana' valley. Instead of then turning along this gently sloping valley the iterations progress at right angles to the preferred direction and just zig-zag back and forth across the valley. Progress along the valley is painfully slow.

One answer to this counter-example is to introduce second-order derivatives into the algorithm. Newton's method does exactly this, and is based on the quadratic model that could be expected to be a successful line of attack. The model follows from a Taylor expansion of the object function about the current point:

$$f(\mathbf{x}^{(k)} + \boldsymbol{\delta}) \sim \mathbf{y}^{(k)}(\boldsymbol{\delta}) = f^{(k)} + \mathbf{g}^{(k)T}\boldsymbol{\delta} + \tfrac{1}{2}\boldsymbol{\delta}^T\mathbf{G}^{(k)}\boldsymbol{\delta} \tag{10.3.8}$$

with $\boldsymbol{\delta} = \mathbf{x} - \mathbf{x}^{(k)}$. In the case when a minimum exists, that is, $\mathbf{G}^{(k)}$ is positive definite, then the step $\boldsymbol{\delta}^{(k)}$ is defined by the equation $\nabla\mathbf{y}^{(k)}(\boldsymbol{\delta}) = 0$. This results in the defining equation

$$\mathbf{G}^{(k)}\boldsymbol{\delta} = -\mathbf{g}^{(k)} \tag{10.3.9}$$

for the step $\boldsymbol{\delta}$ and involves solving a set of $n \times n$ linear equations. The convergence of Newton's method from a point sufficiently close to the minimum can be proved to be second order. The practical problems arise away from the minimum when there is no guarantee of success. If $\mathbf{G}^{(k)}$ is not positive definite, the process breaks down and one suggestion is to revert to steepest descent if this happens. Even with all these safeguards the computational load is still large with the evaluation of the matrix $\mathbf{G}$. This leads to the quasi-Newton methods in which this matrix is estimated.

The rate of convergence is easy to establish by considering the Taylor expansion

$$\mathbf{g}(\mathbf{x}^{(k)} + \boldsymbol{\delta}) = \mathbf{g}(k) + \mathbf{G}(k)\boldsymbol{\delta} + O(||\boldsymbol{\delta}||^2) \tag{10.3.10}$$

Using $\boldsymbol{\varepsilon}^{(k)} = \mathbf{x}^{(k)} - \mathbf{x}^*$ gives

$$\mathbf{0} = \mathbf{g}^* = \mathbf{g}^{(k)} - \mathbf{G}^{(k)}\boldsymbol{\varepsilon}^{(k)} + O(||\boldsymbol{\varepsilon}^{(k)}||^2) \tag{10.3.11}$$

and hence

$$\mathbf{0} = -\boldsymbol{\delta}^{(k)} - \boldsymbol{\varepsilon}^{(k)} + O(||\boldsymbol{\varepsilon}^{(k)}||^2) = -\boldsymbol{\varepsilon}^{(k+1)} + O(||\boldsymbol{\varepsilon}^{(k)}||^2) \tag{10.3.12}$$

Hence from the definition of O, there exists a c such that

$$||\boldsymbol{\varepsilon}^{(k+1)}|| \le c||\boldsymbol{\varepsilon}^{(k)}||^2 \tag{10.3.13}$$

to establish second-order convergence.

The problem of evaluating $\mathbf{G}$ is overcome in the quasi-Newton methods in which only first derivatives are employed. In some way it becomes necessary to estimate $\mathbf{G}$ at each step. The usual approximation is to use a symmetric positive definite matrix $\mathbf{H}^{(k)}$ to replace $[\mathbf{G}^{(k)}]^{-1}$. Now the main iterations include a step to update $\mathbf{H}$. The initial matrix $\mathbf{H}^{(0)}$ is often taken to be the unit matrix, but any other suitable positive definite matrix is feasible. If the descent property is implied by maintaining $\mathbf{H}^{(k)}$ positive definite, the algorithm is of the variable metric type. The rule for passing from $\mathbf{H}^{(k)}$ to $\mathbf{H}^{(k+1)}$ is called the update formula. Such a formula is generated by considering a suitable form for the update which contains some free parameters and then by forcing desirable properties onto the formula to fix the free parameters. One desirable property is the quasi-Newton condition. If

$$\boldsymbol{\delta}^{(k)} = \mathbf{x}^{(k+1)} - \mathbf{x}^{(k)} \tag{10.3.14}$$

$$\boldsymbol{\gamma}^{(k)} = \mathbf{g}^{(k+1)} - \mathbf{g}^{(k)} \tag{10.3.15}$$

then a Taylor expansion gives

$$\boldsymbol{\gamma}^{(k)} = \mathbf{G}^{(k)}\boldsymbol{\delta}^{(k)} + \text{higher order terms} \tag{10.3.16}$$

and hence the rough idea that $\mathbf{H} \sim \mathbf{G}^{-1}$ is encompassed in the quasi-Newton condition

$$\mathbf{H}^{(k+1)}\gamma^{(k)} = \delta^{(k)} \tag{10.3.17}$$

A possible form for the update rule is then

$$\mathbf{H}^{(k+1)} = \mathbf{H}^{(k)} + \alpha\mathbf{u}\mathbf{u}^T \tag{10.3.18}$$

which is a rank-one formula as the added matrix is symmetric and of rank one. Hence using (10.3.17) gives

$$\mathbf{H}^{(k)}\boldsymbol{\gamma}^{(k)} + \alpha\mathbf{u}\mathbf{u}^T\boldsymbol{\gamma}^{(k)} = \boldsymbol{\delta}^{(k)} \tag{10.3.19}$$

which implies that $\mathbf{u}$ is proportional to $\boldsymbol{\delta} - \mathbf{H}^{(k)}\boldsymbol{\gamma}^{(k)}$. Hence fixing the constant of proportionality α gives the update rule

$$\mathbf{H}^{(k+1)} = \mathbf{H}^{(k)} + \frac{(\boldsymbol{\delta} - \mathbf{H}^{(k)}\boldsymbol{\gamma})(\boldsymbol{\delta} - \mathbf{H}^{(k)}\boldsymbol{\gamma})^T}{(\boldsymbol{\delta} - \mathbf{H}^{(k)}\boldsymbol{\gamma})^T\boldsymbol{\gamma}} \tag{10.3.20}$$

This formula suffers from two problems. The most glaring is that the denominator in (10.3.20) may become zero, or small enough to cause numerical difficulties. Furthermore the positive definiteness of $\mathbf{H}$ is not maintained necessarily. In order to give further degrees of freedom which will allow these problems to be overcome, the rank-two formula with the form

$$\mathbf{H}^{(k+1)} = \mathbf{H}^{(k)} + \alpha\mathbf{u}\mathbf{u}^T + \beta\mathbf{v}\mathbf{v}^T \tag{10.3.21}$$

is considered. Now the quasi-Newton condition will not give a unique value for $\mathbf{u}$ and $\mathbf{v}$ and various possibilities can be tried. One fruitful combination is to set $\mathbf{u} = \boldsymbol{\delta}^{(k)}$ and $\mathbf{v} = \mathbf{H}^{(k)}\boldsymbol{\gamma}^{(k)}$ to give the rule

$$\mathbf{H}^{(k+1)} = \mathbf{H}^{(k)} + \frac{\boldsymbol{\delta}\boldsymbol{\delta}^T}{\boldsymbol{\delta}^T\boldsymbol{\gamma}} - \frac{\mathbf{H}^{(k)}\boldsymbol{\gamma}\boldsymbol{\gamma}^T\mathbf{H}^{(k)}}{\boldsymbol{\gamma}^T\mathbf{H}^{(k)}\boldsymbol{\gamma}} \tag{10.3.22}$$

which is usually known as the Davidon–Fletcher–Powell formula and denoted by the letters DFP (Fletcher and Powell, 1963).

It can be shown that for this rule, positive definiteness is maintained, with exact line searches a quadratic object function terminates in, at most, n iterations and convergence has a superlinear order. At one stage this rule was highly used, but with inexact line searches being used, it proves less effective than other more recent alternatives. There is a great gain in not performing line searches exactly, and this gain is lost if exact searches are needed to make the update formula behave correctly.

If the approximate Hessian matrix is denoted by $\mathbf{B}$ so that $\mathbf{H}^{-1} = \mathbf{B}$, then the dual update rule or complementary rule corresponding to (10.3.22) is

$$\mathbf{B}^{(k+1)} = \mathbf{B}^{(k)} + \frac{\gamma\gamma^T}{\gamma^T\delta} - \frac{\mathbf{B}^{(k)}\delta\delta^T\mathbf{B}^{(k)}}{\delta^T\mathbf{B}^{(k)}\delta} \tag{10.3.23}$$

where such rules are formed by replacing $\mathbf{H}$ by $\mathbf{B}$ and δ by γ. If now an $\mathbf{H}$ matrix is generated to force $\mathbf{BH} = \mathbf{I}$, then the formula

$$\mathbf{H}^{(k+1)} = \mathbf{H}^{(k)} + \left(1 + \frac{\gamma^T\mathbf{H}^{(k)}\gamma}{\delta^T\gamma}\right)\frac{\delta\delta^T}{\delta^T\gamma} - \left(\frac{\delta\gamma^T\mathbf{H}^{(k)} + \mathbf{H}^{(k)}\gamma\delta^T}{\delta^T\gamma}\right) \tag{10.3.24}$$

is obtained and this update formula is the BFGS rule suggested by Broyden (1970), Fletcher (1970), Goldfarb (1970) and Shanno (1970). A further parameter can be introduced to give even more generality, producing the Broyden family

$$\mathbf{H}_{\phi}^{(k+1)} = (1 - \phi)\mathbf{H}_{DFP}^{(k+1)} + \phi\mathbf{H}_{BFGS}^{(k+1)} \tag{10.3.25}$$

A different line of approach is furnished by considering methods based on conjugate directions $\mathbf{s}^{(0)}, \mathbf{s}^{(1)}, \ldots, \mathbf{s}^{(n)}$ with respect to a matrix $\mathbf{G}$ which satisfy

$$\mathbf{s}^{(i)T}\mathbf{G}\mathbf{s}^{(j)} = 0 \quad \forall \quad i \neq j \tag{10.3.26}$$

If $\mathbf{G}$ is the Hessian matrix then a conjugate direction method generates conjugate directions when applied to a quadratic. The method can be shown to terminate in at most n exact line searches by considering $\mathbf{g}^{(k+1)T}\mathbf{s}^{(j)}$ for $j = 1, 2, \ldots, k$. If these values are all zero then, as $\mathbf{G}$ is positive definite and the conjugate directions are independent, the result follows. But

$$\mathbf{g}^{(k+1)T}\mathbf{s}^{(j)} = \mathbf{g}^{(j+1)T}\mathbf{s}^{(j)} + \sum_{i=j+1}^{k} (\mathbf{g}^{(i+1)T} - \mathbf{g}^{(i)T})\mathbf{s}^{(j)} \tag{10.3.27}$$

and for each exact line search $\mathbf{g}^{(k+1)T}\mathbf{s}^{(k)} = 0$ and $\mathbf{g}^{(i)} - \mathbf{g}^{(j)} = \mathbf{G}(\mathbf{x}^{(i)} - \mathbf{x}^{(j)})$. These together yield the required zero.

The conjugate gradient method is one way to achieve the above set of directions. By setting

$$\mathbf{s}^{(1)} = -\mathbf{g}^{(1)} \tag{10.3.28}$$

and then letting $\mathbf{s}^{(k+1)}$ be conjugate to the previously generated $\mathbf{s}$'s and be a component of $-\mathbf{g}^{(k+1)}$, gives by Gram–Schmidt (1.2.9)

$$\mathbf{s}^{(k+1)} = -\mathbf{g}^{(k+1)} + \sum_{j=1}^{k} \beta_j \mathbf{s}^{(j)} \tag{10.3.29}$$

and the conjugacy condition is

$$\mathbf{s}^{(i)T} \boldsymbol{\gamma}^{(j)} = 0 \qquad j \neq i \tag{10.3.30}$$

But for quadratic functions $\beta_j = 0$ for $j < k$ and the search directions are given by

$$\mathbf{s}^{(k+1)} = -\mathbf{g}^{(k+1)} + \beta^{(k)} \mathbf{s}^{(k)} \tag{10.3.31}$$

with

$$\beta^{(k)} = \frac{\mathbf{g}^{(k+1)T} \mathbf{g}^{(k+1)}}{\mathbf{g}^{(k)T} \mathbf{g}^{(k)}} \tag{10.3.32}$$

and $\beta^{(0)} = 0$. This method is the Fletcher–Reeves (1964) method. The method is not as efficient as the quasi-Newton methods discussed previously, but requires no matrix operations. From a storage point of view, just three n vectors are required. Hence the importance of this formula is in large systems where storage of the matrices in the quasi-Newton algorithms proves a handicap.

It is not always possible or computationally economical to compute the gradients, and it would be useful to have algorithms which used the advantages of linear searches in conjugate directions without having explicitly to evaluate gradients. The most commonly used method is Powell's method which depends on the following observation. Suppose a direction $\mathbf{d}$ is given and let $\mathbf{x}^*$ minimize a quadratic function

$$f(\mathbf{x}) = \tfrac{1}{2}\mathbf{x}^T \mathbf{G} \mathbf{x} + \mathbf{b}^T \mathbf{x} + c \tag{10.3.33}$$

along $\mathbf{d}$. If $\mathbf{y}$ is a point not on the line $\mathbf{r} = \mathbf{x}^* + s\mathbf{d}$, then if $\mathbf{y}^*$ minimizes $f(\mathbf{x})$ along the parallel line $\mathbf{r} = \mathbf{y} + t\mathbf{d}$, then $\mathbf{y}^* - \mathbf{x}^*$ is conjugate to $\mathbf{d}$. This follows because

$$\mathbf{g}(\mathbf{x}^*) = \mathbf{G}\mathbf{x}^* + \mathbf{b} \tag{10.3.34}$$

$$\mathbf{g}(\mathbf{y}^*) = \mathbf{G}\mathbf{y}^* + \mathbf{b} \tag{10.3.35}$$

and hence

$$\mathbf{d}^T(\mathbf{G}\mathbf{x}^* + \mathbf{b}) = 0 \tag{10.3.36}$$

$$\mathbf{d}^T(\mathbf{G}\mathbf{y}^* + \mathbf{b}) = 0 \tag{10.3.37}$$

and subtracting yields the required condition

$$\mathbf{d}^T \mathbf{G}(\mathbf{y}^* - \mathbf{x}^*) = 0. \tag{10.3.38}$$

Powell's method makes use of this fact in the following algorithm. Initially the directions $\mathbf{d}^{(1)}$ are the co-ordinate directions $\mathbf{e}$. The kth iteration uses $\mathbf{y}_0^{(k)} = \mathbf{x}^k$ and then $\mathbf{y}_r^{(k)}$ is the minimum from $\mathbf{y}_{r-1}^{(k)}$ in the direction $\mathbf{d}_r^{(k)}$ for $r = 1, \ldots, n$. Finally a search is made from $\mathbf{y}_n^{(k)}$ in the direction $\boldsymbol{\delta}^{(k)} = \mathbf{y}_n^{(k)} - \mathbf{x}^{(k)}$ with a minimum at the new current point $\mathbf{x}^{(k+1)}$. The search directions $\mathbf{d}_2^{(k)}, \ldots, \mathbf{d}_n^{(k)}, \boldsymbol{\delta}^{(k)}$ then become the new directions $\mathbf{d}_1^{(k+1)}, \ldots, \mathbf{d}_n^{(k+1)}$. With this algorithm, occasionally the discarding of $\mathbf{d}_1^{(k)}$ will leave a set of directions which will no longer span the complete space, and a breakdown occurs. To guard against this, Powell allows a different direction to be discarded, but pays for this in having to do more than the requisite n iterations to find the minimum of a positive definite quadratic function. The modified algorithm then becomes the following.

(i) Let $\mathbf{y}_0^{(k)} = \mathbf{x}^{(k)}$ and search from $\mathbf{y}_{r-1}^{(k)}$ along the direction $\mathbf{d}_r^{(k)}$ for a minimum at $\mathbf{y}_r^{(k)}$.

(ii) Define

$$\begin{aligned} \Delta &= \max_{r=1,n} [f(\mathbf{y}_{r-1}^{(k)}) - f(\mathbf{y}_r^{(k)})] \\ &= f(\mathbf{y}_{q-1}^{(k)}) - f(\mathbf{y}_q^{(k)}) \end{aligned} \tag{10.3.39}$$

(iii) Compute

$$\begin{aligned} f_1 &= f(\mathbf{y}_0^{(k)}) \\ f_2 &= f(\mathbf{y}_n^{(k)}) \\ f_3 &= f(2\mathbf{y}_n^{(k)} - y_0^{(k)}) \end{aligned} \tag{10.3.40}$$

(iv) If

$$f_3 \geq f_1$$

or

$$(f_1 - 2f_2 + f_3)(f_1 - f_2 - \Delta) \geq \tfrac{1}{2}\Delta(f_1 - f_3)^2$$

then use the old directions $\mathbf{d}_j^{(k)}$ in the $(k+1)$th step and set $\mathbf{x}^{(k+1)} = \mathbf{y}_n^{(k)} = \mathbf{y}_0^{(k+1)}$

(v) otherwise use the Powell unmodified rule but discard $\mathbf{d}_q^{(k)}$ and replace it with $\boldsymbol{\delta}^{(k)}$.

In the development of multivariable optimization algorithms, a great number of test functions have been invented by the various authors of the algorithms. They provide an extensive test-bed for future methods and raise many practical issues. Some of the more important test functions are listed below.

(1) Rosenbrock:

$$f(x) = 100(x_2 - x_1^2)^2 + (1 - x_1)^2 \tag{10.3.41}$$

$$x^{(0)} = (1.2, 1.0), \quad x^* = (1, 1), \quad f(x^*) = 0$$

(2) Wood:

$$\begin{aligned} f(x) = {} & 100(x_1^2 - x_2)^2 + (x_1 - 1)^2 + (x_3 - 1)^2 + 90(x_3^2 - x_4)^2 \\ & + 10.1\{(x_2 - 1)^2 + (x_4 - 1)^2\} + 19.8(x_2 - 1)(x_4 - 1) \end{aligned} \tag{10.3.42}$$

$$x^{(0)} = (-3, -1, -3, -1), \quad x^* = (1, 1, 1, 1), \quad f(x^*) = 0$$

(3) Miele and Cantrell:

$$f(x) = (e^{x_1} - x_2)^4 + 100(x_2 - x_3)^6 + \tan^4(x_3 - x_4) + x_1^8 \tag{10.3.43}$$

$$x^{(0)} = (1, 2, 2, 2), \quad x^* = (0, 1, 1, 1), \quad f(x^*) = 0$$

(4) Powell:

$$f(x) = (x_1 + 10x_2)^2 + 5(x_3 - x_4)^2 + (x_2 - 2x_3)^4 + 10(x_1 - x_4)^4 \tag{10.3.44}$$

$$x^{(0)} = (3, -1, 0, 1), \quad x^* = (0, 0, 0, 0), \quad f(x^*) = 0$$

(5) Fletcher and Powell:

$$f(x) = 100\{(x_3 - 10\theta)^2 + (r - 1)^2\} + x_3^2 \tag{10.3.45}$$

where

$$r = |(x_1 + x_2)^{1/2}|$$

and

$$\theta = \begin{cases} \frac{1}{2\pi} \arctan\left(\frac{x_2}{x_1}\right) & x_1 > 0 \\ \frac{1}{2\pi} \arctan\left(\frac{x_2}{x_1}\right) + \frac{1}{2} & x_1 < 0 \end{cases}$$

$$x^{(0)} = (-1, 0, 0), \quad x^* = (1, 0, 0), \quad f(x^*) = 0$$

(6) Dixon:

$$f(x) = (1 - x_1)^2 + (1 - x_{10})^2 + \sum_{i=1}^{9} (x_i^2 - x_{i+1})^2 \tag{10.3.46}$$

$$x^{(0)} = (-2, \ldots, -2), \quad x^* = (1, \ldots, 1), \quad f(x^*) = 0$$

In each case the starting point $x^{(0)}$ and the final minimum point x^* are given.

To illustrate the behaviour of the methods discussed in this section, they have been applied to these test examples and the results are shown in Table 10.1. The five methods which have been used are firstly Powell's method which uses no

Table 10.1

	Powell			DFP (Powell)			BFGS (Powell)			DFP (Davidon)			BFGS (Davidon)		
	fc	itc	lsc	fc	itc	gc	fc	itc	gc	fc	itc	gc	fc	itc	gc
f_1	436	13	36	527	24	47	601	24	47	268	24	535	273	24	545
f_2	490	13	57	500	42	165	500	42	165	306	41	1221	305	41	1217
f_3	330	7	33	709	61	241	624	61	241	644	23	2573	674	28	2693
f_4	866	20	96	436	30	117	435	30	117	226	23	901	223	23	889
f_5	586	16	61	325	25	73	347	25	73	255	25	763	241	25	721
f_6	4007	39	416	525	45	441	520	45	441	438	47	4371	324	43	3231

derivatives. The three counts in this column are the function count (fc), the iteration count (itc) and the line search count (lsc). The second method is DFP with Powell line searches employed, then BFGS with Powell line searches and finally DFP and BFGS with Davidon line searches. Hence the table will also illustrate how the two line search methods perform, though it would be normal to use Davidon if gradients were available. The gradient count(gc) is triggered for each component of the gradient at each point of evaluation, and therefore rises with the dimension of the problem. In general terms, each gradient component evaluation will be of a similar workload to a function evaluation, and should be counted this way for fairness.

Some of these examples exhibit special difficulties which are highlighted in these results. Example f_6 is the one high-dimension problem with $n = 10$ so high gradient counts are anticipated. Example f_3 has high-order vanishing derivatives near the minimum, and so only restricted accuracy can be expected. This is the sort of example which will cause maximum disruption to the stopping criteria used in the codes. The algorithms were stopped when either two successive f-values differed by less than 10^{-12} or all the components of $\mathbf{x}$ differed from the previous iteration by 10^{-6}. Hence in example f_3 it was the f condition which triggered a halt, as the $\mathbf{x}$ values would still be only accurate to about two figures. Exact line searches to machine accuracy were used throughout with similar stopping conditions to the above. One of the gains in using BFGS is that the line searches can be relaxed with a corresponding gain in efficiency, hence these tables undersell this method.

As the Davidon line searches themselves require gradient evaluation, the gradient counts for the Davidon runs are correspondingly greater than for Powell line searches. In general the number of iterations whether using Powell or Davidon are the same as both are yielding machine accurate results. The Powell method appears much less effective for the larger dimensional problem f_{10}.

10.4 Constrained optimization

The problem of constrained optimization is to minimize the function $f(\mathbf{x})$ of n real variables subject now to the constraints

$$c_i(\mathbf{x}) = 0 \qquad i = 1, p_1 \tag{10.4.1}$$

and

$$c_i(\mathbf{x}) \geq 0 \qquad i = p_1, p \tag{10.4.2}$$

If a point $\mathbf{x}$ satisfies the p constraint conditions then the point is called a feasible point. The set of feasible points defines the feasible region. A constraint is said to be active at a given point $\mathbf{x}$ if the point $\mathbf{x}$ is on the boundary of its feasible region. The set of active constraints $\mathcal{A}$ plays an important role in the subject as the other constraints can be locally ignored, and the inequality constraints treated as equality constraints. The best understood problems arise when both f and the constraints c_i are linear functions. This is known as linear programming. If f is a quadratic function, but the constraints are still linear, then the problem is one of quadratic programming. The general problem is called non-linear programming.

Linear programming is an important topic in operational research and can only be treated briefly here. The problem is the linear version of (10.4.1) and (10.4.2). Hence the object is to maximize (or minimize)

$$z = \mathbf{c}^T\mathbf{x} \tag{10.4.3}$$

subject to

$$\mathbf{Ax} \leq, =, \geq \mathbf{b} \tag{10.4.4}$$

and the non-negativity restriction $\mathbf{x} \geq \mathbf{0}$. The vector $\mathbf{c}$ has components

$$\mathbf{c} = [c_1, c_2, \ldots, c_n] \tag{10.4.5}$$

and is called the price vector. The vector $\mathbf{b}$ is the requirements vector with components $[b_1, b_2, \ldots, b_m]$. By introducing *slack* variables, constraints of the form $g(\mathbf{x}) \leq b$ can be rewritten as $g(\mathbf{x}) + x_q = b$ for q ranging from $n + 1$ as necessary. Equally, constraints of the form $g(\mathbf{x}) \geq b$ can be rewritten as $g(\mathbf{x}) - x_p = b$. The extra variable is called a *surplus* variable. In both cases the non-negativity restriction will apply to the extra variables to maintain the equivalence of the two forms of the constraint equations. Hence the general form of the linear programming problem becomes

$$\text{maximize } z = \mathbf{c}^T\mathbf{x} \tag{10.4.6}$$

with the constraints

$$\mathbf{Ax} = \mathbf{b} \text{ and } \mathbf{x} \geq \mathbf{0} \tag{10.4.7}$$

The matrix $\mathbf{A}$ is partitioned according to

$$\mathbf{A}_0\mathbf{x}_0 + \begin{bmatrix} \mathbf{I}_p & \mathbf{0} \\ \mathbf{0} & -\mathbf{I}_q \\ \mathbf{0} & \mathbf{0} \end{bmatrix} \mathbf{x}_s = \mathbf{b} \tag{10.4.8}$$

where the vector $\mathbf{x}_0 = [x_1, \ldots, x_n]$ is the vector of main variables, $\mathbf{x}$ is the vector

$$\mathbf{x}_s = [x_{n+1}, \ldots, x_{n+p}, x_{n+p+1}, \ldots, x_{n+p+q}] \tag{10.4.9}$$

of p slack and q surplus vectors, and

$$\mathbf{x} = (\mathbf{x}_0, \mathbf{x}_s)^T \quad \text{and} \quad \mathbf{c} = (\mathbf{c}_0, \mathbf{0})^T \tag{10.4.10}$$

The matrix $\mathbf{A}$ is therefore $m \times N$ where $N = n + p + q$.

In two dimensions the constraint equations are just straight lines. Hence the maximum will lie on a boundary because of the uniform slope. Each boundary is limited by its intersection with other boundaries and hence a vertex or point of intersection will be the required maximum. The method for formalizing this process is called the simplex method, and is often laid out in a sequence of tables referred to as *tableaux*. If $N - m$ variables are set equal to zero, and if the remaining corresponding columns of $\mathbf{A}$ are linearly independent then the solution of the resulting set of m simultaneous linear equations is called a basic solution. As any m elements may be taken there are at most ${}^N C_m$ basic solutions. For any such solution the columns of $\mathbf{A}$ are called the basis vectors, and these columns generate the basis matrix. The simplex method involves finding a sequence of basic feasible solutions, each of which gives an improved objective value over the previous solution. This is equivalent in two dimensions to moving from one vertex to another of a convex polygon.

This procedure is first illustrated in an example, after which the general equations involved can be given and finally the tableau layout. Consider the problem:

$$\text{maximize } z = 3x_1 + 6x_2 + 2x_3 \tag{10.4.11}$$

subject to

$$\begin{aligned} 3x_1 + 4x_2 + x_3 &\leq 2 \\ x_1 + 3x_2 + 2x_3 &\leq 1 \\ x_1, x_2, x_3 &\geq 0 \end{aligned} \tag{10.4.12}$$

then the introduction of slack variables gives

$$\begin{aligned} 3x_1 + 4x_2 + x_3 + x_4 &= 2 \\ x_1 + 3x_2 + 2x_3 + x_5 &= 1 \\ x_1, x_2, x_3, x_4, x_5 &\geq 0 \end{aligned} \tag{10.4.13}$$

An obvious basic feasible solution is

$$x_1 = x_2 = x_3 = 0 \quad \text{with} \quad x_4 = 2, \; x_5 = 1$$

Just how to get such a starting point in general will be covered a little later. The object is to choose one of the non-basic variables, that is x_1, x_2 or x_3 which can be changed to allow an increase in z without violating the constraints. The largest element in z is $6x_2$ so this one will become non-zero and one of the basic variables will become zero. How far can x_2 be increased before violating a constraint? The first constraint is just violated when $4x_2 = 2$ or $x_2 > \frac{1}{2}$ and the second when $x_2 > \frac{1}{3}$. This is the first constraint to tip over so that if $x_2 = \frac{1}{3}$ then x_5 becomes non-basic and zero. The new basic variables are x_5 and x_2, and z has risen to 2. The problem is now

$$x_2 = \tfrac{1}{3} - \tfrac{1}{3}x_1 - \tfrac{1}{3}x_5 - \tfrac{2}{3}x_3 \tag{10.4.14}$$

and eliminating x_2 from the first constraint gives

$$x_4 = \tfrac{2}{3} - \tfrac{5}{3}x_1 + \tfrac{4}{3}x_5 + \tfrac{5}{3}x_3 \tag{10.4.15}$$

and z is

$$z = 2 + x_1 - 2x_5 - 2x_3 \tag{10.4.16}$$

again eliminating x_2. The process is now repeated. The pivotal element to give maximum increase in z is the only positive element left, x_1. Constraint (10.4.14) limits x_1 to $\frac{2}{3}/\frac{5}{3}$, and (10.4.15) limits x_1 to $\frac{1}{3}/\frac{1}{3}$. Therefore the first constraint is violated first and hence exchange x_1 for x_4. Substituting out x_1 gives

$$x_1 = \tfrac{2}{5} - \tfrac{3}{5}x_4 + \tfrac{4}{5}x_5 + x_3 \tag{10.4.17}$$

$$x_2 = \tfrac{1}{5} + \tfrac{1}{5}x_4 - \tfrac{3}{5}x_5 - x_3 \tag{10.4.18}$$

and

$$z = \tfrac{12}{5} - \tfrac{3}{5}x_4 - \tfrac{6}{3}x_5 - x_3 \tag{10.4.19}$$

Hence z has a maximum of $\frac{12}{5}$ as increasing any of the remaining variables will decrease z. The maximum occurs when $x_1 = \frac{2}{5}, x_2 = \frac{1}{5}, x_4 = x_5 = x_3 = 0$. This process can be written down in general terms and is conveniently laid out using the tableau approach.

Let the matrix $\mathbf{B}$ be the basis matrix, $\mathbf{x}_B$ be the vector of basic variables, and $\mathbf{c}_B$ be the corresponding price vector. Then the constraint equations are

$$\mathbf{B}\mathbf{x}_B = \mathbf{b} \tag{10.4.20}$$

with the objective function

$$z = \mathbf{c}_B^T\mathbf{x}_B \tag{10.4.21}$$

As in (10.4.14) the next step is to replace one column of $\mathbf{B}$, say $\mathbf{b}_r$ with one of $\mathbf{A}$, say $\mathbf{a}_k$. But $\mathbf{a}_k$ can be expressed as a linear combination the columns $\mathbf{b}_i$ of $\mathbf{B}$ to define β_{ik} as

$$\mathbf{a}_j = \beta_{1j}\mathbf{b}_1 + \cdots + \beta_{mj}\mathbf{b}_m \tag{10.4.22}$$

The process to reach (10.4.14) and (10.4.15) requires solving (10.4.22) for $\mathbf{b}_r$ and substituting out $\mathbf{b}_r$ from (10.4.20). This gives the new basic solution as

$$\sum_{\substack{i=1\\ i\neq r}}^{m}\left(x_{Bi} - \frac{\beta_{ik}}{\beta_{rk}}x_{Br}\right)\mathbf{b}_i + \frac{x_{Br}}{\beta_{rk}}\mathbf{a}_k = \mathbf{b} \tag{10.4.23}$$

This solution must be feasible so that

$$\hat{x}_{Bi} = x_{Bi} - \frac{\beta_{ik}}{\beta_{rk}} x_{Br} \geq 0 \tag{10.4.24}$$

for $i = 1, \ldots, m$ but not $i = r$, where the symbol $\hat{x}$ is used for the new basic variables. At $i = r$ the condition

$$\hat{x}_{Br} = \frac{x_{Br}}{\beta_{rk}} \geq 0 \tag{10.4.25}$$

must hold. Equations (10.4.24) and (10.4.25) show that for $x_{Br} > 0$ then $\beta_{rk} > 0$, and hence either $\beta_{ik} \leq 0$ or

$$\beta_{ik} > 0 \quad \text{and} \quad \frac{x_{Bi}}{\beta_{ik}} \geq \frac{x_{Br}}{\beta_{rk}} \tag{10.4.26}$$

and so the choice of r is that which satisfies

$$\frac{x_{Br}}{\beta_{rk}} = \min_{i=1,\ldots,m} \left[\frac{x_{Bi}}{\beta_{ik}} : \beta_{ik} > 0 \right] \tag{10.4.27}$$

Now the new objective function can be found as in (10.4.16) which develops into

$$\begin{aligned}
\hat{z} &= \sum_{\substack{i=1 \\ i \neq r}}^{m} c_{Bi} \left(x_{Bi} - \frac{\beta_{ik}}{\beta_{rk}} x_{Br} \right) + c_k \frac{x_{Br}}{\beta_{rk}} \\
&= \sum_{i=1}^{m} c_{Bi} \left(x_{Bi} - \frac{\beta_{ik}}{\beta_{rk}} x_{Br} \right) + c_k \frac{x_{Br}}{\beta_{rk}} \\
&= \sum_{i=1}^{m} c_{Bi} x_{Bi} - \frac{x_{Br}}{\beta_{rk}} \sum_{i=1}^{m} c_{Bi} \beta_{ik} + c_k \frac{x_{Br}}{\beta_{rk}} \\
&= z - \frac{x_{Br}}{\beta_{rk}} (z_k - c_k)
\end{aligned} \tag{10.4.28}$$

with

$$z_k = \sum_{i=1}^{m} c_{Bi} \beta_{ik} \tag{10.4.29}$$

In order to make $\hat{z}$ greater than z the choice of k must satisfy

$$z_k - c_k < 0 \tag{10.4.30}$$

If a selection of k satisfy the condition, any of these k may be used.

The new β values follow from eliminating $\mathbf{b}_r$ from (10.4.22) in the forms

$$\mathbf{a}_j = \sum_{\substack{i=1 \\ i \neq r}} \beta_{ij} \mathbf{b}_i + \beta_{rj} \mathbf{b}_r \tag{10.4.31}$$

and

$$\mathbf{a}_k = \sum_{\substack{i=1 \\ i \neq r}} \beta_{ik} \mathbf{b}_i + \beta_{rk} \mathbf{b}_r \tag{10.4.32}$$

The vector $\mathbf{b}_r$ is eliminated in the spirit of (10.4.15) to give

$$\mathbf{a}_j = \sum_{\substack{i=1 \\ i \neq r}} \left(\beta_{ij} - \frac{\beta_{ik}\beta_{rj}}{\beta_{rk}} \right) \mathbf{b}_i + \frac{\beta_{rj}}{\beta_{rk}} \mathbf{a}_k \tag{10.4.33}$$

which yields the update equations

$$\hat{\beta}_{ij} = \beta_{ij} - \frac{\beta_{ik}\beta_{rj}}{\beta_{rk}} \qquad i \neq r \tag{10.4.34}$$

and

$$\hat{\beta}_{rj} = \frac{\beta_{rj}}{\beta_{rk}} \tag{10.4.35}$$

and then

$$\hat{z}_j - c_j = (z_j - c_j) - \frac{\beta_{rj}}{\beta_{rk}} (z_k - c_k) \tag{10.4.36}$$

The tableau layout uses the variables defined above in the scheme shown in Table 10.2.

The expression val() means the value of the expression in the braces.

The rules for updating this tableau are as follows.

(i) If $z_j - c_j \geq 0$ for $j = 1, \ldots N$ then stop.

(ii) If there is one j with $z_j - c_j < 0$ and $\beta_{ik} \leq 0$ for all $i = 1, \ldots, m$ then the solution is unbounded.

(iii) k is chosen from (10.4.30) so that $z_k - c_k = \min_{j=1,\ldots,N}[z_j - c_j : z_j - c_j < 0, \beta_{ij} > 0$ for at least one $i = 1, \ldots, m]$

(iv) r is chosen from (10.4.27).

(v) Update the tableaux values for $\hat{x}_{Bi}$, $\hat{z}$, $\hat{\beta}_{ij}$, and $\hat{z}_j - c_j$ using equations (10.4.24, 10.4.25, 10.4.28, 10.4.34, 10.4.35, and 10.4.36).

Table 10.2

	$\mathbf{c}^T$		c_1	c_2	c_3	c_4
$\mathbf{c}_B$	Basic variables		x_1	x_2	x_3	x_4
val(c_{B1})	x_{B1}	val(x_{B1})	β_{11}	β_{12}	β_{13}	β_{14}
val(c_{B2})	x_{B2}	val(x_{B2})	β_{21}	β_{22}	β_{23}	β_{24}
	$z_j - c_j$	val(z)	$z_1 - c_1$	$z_2 - c_2$	$z_3 - c_3$	$z_4 - c_4$

The earlier example can be laid out in tableau form to give the three Tables (10.3, 10.4 and 10.5).

The minimum $z_j - c_j$ is the entry -6 making $k = 2$ and

$$\min\left(\frac{x_{B1}}{\beta_{12}}, \frac{x_{B2}}{\beta_{22}}\right) = \min\left(\frac{2}{4}, \frac{1}{3}\right)$$

to make $r = 2$ and hence x_5 becomes non-basic and is replaced by x_2. The update formulae then give the second tableau in Table 10.4.

For this tableau $k = 1$ and $r = 1$, from

$$\min\left(\tfrac{2}{3}\tfrac{3}{5}, \tfrac{1}{3}\tfrac{3}{1}\right)$$

making x_4 non-basic to be replaced by x_1. Hence the third tableaux is shown in Table 10.5.

This tableau yields the final result in agreement with the earlier version.

Now that the most basic of constrained problems can be solved, attention turns to more general problems. Suppose the general non-linear programming problem

Table 10.3

	$\mathbf{c}^T$		3	6	2	0	0
$\mathbf{c}_B$	Basic variables		x_1	x_2	x_3	x_4	x_5
0	x_4	2	3	4	1	1	0
0	x_5	1	1	3*	2	0	1
	$z_j - c_j$	0	−3	−6	−2	0	0

Table 10.4

	$\mathbf{c}^T$		3	6	2	0	0
$\mathbf{c}_B$	Basic variables		x_1	x_2	x_3	x_4	x_5
0	x_4	$\frac{2}{3}$	$\frac{5}{3}$*	0	$-\frac{5}{3}$	1	$-\frac{4}{3}$
6	x_2	$\frac{1}{3}$	$\frac{1}{3}$	1	$\frac{2}{3}$	0	$\frac{1}{3}$
	$z_j - c_j$	2	−1	0	2	0	2

Table 10.5

	$\mathbf{c}^T$		3	6	2	0	0
$\mathbf{c}_B$	Basic variables		x_1	x_2	x_3	x_4	x_5
3	x_1	$\frac{2}{5}$	1	0	−1	$\frac{3}{5}$	$-\frac{4}{5}$
6	x_2	$\frac{1}{5}$	0	1	1	$-\frac{1}{5}$	$\frac{3}{5}$
	$z_j - c_j$	$\frac{12}{5}$	0	0	1	$\frac{3}{5}$	$\frac{14}{5}$

defined by

$$\begin{aligned} \text{maximize} \quad & f(\mathbf{x}) \\ \text{subject to} \quad & g_i(\mathbf{x}) \le b_i \quad i = 1, u \\ & g_i(\mathbf{x}) \ge b_i \quad i = u+1, v \\ & g_i(\mathbf{x}) = b_i \quad i = v+1, m \\ & \mathbf{x} \ge 0 \end{aligned} \tag{10.4.37}$$

has a maximum at a point $\mathbf{x}^*$, then $\mathbf{x}^*$ must satisfy the Kuhn–Tucker(1956) necessary conditions. These conditions are now derived and the conditions which are *active* need to be distinguished. Let I_a be the set of indices for which $g_i(\mathbf{x}^*) = b_i$, and J_a be the set for which $\mathbf{x}_j^* = 0$. These are the active constraints. Let I_p and J_p be the indices of the passive constraints. Then problem (10.4.37) can be reformulated in the classical form using Lagrange multipliers to give

$$F(\mathbf{x}, \boldsymbol{\lambda}) = f(\mathbf{x}) + \sum_i \lambda_i (b_i - g_i(\mathbf{x})) - \sum_j \lambda_{m+j} x_j \tag{10.4.38}$$

with Lagrange multipliers $\lambda_1, \ldots, \lambda_{m+n}$. Classical results then yield, after the addition of slack and surplus variables, that there is a constrained local maximum of (10.4.38) when there exists $(m+n)$ Lagrange multipliers λ_i^* and

$$\frac{\partial F(\mathbf{x}^*, \boldsymbol{\lambda}^*)}{\partial x_j} = \frac{\partial f(\mathbf{x}^*)}{\partial x_j} - \sum_i \lambda_i^* \frac{\partial g_i(\mathbf{x}^*)}{\partial x_j} - \lambda_{m+j}^* = 0 \tag{10.4.39}$$

Hence the conditions

$$\frac{\partial f(\mathbf{x}^*)}{\partial x_j} - \sum_i \lambda_i^* \frac{\partial g_i(\mathbf{x}^*)}{\partial x_j} \le 0 \quad (j \in J_a) \tag{10.4.40}$$

$$= 0 \quad (j \in J_p) \tag{10.4.41}$$

$$\lambda_i^* \ge 0 \quad (i = 1, \ldots, u) \tag{10.4.42}$$

$$\le 0 \quad (i = u+1, \ldots, v) \tag{10.4.43}$$

and for $i = v+1, \ldots, m$, λ_i^* is restricted in sign. This result allows the necessary conditions for a local constrained maximum to be written down. These conditions are called the Kuhn–Tucker necessary conditions and are as follows.

(i) Using (10.4.40) and (10.4.41)

$$\boldsymbol{\nabla} f(\mathbf{x}^*) - \sum_i \lambda_i^* \boldsymbol{\nabla} g_i(\mathbf{x}^*) \le 0 \tag{10.4.44}$$

with strict equality for $j \in J_p$.

(ii)

$$\sum_j \left[\frac{\partial f(\mathbf{x}^*)}{\partial x_j} - \sum_i \lambda_i^* \frac{\partial g_i(\mathbf{x}^*)}{\partial x_j} \right] x_j^* = 0 \qquad (10.4.45)$$

as the expression in [ ] is zero for $j \in J_a$ and x_j^* is zero for $j \in J_b$ from (10.4.41).

(iii) The third condition consists of the constraints (10.4.37) and hence $b_i - g_i(\mathbf{x}^*)$ is non-negative for $i = 1, u$, is non-positive for $i = u + 1, v$ and is zero for the rest.

(iv) Each constraint is either active or passive and hence

$$\sum_i (b_i - g_i(\mathbf{x}^*))\lambda_i^* = 0 \qquad (10.4.46)$$

as either $b_i - g_i(\mathbf{x}^*) = 0$ for $i \in I_a$ or $\lambda_i^* = 0$ for $i \in I_p$.

All these conditions rely on the existence of $\boldsymbol{\lambda}^*$ and the *constraint qualification* ensures this requirement. This condition says that the vector $\nabla f(\mathbf{x}^*)$ must lie within or on the convex cone generated by the vectors

$$\nabla g_i(\mathbf{x}^*) \quad i = 1, u \quad i \in I_a \qquad (10.4.47)$$

$$-\nabla g_i(\mathbf{x}^*) \quad i = u + 1, v \quad i \in I_a \qquad (10.4.48)$$

$$-\mathbf{e}_j \quad j \in J_a \qquad (10.4.49)$$

In terms of these conditions it is possible to solve the quadratic programming problem. This problem can be solved using the Kuhn–Tucker conditions to reduce the problem to a linear programming problem. A brief derivation of how this is done follows. The general quadratic problem is

$$\begin{gathered} \text{maximize } \mathbf{x}^T\mathbf{D}\mathbf{x} + \mathbf{c}^T\mathbf{x} \\ \text{subject to } \mathbf{A}\mathbf{x} \leq, =, \geq \mathbf{b} \\ \mathbf{x} \geq 0 \end{gathered} \qquad (10.4.50)$$

where $\mathbf{D}$ is assumed symmetric. The Kuhn–Tucker conditions give the constraints:

$$2\mathbf{D}\mathbf{x} - \mathbf{A}^T\boldsymbol{\lambda} + \mathbf{v} = -\mathbf{c} \qquad (10.4.51)$$

where $\mathbf{v}$ is an n component slack variable, and

$$\mathbf{A}\mathbf{x} \leq, =, \geq \mathbf{b} \qquad (10.4.52)$$

The second and third Kuhn–Tucker conditions give the non-linear conditions

$$\mathbf{v}^{*T}\mathbf{x}^* = 0 \qquad (10.4.53)$$

or

$$v_j^* x_j^* = 0 \qquad j = 1, \ldots, n \qquad (10.4.54)$$

and

$$\lambda_i^* x_{n+i}^* = 0 \tag{10.4.55}$$

Hence using phase I of the two-phase linear programming approach, z is taken to be the sum of all the artificial variables in a minimization problem or negative of this for a maximizing problem. Hence we have an equivalent linear programming exercise.

The above methods of solution are moving away from the usual areas of numerical analysis and the reader is referred to the bibliography for further reading on various forms of 'programming'. There are a number of methods to deal with the completely general programming problem using the type of optimization procedure discussed in the earlier part of the chapter.

The full non-linear programming problem is still subject to much current research. An early method called hemstitching used a conventional unconstrained minimization method. If any new point of the optimization procedure violates a constraint, then a step is taken perpendicular to the constraint boundary to give a point within the constraint region, that is a feasible point.

Another approach is to modify the objective function so that a penalty is incurred as the current point approaches the constraint boundary. The problem is then reduced to an unconstrained optimization. Hence for the constrained problem of minimizing $f(\mathbf{x})$ subject to $c_i(\mathbf{x}) = 0$, for $i = 1, m$, consider the associated problem of minimizing

$$F(\mathbf{x}, \rho) = f(\mathbf{x}) + \rho \sum_{i=1}^{m} c_i^2(\mathbf{x}) \tag{10.4.56}$$

The added sum will be large when the current point is away from the required constraints. The level of penalty can be varied with changes to the parameter ρ. One algorithm is to choose a ρ, say $\rho = 1$ and complete an unconstrained minimization on $F(\mathbf{x})$. Then increase ρ in stages, possibly, $\{10, 10^2, 10^3, \ldots\}$ until the final constraint conditions are satisfied to machine accuracy. It can be shown that this process converges under quite general conditions, but hidden behind this proof are considerable practical difficulties. These arise in the unconstrained minimization which becomes increasingly difficult as the penalty function is scaled up for large ρ.

The inequality problem with constraints $c_i(\mathbf{x}) \geq 0$ can also be solved in this way using for example

$$F(\mathbf{x}, \rho) = f(\mathbf{x}) + \rho \sum_{i=1}^{m} [\min(c_i(\mathbf{x}), 0)]^2 \tag{10.4.57}$$

but similar problems arise in this case.

Inequality constraints can also be dealt with by adding on a barrier function which is infinite on the constraint boundary. Two typical barrier functions are

$$F(\mathbf{x}, \rho) = f(\mathbf{x}) + \rho \sum_{i=1}^{m} [c_i(\mathbf{x})]^{-1} \tag{10.4.58}$$

and

$$F(\mathbf{x}, \rho) = f(\mathbf{x}) - \rho \sum_{i=1}^{m} \ln[c_i(\mathbf{x})] \tag{10.4.59}$$

Now the scaling term ρ is reduced in the sequence of unconstrained problems.

Further reading

Fletcher R. (1987) *Practical Optimization*, John Wiley and Sons, Chichester.
Walsh G. R. (1971) *An Introduction to Linear Programming*, Holt, Rinehart and Winston Ltd, London.
Walsh G. R. (1975) *Methods of Optimization*, John Wiley and Sons, London.

Exercises

(10.1). Apply the direct search method of Rosenbrock to the function

$$y = 1 - x\,e^{-x}$$

to find its minimum. If you are using a computer programme, verify the attainable accuracy in $x_{\min}$ as being only half the full precision used. What is the precision of $y_{\min}$?

(10.2). Apply the golden section search to the above problem. Compare efficiency with the above.

(10.3). Use a Fibbonaci search on the example in exercise (10.1).

(10.4). Apply Powell's quadratic method to the above problem.

(10.5). Apply the above direct search methods to

$$y = 1 - \frac{1}{5x^2 - 6x + 5}$$

using $0 \le x \le 1$ as a starting region.

(10.6). Consider the above methods applied to

$$y = (x - 1)^2(x - 3)$$

with initial point $x = 1.5$ or bracketing interval $1 \le x \le 3$.

(10.7). Use Powell's quadratic and Davidon's cubic methods on

$$y = x^4 - 4x^3 - 6x^2 - 16x + 4$$

with starting points $x_1 = 0$, $x_2 = 1$ and $x_3 = 2$ to make the first quadratic fit. Use a suitable set of fitting points for the cubic.

(10.8). Use Powell's quadratic and Davidon's cubic methods for

$$y = x^2 + 2\,e^{-x}$$

with $0 \le x_{\min} \le 2$.

(10.9). Attempt to take at least two steps of the method of steepest descent from each of the points $(2,\ 0)^T$ and $(0,\ 0)^T$ on the function

$$4x_1^2 + x_2^2 - 16x_1 - 4x_2.$$

Determine explicitly the contours of the function, sketch them and use your diagram to illustrate the results of the steps of steepest descent.

(10.10). Attempt to take three steps of both steepest descent and Newton's method from $x^{(0)} = (1, 1)^T$ on the function $x_1^6 + x_2^6$. Comment critically on your result.

(10.11). Find the stationary points of

$$f = x_1^2 + x_2^2 - x_1^2 x_2$$

and determine their nature. Plot the contours of f.

(i) Find the value of f after taking a step of the basic Newton optimization method from $x^{(0)} = (1,\ 1)^T$.

(ii) Find f after taking a step of Newton's method with line search from $x^{(0)}$.

(iii) Find the smallest integer r^* for which $G(x^{(0)}) + r\, I_2$ is positive definite and repeat (i) and (ii) with $G(x^{(0)}) + r^*\, I_2$ instead of $G(x^{(0)})$. Comment critically on your results.

(10.12). Verify that the DFP and BFGS methods have finite termination on the quadratic $x_1^2 + 2x_2^2$. Start from $x^{(0)} = (1,\ 1)^T$ with $H^{(0)} = I_2$. Verify that $H^{(2)} = G^{-1}$ (i.e. the inverse Hessian of the quadratic). Use exact arithmetic.

(10.13). Show that the FR method has finite termination on the quadratic $2x_1^2 + 3x_2^2$. Start with $x^{(0)} = (1,\ 1)^T$ and use exact arithmetic.

(10.14). Show by induction that Newton's method when applied to the function $x_1^6 + x_2^4$ from $x(0) = (1,\ 1)^T$ generates the sequence

$$\left\{ \left(\frac{4}{5}\right)^k,\ \left(\frac{2}{3}\right) \right\}^T, \quad k = 1, 2, \ldots$$

and that convergence is linear with asymptotic error constant $\frac{4}{5}$. Why is convergence not quadratic?

(10.15). Show that the directions $(1,\ 0)^T$ and $(\frac{1}{5},\ 1)^T$ are conjugate with respect to the quadratic

$$5x_1^2 - 2x_1x_2 + 2x_2^2 - 5x_1 - 2x_2$$

and use them to find its minimizer. Verify the result analytically.

(10.16). Find the first value in the sequence $(0.1)^k,\ k = 0, 1, \ldots,$ for which the Goldstein–Armijo conditions are satisfied for the function

$$100(x_1^2 - x_2)^2 + (1 - x_1)^2$$

with the starting point $x^{(0)} = (-1.2,\ 1)^T$ and the search direction $(1,\ 0)^T$. Take the scalar parameter ρ in the Goldstein–Armijo conditions equal to 0.1.

(10.17). Without using calculus show that the function

$$\Phi = (x^2 + y^2 - 1)^2 + (x + y - 1)^2$$

has minimizers at the points $(1,\ 0)^T$ and $(0,\ 1)^T$. Prove this result using calculus and show also that Φ has a saddle point at $\{(\frac{1}{4})^{1/3},\ (\frac{1}{4})^{1/3}\}^T$. Give a rough sketch of the contours of Φ.
Take three steps of Newton's method of optimization on Φ from $x^{(0)} = (\frac{1}{2},\ 0)^T$ and comment on the results. What points would you expect Newton's method to converge to, from $x^{(0)} = (0,\ \frac{1}{2})^T$ and $(\frac{1}{2},\ \frac{1}{2})^T$ respectively?

(10.18) Maximize $f = 15x_1 + 5x_2$ subject to the conditions

$$2x_1 + 3x_2 \le 54$$

$$4x_1 + 2x_2 \le 40$$

$$x_1,\ x_2 \ge 0$$

Appendix A Answers, Notes and Hints for Selected Exercises

Chapter 1

(1.1) Put $x = \cos\theta$ to give $T_r(x) = \cos r\theta$ and use trigonometric formulae for $\cos r\theta \cos s\theta$.

(1.2) Put $x = \cos\theta$ again.

(1.3) As in the text, use (1.2.15) after expanding $\sin(\pi is/N)\sin(\pi js/N)$ as a sum of cosines.

(1.4) $x^4 - x = 10$.
Start at $x = 0.0$ to get $-1.697\,471$ and at 1.0 to get 1.855 584.

(1.7) (a) diverges and (b) converges.

(1.8) Converges to 0.963.

(1.9) (i) Same problem as exercise (1.4).
(ii) $x_0 = 0$ gives 0.588 532 7434.
(iii) $x_0 = 1.0$ gives 1.497 300 389 09.

(1.10) The series is a Taylor expansion of $J_0(2\sqrt{x})$. $J_0(y)$ has its smallest zero at 2.404 82. (Abramowitch and Stegun, p409). Hence $x = 2.404\,82^2/4$.

(1.11) Aim for a non-linear equation in θ, the angle subtended at the circumference by the arc of the tether. The tether radius is 1.1587 times the field radius.

(1.12) (a) 0.739 085 1332.
(b) as 1.9(ii).

(1.13) Two real roots are 1.200 858 006 and 1.566 565 928.
Divide these out to get the quadratic factor for the other two roots.

(1.16) The factor from the given trial is $z^2 + 4.5049z + 1.5226$.

(1.18) Put $x = z - 1$ to avoid zero coefficients and solve

$$z^6 - 6z^5 + 11z^4 - z^3 - 18z^2 + 23z - 11 = 0.$$

The 100th iteration gives the rows:

4.414 99	0.407 84	−1.366 59	1.242 60	0.680 88	0.614 0968
3.435 51	1.387 292	−1.366 78	1.242 78	−2.378 48	1.539 033

Hence col 3 gives the root −1.3667, col 4 the root 1.2428, and cols 1 and 2 the factor $x^2 - rx - s = 0$ with $r = 3.435\,51 + 1.3872$ and $s = 4.414\,99 \times 1.3872$. Similarly cols 5 and 6 give $r = -2.378\,48 + 1.539\,033$ and $s = 0.687\,088 \times 1.539\,033$.

(1.19) As above the QD rows are

−3.461 79	−1.207 585	1.949 422	−2.055 304	1.176 9202	0.511 507
−3.461 79	12.873 27	−11.044 606	−0.466 349	−0.412 035	0.511 507

with col 1 and col 5 converging directly, cols 2 and 3 giving $r = 12.873\,27 - 11.044\,606$ and $s = (-1.207\,585) \times (-11.044\,606)$ and cols 4 and 5 giving $r = -0.466\,349 - 0.412\,035$ and $s = (-2.055\,304) \times (-0.412\,035)$.

(1.20) Only five partial products with values 0.9375, 0.922 851 56, 0.916 442 87, 0.912 8630 and 0.910 580 86, are required to give $B_{22} = 0.905\,629\,14$.

(1.21) Only three cycles with values 0.019 688 755, 0.000 060 892 543 and 0.019 098 663 are needed to give $B_{11} = -0.009\,999\,0005$. The true value is −0.009 999 0001.

(1.22) $x = y^2$ gives easy analytic integrals. Five sequence elements (0.477 302 44, 0.668 618 12, 0.821 842 68, 0.940 174 89 and 1.029 2349) give $B_{22} = 1.257\,867$. The true value is 1.264 2411.

Chapter 2

(2.1)

x		f		Δ	Δ^2	Δ^3	Δ^4
x_1	1.0	f_1	0.8415				
				0.0253			
x_2	1.1	f_2	0.8668		−0.0030		
				0.0223		−0.0003	
x_3	1.2	f_3	0.8891		−0.0027		−0.0001
				0.0196		−0.0002	
x_4	1.3	f_4	0.9087		−0.0025		
				0.0171			
x_5	1.8	f_5	0.9258				

With $s = 0.5$,

$$f(1.05) = 0.8415 + 0.5(0.0253) + \frac{(0.5)(-0.5)}{1.2}(-0.0030) + \cdots$$

to give $f(1.05) = 0.85455$

(2.2)

x		f		Δ	Δ^2	Δ^3
x_1	0.0	f_1	1.0000			
				0.1052		
x_2	0.1	f_2	1.1052		0.0110	
				0.1162		0.0013
x_3	0.2	f_3	1.2214		0.0123	
				0.1285		
x_4	0.3	f_4	1.3499			

$$f'(0.0) = \frac{1}{0.1}\left(0.1052 - \frac{1}{2}(0.0110) + \frac{1}{3}(0.0013)\right) = 1.0013$$

and

$$f'(0.3) = \frac{1}{0.1}\left(0.1285 + \frac{1}{2}(0.123) + \frac{1}{3}(0.0013)\right) = 1.3508$$

(2.3) $f^{(n+1)} \leq e$ on [0, 1] to give $h < 0.0017$ in the linear case. $\Pi(x) = x(x - h)$ has its maximum at $x = h/2$. In the quadratic case $h < 0.025$ with a maximum of $|\Pi(x)|$ at $0.4226h$ and $1.577h$.

(2.13) $a = 1.2994$ and $b = 0.5987$.

(2.14) $a = -9.7232$ and $b = 2.8353$.

Chapter 3

(3.3) The characteristic equation is

$$\lambda^3 - 6\lambda^2 + 11\lambda - 6 = 0$$

which factorizes into

$$(\lambda - 1)(\lambda - 2)(\lambda - 3) = 0.$$

(3.5) The eigenvalues of **A** are 0.01015, 3.85806, 30.2886 and 0.8431. Hence

$$\kappa(\mathbf{A}) = 30.2886/0.010\,15.$$

The characteristic equation is

$$\lambda^4 - 35\lambda^3 + 146\lambda - 100\lambda + 1 = 0.$$

(3.6) $x_1 = x_2 = x_3 = x_4 = 1.$

(3.8) The first set of equations gives $x_1 = x_2 = x_3 = 1$. The second demonstrates how small changes in $\mathbf{b}$ affect $\mathbf{x}$.

(3.9) Characteristic equation is

$$\lambda^3 - \lambda^2(\alpha + 101) + \lambda(101\alpha + 94) - 99\alpha + 108 = 0$$

Hence $\kappa(\mathbf{A}) = \infty$ when $\alpha = 108/99$ to give a zero root. For large α, $\kappa(\mathbf{A}) \sim \alpha^2/100$.

(3.10) Characteristic equation is

$$\lambda^3 - (3 + c)\lambda^2 + (3c - 35)\lambda - c + 22 = 0$$

with $\kappa(\mathbf{A}) = \infty$ when $c = 22$.
See also exercise (3.13) with $c = 24$.

(3.11)

$$\mathbf{D}^{-1}(\mathbf{L} + \mathbf{U}) = \begin{bmatrix} 0 & 1/2 & 1/2 \\ -1/3 & 0 & 1/3 \\ 0 & -1/4 & 0 \end{bmatrix}$$

Its eigenvalues are 0.268 28 and $-0.134\,14 \pm 0.370\,56i$, which has modulus 0.394 09. Hence the error falls off as $(0.394)^p$ which is roughly three iterations per decimal place.

(3.13) This is the Vandermonde matrix for Laguerre polynomials (Chapter 5) and Gauss–Laguerre quadrature (Chapter 6).

(3.16) Expect an eigenvalue around 100 and another around 0.1 from the diagonally dominant nature of the matrix. Because x_1 turns out to be small (and multiplies a big coefficient in the first row) and x_3 is larger, the first row for example exhibits cancellation in the Jacobi process.

(3.17) $x_1 = 1.246\,988$

$x_2 = -2.433\,7349$

$x_3 = 0.909\,638\,55$

(3.18) $y_1 = 1.420\,1681$

$y_2 = 0.920\,16807$

$y_3 = 0.340\,336\,14$

$y_4 = 0.121\,848\,74$

(3.21) Eigenvalues of **A** satisfy

$$-\lambda^3 + 18\lambda^2 - 102\lambda + 180 = 0$$

with factors

$$(\lambda - 6)(\lambda^2 - 12\lambda - 30)$$

to give $\lambda_1 = 6$, $\lambda_2 = 8.45$ and $\lambda_3 = 3.55$ to confirm **A** is positive definite.

(3.22) $x_1 = 0.363\,636\,36$

$x_2 = 0.454\,545\,46$

$x_3 = 0.454\,545\,46$

$x_4 = 0.363\,636\,36$

Chapter 4

(4.1) $\lambda_1 = 19.286\,08$ $\quad \mathbf{x}_1 = (1.495\,95, 1, 0.6734)^T$.

$\lambda_2 = -7.077\,4322$ $\quad \mathbf{x}_2 = (-1.109\,72, 1, 0.9852)^T$.

$\lambda_3 = 4.791\,35$.

(4.2) $\lambda_1 = 10.321\,8298$.

$\lambda_2 = 5.851\,7950$.

$\lambda_3 = 7.826\,3752$.

(4.3) $\lambda_1 = 13.9999$ $\quad \mathbf{x}_1 = (0.199\,990\,42, -0.9, 1)^T$.

$\lambda_2 = -7.999\,2588$ $\quad \mathbf{x}_2 = (-0.6, 1, 0.4)^T$.

(4.4) Direct iteration gives $\lambda_1^{(1)} = 10.0$, $\lambda_1^{(2)} = 10.4$ and $\lambda_1^{(3)} = 10.201$. The other eigenvalues are $1 \pm i\sqrt{2}$ with $|\lambda| \sim 1.7$. Hence rate of convergence is $(1.7/10.0)^p$, to give roughly one figure per iteration.

(4.6) $\lambda_1 = -3.131\,605\,46$

$\lambda_2 = 2.350\,622\,95$

$\lambda_3 = 9.780\,9825$

(4.8) $\lambda_1 = 15.020\,0439$

$\lambda_2 = 14.999\,956$

$\lambda_3 = 15.010\,000$

These eigenvalues are too close for a practical rate of convergence by direct iteration.

(4.12)

$$\mathbf{B}^{-1} = \begin{bmatrix} 1 & 2 & 15 \\ -2 & -3 & -24 \\ -1 & -3 & -19 \end{bmatrix}$$

and

$$\mathbf{B}^{-1}\mathbf{A} = \begin{bmatrix} 10 & 1 & -12 \\ -1 & -1 & 19 \\ -1 & -4 & 15 \end{bmatrix}$$

to give $\lambda_1 = 12.357$.

(4.13) The corresponding eigenvector is $(-1, 2, 1)$. Deflation gives

$$\begin{bmatrix} 8 & 3 \\ -2 & -8 \end{bmatrix}$$

which yields the eigenvalues ± 1 correctly.

(4.16) The upper Hessenberg form is

$$\begin{bmatrix} \frac{1}{\sqrt{13}} & \frac{1}{\sqrt{13}} & \frac{8}{\sqrt{13}} \\ \sqrt{13} & \frac{41}{13} & \frac{93}{13} \\ 0 & \frac{24}{13} & \frac{24}{13} \end{bmatrix}$$

(4.21) The upper Hessenberg form is

$$\begin{bmatrix} \frac{17}{2} & 0.032\,686 & 8.3349 \\ -7.6485 & 2.538\,46 & -12.6923 \\ 0 & -1.525\,64 & -1.038\,46 \end{bmatrix}$$

with eigenvalues 10, 3 and -2. The first QR step gives

$$\begin{bmatrix} 9.6195 & -4.1659 & -15.4703 \\ -1.6345 & 2.8966 & 0.7146 \\ 0 & -2.0102 & -2.515 \end{bmatrix}$$

Chapter 5

(5.6) $J_0(z) \sim 0.822\,03 - 0.058\,695x^2$

$$= 0.880\,724T_0 - 0.117\,39T_2$$

$J_0(1) = 0.76519$

$J_0(1) \sim 0.76333$ by economized series

$J_0(1) \sim 0.75$ by two terms of Taylor

(5.7) $y = (\sqrt{2} - 1)/2$

(5.8) The degree 2 result has the $n+2$ turning points 0.0, 0.192 8188, 0.584 59819, $\pi/4$. The polynomial is $-0.188\,28179x^2 + 1.054\,0786x - 0.002\,309\,8048$

(5.9) The turning points are 0.0, 0.014, 0.558, 4.1799 and 10.0. The rational approximation is

$$\frac{0.117\,8852x^2 + 1.095\,4229x + 0.014\,280\,479}{x + 0.173\,091\,31}$$

(5.11) See (5.7)

(5.13)

$$\sin x \sim \frac{x - \frac{7}{78}x^3}{1 + \frac{1}{13}x^2 + \frac{11}{5880}x^4}$$

Maximum error at $\pi/4$.

(5.14)

$$\frac{z - \frac{29\pi^2}{2160}z^5}{1 + \frac{5\pi^2}{432}z^4}$$

Taylor $C(1) = 0.78\,144$ to $z^9 K$
Padé $C(1) = 0.778\,556$
True $C(1) = 0.779\,8934$

Chapter 6

(6.3) $A_1 - \frac{17}{36}$, $A_2 = \frac{5}{9}$ and $A_3 = -\frac{1}{36}$
$E = \frac{11}{864} f'''(0)$
Gauss 3pt gives 0.0626.

(6.4) $A = \pi/4$, $C = \pi/4$ and $B = \pi/2$.
$E = (\pi/272) f^{(iv)}(0)$
The rule gives

$$\int_{-1}^{1} \frac{\tan x^2}{\sqrt{1 - x^2}}\,dx = \frac{\pi}{2}\tan 1 = 2.446\,37$$

Substitute $x = \cos\theta$ gives

$$\int_0^{\pi} \tan(\cos^2\theta)\,d\theta$$

3pt Simpson gives $(\pi/3)\tan(1) = 1.630\,91$.

(6.5) $A_1 = A_2 = A_3 = \pi/3$ and $x_1 = \sqrt{3}/2$, $x_2 = 0$ and $x_3 = -\sqrt{3}/2$. By the quadrature rule:

$$\int_{-1}^{1} \frac{\arccos(x)}{\sqrt{1-x^2}}\,dx = \frac{\pi^2}{2}$$

which is also the correct value. This is because all the equations of the form

$$A_1 x_1^{2n+1} + A_2 x_2^{2n+1} + A_3 x_3^{2n+1} = 0$$

are satisfied exactly. Hence as arccos expands in odd powers the rule integrates the example exactly.

(6.14)

$$\int_0^{\pi/2} f(x)\sin x\,dx = 0.159\,1448 f(0.323\,5276) + 0.476\,842 f(0.904\,6135) + 0.364\,0129 f(1.420\,7031)$$

The Vandermonde matrix gives the polynomial

$$x^3 - 2.648\,8441x^2 + 2.037\,4912x - 0.415\,7935 = 0$$

whose roots are the above abscissae.

$$\int_0^{\pi/2} \sim 2.905\,227\,106$$

and the true value is $(e^{\pi/2} + 1)/2 = 2.905\,238\,69$.

(6.16) There is only one extended unknown which satisfies $c_1 P_3(x_1) = -P_5(x_1)$ to give $c_1 = -\frac{1}{4}$. Hence

$$G_{n+p} = \tfrac{1}{8}(63x^5 - 70x^3 + 15x) - \tfrac{1}{8}(5x^3 - 3x)$$

with roots 0 and $\sqrt{1/3}$ which are the original abscissae, and the extended roots $\sqrt{54/63}$ and $\sqrt{6/7}$. The corresponding weights are $W_1 = 98/495$, $W_2 = 12/55$ and $W_3 = 578/495$.

(6.19) Let I_1 be the integral over $[1/4, 1]$, I_2 over $[1/8, 1]$ and I_3 over $[1/16, 1]$. Then $I_1 = 0.537\,5204$ (Simpson 5pt.) (True $= 0.537\,574\,56$), $I_2 = 0.610\,063\,58$ (True $= 0.610\,1228$) and $I_3 = 0.620\,539\,56$ to give $B_{1n} = 0.620\,539\,56$. Putting $x = t^2$ gives

$$\int_0^1 \sin t^2\,dt \sim 0.619\,887$$

using 5pt Simpson.

(6.27) Split the interval of integration into $[0, \pi/10]$, $[\pi/10, \pi/5]$, $[\pi/5, 3\pi/10]$ and so on to give $A_0 = \pi/100, A_1 = -\pi/50, A_2 = 3\pi/100, A_3 = -2\pi/50$, and $A_4 = \pi/20$. Levin's $v_{21} = 0$ which is exact! Indeed the Abel limit

$$\lim_{\alpha \to 0} \int_0^\infty e^{-\alpha x} x \sin 10x \, dx = 0$$

(6.28) The first zeroes of $J_0(kx)$ occur at 0.240 483, 0.552 008 and 0.865 372. Hence integrate over the ranges $[0, 0.240\,483]$, $[0.240\,483, 0.552\,008]$ and $[0.552\,008, 0.865\,372]$ to get the sequence 0.075 9322, -0.044 1963, 0.025 2733 and -0.018 9655. Levin's transformation v_{21} gives 0.053 014 and Shanks 0.047 81. For increased accuracy the sequence requires extension.

(6.29)

$$\begin{aligned}\int_0^1 \int_0^1 \ln(xy) f(x, y) \, dx \, dy = &-0.8125 f(0.1, 0.1) \\ &- 0.5(f(0.1, 0.9) + f(0.9, 0.1)) \\ &- 0.1875 f(0.9, 0.9)\end{aligned}$$

and the example yields -1.6615 with this quadrature rule.

(6.30) $W_1 = W_3 = W_4 = W_2 = \pi/3$ with

$$E \sim \frac{5\pi}{84}(f_{xx} + f_{yy})$$

and the true value of the double integral is $16\pi^3/9$, the formula gives $7\pi^3/4\sqrt{2}$.

(6.31) The transformed integral is

$$\int_{-\infty}^{\infty} \int_{-\infty}^{\infty} \frac{\operatorname{sech}^2 t \operatorname{sech}^2 s \, ds \, dt}{\sqrt{1 - \tanh^2 t \tanh^2 s}}$$

which using points $s = 0$, $s = 2$ and $s = 4$ with all the combinations of the same points in t gives 4.633 08. True value is $2\pi \ln 6 = 4.355\,17$.

Chapter 7

(7.1) $h = 0.1$ gives $y = 0.292\,542\,11$ at $x = 1$.
$h = 0.05$ gives $y = 0.320\,211\,63$ at $x = 1$.

(7.2) $y^{[3]} = (x^3/3) + (x^7/63) + (2x^{11}/2079)$

(7.3)

$$y = 1 + 0.1x + \frac{0.02x^2}{2!} + \frac{0.006x^3}{3!} + \frac{0.6024x^4}{4!} + \frac{0.120\,5016x^5}{5!} + \frac{0.084\,58x^6}{6!} + \cdots$$

(7.4) Runge–Kutta gives $y = 0.350\,233$ with $h = 0.1$, and $y = 0.350\,232$ with $h = 0.05$.

(7.5) Euler with $h = 0.05$ gives $y = 0.640\,18$ and $x = 0.728\,72$. Runge–Kutta with $h = 0.05$ gives $y = 0.617\,58$ and $x = 0.660\,59$. These values agree with those for $h = 0.01$.

(7.6) The small-amplitude period is $2\pi\sqrt{l/g} = 4 \times 0.2867$. One quarter cycle numerically by Runge–Kutta with $h = 0.01$ gives $\theta = 0.0346$ at $t = 0.29$ and $\theta = -0.0073$ at $t = 0.30$. Linear interpolation gives a time for $\theta = 0$ of 0.2982.

(7.9) At $\theta = 50°$, the range, R, is 27.26m using Runge–Kutta with a step $h = 0.05$. At $\theta = 42°$, $R = 27.351$; at 43°, $r = 27.421$; at 44°, $R = 27.457$ and at 45°, $R = 27.328$. The cubic

$$R = -0.021\,83\theta^3 + 2.7996\theta^2 - 119.577\theta + 1728.731$$

fits these points, and has a maximum at $\theta = 43.84$.

(7.10) (i) Using $y'(0) = 0$ gives $y(\pi/2) = 0.451\,006\,87$ and $y'(0) = 0.4$ gives $y(\pi/2) = 0.851\,0069$. Linear interpolation gives the required value of $y'(0) = 0.548\,993\,14$

(ii) As above, $y'(0) = 0$ gives $y(1) = 0.162\,734\,33$, and $y'(0) = 0.4$ gives $y(1) = 0.543\,0103$. Linear interpolation then gives $y'(0) = 0.880\,692\,59$ which leads to $y(1) = 1$ correctly.

(7.12) These examples require a sequence of 'shots'. For the results here linear interpolation is used on two shots. The worst shot is rejected and the process repeated to convergence.

(i) $y'(0) = 0$ gives $y(1) = 0.150\,46$, and the following pairs of $y'(0)$ and $y(1)$ are $(0.1, 0.234\,11)$, $(0.488\,77, 0.548\,79)$, $(0.499\,55, 0.558\,046)$, and $(0.5, 0.5584)$ as required.

(ii) The pairs as above are: $(0.0, 0.164\,23)$, $(0.1, 0.2576)$, $(0.2413, 0.3823)$, $(0.2497, 0.3894)$ and $(0.25, 0.389\,76)$.

(iii) Again the pairs are: $(0.0, 0.048\,548)$, $(0.1, 0.135\,46)$, $(0.5917, 0.475\,14)$, $(0.718\,655, 0.546\,096)$, $(0.748\,59, 0.562\,08)$ and $(0.75, 0.562\,83)$.

(iv) The guessed value is now $y''(0)$ with $y(1)$ being the target to give pairs (0.0, 2.025 64), (0.1, 2.070 303), (0.55, 2.27 108), (0.560 886, 2.2735) and (0.560 945, 2.2736)

(7.13) Eigenvalues at $k = 1 + n^2\pi^2$.

(7.14) (i) $k = n\pi$
(ii) Eigenvalues occur at zeroes of $J_1(x)$. (Abramowitch and Stegun, p 370).

(7.20) Empty region of absolute stability: squeezed out by forcing the formula to be optimal.

(7.22) (i) $(-1, 0)$
(ii) $(-3, 0)$

(7.23) $\Pi = r^2 - (4\bar{h}/3)r - 1 - 2\bar{h} - 2\bar{h}/3$ to give the interval of absolute stability of $(-1, 0)$.

(7.24) Corrector is used in the form

$$y_{n+4} - \tfrac{9}{8}y_{n+3} + \tfrac{1}{8}y_{n+1} = \tfrac{3}{8}h(f_{n+4} + 2f_{n+3} - f_{n+2})$$

to give

$$\Pi = r^4 - r^3\left(\frac{9}{8} + \frac{3\bar{h}}{4} + \bar{h}^2\right) + r^2\left(\frac{3\bar{h}}{8} + \frac{\bar{h}^2}{2}\right) + r\left(\frac{1}{8} - \bar{h}^2\right) - \frac{3\bar{h}}{8}$$

and $\bar{h} \in (-\frac{13}{6}, 0)$.

(7.28) $y_{n+1} = y_n + (h/4)(3f_{n+2/3} + f_n)$

(7.31) Order is 4.

Chapter 8

(8.9)

$$g(x) = \frac{x^2}{2} - x + 1 - \frac{3e}{2(e-1)}e^{-x} + \frac{e+2}{2(e-1)}$$

with

$$K(x, t) = \begin{cases} e^{-x} - 1 + (1 - e^{-x})\left(\frac{1}{1-e^{-1}} - \frac{1}{e-1}\right) & x < t \\ (1 - e^{-x})\left(\frac{1}{1-e^{-1}} - \frac{1}{e-1}\right) & x > t \end{cases}$$

(8.10) $f^{[1]} = 1 + \frac{1}{2}\sinh^{-1}x + (x/2)(1+x^2)^{1/2}$ for the first iterate and Simpson's rule gives $f(0.2) \sim 1.25$.

(8.11)

$$y^2(x) = x + \int_0^1 K(t, x, y)\,\mathrm{d}t$$

with

$$K(t, x, y) = \begin{cases} 2t(1-x)(y^3(t) - t) & t \le x \\ 2x(1-t)(y^3(t) - t) & t \ge x \end{cases}$$

and δf satisfies

$$2y^{(n)}(x)\delta f(x) = x - {y^{(n)}}^2(x) + \int_0^1 ({y^{(n)}}^3(t) - t)G(x, t)\,\mathrm{d}t + \int_0^1 {y^{(n)}}^2(t)\delta f(t)G(x, t)\,\mathrm{d}t$$

with

$$G(x, t) = \begin{cases} 2t(1-x) & t \le x \\ 2x(1-t) & t \ge x \end{cases}$$

(8.20) $\Lambda = c + 1 + 1/4c$ with a minimum at $c = \frac{1}{2}$ and $\Lambda = 2$. With two parameters the extremum are at $\lambda_1 \sim 2.1397$ and $\lambda_2 \sim -7.9853$

Chapter 9

(9.5) A step of $k = \frac{1}{4}$ and $h = \frac{1}{4}$ gives $u(\frac{1}{4}, \frac{1}{4}) = 1.209$, $u(\frac{1}{4}, \frac{1}{2}) = 1.116$ and $u(\frac{1}{4}, \frac{3}{4}) = u(\frac{1}{4}, 1) = 0.744$ with the expected symmetry.

(9.7) Characteristics are $y - 2x = A$ and $y + x = B$. Those through $(0, 0)$ and $(0.1, 0)$ intersect at $R = (0.1/3, 0.2/3)$ and yield $p_R = 1$, $q_R = 1$ and $z_R \sim 0.1$.

(9.8) $y = x^2 + A$ and $y^2 = 4x + B$.

(9.9) Equation is parabolic with the characteristic $3y - 4x = \text{const}$. Hence set $3y - 4x = \xi$ and $x = \eta$ to get

$$\frac{\partial^2 u}{\partial \eta^2} = \frac{\partial u}{\partial \xi} + u^2$$

and boundary conditions $u = 0$ on $\xi = 0$ and $u = 10$ on $\eta = 0$ and $\eta = 1$. Now solve by conventional Crank–Nicolson with the term $u_{i,j}^2$ added. This causes no problems.

Chapter 10

(10.1) Minimum at $x = 1$

(10.5) Minimum at $x = \frac{4}{5}$

(10.6) Minimum at $x = \frac{7}{3}$

(10.7) $y' = 4x^3 - 12x^2 - 12x - 16$, Solve by Newton as a check, to give $x = 4$

(10.8) $y' = 2(x - e^{-x})$, as (10.7), with $x = 0.567\,143\,29$.

(10.18) Maximum at $x_1 = 10$, $x_2 = 0$ and $f = 150$.

Appendix B Gauss Quadrature Weights and Abscissae

The most commonly used Gaussian quadrature rule, namely Gauss–Legendre is presented in Table B1, for

$$\int_{-1}^{1} f(x)\,\mathrm{d}x = \sum_{i=1}^{N} W_i f(x_i) \qquad B1$$

The second group are the Gauss–Laguerre formulae for the integral

$$\int_{0}^{\infty} e^{-x} f(x)\,\mathrm{d}x = \sum_{i=1}^{N} W_i f(x_i) \qquad B2$$

and are presented in Table B2.

Table B3 gives low-order weights and abscissae for the Gauss–Hermite rules

$$\int_{-\infty}^{\infty} e^{-x^2} f(x)\,\mathrm{d}x = \sum_{i=0}^{N} W_i f(x_i) \qquad B3$$

Further values for Tables B1, B2 and B3 may be found in Abramowich and Stegun (1965) and Stroud and Secrest (1966).

The final two tables, B4 and B5, are the coefficients for the trigonometric rules:

$$\int_{-\pi/2}^{\pi/2} f(x) \cos x \mathrm{d}x \simeq \sum_{i=1}^{N} W_i f(x_i) \qquad (B4)$$

and

$$\int_{0}^{\pi/2} f(x) \cos x \mathrm{d}x \simeq \sum_{i=1}^{N} W_i f(x_i) \qquad (B5)$$

The symmetric interval $[-\pi/2, \pi/2]$ is chosen for convenience in tabulation. If the coefficients for the weight function $\sin t$ on $[0, \pi]$ are required, then $\pi/2$ is simply added to the abscissae t_i, the weights w_i remaining the same.

Table B1 Gauss–Legendre coefficients

N	$\pm x_i$	W_i
2	0.577 350 2692	1.000 000 0000
3	0.000 000 0000	0.888 888 8889
	0.774 596 6692	0.555 555 5556
4	0.339 981 0436	0.652 145 1549
	0.861 136 3116	0.347 854 8451
5	0.000 000 0000	0.568 888 8889
	0.534 869 3101	0.478 628 6705
	0.906 179 8459	0.236 926 8851
6	0.238 619 1861	0.467 913 9346
	0.661 209 3865	0.360 761 5730
	0.932 469 5142	0.171 324 4924
7	0.000 000 0000	0.417 599 1837
	0.405 845 1514	0.381 830 0505
	0.741 531 1856	0.279 705 3915
	0.949 107 9123	0.129 484 9662
8	0.183 434 6425	0.362 683 7834
	0.525 532 4099	0.313 706 6459
	0.796 666 4774	0.222 381 0345
	0.960 289 8566	0.101 228 5363
9	0.000 000 0000	0.330 239 3550
	0.324 253 4234	0.312 347 0770
	0.613 371 4327	0.260 610 6964
	0.836 031 1073	0.180 648 1607
	0.968 160 2395	0.081 274 3884
10	0.148 874 3390	0.295 524 2247
	0.433 395 3941	0.269 266 7193
	0.679 409 5682	0.219 086 3625
	0.865 063 3667	0.149 451 3492
	0.973 906 5285	0.066 671 3443

Table B2 Gauss–Laguerre coefficients

N	x_i	W_i
2	0.585 786 4376	8.535 533 9059(−1)
	3.414 213 5623	1.464 466 0941(−1)
3	0.415 774 5568	7.110 930 0993(−1)
	2.294 280 3603	2.781 577 3357(−1)
	6.289 945 0829	1.038 925 6502(−2)
4	0.322 547 6896	6.031 541 0434(−1)
	1.745 761 1012	3.574 186 9244(−1)
	4.536 620 2969	3.888 790 8515(−2)
	9.395 070 9123	5.392 947 0556(−4)
5	0.263 560 3197	5.217 556 1058(−1)
	1.413 403 0591	3.986 668 1108(−1)
	3.596 425 7710	7.594 244 9682(−2)
	7.085 810 0059	3.611 758 6799(−3)
	12.640 800 8443	2.336 997 2386(−5)
6	0.222 846 6042	4.589 646 7395(−1)
	1.188 932 1017	4.170 008 3077(−1)
	2.992 736 3261	1.133 733 8207(−1)
	5.775 143 5691	1.039 919 7453(−2)
	9.837 467 4183	2.610 172 0282(−4)
	15.982 873 9806	8.985 479 0643(−7)

Table B3 Gauss–Hermite coefficients

N	$\pm x_i$	W_i
2	0.707 106 7812	8.862 269 2545(−1)
3	0.000 000 0000	1.181 635 9006(0)
	1.224 744 8714	2.954 089 7515(−1)
4	0.524 647 6233	8.049 140 9001(−1)
	1.650 680 1239	8.131 283 5447(−2)
5	0.000 000 0000	9.453 087 2048(−1)
	0.958 572 4646	3.936 193 2315(−1)
	2.020 182 8704	1.995 324 2059(−2)
6	0.430 774 119	7.246 295 9522(−1)
	1.335 849 0740	1.570 673 2032(−1)
	2.350 604 9736	4.530 009 9055(−3)

Table B4

N	$\pm x_i$	W_i
2	6.836 673 9009(−1)	1.000 000 0000(0)
3	0.000 000 0000(0)	1.088 319 1839(0)
	1.012 601 2400(0)	4.558 404 0804(0)
4	4.392 874 6686(−1)	7.759 293 8187(−1)
	1.190 676 5639(0)	2.240 706 1813(−1)
5	0.000 000 0000(0)	7.522 109 7881(−1)
	7.259 867 3794(−1)	5.037 825 1239(−1)
	1.296 440 2800(0)	1.201 119 9821(−1)
6	3.238 521 1421(−1)	6.058 137 0012(−1)
	9.197 906 6552(−1)	3.247 985 5138(−1)
	1.363 911 3021(0)	6.938 774 8500(−2)
7	0.000 000 0000(0)	5.755 404 0031(−1)
	5.635 019 6618(−1)	4.560 738 8993(−1)
	1.055 539 9634(0)	2.135 301 5976(−1)
	1.409 416 8673(0)	4.262 575 0151(−2)
8	2.564 965 0742(−1)	4.919 957 9660(−1)
	7.434 686 4788(−1)	3.362 644 7785(−1)
	1.153 725 6455(0)	1.442 040 9203(−1)
	1.441 490 5402(0)	2.753 563 3514(−2)
9	0.000 000 0000(0)	4.663 064 1290(−1)
	4.598 098 5871(−1)	4.004 335 5770(−1)
	8.809 816 6893(−1)	2.477 678 2327(−1)
	1.226 760 7474(0)	1.001 047 2234(−1)
	1.464 915 0799(0)	1.854 069 0233(−2)
10	2.123 428 8707(−1)	4.127 123 7189(−1)
	6.221 414 5705(−1)	3.189 730 7574(−1)
	9.878 852 8562(−1)	1.840 793 4564(−1)
	1.282 421 8393(0)	7.131 163 1869(−2)
	1.482 529 4562(0)	1.292 357 4871(−2)

Table B5

N	x_i	W_i
2	2.658 738 8056(−1)	6.0362553281(−1)
	1.035 152 6093(0)	3.963 744 6719(−1)
4	9.566 938 9197(−2)	2.378 307 1420(−1)
	4.524 090 2327(−1)	4.026 569 6624(−1)
	9.318 505 7672(−1)	2.868 173 7949(−1)
	1.356 443 9600(0)	7.269 495 1083(−2)
6	4.833 720 2961(−2)	1.223 313 0510(−1)
	2.415 743 8682(−1)	2.489 764 3030(−1)
	5.424 156 9594(−1)	2.866 621 2347(−1)
	8.888 823 5763(−1)	2.187 192 2991(−1)
	1.214 562 5204(0)	1.025 086 6521(−1)
	1.457 917 6907(0)	2.080 224 6015(−2)
8	2.902 372 9769(−2)	7.390 899 8095(−2)
	1.482 852 4405(−1)	1.600 299 3702(−1)
	3.453 111 1152(−1)	2.144 443 4342(−1)
	5.944 769 6798(−1)	2.197 958 1269(−1)
	8.653 838 0686(−1)	1.758 116 4478(−1)
	1.126 307 6093(0)	1.056 044 8025(−1)
	1.347 015 0460(0)	4.248 549 7299(−2)
	1.501 560 3622(0)	7.919 286 4406(−3)
10	1.932 446 8597(−2)	4.934 921 0907(−2)
	9.979 921 7690(−2)	1.098 219 5774(−1)
	2.357 222 3071(−1)	1.569 244 1652(−1)
	4.179 775 7474(−1)	1.814 218 5381(−1)
	6.282 707 3435(−1)	1.773 5663121(−1)
	8.505 747 8126(−1)	1.467 161 6225(−1)
	1.067 343 0163(0)	1.004 666 1487(−1)
	1.261 602 1055(0)	5.407 427 2678(−2)
	1.418 036 7416(0)	2.024 647 7726(−2)
	1.524 103 4398(0)	3.622 402 2909(−3)

Appendix C Patterson Quadrature Weights and Abscissae

Tables C1, C2, C3 and C4 contain the weights and abscissae for the Patterson quadrature rules of point numbers 3, 7,15, 31, 63 and 127 with the three-point rule being the normal Gauss–Legendre rule.

Table C1

$\pm x_i$	W_i
0.774 596 669 241 483 377 04	0.555 555 555 555 555 555 556
0.0	0.888 888 888 888 888 888 889
0.960 491 268 708 020 283 42	0.104 656 226 026 467 265 194
0.774 596 669 241 483 377 04	0.268 488 089 868 333 440 729
0.434 243 749 346 802 558 00	0.401 397 414 775 962 222 905
0.0	0.450 916 538 658 474 142 345
0.993 831 963 212 755 022 21	0.017 001 719 629 940 260 3390
0.960 491 268 708 020 283 42	0.051 603 282 997 079 739 6969
0.888 459 232 872 256 998 89	0.092 927 195 315 124 537 6859
0.774 596 669 241 483 377 04	0.134 415 255 243 784 220 360
0.621 102 946 737 226 402 94	0.171 511 909 136 391 380 787
0.434 243 749 346 802 558 00	0.200 628 529 376 989 021 034
0.223 386 686 428 966 881 63	0.219 156 858 401 587 496 404
0.0	0.225 510 499 798 206 687 386
0.999 098 124 967 667 597 66	0.002 544 780 791 561 874 41540
0.993 831 963 212 755 022 21	0.008 434 565 739 321 106 24631
0.981 531 149 553 740 106 87	0.016 446 049 854 387 810 9338
0.960 491 268 708 020 283 42	0.025 807 598 096 176 653 5646
0.929 654 857 429 740 056 67	0.035 957 103 307 129 322 0968
0.888 459 232 872 256 998 89	0.046 462 893 261 757 986 5414
0.836 725 938 168 868 735 50	0.056 979 509 494 123 357 4122
0.774 596 669 241 483 377 04	0.067 207 754 295 990 703 5404
0.702 496 206 491 527 078 61	0.076 879 620 499 003 531 0427
0.621 102 946 737 226 402 94	0.085 755 920 049 990 351 1542
0.531 319 743 644 375 623 97	0.093 627 109 981 264 473 6167
0.434 243 749 346 802 558 00	0.100 314 278 611 795 578 771

(*continued overleaf*)

Table C1 *(continued)*

$\pm x_i$	W_i
0.331 135 393 257 976 833 09	0.105 669 893 580 234 809 744
0.223 386 686 428 966 881 63	0.109 578 421 055 924 638 237
0.112 488 943 133 186 625 75	0.111 956 873 020 953 456 880
0.0	0.112 755 256 720 768 691 607

Table C2

$\pm x_i$	W_i
0.999 872 888 120 357 611 94	0.000 363 221 481 845 530 659694
0.999 098 124 967 667 597 66	0.001 265 156 556 230 068 01137
0.997 206 259 372 221 959 08	0.002 579 049 794 685 688 27243
0.993 831 963 212 755 022 21	0.004 217 630 441 558 854 83908
0.988 684 757 547 429 479 94	0.006 115 506 822 117 246 33968
0.981 531 149 553 740 106 87	0.008 223 007 957 235 929 66926
0.972 182 874 748 581 796 58	0.010 498 246 909 621 321 8983
0.960 491 268 708 020 283 42	0.012 903 800 100 351 265 6260
0.946 342 858 373 402 905 15	0.015 406 750 466 559 497 8021
0.929 654 857 429 740 056 67	0.017 978 551 568 128 270 3329
0.910 371 156 957 004 292 50	0.020 594 233 915 912 711 1492
0.888 459 232 872 256 998 89	0.023 231 446 639 910 269 4433
0.863 907 938 193 690 477 15	0.025 869 679 327 214 746 9108
0.836 725 938 168 868 735 50	0.028 489 754 745 833 548 6125
0.806 940 531 950 217 611 86	0.031 073 551 111 687 964 8799
0.774 596 669 241 483 377 04	0.033 603 877 148 207 730 5417
0.739 756 044 352 694 758 68	0.036 064 432 780 782 572 6401
0.702 496 206 491 527 078 61	0.038 439 810 249 455 532 0386
0.662 909 660 024 780 595 46	0.040 715 510 116 944 318 9339
0.621 102 946 737 226 402 94	0.042 877 960 025 007 734 4929
0.577 195 710 052 045 814 84	0.044 914 531 653 632 197 4143
0.531 319 743 644 375 623 97	0.046 813 554 990 628 012 4026
0.483 618 026 945 841 027 56	0.048 564 330 406 673 198 7159
0.434 243 749 346 802 558 00	0.050 157 139 305 899 537 4137
0.383 359 324 198 730 346 92	0.051 583 253 952 048 458 7768
0.331 135 393 257 976 833 09	0.052 834 946 790 116 519 8621
0.277 749 822 021 824 315 07	0.053 905 499 335 266 063 9269
0.223 386 686 428 966 881 63	0.054 789 210 527 962 865 0322
0.168 235 251 552 207 464 98	0.055 481 404 356 559 363 9878
0.112 488 943 133 186 625 75	0.055 978 436 510 476 319 4076
0.056 344 313 046 592 789 972	0.056 277 699 831 254 301 2726
0.0	0.056 377 628 360 384 717 3877

Table C3

$\pm x_i$	W_i
0.999 982 430 354 891 598 58	0.000 050 536 095 207 862 517 6247
0.999 872 888 120 357 611 94	0.000 180 739 564 445 388 357 820
0.999 598 799 671 910 683 25	0.000 377 746 646 326 984 660274
0.999 098 124 967 667 597 66	0.000 632 607 319 362 633 544219
0.998 316 635 318 407 392 53	0.000 938 369 848 542 381 500794
0.997 206 259 372 221 959 08	0.001 289 524 082 610 417 39210
0.995 724 104 698 407 188 51	0.001 681 142 865 421 469 90631
0.993 831 963 212 755 022 21	0.002 108 815 245 726 632 87933
0.991 495 721 178 106 132 40	0.002 568 764 943 794 020 37313
0.988 684 757 547 429 479 94	0.003 057 753 410 175 531 13613
0.985 371 499 598 520 371 11	0.003 572 892 783 517 299 64938
0.981 531 149 553 740 106 87	0.004 111 503 978 654 693 04717
0.977 141 514 639 705 714 16	0.004 671 050 372 114 321 74741
0.972 182 874 748 581 796 58	0.005 249 123 454 808 859 12513
0.966 637 851 558 416 567 09	0.005 843 449 875 835 639 50756
0.960 491 268 708 020 283 42	0.006 451 900 050 175 736 92280
0.953 730 006 425 761 136 41	0.007 072 489 995 433 555 46805
0.946 342 858 373 402 905 15	0.007 703 375 233 279 741 84817
0.938 320 397 779 592 883 65	0.008 342 838 753 968 157 70558
0.929 654 857 429 740 056 67	0.008 989 275 784 064 135 72328
0.920 340 025 470 012 420 73	0.009 641 177 729 702 536 69530
0.910 371 156 957 004 292 50	0.010 297 116 957 956 355 5237
0.899 744 899 776 940 036 64	0.010 955 733 387 837 901 6480
0.888 459 232 872 256 998 89	0.011 615 723 319 955 134 7270
0.876 513 414 484 705 269 74	0.012 275 830 560 082 770 0870
0.863 907 938 193 690 477 15	0.012 934 839 663 607 373 4547
0.850 644 494 768 350 279 76	0.013 591 571 009 765 546 7896
0.836 725 938 168 868 735 50	0.014 244 877 372 916 774 3063
0.822 156 254 364 980 407 37	0.014 893 641 664 815 182 0348
0.806 940 531 950 217 611 86	0.015 536 775 555 843 982 4399
0.791 084 933 799 848 361 43	0.016 173 218 729 577 719 9419
0.774 596 669 241 483 377 04	0.016 801 938 574 103 865 2709

Table C4

$\pm x_i$	W_i
0.757 483 966 380 513 63793	0.017 421 930 159 464 173 7472
0.739 756 044 352 694 75868	0.018 032 216 390 391 286 3201
0.721 423 085 370 098 91548	0.018 631 848 256 138 790 1863
0.702 496 206 491 527 07861	0.019 219 905 124 727 766 0193
0.682 987 431 091 079 22809	0.019 795 495 048 097 499 4880
0.662 909 660 024 780 59546	0.020 357 755 058 472 159 4669
0.642 276 642 509 759 51377	0.020 905 851 445 812 023 8522
0.621 102 946 737 226 40294	0.021 438 980 012 503 867 2465
0.599 403 930 242 242 89297	0.021 956 366 305 317 824 9393
0.577 195 710 052 045 81484	0.022 457 265 826 816 098 7071
0.554 495 132 631 932 54887	0.022 940 964 229 387 748 7608
0.531 319 743 644 375 62397	0.023 406 777 495 314 006 2013
0.507 687 757 533 716 60215	0.023 854 052 106 038 540 0804
0.483 618 026 945 841 02756	0.024 282 165 203 336 599 3580
0.459 130 011 989 832 33287	0.024 690 524 744 487 676 9091
0.434 243 749 346 802 55800	0.025 078 569 652 949 768 7068
0.408 979 821 229 888 67241	0.025 445 769 965 464 765 8126
0.383 359 324 198 730 34692	0.025 791 626 976 024 229 3884
0.357 403 837 831 532 15238	0.026 115 673 376 706 097 6805
0.331 135 393 257 976 83309	0.026 417 473 395 058 259 9310
0.304 576 441 556 714 04334	0.026 696 622 927 450 359 9062
0.277 749 822 021 824 31507	0.026 952 749 667 633 031 9634
0.250 678 730 303 483 17661	0.027 185 513 229 624 791 8192
0.223 386 686 428 966 88163	0.027 394 605 263 981 432 5161
0.195 897 502 711 100 15392	0.027 579 749 566 481 873 0349
0.168 235 251 552 207 46498	0.027 740 702 178 279 681 9939
0.140 424 233 152 560 17459	0.027 877 251 476 613 701 6085
0.112 488 943 133 186 62575	0.027 989 218 255 238 159 7038
0.084 454 040 083 710 883710	0.028 076 455 793 817 246 6068
0.056 344 313 046 592 789972	0.028 138 849 915 627 150 6363
0.028 184 648 949 745 694339	0.028 176 319 033 016 602 1307
0.0	0.028 188 814 180 192 358 6938

Appendix D Chebyshev Polynomials

Chebyshev polynomials are defined by

$$T_r(x) = \cos(r \arccos x) = \cos r\theta \tag{1.2.6}$$

where $x = \cos\theta$. The following is a list of the first twelve such polynomials. Hence

$$\begin{aligned}
T_0 &= 1 \\
T_1 &= x \\
T_2 &= 2x^2 - 1 \\
T_3 &= 4x^3 - 3x \\
T_4 &= 8x^4 - 8x^2 + 1 \\
T_5 &= 16x^5 - 20x^3 + 5x \\
T_6 &= 32x^6 - 48x^4 + 18x^2 - 1 \\
T_7 &= 64x^7 - 112x^5 + 56x^3 - 7x \\
T_8 &= 128x^8 - 256x^6 + 160x^4 - 32x^2 + 1 \\
T_9 &= 256x^9 - 576x^7 + 432x^5 - 120x^3 + 9x \\
T_{10} &= 512x^{10} - 1280x^8 + 1120x^6 - 400x^4 + 50x^2 - 1 \\
T_{11} &= 1024x^{11} - 2816x^9 + 2816x^7 - 1232x^5 + 220x^3 - 11 \\
T_{12} &= 2048x^{12} - 6144x^{10} + 6912x^8 - 3584x^6 \\
&\quad + 840x^4 - 72x^2 + 1
\end{aligned}$$

It is also useful to have the powers of x expressed as series of Chebyshev polynomials, in particular in the process of economization of series:

$$\begin{aligned}
1 &= T_0 \\
x &= T_1
\end{aligned}$$

$$x^2 = \tfrac{1}{2}[T_0 + T_2]$$

$$x^3 = \tfrac{1}{4}[3T_1 + T_3]$$

$$x^4 = \tfrac{1}{8}[3T_0 + 4T_2 + T_4]$$

$$x^5 = \tfrac{1}{16}[10T_1 + 5T_3 + T_5]$$

$$x^6 = \tfrac{1}{32}[10T_0 + 15t_2 + 6T_4 + T_6]$$

$$x^7 = \tfrac{1}{64}[35T_1 + 21T_3 + 7T_5 + T_7]$$

$$x^8 = \tfrac{1}{128}[35T_0 + 56T_2 + 28T_4 + 8T_6 + T_8]$$

$$x^9 = \tfrac{1}{256}[126T_1 + 84T_3 + 36T_5 + 9T_7 + T_9]$$

$$x^{10} = \tfrac{1}{512}[126T_0 + 210T_2 + 120T_4 + 45T_6 + 10T_8 + T_{10}]$$

$$x^{11} = \tfrac{1}{1024}[462T_1 + 330T_3 + 165T_5 + 55T_7 + 11T_9 + T_{11}]$$

$$x^{12} = \tfrac{1}{2048}[462T_0 + 792T_2 + 495T_4 + 220T_6 + 66T_8 + 12T_{10} + T_{12}]$$

References

Abramowitch M. and Stegun I. A. (1965) *Handbook of Mathematical Functions*, Dover, New York.

Acheson D. J. (1990) *Elementary Fluid Dynamics*, Clarendon Press, Oxford.

Ahlberg J., Nilson E. and Walsh J. (1967) *The Theory of Splines and Their Application*, Academic Press, New York.

Alaylioglu A., Evans G. A. and Hyslop J. (1973) The evaluation of integrals with infinite limits, *J. Comp. Phys.*, **13**, 433-438.

Alaylioglu A., Evans G. A. and Hyslop J. (1976) The use of Chebyshev series for the evaluation of oscillatory integrals, *C. J.*, **19**, No. 3, 258-267.

Ames W. F. (1969) *Numerical Methods for Partial Differential Equations*, Nelson, London.

Apostol T. M. (1974) *Mathematical Analysis*, Addison-Wesley, Reading, Massachusetts.

Atkin R. H. (1962) *Theoretical Electromagnetism*, Heinemann, London.

Bairstow L. (1914) The solution of algebraic equations with numerical coefficients, *Rep. Memor. Adv. Comm. Aero., London*, No. 154, 51-64.

Baker G. A. and Graves-Morris P. (1981) Padé Approximants, *Encyclopaedia of Mathematics*, **13**, **14**.

Bourne D. E. and Kendall P. C. (1977) *Vector Analysis and Cartesian Tensors*, Nelson, Sunbury-on-Thames.

Broyden C. G. (1965) A class of methods for solving non-linear simultaneous equations, *Math. Comp.*, **19**, 577-593.

Broyden C. G. (1970) The convergence of a class of double rank minimization algorithms, parts I and II, *J. Inst. Maths. Appl.*, **6**, 76-90; 222-231.

Butcher J. C. (1965) A modified multi-step method for the numerical integration of ordinary differential equations, *J. Assoc. Comput. Mach.* **12**, 124-135.

Butcher J. C. (1987) *The numerical analysis of ordinary differential equations: Runge-Kutta and general linear methods*, John Wiley and Sons, Chichester.

Calio F., Gautschi W. and Marchetti E. (1986) On computing Gauss-Kronrod quadrature formulae, *Math. Comp.* **47**, No. 176, 639-650.

Chisholm J. S. R., Genz A. and Rowlands G. E. (1972) Accelerated convergence of sequences, *J. Comp. Phys.* **10**, 284-307.

Clenshaw C. W. (1957) The numerical solution of linear differential equations in Chebyshev series, *Proc. Camb. Phil. Soc.* **53**, 134-149.

Clendenin W. W. (1966) A method for numerical calculation of Fourier integrals, *Num. Math.* **8**, 422-436.

Clenshaw C. W. and Curtis A. R. (1960) A method for numerical integration on an automatic computer, *Num. Math.* **12**, 197-205.

Cohen A. M. (1980) Cautious Romberg extrapolation, *Int. J. Comp. Maths.* **8**, 137-147.

Cohen A. M. and Gismalla D. A. (1985) The construction of quadrature rules by parameter optimization, *Int. J. Comp. Math.* **17**, 203-214.

Crane R. L. and Klopfenstein R. W. (1965) A predictor-corrector algorithm with an increased range of absolute stability, *J. Assoc. Comput. Mach.* **12**, 227-241.

Crank J. and Nicolson P. (1947) A practical method for numerical evaluation of solutions of partial differential equations of the heat conduction type, *Proc. Camb. Phil. Soc.* **43**, 50–67.

Dahlquist G. (1963) A special stability problem for linear multi-step methods, *B. I. T.* **3**, 27–43.

Davidon W. C. (1959) Variable metric method for optimization, *A. E. C. R and D report, ANL-5990(Rev), Argonne National Laboratory*.

Davies A. J. (1980) *The Finite Element Method: a First Approach*, Oxford University Press, Oxford.

Davis P. J. (1967) A construction of non-negative approximate quadratures, *Math. Comp.* **21**, 578–582.

Davis P. J. (1969) *Approximate Integration Rules with Non-negative Weights; Lectures in Differential Equations*, Van Nostrand Reinhold, New York.

Davis P. J. and Rabinowitz P. (1984) *Methods of Numerical Integration*, Academic Press, New York.

Dekker K. and Verwer J. G. (1984) *Stability of Runge–Kutta Methods for Stiff Non-linear Differential Equation*, North Holland, Amsterdam.

Delves L. M. and Walsh J. (1974) *Numerical Solution of Integral Equations*, Clarendon Press, Oxford.

Dennis J. E. and Goldstein A. A. (1969) Cubature and the Tchakaloff cone, *J. Comp. Syst. Sci.* **3**, 218–220.

Duff I. S. (1986) The influence of vector and parallel processors on numerical analysis, *AERE-R 12329*, Harwell laboratory, Oxford.

Duff I. S., Erisman A. M. and Reid J. K. (1986) *Direct Methods for Sparse matrices*, Oxford University Press, Oxford.

Ehle B. L. (1969) On Padé approximations to the exponential functions and A-stable methods for the numerical solution of initial value problems, *University of Waterloo, Department of Applied Analysis and Computer Science, Research Report No. CSRR 2010*.

Evans D. J. and Abdullah A. R. B. (1983) Group explicit methods for parabolic equations, *Int. J. Comp. Math.* **14**, 73–105.

Evans G. A. (1979) A numerical differentiator, *J. Comp. Phys.* **32**, 247–253.

Evans G. A. (1993) *Practical Numerical Integration*, John Wiley and Sons, Chichester.

Evans G. A. (1994a) Two robust methods for irregular oscillatory quadrature *App. Num. Math.* **14**.

Evans G. A. (1994b) An alternative method for irregular oscillatory integrals over a finite range *Int. J. Comp. Math.* **53**.

Evans G. A. (1995) An efficient stable implementation for the integration of oscillatory integrals using Chebyshev polynomials, *I. M. A. J. Num. Anal.* in press.

Evans G. A., Forbes R. C. and Hyslop J. (1983) Polynomial transformations for singular integrals, *Int. J. Comp. Math.* **14**, No. 4, 157–170.

Evans G. A., Hyslop J., and Morgan A. P. G. (1982) The iterative solution of a non-linear Fredholm integral equation arising from a chemical reactor problem, *Int. J. Comp. Math.* **11**, 73–83.

Evans G. A., Hyslop J. and Morgan A. P. G. (1983) An extrapolation procedure for the evaluation of singular integrals, *Int. J. Comp. Math.* **12**, 251–265.

Filon L. N. G. (1928) On a quadrature formula for trigonometric integrals, *Proc. Roy. Soc. Edinburgh* **49**, 38–47.

Finkbeiner D. T. (1960) *Introduction to Matrices and Linear Transformations*, W.H. Freeman, San Francisco.

Fletcher R. (1970) A new approach to variable metric algorithms, *C. J.* **13**, 317–322.

Fletcher R. (1987) *Practical Optimization*, John Wiley and Sons, Chichester.

Fletcher R. and Powell M. J. D. (1963) A rapidly convergent descent method for minimization, *C. J.* **6**, 163–168.

Fletcher R. and Reeves C. M. (1964) Function minimization by conjugate gradients, *C. J.* **7**, 149–154.

Flinn E. A. (1960) A modification of Filon's method for numerical integration, *J. Assoc. Comput. Mach.* **7**, 181–184.

Fox L. (1968) *Introduction to Numerical Linear Algebra*, Oxford University Press, Oxford.

Fox L. and Parker I. (1968) *Chebyshev Polynomials in Numerical Analysis*, Oxford Press, London.

Fraboul F. (1962) Un critère de stabilité pour l'intégration numérique des équations différentielles, *Chiffres* **1** 55–63.

Francis J. G. F. (1961) The QR transformation, Part 1, *Comp. J.* **4**, 265–271.

Francis J. G. F. (1962) The QR transformation, Part 2, *Comp. J.* **4**, 332–345.

Gauss C. F. (1886) Methodus nova integralium valores per approximationem inveniendi, *C.F. Gauss Werke, Göttingen: Königlichen Gesellschaft der Wissenschaften* **3**, 163–196.

Gautschi W. (1968) Construction of Gauss–Christoffel quadrature formulas, *Math. Comp.* **22**, 251–270.

Gautschi W. (1970) Tables of Gaussian quadrature rules for the calculation of Fourier coefficients, *Maths. Comp.* **24**, microfiche.

Gautschi W. and Notaris S. E. (1988) Newton's method and Gauss–Kronrod quadrature, *Int. Series of Num. Math.* **85**, Birkhauser Verlag, Basel, 60–73.

Gear C. W. (1965) Hybrid methods for initial value problems in ordinary differential equations, *SIAM J. Num. Anal.* **2**, 69–86.

Gear C. W. (1971) *Numerical Initial Value Problems in Ordinary Differential Equations*, Prentice-Hall, Englewood Cliffs, New Jersey.

Gismalla D. A., Jenkins L. D. and Cohen A. M. (1988) Acceleration of convergence of certain multiple integrals, *Int. J. Comp. Math.* **24**, 155–168.

Givens J. W. (1954) Numerical computation of the characteristic values of a real symmetric matrix, *Oak Ridge National Laboratory, ORN 1-1574.*

Goldfarb D. (1970) A family of variable metric methods derived by variational means, *Math. Comp.* **24**, 23–26.

Golub G. H. and Welsch J. H. (1969) Calculation of Gauss quadrature rules, *Math. Comp.* **23**, 221–230.

Gragg W. B. and Stetter H. J. (1964) Generalized multi-step predictor–corrector methods, *J. Assoc. Comput. Mach.* **11**, 188–209.

Gradshteyn I. S. and Ryzhik I. M. (1965) *Tables of Integrals, Infinite Series and Products*, Academic Press, New York.

Greville T. N. E. (Ed) (1969) *Theory and Application of Spline Functions*, Academic Press, New York.

Hasting C. Jr. (1955) *Approximations for Digital Computers*, Princeton University Press, Princeton, New Jersey.

Henrici P. (1958) The quotient-difference algorithm, *Appl. Math. Ser. Nat. Bur. Stand.* **49**, 23–46.

Henrici P. (1963) Some applications of the quotient-difference algorithm, *Proc. Symp. Appl. Math.* **15**, 159–183.

Heun K. (1900) Neue Methode zur approximativen Integration der Differentialgleichungen einer unabhängigen Veränderlichen, *Z. Math. Phys.* **45**, 45–38.

Hildebrand F. (1958) *Introduction to Numerical Analysis*, McGraw-Hill, New York.

Hlawka E. (1962) Zur angenäherten Berechnung mehrfacher Integrale, *Monatsh. Math.* **66**, 140–151.

Holladay J. C. (1957). A smoothest curve approximation, *Math. Tab. and Other Aids to Comp.* **11**, 233–243.

Householder A. S. (1958) The approximate solution of matrix problems, *J. Ass. Comp. Mach.* **5**, 205–243.

Householder A. S. (1970) *The Numerical Treatment of a Single Non-linear Equation*, McGraw-Hill, New York.

Hyman M. A. (1957) Eigenvalues and eigenvectors of general matrices, *Twelfth National Meeting A. C. M.*, Houston, Texas.

Jacobi C. G. J. (1846) Über ein leichtes Verfahren die in der Theorie der Säculärstörungen vorkommenden Gleichungen numerisch aufzulösen, *Crelle's J.* **30**, 51–94.

Kahaner D. K. (1971) *Comparison of Numerical Quadrature Formulae in Mathematical Software (Rice J. R. ed)*, Academic Press, ACM Monograph series.

Klopfenstein R. W. and Millman R. S. (1968) Numerical stability of a one evaluation predictor-corrector algorithm for numerical solution of ordinary differential equations, *Math. Comp.* **22**, 557–564.

Knopp K. (1928) *Theory and Application of Inifinite series*, Blackie and Son Ltd., London and Glasgow.

Korobov N. M. (1963) Number theoretic methods in approximate analysis, *G. I. F. L., Moscow.*

Kronrod A. S. (1965) *Nodes and Weights of Quadrature Formulae*, Consultants Bureau, New York.

Kuhn H. W. and Tucker A. W. (1956) *Non-linear programming*, in Proceedings of the 2nd. Berkeley Symposium on Mathematical Statistics and Probability. (Ed Neyman J.), University of California Press.

Kutta W. (1901) Beitrag zur näherugsweisen Integration totaler Differentialgleichungen, *Z. Math. Phys.* **46**, 435–453.

Lambert J. D. (1973) *Computational Methods in Ordinary Differential Equations*, John Wiley and Sons, London.

Lambert J. D. (1991) *Numerical Methods for Ordinary Differential Systems*, John Wiley and Sons, Chichester, England.

Levin D. (1973) Development of non-linear transformations for improving convergence of sequences, *Int. J. Comp. Maths.* **3**, 371–388.

Levin D. (1982) Procedures for computing one and two dimensional integrals of functions with rapid irregular oscillations, *Math. Comp.* **38**, No. 158, 531–538.

Littlewood R. K. and Zakian V. (1976) Numerical evaluation of Fourier integrals, *J. Inst. Maths. Applics.* **18**, 331–339.

Lorentz G. G. (1966) *Approximation of Functions*, Holt, Rinehart and Winston, New York.

Markowitz H. M. (1957) The elimination form of the inverse and its application to linear programming, *Management Sci.* **3**, 255–269.

Milne-Thomson L. M. (1951) *The calculus of finite differences*, Macmillan, London.

Monegato G. (1976) An overview of results and questions related to Kronrod schemes *Int, series of Num. Math.* **45**, Birkhauser Verlag, Basel, 231–240.

Morrow C. R. (1977) *A Study of some Numerical Quadrature and Cubature Processes*, Ph. D. thesis, Belfast.

Morse P. M. and Feshbach H. (1953) *Methods of Theoretical Physics, Part II*, McGraw-Hill, New York.

O' Brien C. G., Hyman M. A. and Kaplan S. (1951) A study of the numerical solution of partial differential equations, *J. Math. Phys.* **29**, 223–251.

O'Hara H. and Smith F. H. (1968) Error estimation in the Clenshaw–Curtis quadrature formulae, *C. J.* **11**, 213–219.

Ortega J. M. (1972) *Numerical Analysis — a Second Course*, Academic Press, New York.

Ortega J. M. and Rheinboldt W. (1970) *Iterative Solution of Non-linear Equations in Several Variables*, Academic Press, New York.

Pakes H. W. and Storey C. (1967) Solution of the equations for a tubular reactor with axial diffusion by a variational technique, *The Chemical Engineer*, **208**, 96–99.

Patterson T. N. L. (1968) The optimum addition of points to quadrature formulae, *Math. Comp.* **22**, 847–856.

Patterson T. N. L. (1976) On high precision methods for the evaluation of Fourier integrals with finite and infinite limits, *Numer. Maths.* **27**, 41-52.

Peaceman D. W. and Rachford H. H. (1955) The numerical solution of parabolic and elliptic differential equations, *J. Soc. Ind. App. Math.* **3**, 28.

Piessens R. (1970a) Gaussian quadrature formulae for the integration of oscillating functions, *Z. A. M. M.* **50**, 698-700.

Piessens R. (1970b) Gaussian quadrature formulae for the integration of oscillating functions, *Maths. Comp.* **24**, microfiche.

Piessens R. (1970c) Numerical inversion of the Laplace transform, *Ph. D. thesis, Univ. of Leuven, Belgium.*

Piessens R. and Poleunis F. (1971) A numerical method for the integration of oscillatory functions, *B. I. T.* **11**, 317-327.

Powell M. J. D. (1964) An iterative method for finding stationary values of a function of several variables without calculating derivatives, *C. J.* **7**, 155-162.

Radon J. (1948) Zur mechanische Kubatur, *Monatsh. Math.* **52**, 286-300.

Roach G. F. (1970) *Green's Functions*, Van Nostrand Reinhold Company, London.

Romberg W. (1955) Vereinfachte numerische Integration, *Det. Kong. Norske Videnskabers Selskab Forhandlinger, Trondheim*, **28**, No. 7, 30-36.

Runge C. (1895) Über die numerische huflösung von Differentialgleichungen, *Math. Ann.* **46**, 167-178.

Rutishauser H. (1958) Solution of eigenvalue problems with the LR transformation, *Appl. Math. Ser. Nat. Bur. Stand.* **49**, 47-81.

Sack R. A. and Donovan A. F. (1972) An algorithgm for Gaussian quadrature given modified moments, *Num. Math.* **18**, 465-478.

Saul'yev V. K. (1964) *Integration of Equations of Parabolic Type by the Method of Nets*, Pergamon Press, New York.

Schur J. (1916) Über Potenzreihen die im Innern des Einheitskreises beschränkt sind, *J. Reine. Angew. Math.* **147**, 205-232.

Schwartz C. (1969a) Numerical integration of analytic functions, *J. Comp. Phys.* **4**, 19-29.

Schwartz C. (1969b) Variational principles for integrals, *J. Comp. Phys.* **3**, 512-522.

Shanks D. (1955) Non-linear transformations of divergent and slowly convergent sequences, *J. Math. Phys.* **34**, 1-42.

Shanno D. F. (1970) Conditioning of quasi-Newton methods for function minimization, *Math. Comp.* **24**, 647-656.

Smith G. D. (1978) *Numerical Solution of Partial Differential Equations: Finite Difference Method*, Oxford University Press, Oxford.

Squire W. (1976) A quadrature method for finite intervals, *Int. J. Num. Meth. Engng.* **10**, 708-712.

Squire W. (1979) A comment on quadrature in the presence of end-point singularities, *Int. J. Comp. Maths.* **7**, 239-241.

Sneddon I. N. (1957) *Elements of Partial Differential Equations*, McGraw-Hill, New York.

Steffensen J. F. (1927) *Interpolation*, The Williams and Wilkins Co., Baltimore.

Stenger F. (1973) Integration formulae based on the trapezoidal formula, *J. Inst. Maths. Applic.* **12**, 102-114.

Stroud A. H. (1971) *Approximate Calculation of Multiple Integrals*, Prentice-Hall, Englewood Cliffs, New Jersey.

Stroud A. H. and Secrest D. (1966) *Gaussian Quadrature Formulae*, Prentice-Hall, Englewood Cliffs, New Jersey.

Sugihara M. (1987) Method of good matrices for multi-dimensional integrations — an extension of the method of good lattice points, *J. Comp. Math.* **17**, 197-213.

Takahasi H. and Mori M. (1973) Quadrature formulas obtained by variable transformation, *Numer. Math.* **21**, 206-219.

Tchakaloff V. (1957) Formules de cubature méchanique à coefficients non négatifs, *Bull. Sci. Math.[2]* **81**, 123-134.

Tricomi F. G. (1961) *Differential Equations*, Blackie, London.

Walsh G. R. (1971) *An Introduction to Linear Programming*, Holt, Rinehart and Winston Ltd., London.

Walsh G. R. (1975) *Methods of Optimization*, John Wiley and Sons, London.

Wendroff B. (1966) *Theoretical Numerical Analysis*, Academic press, New York.

Widlund O. D. (1967) A note on unconditionally stable linear multi-step methods, *B. I. T.* **7**, 65-70.

Wilkinson J. H. (1965) *Algebraic Eigenvalue Problem*, Oxford University Press, Oxford.

Wilson M. W. (1969) A general algorithm for non-negative quadrature formulae, *Math. Comp.* **23**, 253-258.

Wynn P. (1956) On a device for computing the $e_m(s_n)$ transformation, *Maths. Comp.* **10**, 91-96.

Zienkiewicz O. C. (1977) *The Finite Element Method*, McGraw-Hill, Maidenhead, England.

Author Index

Subject Index